IFIP Advances in Information and Communication Technology

The IFIP AICT series publishes state-of-the-art results in the sciences and technologies of information and communication. The scope of the series includes: foundations of computer science; software theory and practice; education; computer applications in technology; communication systems; systems modeling and optimization; information systems; ICT and society; computer systems technology; security and protection in information processing systems; artificial intelligence; and human-computer interaction.

Edited volumes and proceedings of refereed international conferences in computer science and interdisciplinary fields are featured. These results often precede journal publication and represent the most current research.

The principal aim of the IFIP AICT series is to encourage education and the dissemination and exchange of information about all aspects of computing.

More information about this series at https://link.springer.com/bookseries/6102

Matthias Thürer · Ralph Riedel ·
Gregor von Cieminski ·
David Romero
Editors

Advances in Production Management Systems

Production Management Systems for Volatile, Uncertain, Complex, and Ambiguous Environments

43rd IFIP WG 5.7 International Conference, APMS 2024
Chemnitz, Germany, September 8–12, 2024
Proceedings, Part VI

Springer

Editors
Matthias Thürer
Chemnitz University of Technology
Chemnitz, Germany

Gregor von Cieminski
ZF Friedrichshafen AG
Friedrichshafen, Germany

Ralph Riedel
West Saxon University of Applied Sciences Zwickau
Zwickau, Germany

David Romero
Tecnológico de Monterrey
Mexico City, Mexico

ISSN 1868-4238 ISSN 1868-422X (electronic)
IFIP Advances in Information and Communication Technology
ISBN 978-3-031-71644-7 ISBN 978-3-031-71645-4 (eBook)
https://doi.org/10.1007/978-3-031-71645-4

This Springer imprint is published by the registered company Springer Nature Switzerland AG
The registered company address is: Gewerbestrasse 11, 6330 Cham, Switzerland

Preface

The year 2024 highlighted that the world is getting more and more volatile, uncertain, complex, and ambiguous. There are no simple solutions to complex problems, which require new ways to solve and manage them. New technology opens new ways as does the continuous progress of science. But to bring real benefits to people on the ground who are suffering from the consequences of conflicts and climate change, just to name two, more research is needed on how to apply and further advance this technology and knowledge. A scientific conference plays an important part in this. It is a symbol of aiming for a better future. We create new knowledge and solutions, we share all our achievements, and we meet to get to know people from all over the world, creating new ties and making new friends.

The International Conference on "Advances in Production Management Systems" (APMS) is the leading annual event of the IFIP Working Group (WG) 5.7 of the same name. At APMS 2024 in Chemnitz, Germany, hosted by Chemnitz University of Technology and West Saxon University of Applied Sciences Zwickau, more than 200 papers were presented and discussed. This is a significant step up from the first APMS Conference in 1980, which assembled just a few participants. The IFIP WG 5.7 was established in 1978 by the General Assembly of the International Federation for Information Processing (IFIP) in Oslo, Norway. Its first meeting was held in August 1979 with all its seven members present. The WG has since grown to 124 full and candidate members as well as 27 honorary members.

APMS was held for the first time in East Germany, which experienced specific challenges and disruptions, but also hopes and new futures. The Ore Mountains are a traditional industrial region, and it can be argued that the first textbook on operations management was written on the mining industry in this region. Mining always brought its own challenges in terms of sustainability, quality of work life, and process efficiency, but also major technological advancements and highly skilled labour. Many of the papers presented address similar challenges as 500 years ago. The problems did not change but are more volatile, uncertain, complex, and ambiguous today given the interconnectedness of society and production. No other technology can show a more rapid development and impact in industry and society than Information and Communication Technology (ICT).

The APMS 2024 program shows that the IFIP WG 5.7 can still make, and will continue to make, a significant contribution to production and production management disciplines. In 2024, the international review board for APMS included 95 recognized experts working in the disciplines of production and production management systems. For each paper, an average of 2.5 single-blind reviews were provided. Over two months, each submitted paper went through two rigorous rounds of reviews to allow authors to revise their work after the first round of reviews to guarantee the highest scientific quality of the papers accepted for publication. Following this process, 201 full papers were selected for inclusion in the conference proceedings from a total of 224 submissions.

APMS 2024 brought together leading international experts from academia, industry, and government in the areas of production and production management systems to discuss how to achieve responsible manufacturing, service, and logistics futures. This included topics such as innovative manufacturing, service, and logistics systems characterized by their agility, circularity, digitalization, flexibility, human-centricity, resilience, and smartification contributing to more sustainable industrial futures that ensure that products and services are manufactured, servitized, and distributed in a way that creates a positive effect on the triple bottom line.

The APMS 2024 conference proceedings are organized into 6 volumes, covering a large spectrum of research addressing the overall topic of the conference "Production Management Systems for Volatile, Uncertain, Complex, and Ambiguous Environments". We would like to thank all contributing authors for their high-quality research work and their willingness to share their findings with the APMS and IFIP WG 5.7 community. We are equally grateful for the outstanding work of all the International Reviewers, the Program Committee Members, and the Special Sessions Organizers.

September 2024

Matthias Thürer
Ralph Riedel
Gregor von Cieminski
David Romero

Organisation

Conference Co-chairs

Matthias Thürer	Chemnitz University of Technology, Germany
Ralph Riedel	West Saxon University of Applied Sciences Zwickau, Germany
Gregor von Cieminski	ZF Friedrichshafen AG, Germany

Honorary Co-chair

Egon Müller	Chemnitz University of Technology, Germany

Program Co-chairs

Philipp Wilsky	Chemnitz University of Technology, Germany
David Romero	Tecnológico de Monterrey, Mexico

Organizing Committee Co-chairs

Luise Weißflog	Chemnitz University of Technology, Germany
Felix Franke	Chemnitz University of Technology, Germany
Julia Sprigode	West Saxon University of Applied Sciences Zwickau, Germany

Doctoral Workshop Chairs

Marina Ivanova	Chemnitz University of Technology, Germany
Pierre Grzona	Chemnitz University of Technology, Germany
David Romero	Tecnológico de Monterrey, Mexico

Program Committee

Federica Acerbi
Lorenzo Agbomemewa
Natalie Cecilia Agerskans
Carla Susana Agudelo Assuad
Alireza Ahmadi
Kartika Nur Alfina
Erlend Alfnes
Farhad Ameri
Bjorn Andersen
Yevheniia Andriichenko
Oliver Antons
Veronica Arioli
Doris Aschenbrenner
Aylin Ates
Nestor Fabián Ayala
Jannicke Baalsrud Hauge
Elena Beducci
Monica Bellgran

Frederique Biennier
Alessia Bilancia
Vlad Bocanet
Jos A. C. Bokhorst
Frank Börner
Magali Bosch-Mauchand
Hafida Bouloiz
Alexandros Bousdekis
Greta Braun
Gerald Bräunig
Jessica Bruch
Adauto Bueno
Lien Bui
Guillherme Hörner Bussolo
Mario Henrique Callefi
Davide Castellano
Perawit Charoenwut
Ferdinando Chiacchio
Steve Childe
Chiara Cimini
Beatrice Colombo
Jelena Crnobrnja
Catherine da Cunha
Patrick Dallasega
Mélanie Despeisse
Mario Di Nardo
Alexandre Dolgui
Heidi Dreyer
Milos Drobnjakovic
Malin Elvin
Adrodegari Federico
Jannick Fiedler
Daniel Fischer
Martin Folz
Florian Förster
Chiara Franciosi
Jana Frank
Susanne Franke
Stefano Frecassetti
Mike Freitag
Jan Frick
Paolo Gaiardelli
Mattia Galimberti
Maryam Gallab
Mosè Gallo
Fernando José Gómez Paredes
Clarissa A. González Chávez
Jon Gosling
Pierre Grzona
Christopher Gustafsson
Harumi Haraguchi
Stefanie Hatzl
Zhiliang He
Dominik Hertel
Jan Holmström
Djerdj Horvat
Ryuichiro Ishikawa
Dmitry Ivanov
Marina Ivanova
Niloofar Jafari
Tim Maximilian Jansen
Yongkuk Jeong
HongBae Jun
Toshiya Kaihara
Matthias Kalverkamp
Kai Kang
Ibrokhimjon Khamidov
Maria Kollberg Thomassen
Juhoantti Viktor Köpman
Alexandra Lagorio
Beñat Landeta
Nicolas Leberruyer
Jan-Peter Lechner
Liu Lei
Yee Yeng Liau
Ming Lim
Ma Lin
Francesco Lolli
Damien Lovato
Egon Lüftenegger
Sivanilza Teixeira Machado
Ugljesa Marjanovic
Larisa Markov
Melissa Marques-McEwan
Antonio Masi
Sandra Mattsson
Gokan May
Nayara Cardoso Medeiros
Khaled Medini
Kunruthai Meechang
Joan Gilberto Mendes dos Reis
Nicola Mercogliano

Hajime Mizuyama
Katharina Moitzi
Elias Montini
Eiji Morinaga
Francesco Moroni
Sobhan Mostafayi Darmian
Mohamed Naim
Torbjørn Netland
Phu Nguyen
Nariaki Nishino
Sang Do Noh
Tomomi Nonaka
Maniva Oliva
Margherita Pero
Claudia Piffari
Marta Pinzone
Fabiana Pirola
Daryl Powell
Vittaldas Prabhu
Hiran Harshana Prathapage
Anna Presciuttini
Moritz Quandt
Ricardo Rabelo
Slavko Rakic
Mario Rapaccini
R. M. Chandima Ratnayake
Daniel Resanovic
Ralph Riedel
David Romero
Anita Romsdal
Christoph Roser
Martin Rudberg
Tamas Ruppert
Roberto Sala
Omkar Salunkhe
Jan Salzwedel
Adriana Saraceni
Claudio Sassanelli
Mizuho Sato
Bennet Schulz
Daniel Seifert
Marcia Terra Silva
Katrin Singer-Coudoux
Dragana Slavic
Selver Softic
Per Solibakke
Endre Sølvsberg
Dongping Song
Kenn Steger-Jensen
Oliver Stoll
Jo Wessel Strandhagen
Shota Suginouchi
Yoshitaka Tanimizu
Takashi Tanizaki
Moreira Tavares
Klaus-Dieter Thoben
Peter Thorvald
Matthias Thürer
Rodrigo Carlo Toloi
Ivan Tomašević
Mario Tucci
Ioan Turcin
Ali Turkyilmaz
Hendrik Unger
Kenneth Vidskjold
Gregor von Cieminski
Sabine Waschull
Kasuni Vimasha Weerasinghe
Shaun West
Stefan Alexander Wiesner
Joakim Wikner
Philipp Wilsky
Heiner Winkler
Thorsten Wuest
Mohammed Yaqot
Mingze Yuan
Lara Popov Zambiasi
Matteo Zanchi
Anton Zitnikov
Anne Zouggar

Contents – Part VI

Modelling Supply Chain and Production Systems

Management of Measuring Equipment for Quality Assurance in Manufacturing Processes: A Decision-Making Support System 3
Isabel Lopes, Cláudia Pires, and Tiago Alão

Understanding Coopetition Dynamics in Manufacturing Value Networks: A System Dynamics Based Causal Loop Diagram (CLD) Modelling Approach 19
Aziz Kemal Konyalioglu, Aylin Ates, Steve Paton, and Tuğçe Apaydın

De-politicised in Court: The Interaction of Democracy with Innovation Projects 34
Atle M. Christiansen, Kenn Steger-Jensen, and Kenneth Vidskjold

Dynamic Dispatching of DDMRP Replenishment Orders 49
Lucas Alves, Maha Benali, and Pierre Baptiste

Assembly Line Design for Industrialized Electrolyser Production 65
Sven-Vegard Buer and Lars Skjelstad

Empirical Grounded Simulation Models for Make-To-Order (MTO) Supply Chains: An Application in the Furniture Industry 78
Alexandra Lagorio, Giulio Mangano, Roberto Pinto, Carlo Rafele, and Giovanni Zenezini

Modeling of Logistics Objectives at Rail-Road Inland Transshipment Terminals 92
Volodymyr Alieksieiev, Rafat Beigpoor Shahrivar, Carlos Jahn, and Hermann Lödding

Resilience Management in Supply Chains

Forging Resilience Through Supply Chain Collaboration: Insights from the Chinese Automotive Industry 111
Weiyuan Li, Ming K. Lim, and James Wilson

Anticipating VUCA by Utilizing the Potential of Technological and Logistical Degrees of Freedom 123
Julian B. Maier and Hans-Hermann Wiendahl

Material Shortages Propagation: Using Network Science to Evaluate Inventory Efficacy . . . 138
Michele Martignago, Martina Calzavara, and Daria Battini

Digital Twin Concepts in Production and Services

A Digital Twin Framework for Flexible Manufacturing System. . . . 155
Mokhtar Nizar Sid-Lakhdar, Hichem Haddou Benderbal, and Mehdi Souier

A Dynamic Fit-Out Scheduling Framework for Digital Twin-Enabled Modular Integrated Construction . . . 168
Qiqi Chen, Jiyuchen Ding, Mingyue Sun, Zhiheng Zhao, and George Q. Huang

Comparing Digital Twins and Virtual Engineering in Buyer Supplier Relationships for Complex Production Facilities . . . 180
Luca Janecki, Oliver Antons, Daniel Reh, and Julia C. Arlinghaus

Optimization

A Cost Effective Two-Step Exploration for Constrained Two-Dimensional Packing Problems . . . 197
Hyeongoo Pyeon and SookYoung Son

Energy Conscious Bi-objective Job Shop Scheduling: A New Formulation and Augmented -Constraint Method . . . 213
S. Hocine Bouzegag and Saadettin Erhan Kesen

Routing Heuristics for in-House Transportation in Assembly Systems . . . 229
Ebenezer Olatunde Adenipekun, Veronique Limère, and Nico André Schmid

A Flexible Job Shop Scheduling Problem Involving Reconfigurable Machine Tools Under Industry 5.0 . . . 244
Hessam Bakhshi-Khaniki, Reza Tavakkoli-Moghaddam, Zdenek Hanzalek, and Behdin Vahedi-Nouri

Makespan Minimisation in Hybrid Flexible Flowshops with Buffers and Machine-Dependent Transportation Times . . . 258
Stavros Vatikiotis, Ilias Mpourdakos, Dimitrios Papathanasiou, and Ioannis Mourtos

Multi-depot Electric Vehicle Routing Problem with Half-Open Routes and Rotations: A Mathematical Formulation 274
Hakan Erdeş and Saadettin Erhan Kesen

The Generalized One-to-One Pickup and Delivery Vehicle Routing Problem 291
Nurşah Yilmaz and İsmail Karaoğlan

An AUGMECON2VIKOR Algorithm for a Multi-objective Model in a Sustainable Manufacturing System Under Reliable Constraints 307
Reza Shahabi-Shahmiri, Reza Tavakkoli-Moghaddam, Alexandre Dolgui, Mohammad Ghasemi, and Seyed-Ali Mirnezami

Additive Manufacturing

An Accessibility Assessment Algorithm for Support Structure Removal in Parts Produced by Powder Bed Fusion of Metal Using a Laser Beam 325
Leonie Pauline Pletzer-Zelgert, Lukas Unruh, Martin Iza-Mendez, and Johannes Henrich Schleifenbaum

Conceptualization of an Operational Capability Development Framework to Facilitate the Integration of Additive Manufacturing Based on a Use Case Perspective 338
Christopher Gustafsson and Christer Nygren

Operations Management of Additive Manufacturing 353
Siavash Haghighat Khajavi, Mika Salmi, and Jan Holmström

Advances in Production Management Systems

A Worker-Centric Order Release Method Based on Workload Control: An Assessment by Simulation 381
Lin Ma, Mingze Yuan, Ting Qu, Lei Liu, Kai Zhang, Congdong Li, and Matthias Thürer

Understanding Part Complexity: A Novel Approach for the Identification of Complexity-Influencing Part Characteristics 394
Greta Tjaden, Nick Große, and Anne Meyer

Covariance Structure Analysis of the Collaboration Between Local Enterprises and Local Governments for the Creation and Communication of Value in Local Brands 410
Yu Fujii, Tomomi Nonaka, Fumihito Oura, and Seiko Shirasaka

Developing 3D Production Simulation Models in Industrial Production Systems . . . 425
Akshay Goyal, Yuji Yamamoto, and Ioanna Aslanidou

From Product-Service Systems to Smart Product-Service Systems. A Survey About German Companies and Practical Implications . . . 437
Christian Schiller, Mike Freitag, and Oliver Hämmerle

The Impact of Blockchain Implementations on Supply Chain Collaboration . . . 449
Yutong Bai and Ming K. Lim

An Overview of Cloud-Based Services for Smart Production Plants . . . 461
Auday Al-Dulaimy, Leo Hatvani, Moris Behnam, Anas Fattouh, and Koteshwar Chirumalla

Sustainability Evaluation of Electronic Components: A Case Study of a Swedish Temperature-sensing Solutions Manufacturer . . . 476
Trupti Borade, Zuhara Zemke Chavez, and Emanuel Bengtsson

Developing a Robust Multi-objective Optimization Model for Reverse Logistics of Electric Vehicle Batteries . . . 492
Roozbeh Yaghoubi, Aysan Mahboubi, Mina Kazemi Miyangaskary, and Samira Keivanpour

Construction Logistics Conceptualization: A Comprehensive Framework and Existing Challenges . . . 508
Uyen Ngo, Erlend Alfnes, Fabio Sgarbossa, and Marco Semini

Integrated Dynamic Flexible Job Shop and AIV Scheduling: Deep Reinforcement Learning Approach Considering AIV Charging and Capacity Constraints . . . 522
Arman Hosseini, Mohammad Feizabadi, and Zakaria Yahouni

Multi-project Scheduling Problem with Limited Preemption and Ordinary Feeding Precedence Relations . . . 537
Ali Fradi and Maha Benali

Author Index . . . 553

Modelling Supply Chain and Production Systems

Management of Measuring Equipment for Quality Assurance in Manufacturing Processes: A Decision-Making Support System

Isabel Lopes[1(✉)], Cláudia Pires[2], and Tiago Alão[3]

[1] ALGORITMI Research Centre/LASI, University of Minho, Guimarães, Portugal
ilopes@dps.uminho.pt

[2] Technological Centre for the Metal Working Industry (CATIM), Porto, Portugal

[3] Egitron, Mozelos, Portugal

Abstract. This paper aims to present a decision-making support system for the efficient management of measuring equipment in Manufacturing plants, providing support for condition analysis and calibration interval definition, a missing functionality in current computerized systems. For the same equipment, existing methods for defining the calibration/verification interval give very different results. The proposed solution for management of measuring equipment includes a developed approach for the selection of the method that best suits the equipment, considering its characteristics and condition of use. Since unpredictable situations, such as falls, shocks or misuse can change the behavior of measuring equipment, an additional approach for unpredictable behavior identification is included, using EWMA and I-MR charts. The approaches were developed and tested with data of measuring equipment of a cork company. This solution brings cost savings to companies since, by helping to adjust calibration or verification intervals, allows avoiding the use of inapt equipment for quality control of manufactured products.

Keywords: Calibration and verification · decision-making support system · EWMA chart · I-MR charts · quality assurance

1 Introduction

The measurement processes play an essential role in the whole operational process chain of products. One of the important elements of providing the customer with products of the required quality is correct measurements [1]. Quality assurance depends on the reliability of measuring systems, making the management of this equipment an important factor for organizational performance.

The international quality management standard ISO 9001:2015 requires that organizations determine and provide the necessary resources to ensure valid and reliable results [2]. When measurement traceability is considered by organizations to be an essential part of providing confidence in the validity of measurement results, it also requires that measurement equipment is calibrated or verified, or both, at specified intervals or prior

Published by Springer Nature Switzerland AG 2024
M. Thürer et al. (Eds.): APMS 2024, IFIP AICT 733, pp. 3–18, 2024.
https://doi.org/10.1007/978-3-031-71645-4_1

to use. As every measuring equipment has errors, calibration is a fundamental requirement to determine the magnitude of errors, for its approval and application in production control.

Regarding the periodicity of calibration, the management quality standard delegates this decision to the user of the measuring instrument. Calibration periodicity is a determining factor in metrological reliability. Short intervals involve high costs, while long ones may originate faulty measurements ([3, 4]). According to the ISO 10012:2003 standard, calibration intervals shall be reviewed and adjusted and historical data relating to previous calibrations and confirmations, as well as progress in knowledge and technology, may be used to determine confirmation intervals [5].

In practice, the most adopted calibration frequency is once a year [1]. Some methods are also adopted for defining or adjusting the calibration frequency [6]: adjustment of the interval by drift; adjustment of the interval based on the last three calibrations; and the Schumacher Method. However, there is no standardized method that presents the best efficiency in all cases ([6, 7]). The selection of the most appropriate method may be conditioned by certain factors associated with the characteristics of measuring equipment, its condition and use environment. Being a little explored problem in the industrial environment, one of the biggest challenges is also, currently, the implementation of the methods.

According to Wang and Huang [8], several information systems that assist in the effective management of monitoring and measuring equipment have been built. However, in general, these systems focus on managing standing information and periodical verification, without connection with production process management, which makes it difficult for these systems to guarantee the integrity and authenticity of the product data and to evaluate the condition and influence of the measuring equipment.

Therefore, in this paper a computer system, designated by System for Intelligent Management of Measuring Equipment (SIMME), is presented. This system includes a new approach for the selection of the most appropriate calibration interval definition method for each equipment. Furthermore, since the metrological properties of equipment undergo changes over time, caused by the aging of the equipment itself, as well as by unpredictable situations such as falls or shocks, the computer solution also includes an approach for monitoring measuring equipment, allowing through Exponential Weighted Moving Average (EWMA) and Individual Moving Range (I-MR) charts, to prevent them from being used inappropriately.

The paper is structured as follows. Section 2 presents a literature review on measuring equipment calibration and its management. Section 3 reports the research methodology. The proposed computer system with the new developed approaches is presented in Sects. 4 and in 5 the conclusions are drawn.

2 Literature Review

2.1 Equipment Calibration

Metrological confirmation is required to manufacturing companies to assure conformance to specifications of the manufactured products. According to the ISO 10012:2003 standard, metrological confirmation includes calibration and verification, any necessary adjustment or repair and subsequent recalibration.

The calibration consists of the process of comparing the accuracy of an instrument measurements to a known standard. The International Vocabulary of Metrology (VIM) defines calibration as "operation that, under specified conditions, in a first step, establishes a relation between the quantity values with measurement uncertainties provided by measurement standards and corresponding indications with associated measurement uncertainties and, in a second step, uses this information to establish a relation for obtaining a measurement result from an indication" [9].

On the other hand, the verification process, which can be carried out by the operator or technician responsible for the equipment, aims to verify if there has been any significant change in the characteristics of an equipment during the period of use since the last calibration and to establish a qualification of the equipment as a whole. Thus, it is a quick and effective means of verifying whether a given instrument is working within the tolerance limits and, although it requires a documented record, it does not require the issuance of a certificate as do a calibration process.

Calibrations or verifications need to be repeated periodically at appropriate intervals, the length of which depends on different variables (uncertainty required, frequency of use, type of use, equipment stability, etc.) [10]. Concerning the frequency of calibration or verification, there is not a universally applicable single best practice [7]. The officially published documents only provide general guidance [11]. Some criteria are recommended to be considered when establishing the calibration or verification interval: type of instrument, tendency to wear and drift, extent and severity of use, environmental conditions (climatic conditions, vibration, ionizing radiation, etc.), trend data obtained from previous calibration records, frequency of cross-checking against other reference standards or measuring devices, transportation arrangements and risk, etc. Portella and Frota [12] summarize some criteria identified in the literature to establish the calibration interval and define a method for the pharmaceutical industry where criteria such as overall impact of equipment non-compliance, intensity of use and stability stand out.

The Drift-based Adjustment method, the method "adjustment of the interval based on the last three calibrations" and the Schumacher method use historical calibration or verification data to define the next interval [6]. Some new methods can be found in the literature ([1, 4, 7]). Krndija et al. [4] present an algorithm to calculate the initial calibration interval. According to the authors, since there are often no instructions or recommendations defined by the standards or given by the manufacturer, the development of such models is of utmost importance. The methods of Pashnina [7] and Pacana [1] are both based on risk, complying with the ISO 9001 standard, which in 2015 included the risk-based approach. Pashnina [7] details the implementation of the risk-based approach to determine the optimal calibration interval by balancing financial exposure against measurement costs. Pashnina [7] concluded that the process of establishing the optimal calibration interval can be a complex endeavour requiring the use of mathematical and statistical processes and demanding a sufficient quantity of data including calibration results and registered costs. Pacana [1] proposed a five-stage methodology for implementing a risk-based approach to the management of measuring equipment. The matrix method was used to assess the equipment risk, including the two elements: probability and potential effects of the risk occurrence. Taking into account the risk associated with the equipment, a fixed period of calibration is assigned.

Once there are several known statistical methods for defining the calibration or verification interval, Oliveira and Jesus [6] used a modified SWOT (Strengths-Weaknesses-Opportunities-Threats) analysis, with weightings for each indicator, to define the method with the greatest potential for re-evaluating the calibration intervals of the instruments for temperature and static pressure utilized in natural gas networks in Brazil. This approach focuses exclusively on the intrinsic characteristics of each method (namely, facility of application, time of sample collection, robustness, etc.), disregarding the influence of a set of factors related to the characteristics of the measuring equipment, as well as to the condition of use.

In this paper, an approach developed to be included in a measuring equipment management system, aims to assist in identifying the most appropriate interval determination method for a given equipment such as in [6]. However, in addition of considering the intrinsic characteristics of the methods, it also takes into account the factors related to the measuring equipment characteristics, condition and environment of use.

Since during use measuring equipment may suffer unexpected behavior due to shocks, falls or misuse, for instance, monitorization of calibration or verification results is helpful to adjust the interval. With this purpose, Gaber [13] used the results of intermediate checks and control charts to monitor the need for recalibration of electronic balances and micropipettes used in a laboratory for environmental quality monitoring. This procedure allows for shortening or lengthened calibration intervals. The author concluded that laboratory equipment that requires certified calibration does not necessarily have to be done annually or at fixed time intervals. Equipment should be tracked through its control charts, and in case the device goes out of control, calibration is performed. The author used Xbar and R charts with subgroups of five observations which consists of the values obtained in one week. Baldo et al. [14] use individual and moving range charts (I-MR) for subgroups of three observations, to ascertain the consistency of the computer tomography measurement process.

In the present work, since usually only one observation is available concerning calibration or verification results, the EWMA and I-MR charts with subgroups of one observation are used to monitor the measuring equipment and to ascertain its stability and degree of conformance with the acceptance criterion.

2.2 Monitoring and Measuring Equipment Management Systems

Several recent publications related to measuring equipment management are focused on digitalization ([10, 17–21]).

Gadelrab and Abouhogail [10] present a thorough analysis of the calibration certificate and the calibration process to establish a bridge between the metrological concepts and information technology concepts addressed to those involved in digital transformation in metrology and specially developing digital calibration certificates.

Oppermann, [18] and [22], present a distributed software architecture that aims the digital transformation of paper-based procedures within legal metrology. In legal metrology, the calibration interval is predetermined and legally imposed. Such as the architecture proposed by Oppermann, the conceptual model of Melo Jr and Carmo [17] is also focused on legal metrology and intends to implement measuring instruments in a distributed blockchain-based architecture. Peters et al. [19] also refer to the application of blockchain solutions to measuring instruments within the scope of legal metrology. In Takatsuji et al. [20], blockchain technology is proposed to allow for the owner of the calibration certificate to picture the chain of calibrations to the primary standard.

In Wu et al. [15], a vehicle-mounted metrology laboratory information management system for military equipment is designed and developed to realize the information recording and traceability of the whole execution process of metrology task, which includes metrology task preparation, metrology task implementation and metrology task reporting. The system is designed to be used as a tool to support metrology tasks by a metrology laboratory and, is then oriented to record metrology results and accompany the metrology process, from task creation to the issuing of the calibration certificate.

With the aim of guaranteeing metrological quality to achieve product quality in a machinery manufacturing plant, Leonov and Shkaruba [16] defined processes according to the ISO 9001 standard and developed a functional model for the management system of metrological assurance of measurements. The determination of the calibration or verification interval was not included.

In addition to this system, several computer systems have been developed for manufacturing companies to manage their measuring equipment. However, despite there being some diversity of solutions that support the management of measuring equipment, they have essentially record functions and none were found to support decisions related to the definition of calibration interval and its adjustment depending on the measurement status, to ensure quality of manufactured products. In this sense, the decision support system proposed in this paper is considered innovative since it encompasses a difficulty faced by companies in this field and translates it into functions that allow for effective management of measuring equipment. Two new approaches are included in the computer system for supporting calibration and verification planning. The first allows defining the most adequate method for defining the calibration interval and the second allows monitoring the measuring equipment through calibration or verification results and identifying abnormal behaviors.

3 Research Method

A Portuguese cork company that produces stoppers and has more than a hundred measuring and monitoring equipment to manage, participated in the study to provide the necessary data and assist in the validation of the proposed approaches. Data that consists of the verification results of the company measuring equipment over a period of three years was used. This data was recorded in an Excel file which is the only tool in the company for measuring equipment management. The first step of the study was to apply the three best known and most used methods in practice, namely the Drift-based Adjustment method, the method "adjustment of the interval based on the last three calibrations" and the Schumacher method, to determine the interval for the next equipment verification to a significant equipment sample to understand the differences between the methods.

The Drift-based Adjustment method (method 1) is based on the drift of the history over determined periods. It determines the maximum interval until reaching the acceptance criterion limit taking into account the average drift. The method "adjustment of the interval based on the last three calibrations" (method 2) defines an increase or decrease relative to the previous interval. The intensity of the increase or decrease depends on the proximity of the error from the acceptance criterion in the last three calibrations. Considering the previous calibration interval, Schumacher method (method 3) suggests a new interval that is dependent on the conditions the measuring equipment is found in the previous calibrations [6]. This first step gave rise to drawing some conclusions about the methods, that were the basis for the definition of the selection approach.

Subsequently, bearing in mind that the equipment behavior can change due to abnormal conditions, the application of control charts to the verification data was tested. The first chart to be tested was EWMA (Exponentially Weighted Moving Average) chart since it is sensitive to small variations. The I-MR (Individuals and Moving Range) chart was also tested since it has the advantage of plotting the original values, the errors, making it possible to include in the chart the predefined acceptance criteria. This allows analyzing if the stability of the equipment errors is important to maintain the compliance of the equipment with the acceptance criteria. With the construction of these charts, the aim was to understand if they would bring relevant information about the behavior of the instruments, that is, to understand if the measuring equipment is stable and if there is a need to anticipate the next calibration or verification period.

After these approaches have been defined, the computer system was developed with the objective, on one hand, to include functionalities that allow recording relevant data about the measuring equipment management process and issuing reports, and on another, to aggregate the new functionalities to support decision-making.

4 The Proposed Decision-Making System

4.1 The Selection Approach for Interval Definition Method

The results of the application of the three different methods to a sample of 30 measuring equipment are reported in Table 1. For method 1, the new interval value is indicated. For method 2 and method 3, an increase or a decrease compared to the previous interval is specified.

Concerning the suggested interval for each method, there are great differences between the intervals proposed by method 1 and the intervals proposed by methods 2 and 3, although in most cases the three methods are aligned with regard to increasing or decreasing the periodicity. While, in comparison with the already defined interval, method 2 proposes a maximum increase of 1,2 months (an increase of 10%) and method 3 a maximum increase of 0.8 months (about 27%), method 1 even proposes a maximum periodicity between verifications of 60 months (5 years), corresponding to an increase of 57 months. Furthermore, it is observed that the decisions of methods 2 and 3 tend to be close, both with small increases.

The fact that it is not based on the current verification interval, allows explaining the great differences verified between the verification periodicity suggested by method 1 and that suggested by methods 2 and 3. Therefore, comparatively with the other two methods, adopting the interval indicated by method 1 can imply a bigger change compared to the predefined previous interval which is not always advisable. The calibration intervals defined by method 1 in the selected sample, consider that the drift remains stable. However, increases in drift may occur due to factors that are not intrinsic to the equipment, making it very likely that at the end of the interval the acceptance criteria will not be respected. On the other hand, for both method 2 and method 3, if the initial interval is too large or small, taking as reference the suggested increase or decrease, only after several verifications the appropriate interval will be defined since slight increase or decrease are suggested each time the method is applied.

Finally, the differences between the intervals defined by method 1 and by methods 2 and 3 may suggest that the interval initially specified by the organization was poorly defined. Therefore, the application of the Drift-based Adjustment method, method 1, can be suggested to analyze whether or not the initial interval was poorly defined.

Identification of Factors that May Influence the Selection. When selecting the method that most suits each equipment and its context of use, in addition to the intrinsic characteristics of each method, other factors that affect this decision should be considered. This includes the extent and severity of use and, the susceptibility of the equipment to the environmental conditions where it is used and the tendency to wear and drift [11]. To take into account these factors, a decision tree was developed to select the most appropriate method from among the three most used methods indicated above, including deciding whether the interval should be set based on calendar time or time in use.

Five criteria, explained below, were considered. The criteria Regularity of Use and Ease of Recording Measurements aim to understand if it is recommended to define the calibration interval based on time in use or on calendar time. The criterion Criticality of Use aims to take into account the impact of using the equipment when it is not in conformity, leading to greater or lesser caution in defining the calibration interval. The susceptibility of the equipment to the environmental conditions as well as the tendency to wear and drift is reflected in the criterion Equipment Stability which is defined based on equipment history. The criterion Calibration Costs aim to consider the financial aspect and is the last to be considered in the decision tree.

Regularity of use: A measuring equipment can be used more or less frequently, which influences the wear to which the instrument is subject. However, more important than knowing whether there is high wear or not, is whether or not this wear is steady. For equipment with regular use, the method could be based on the calendar time, and for equipment with variable use, the method should be based on the time in use.

Ease of recording measurements: Monitoring the use time of a piece of equipment or the number of measurements carried out by a piece of equipment can involve unnecessary costs when measurement record is not automatic. Therefore, before analyzing the regularity of use of the equipment, this criterion must be taken into account, since

Table 1. Verification interval determined using different methods

Equipment	Initial interval (month)	Method 1	Method 2	Method 3
Medcork 47	12	41,8	1,2 incr	keep the same
Medcork 22	12	30	1,2 incr	keep the same
Comparator table 9	3	48,5	0,3 incr	0,8 incr
Comparator table 8	3	35	0,3 incr	0,8 incr
Comparator table 7	3	63	0,3 incr	0,8 incr
Comparator table 6	3	63	0,3 incr	0,8 incr
Comparator table 4	3	63	0,3 incr	0,8 incr
Comparator table 3	3	31,5	0,3 incr	0,8 incr
Comparator table 2	3	31,5	0,3 incr	0,8 incr
Comparator table 1	3	31,5	0,3 incr	0,8 incr
Scale 18	3	22,8	0,3 incr	0,8 incr
Scale 17	3	15,6	0,3 incr	0,8 incr
Scale 6	3	8,3	0,3 incr	0,8 incr
Scale 1	3	6,2	0,3 incr	Keep the same
Scale 15	3	10,7	0,3 incr	Keep the same
Scale 14	3	8	0,7 decr	0,3 decr
Scale 12	3	18,7	0,3 incr	0,8 incr
Scale 10	3	14,9	0,3 incr	0,8 incr
Scale 9	3	37,4	0,3 incr	0,8 incr
Caliper 41	3	60	0,3 incr	0,8 incr
Caliper 40	3	15,8	0,3 incr	0,8 incr
Caliper 36	3	31,5	0,3 incr	0,8 incr
Caliper 35	3	21	0,3 incr	Keep the same
Caliper 32	3	15,8	0,3 incr	0,8 incr
Caliper 28	3	60	0,3 incr	0,8 incr
Caliper 27	3	31,5	0,3 incr	0,8 incr
Caliper 21	3	15,8	0,3 incr	0,8 incr
Caliper 17	3	21	0,3 incr	0,8 incr
Caliper 12	3	15,8	0,3 incr	0,8 incr
Caliper 7	3	31,5	0,3 incr	0,8 incr

in cases where this monitoring is not automatic, it is not feasible to use the time in use even if the use of the equipment is variable.

Criticality of use: The more critical the instrument is in terms of use, the more rigorous its control must be. Thus, knowing that the measuring instrument has to measure high critical quality characteristics affects the decision of which method is the most appropriate, since its fitness for use is even more important. For this study, a given equipment will be considered critical if its measurements determine whether the product should be rejected or not or if it is used to measure high critical quality characteristics.

Equipment stability: According to the International Vocabulary of Metrology [9], stability is the "property of a measuring instrument whereby it maintains its metrological properties constant over time" and can be expressed quantitatively in different ways, for example, by the variation of a metrological characteristic during a given time interval. More unstable equipment requires more regular and rigorous control. The stability of a measuring equipment can be described, for example, by its susceptibility to the environmental conditions where it is used. The instability of a given piece of equipment, which may be caused by the factors described (susceptibility to the surrounding environment or condition of use), may be reflected in its verification/calibration history.

Calibration costs: Costs involved in the calibration and/or verification processes have a significant weight with regard to the decision on the frequency of these procedures. In this way, it is necessary to balance the regularity of calibration/verification taking into account the costs involved.

The Approach. Based on the previously defined criteria and considering first the criteria that are more decisive in the selection of the interval between calibration determination method, the decision tree was developed (Fig. 1).

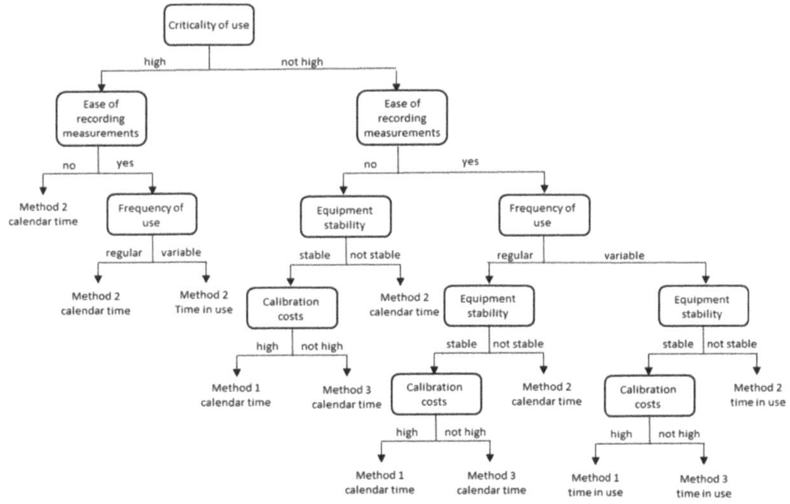

Fig. 1. Approach for interval between calibration determination method

When developing the decision tree, the following were considered:

- Whenever the equipment is used in a final inspection of the product or to measure a highly critical quality characteristic, it is not intended to significantly change the calibration interval compared to the previously defined interval. It is assumed that the

initial interval was defined with high caution, which can be considered whenever the interval defined by the manufacturer is adopted. Method 2 (adjustment of the interval based on the last three calibrations) is the one that causes the least change compared to the previous interval, which is why it is the chosen interval.
- Method 2 is also chosen whenever the equipment is not stable for the same reason.
- Time of use is adopted, instead of calendar time, whenever the equipment allows automatic data collection and the frequency of use of the equipment is variable.
- Method 1 (Drift-based adjustment method), which can originate high changes compared to the previous intervals, is selected when the equipment is not used in the final inspection or to measure a high critical quality characteristic, the equipment is stable and the calibration costs are high.
- Method 3 (Schumacher method), which in terms of interval change is positioned between method 1 and method 2, is adopted in the other cases.

The approach was tested and approved by the cork company after being applied to an equipment sample.

4.2 Approach for Equipment Stability Analysis

The I-MR is a Shewhart chart that is used to monitor the mean and variation of a process based on data that are individual observations [23]. This chart was applied in this work since the results of sequential calibrations are individual values and it is intended to observe the evolution of the measurement error over time. The error was used instead of the measuring value since it is possible to add the acceptance criteria (or maximum allowed error) to the charts and, consequently, to understand if it is being fulfilled and if there is a significant slack, allowing a less tight equipment control in this case. However, the application of I-MR chart requires normally distributed data to set control limits that reflect the process true variation. Therefore, the Kolmogorov-Smirnov test was used to verify if the set of data comes from a Normal distribution.

In addition to the I-MR chart, the EWMA chart was also built, on the one hand, to monitor data with non-normal distribution and, on the other hand, to allow faster detection of small variations in average measurements. The EWMA chart is used in statistical process control to monitor continuous or discrete variables (attributes) and can be used with individual samples or subgroups. Instead of Shewhart charts, EWMA chart has the particularity of using the entire history of observations. For each observation, a given weight is given, resulting in the current value of EMWA that is plotted on the chart. This gives this chart the capacity to detect small variations faster [24]. The weighting constant λ, which takes values between 0 and 1, defines the amount of influence that previous observations have on the current EWMA value. Values of λ close to 1 give more influence on the recent observations. For values of λ close to 0, the recent observations have less influence. Usually, the central line (CL) of this chart is the desired or target value of the process average. In this work, the observations are the values obtained sequentially in the verification process when the equipment is measuring the same reference standard and, the average value in the steady state is considered as the central line. Assuming that the error is always present, the aim was not to know if the equipment was measuring the true value of the reference standard, but instead to know if this error remains stable.

Figure 2 shows both charts for one test point of each of two measuring equipment, a scale and a comparator table. For the scale, the test point is 1000,00 g and for the comparator table, it is 15,00 mm and the corresponding acceptance criteria (CA) are 0,33 g and 0,07 mm, respectively. For obtaining the EWMA charts, the chart parameters, the constants λ and k, which can influence the chart efficiency, appraised through type I and type II errors, have to be defined. In this work, to detect small variations faster, a lower λ equal to 0,1 and a k of 2,7 were selected, providing in this way the same level of confidence as Shewhart charts.

In Fig. 2, the three charts corresponding to the scale show the observations 22, 23 and 24 outside the control limits (UCL- Upper Control Limit; LCL – Lower Control Limit) which means that the measuring equipment did not remain stable. The application of the Kolmogorov-Smirnov test allows concluding that the verification data follows a Normal distribution (in a stable situation) which validates the conclusions obtained with the I-MR chart. In this case, the I-MR chart and the EWMA chart detected the out-of-control situation at the same time. However, additional and valuable information is provided by the I chart. The central limit of the graph is not at the centre defined by the acceptance interval and the lower control limit of the chart is outside this interval, which means that even if the equipment remains stable, there is a possibility that it will become unfit for use. Therefore, the chart indicates the need for an adjustment even before the 22nd verification. Using the standard deviation obtained for the in-control situation, a capacity index of Cp = 1,23 was obtained. This value shows the need for frequent control of this equipment.

Fig. 2. I-MR and EWMA charts for a test point of a scale and comparator table

Regarding the comparator table, Fig. 2 shows a stable situation. However, for this equipment, the Kolmogorov-Smirnov test revealed that the verification data cannot be considered to follow a Normal distribution due to the little variability (i.e., there are several identical values). In the I chart, it can be observed that compared to the present variability, the acceptance interval is large which gives some assurance that in these conditions and if no abnormal situation occurs, the equipment will remain fit for use. The recommended interval definition method obtained by the application of the described approach is the Schumacher method which suggests an increase in the interval of 0,8 month (interval of 3,8 month). The Drift-based adjustment method recommends a new interval of 31,5 months. This information combined with what is observed in the charts provides certainty in adopting the change that the Schumacher method proposes.

The charts were applied to the results of the verification process, which are one error for each verification made. In the calibration process, in addition to the systematic error, uncertainty or random error is determined. Therefore, to build the charts for calibration data, whenever the error is positive, the random error should be added to the systematic error and, whenever the error is negative, the random error should be subtracted. In this way, the worst scenario is represented.

4.3 The Decision-Making System Architecture

The system architecture is represented in Fig. 3. Four main functions were defined: Master data management, Calibration, Decision support and Integration. The main function "master data management" aims to record all data characterizing the measuring equipment, including context data, which provides information to support the selection of the interval definition method. The Calibration function allows the record of calibration results and has the purpose of supporting the user in the definition of the next calibration date through the determination of the calibration interval. Furthermore, it allows the management of calibration plans to monitor and ensure their implementation. The decision support main function includes the approaches presented previously in this paper and also the application of the Repeatability and Reproducibility (R&R) study. The results of the application of the approaches and also of R&R study will be shown associated with the Calibration main function for identifying the need for calibration and guiding the user in the definition of the calibration interval. A R&R study is a systematic approach to assess the reliability and capability of a measurement system, which involves a measuring equipment and may involve different operators [25].

The Integration main function has the purpose of the integration of the application with other computer systems or applications allowing the registration of equipment from lists received, the indication of equipment availability (period not assigned to a production order) from other systems and reservation of the equipment for calibration.

Figures 4 and 5 show interfaces of the system prototype. Figure 4 shows a defined calibration/verification plan for a comparator table, including the criteria used to define the interval definition method which values or levels are introduced by the user, the method for interval definition and the respective calibration interval indicated automatically by the system. Figure 5 shows a dashboard returned by the system for a measuring device that includes the results of the past calibrations/verifications for the different test

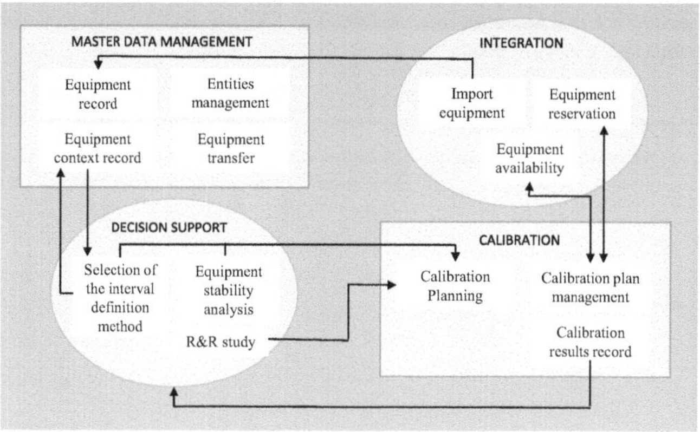

Fig. 3. System architecture

points, the method used for the interval definition and the visualization of the EWMA and I-MR charts to the corresponding equipment.

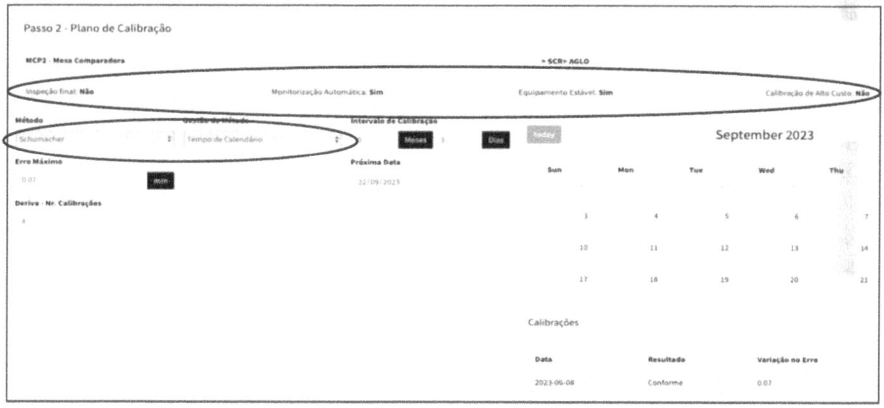

Fig. 4. System interface to define an equipment calibration plan

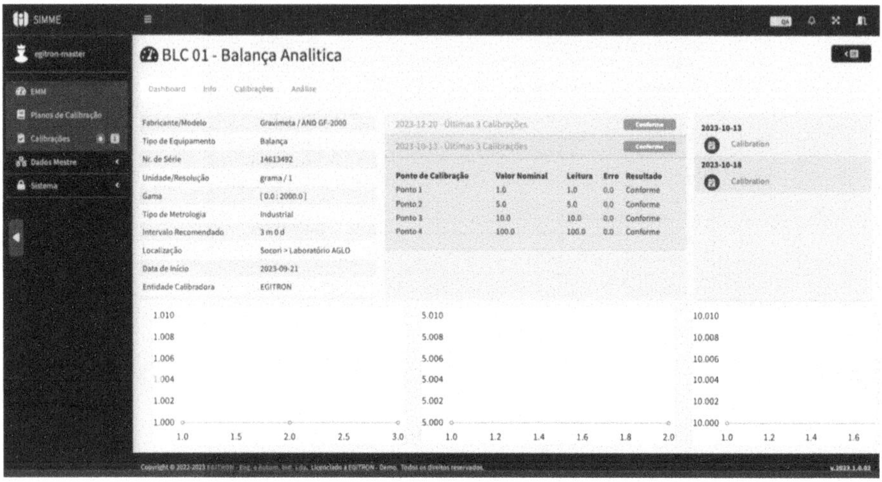

Fig. 5. System dashboard for a measuring device

5 Conclusion

The management of measuring equipment represents an essential tool for controlling the quality of any process or product, being a requirement of several standards including the ISO 9001 standard (7.1.5 requirement). The system proposed in this paper is intended to ensure the conformity of the measuring equipment for the intended use, thus ensuring the rigor of the measurements carried out for assessing the product conformity and for monitoring the manufacturing process, minimizing the damage resulting from its incorrect use. For this purpose, two approaches to help define the calibration interval were defined. Defining calibration intervals is considered one of the main difficulties for organizations. And, although this topic is addressed in international references and some methods are proposed, they lack guidelines for selecting the most appropriate method for each situation. The new approach defined and included in the system combines the characteristics of each method with a set of factors related to the measuring equipment characteristics, condition and environment of use, which makes the selection of the method for defining calibration intervals more reliable. Through the proposed system, the selection of the interval definition method and the calculation of this interval are automatic.

Besides, since measuring equipment may suffer damage or premature deterioration, the system also allows analysing calibration or verification results through control charts to detect abnormal behaviours. This second approach provides useful information for calibration management since it allows anticipating future failures in meeting acceptance criteria. The charts, the I-MR and the EWMA charts, are directly applied to the individual data generated in the successive calibration or verification tasks. The charts have the same function of detecting abnormal behaviour. However, the I-MR charts require the data to follow the Normal distribution. With data following the Normal distribution, both charts are still useful as the I-MR charts allow analysing error variability compared to the acceptance criteria.

These two approaches give this measuring equipment management system the ability to support decision-making. New functionalities will be added in the future, including the assignment of equipment to the production order based on its characteristics and availability.

Acknowledgments. This work has been supported by FCT – Fundação para a Ciência e Tecnologia within the R&D Units Project Scope: UIDB/00319/2020.

Disclosure of Interests. The authors have no competing interests to declare that are relevant to the content of this article.

References

1. Pacana, A.: Supervision of measuring equipment based on risk management and ISO 9001. Prod. Eng. Arch. **21**(21), 8–11 (2018)
2. ISO: ISO 9001 - Quality management systems — Requirements. International Organization for Standardization (ISO) (2015)
3. Morello, R., Paciello, V.: A decisional model for estimating reliability curve and maintenance intervals of instrumentation. Instrum Sci. Technol. **42**(4), 446–457 (2014)
4. Krndija, M., Latinović, M., Broćeta, G., Savić, G.: Measuring equipment calibration and determination of the initial calibration interval. Савремена Теорија И Пракса У Градитељству **14**(1), 411–421 (2020)
5. ISO: ISO 10012 - Measurement management systems — Requirements for measurement processes and measuring equipment. International Organization for Standardization (ISO) (2003)
6. Oliveira, E.C., Jesus, V.C.B.M.: Management of calibration intervals for temperature and static pressure transmitters applied to the natural gas industry. J. Nat. Gas Sci. Eng. **24**, 178–184 (2015)
7. Pashnina, N.: Determination of optimal calibration intervals by balancing financial exposure against measurement costs. Flow Meas. Instrum. **60**, 115–123 (2018)
8. An, H., Wang, M., Huang, M.: Research on the Measuring Equipment Management Methodology for the manufacturing process. In proceedings of the 2015 International Conference on Education, Management and Computing Technology, vol. 30, pp. 1630–1636 (2015)
9. JCGM 200: International vocabulary of metrology (VIM) – basic and general concepts and associated terms. In: 3rd edition, Joint Committee for Guide in Metrology (JCGM) (2012)
10. Gadelrab, M.S., Abouhogail, R.A.: Towards a new generation of digital calibration certificate: analysis and survey. Measurement **181**, 109611 (2021)
11. ILAC-G24, OIML D10: Guidelines for the Determination of Calibration Intervals of Measuring Instruments. Edition 2007 (E). Organisation Internationale de Metrologie Legale (OIML) (2007)
12. Portella, W. Frota, M. N.: Frequency of calibration: a critical issue for quality control in the pharmaceutical industry. In: 18th IMEKO World Congress 2006: Metrology for a Sustainable Development, vol. 3, pp. 1897–1899 (2006)
13. Gaber, S.E.: Adjusting external calibration intervals for auxiliary devices in testing laboratories according to intermediate checks results. Water Sci. **35**(1), 82–88 (2021)
14. Baldo, C.R., Probst, G.M., Dewulf, W.: Process behavior charts used to check the consistency of CT instruments. Procedia CIRP **96**, 330–335 (2020)

15. Wu, B., Xiao, Y., Ma, C.: Design and implementation of mobile metrology information management system. In: Proceedings of the 7th international Conference on Cyber Security and Information Engineering, pp. 657–660 (2022)
16. Leonov, A., Shkaruba, N.: Development of the management system for metrological assurance of measurements. J. Phys: Conf. Ser. **1515**(3), 032010 (2020)
17. Melo Jr, W., Carmo, L.F.R.C.: How blockchains can improve measuring instruments regulation and control. In: 2018 IEEE International Instrumentation and Measurement Technology Conference (I2MTC), pp. 1–6 (2018)
18. Oppermann, A., et al.: Digital transformation transformation in metrology: building a metrological service ecosystem. Procedia Comput. Sci. **200**, 308–317 (2022)
19. Peters, D., Wetzlich, J., Thiel, F., Seifert, J.: Blockchain applications for legal metrology. In: 2018 IEEE International Instrumentation and Measurement Technology Conference (I2MTC), pp. 1–6 (2018)
20. Takatsuji, T., Watanabe, H., Yamashita, Y.: Blockchain technology to visualize the metrological traceability. Precis. Eng. **58**, 1–6 (2019)
21. Barbosa, C.R.H., Sousa, M.C., Almeida, M.F.L., Calili, R.F.: Smart manufacturing and digitalization of metrology: a systematic literature review and a research agenda. Sensors **22**(16), 6114 (2022)
22. Oppermann, A., Eickelberg, S., Exner, J.: Toward digital transformation of processes in legal metrology for weighing instruments. In: 15th Conference on Computer Science and Information Systems (FedCSIS), pp. 559–562 (2020)
23. Rodríguez-Álvarez, J.L., et al.: Analysis and control of the paper moisture content variability by using fuzzy and traditional individual control charts. Chemom. Intell. Lab. Syst. **208**, 104211 (2021)
24. Montgomery, D.C.: Introduction to Statistical Quality Control, 6th edn. Wiley, New York (2009)
25. Zanobini, A., Sereni, B., Catelani, M., Ciani, L.: Repeatability and reproducibility techniques for the analysis of measurement systems. Measurement **86**, 125–132 (2016)

Understanding Coopetition Dynamics in Manufacturing Value Networks: A System Dynamics Based Causal Loop Diagram (CLD) Modelling Approach

Aziz Kemal Konyalioglu[1(✉)], Aylin Ates[1], Steve Paton[2], and Tuğçe Apaydın[3]

[1] University of Strathclyde, Hunter Center for Entrepreneurship, 199 Cathedral Street, Glasgow, UK
aziz.konyalioglu@strath.ac.uk

[2] University of Strathclyde, Management Science, 199 Cathedral Street, Glasgow, UK

[3] Industrial Engineering Department, Istanbul Kultur University, Istanbul, Turkey

Abstract. The concept of "coopetition" in manufacturing value networks involves firms engaging in both collaboration and competition simultaneously This approach, while aiming to leverage the benefits of both, inherently introduces a paradox concerning value creation and capture. Within a value network, coopetition involves various entities such as suppliers, distributors, subcontractors, and even competitors working together to enhance overall value. Despite a surge in research on coopetition, there remains a disjointed understanding, with limited exploration of its dynamics within manufacturing contexts. To address this gap, our study constructs a system dynamics model using causal loop diagrams (CLD) derived from an in-depth literature review within the manufacturing sector. Our aim is to comprehensively elucidate the factors influencing co-opetitive relationships and dynamics in manufacturing value networks and business ecosystems. Furthermore, existing literature emphasizes the need for a multifaceted perspective on coopetition in manufacturing. Our model, developed through consultation of extensive manufacturing and business literature and CLD application, identifies key factors driving coopetition dynamics in manufacturing and business ecosystem contexts. It represents the first comprehensive content analysis of coopetition dynamics within manufacturing and business ecosystems, serving as a valuable resource for scholars and professionals in the manufacturing and management field. By examining interconnected elements in a causal loop framework specific to manufacturing and business ecosystems, our study reveals how dynamic factors influence co-opetitive outcomes in these ecosystems. We explore various manufacturing-related aspects, such as supply chain dynamics in coopetition, technological innovations, and market dynamics, all impacting co-opetitive interactions. This comprehensive approach fills a literature gap, offering insights into critical factors affecting the co-opetitive process within manufacturing and business value networks. Our study's methodology, employing causal loop diagrams tailored to the manufacturing domain, stands out in the literature, providing a unique perspective on coopetition dynamics within manufacturing and management contexts.

Keywords: Coopetition · Dynamics · Manufacturing · Business · Innovation

Published by Springer Nature Switzerland AG 2024
M. Thürer et al. (Eds.): APMS 2024, IFIP AICT 733, pp. 19–33, 2024.
https://doi.org/10.1007/978-3-031-71645-4_2

1 Introduction

Nalebuff and Brandenburger (1996) have embraced the concept of 'coopetition,' which involves the intricate interplay of cooperation and competition within relationships [1]. Drawing upon principles of game theory [2, 3], Nalebuff and Brandenburger (1996) argue that coopetition significantly enhances firm performance as a strategy [1]. Their rationale is grounded in the idea that organizations operate within a complex ecosystem encompassing manufacturer, suppliers, competitors, clients, and complementors. These firms adopt various roles in this business landscape, allowing them to employ both collaborative and competitive strategies to achieve success. In this network, coopetition can manifest both horizontally and vertically among two or more companies [3–5]. Some authors pointed out that coopetition should be separately investigated based on two different logics cooperation and competition, which are interacting with each other, to have an understandable coopetition relationship [6–8]. Conceptualized by two different definitions, coopetition has been treated as a dyadic relationship since factors in coopetition strategy are complex and bilateral [7, 8].

On the other hand, recognizing their potential for coexistence and mutual benefit, they should be seen as interdependent constructs [6]. When the interplay between cooperation and competition (referred to as coopetition) is viewed as a continuum, it means that the phenomenon ranges from intense competition to robust cooperation. In this scenario, an increase in cooperation typically leads to a decrease in competition, and vice versa [8]. Furthermore, Bengtsson and Kock (2000) categorize co-opetitive relationships into three types depending on the emphasis placed on each behaviour: cooperation-driven coopetition, competition-driven coopetition, and balanced or equally weighted coopetition [6]. However, the dominance in a coopetitive relationship can be shaped by the types of activities carried out by the firms involved [9].

Coopetition dynamics feature firms interconnected by contracts within networks to achieve shared goals in supply chains, manufacturing, and business ecosystems [10]. These interconnections create unique relational strategies, influencing firm interactions in production and service networks [11–13]. The literature debates critical success factors of coopetition, noting its tension-filled and risky nature, potentially leading to negative outcome [14–16]. Understanding these tensions is crucial as they occur at all hierarchical levels [6]. Key success factors depend on multiple elements, including supply chain performance, analytics capabilities, process management, and manufacturing ability [17–19]. Studies categorize these factors into inter-relationship and internal factor and the evaluation methods include the Analytic Hierarchy Process and statistical tests [19, 20]. In this context, the scholars stated that trust, commitment, inter-organizational factors, learning experiences, organizational culture, and external factors are pivotal for coopetition effectiveness [20, 21].

While coopetition is inherent in business and manufacturing ecosystems and has been acknowledged as significant by scholars, it has not been comprehensively incorporated into the existing body of literature on ecosystems [22, 23]. Regrettably, there is a conspicuous lack of extensive research addressing the complex dynamics of coopetition among organizations and firms. In many instances to identify the dynamics of strategies, the scenario falls somewhere between the pure cooperation and pure competition, which is commonly referred to as "coopetition." [24]. Thus, there is a broad consensus in the literature to comprehensively elucidate the dynamics of coopetition or to employ a multifaceted perspective on coopetition in order to attain a deeper and more comprehensive grasp of its dynamics [12, 25, 26].

In this study, thus, we will try to investigate two research questions centred around coopetition dynamics.

RQ1. What are the dynamics of coopetition in manufacturing and business ecosystems?
RQ2. How can the dynamics of coopetition be modelled by using Causal Loop Diagrams (CLD), and how can one factor in coopetition strategy affect another to observe the overall system behaviour?

Based on the research questions given above that we will address; this study aims to create a system dynamics model using causal loop diagrams (CLD) to elucidate the dynamic nature of coopetition. The model we've developed comprises two phases: one involving an extensive literature review on coopetition dynamics and the other utilizing CLD to analyse coopetition literature. In our research, we've identified the pivotal factors that drive the dynamics of coopetition through a comprehensive literature review, followed by the application of the CLD approach to comprehend the components of coopetition. Notably, our research represents one of the first in-depth analysis of coopetition dynamics, drawing from existing coopetition studies in the literature, making it a valuable resource for both academics and practitioners seeking a deeper understanding of this field.

The reminder of our paper is as follows. First, we introduce the materials and method to understand the dynamics of coopetition used in our CLD modelling. Second, we present CLD methodology, which will be involved in our modelling methodology and lastly the discussion around our modelling and the dynamics of coopetition.

2 Materials and Method

We have conducted an extensive systematic literature review on coopetition, employing Latent Dirichlet Allocation Algorithm and bibliometric analysis with a corpus of 1802 articles sourced from various databases. Thus, this integration of methods to conduct a literature review provides us the dynamics and key success factors of coopetition strategy. Our integrated method also gives clusters to identify each factor affecting coopetition dynamics in manufacturing and business ecosystems. These clusters have been used to understand coopetition dynamics used in CLD modelling. The subsequent phase we apply in this study is to use these dynamics to develop a causal loop diagram integrated with system dynamics model approach by the aid of the results of our literature review methodology.

2.1 System Dynamics and Causal Loop Diagrams

Causal loop diagrams serve to pinpoint the key feedback loops within systems. They are instrumental in illustrating the fundamental causal mechanisms that are theorized to produce the typical behaviour pattern of the system as time progresses [27]. A feedback loop consists of two or more interlinked variables, causing a self-reinforcing or self-regulating effect. The connection between variables in the loop can be either positively reinforcing or negatively counteracting [28, 29]. The arrows in the diagram represent the direction of cause-and-effect relationships. The plus (+) sign next to an arrow signifies that a change in the variable at the tail of the arrow will lead to a change in the variable at the head of the arrow in the same direction. Conversely, a minus (-) sign indicates that a change in the variable at the tail of the arrow will result in a change in the variable at the head of the arrow in the opposite direction [30].

One way to refine the traditional definition of a positive influence in a causal-loop diagram, as hinted by the previous observations on curvature, is as follows: A has a positive influence on B when an increase (or decrease) in A leads to a value of B that is higher (or lower) than it would have been if A had not changed. A similar definition can be devised for a negative influence. In the example of the beer glass, when the flow rate increases, it causes a higher beer level in the glass compared to what would have happened if the rate had remained constant. Likewise, a decrease in the flow rate results in a lower beer level compared to a scenario with a constant rate [31, 32].

Figure 1 mainly shows the CLD approach steps to understand value creation in manufacturing and business ecosystems.

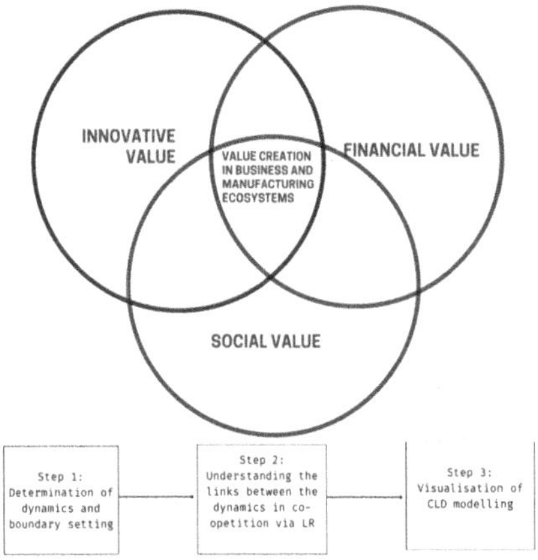

Fig. 1. CLD methodology to understand coopetition dynamics in manufacturing and business ecosystems.

3 Modelling Coopetition Dynamics Via Causal Loop Diagrams

In this part of the study, the affecting factors, thus, the dynamics of coopetition are identified to construct a CLD model. Our study embarks on a comprehensive exploration of the intricate dynamics governing coopetition within manufacturing value networks, employing a robust system dynamics methodology anchored in seminal contributions by Forrester (1977) and Sterman (2006) [33, 34]. System dynamics based CLD offers a structured framework for dissecting the multifaceted relationships among various factors that influence co-opetitive performance. By following a systematic process encompassing model conceptualization, formulation, CLD visualisation, and analysis, this approach facilitates a nuanced understanding of how coopetition strategies shape performance outcomes over time [31–33].

The Causal Loop Diagram (CLD) developed herein serves as a powerful visual representation elucidating the complex interplay among managerial, financial, social, and innovative dynamics underpinning coopetition. Within the realm of managerial and organizational dynamics in coopetition, pivotal factors such as trust and loyalty emerge as foundational elements, echoing the assertions of Osarenkhoe (2010), further substantiated by Akdoğan and Cingoz (2012) and Lascaux (2020) [24, 35, 36]. These factors foster an environment conducive to commitment and collaboration, thereby positively influencing co-opetitive performance. Similarly, the pursuit of common economic and social goals, as advocated by Schmoltzi and Wallenburg (2012), fosters resource pooling and alignment of objectives, thereby enhancing manufacturing and supply chain performance [37, 38]. Effective conflict management capabilities, leadership styles, networking strategies, and organizational culture further augment the stability and efficacy of co-opetitive relationships, as expounded upon by prior research [39–42].

Financial dynamics constitute another critical dimension in shaping coopetition strategies. Equitable profit-sharing agreements, robust investment capacities, and efficient cost-sharing mechanisms emerge as indispensable determinants of co-opetitive success, as underscored by the scholarly discourse [43–45]. Nevertheless, financial risks associated with unintended knowledge transfer, opportunistic behaviours, and escalating operational costs introduce complexities into the co-opetitive landscape, as noted by extant literature [46, 47].

Social and innovative dynamics complement these financial considerations, with factors such as social cohesion, mutual sharing in business ecosystems, entrepreneurial orientation, knowledge acquisition and transfer capabilities, and analytics capabilities exerting significant influence on co-opetitive outcomes, as delineated in scholarly investigations [47–53].To provide a broader aspect, for instance, Strese et al. (2016) state that trust can be a contributing factor for the management of tensions [54], which affects leadership capabilities also affecting co-opetitive performance [54, 55]. Combining these effects via literature review provides us to construct our CLD modelling as seen in Fig. 2. Also, exemplifying CLD approach in the context of coopetition dynamics helps us address our research questions in a visualised way to understand the overall behaviour. Figure 2 shows the CLD modelling and the effects of dynamics on coopetition performance as well as the relationships. Stella software (https://www.iseesystems.com/store/products/stella-architect.aspx) has been used to create our CLD.

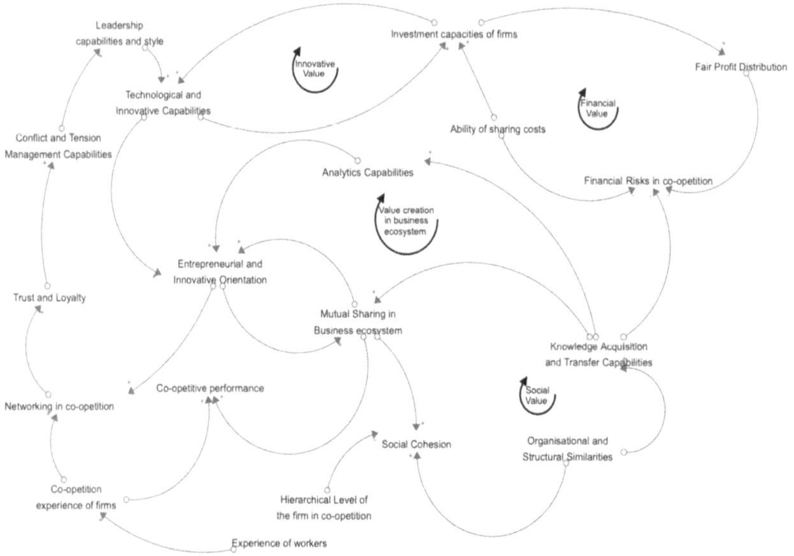

Fig. 2. CLD modelling for coopetition dynamics.

4 Results and Discussion

Causal loop diagrams unveil intricate interactions within systems and expose latent systemic issues [33]. Positive loops are depicted by lines marked with plus signs (+), while negative loops are represented by lines adorned with minus signs (−) [34] regarding coopetition dynamics in manufacturing and business ecosystems, as illustrated in Fig. 2. Based on the results obtained by the literature integrated CLD modelling, which includes 21 system variables, we have seen that there are 3 main values created by coopetition dynamics which are innovative, social and financial value respectively.

We found that value creation in a business and manufacturing ecosystems is dynamic and centred around mutual sharing. This mutual sharing activity creates financial value, innovative value, and social value. This mutual sharing among actors creates social cohesion, co-opetitive performance and entrepreneurial and innovative orientation. Next, we will explain our findings as presented in Fig. 2 in detail.

4.1 Innovative Value in Coopetition

Opting to initiate the innovation process within the firm yields certain advantages by facilitating access to complementary knowledge sources. However, concurrently, it subjects the firm to the potential hazard of opportunistic behaviour from external partners. Consequently, inter-organizational R&D agreements are distinguished by a blend of competitive and collaborative actions including innovation, named as "coopetition" [56, 57]. The significance of coopetition appears particularly in the realm of Small and Medium-Sized Enterprises (SMEs) regarding the innovative value created during coopetition process. In the face of escalating technological competitions and the increasing complexity of technologies, SMEs encounter a myriad of challenges, including escalating research and development costs, heightened risk and uncertainty in technological advancements, and limited resources for undertaking expansive innovation projects [58, 59].Typically, this challenge is addressed by acquiring the requisite knowledge for innovation from external sources through inter-organizational agreements and partnerships and coopetition proves especially advantageous for innovations within high-technology contexts [58, 60, 61]. As seen in Fig. 2 designed by CLD and the literature, coopetition has demonstrated its efficacy in facilitating joint investments, pooling resources, and collaborative Research and Development (R&D) projects, which enhances innovative capabilities in a firm through coopetition activities [62, 63].

Furthermore, there is a direct relationship between entrepreneurial and innovative orientation and coopetition activities. The innovation capacity in a firm serves as a mediating factor in the association between entrepreneurial orientation and firm performance within the creative industries [64]. This knowledge and capacity can guide them in leveraging their collective learning, resources, and skills to achieve crucial performance advantages, including innovation, entrepreneurial orientation, and operational excellence. These advantages, in turn, can confer a competitive edge and leadership positions. This advantage can be knowledge sharing obtained by coopetition process affecting entrepreneurial orientation positively [65, 66]. On the other hand, from supply chain perspective, Li et al. (2011) states that the entrepreneurial orientation of a distributor plays a positive moderating role in the connections between coopetition factors and a manufacturer's knowledge acquisition. This suggests that enhancing the distributor's entrepreneurial orientation can enhance the effectiveness of coopetition, subsequently influencing the knowledge acquisition of the manufacturer [10]. This underscores the significance of integrated analysis spanning the domains of supply chain management and entrepreneurship [10]. Additionally, analytical capabilities stand out as positive contributors to firm performance within the context of coopetition and innovation, in particular, the capability in big data analytics exerts a moderating influence on the correlation between cooperation within the innovation ecosystem and coopetition in terms of coopetition performance [67, 68].

4.2 Social Value in Coopetition

The second part that we investigated in our CLD modelling is to understand social value created by coopetition process. Coopetition is regarded as an essential process for enhancing market performance, emerging as a natural strategy stemming from the activities of social enterprises. The behaviour and context of social enterprises contribute to the creation of a novel framework for the formation of coopetition to create a social value [69]. In certain cases, coopetition has the potential to generate socio-environmental value. While socio-environmental value is inherently a public benefit and cannot be directly appropriated by coopeting parties, it functions as a catalyst for creating value for the collaborating firms [70]. To evaluate the social value in coopetition, knowledge transfer efficiency can be a measure since the transfer of knowledge, conceptualized as the mobilization, assimilation, and utilization of knowledge resources, serves as a mediator between intra-organizational social capital among organization members and the organizational performance outcomes of growth and innovation performance, thereby influencing social value within the organization [71, 72]. From the perspective of partnerships, efficient knowledge sharing constitutes a crucial component of coopetitive ecosystem partnerships [19]. In the context of coopetitive partnerships, the presence of explicit knowledge, strong relationship ties, and a shared intent for learning is imperative. Therefore, firms should implement mechanisms that foster these conditions within their partnerships [73].

However, to create a social value, a characteristic of co-opetitive relationships is the mutual adaptation of processes and products by the involved parties to enhance alignment. This involves the sharing of information and experiences, with a concerted effort to eliminate or minimize sources of insecurity and uncertainty [74]. In that sense, organizations possessing extensive prior experience tend to enjoy a more favourable reputation in their engagement and execution of successful R&D funded projects creating a social value additionally [75, 76]. Consequently, such organizations can readily recognize one another and make informed decisions to leverage their respective expertise by participating in the same consortium, even when it entails forming partnerships with potential competitors [75]. However, it's also worthy to note that tensions may arise within coopetition scenarios and they may originate from human-based disagreements, as the intricate relationships among individuals or firms can be a contributing factor [76, 77]. The duration of tenures in competitors within firms is presumed to be a factor leading to involvement in tensions, potentially stemming from the hierarchical levels established by the firm [78]. When it comes to the individual level, the dynamics of coopetition are significantly influenced by factors such as organizational culture, motivation, and the established rules for interaction due to the possible tensions [79, 80].

Corporate responsibility is considered as a social value in coopetition process [76]. Even in multi-national firms, coopetition among multinational corporations has the potential to catalyse subsequent coopetition among other firms, fostering advanced technological development. Embracing coopetition may be regarded as a viable strategy when navigating the intricate landscape of sustainability and addressing a myriad of public policy concerns [76, 81]. For instance, Ritala (2012) states that innovation performance sees improvement through coopetition alignment, implying that the sampled firms may utilize coopetitive agreements to collaboratively generate more valuable and innovative solutions for their customers socially [82].

4.3 Financial Value in Coopetition

Experience in coopetition consistently proves advantageous, suggesting a contingency perspective on the correlation between coopetition experience and firm profitability [83]. In that sense, value appropriation in coopetition is delineated as the individual portion of the value that a firm can seize. Rational, profit-seeking firms aim to secure the profits stemming from the innovations they generate [84] For this value appropriation in coopetition, Cohen and Zhang (2022) states that co-opetitive partnerships prove advantageous for both platforms, particularly in the face of intense demand-side competition financially and economically. It is emphasized that riders and drivers can also experience benefits from coopetition financially, highlighting that well-crafted coopetition terms can result in mutual advantages for all parties involved [43].

On the other hand, Ritala states that (2012) coopetition is perceived as a mechanism for implementing price discrimination, primarily enhancing profitability, and diminishing the incentive to innovate [82]. However, there is a big consensus in the literature that a positive correlation is present between coopetition and financial performance, aligning with findings from prior research [82, 83, 85]. This financial performance via coopetition strategy has been used in our CLD modelling illustrated in Fig. 2. This outcome is attributed to decision-makers leveraging cooperation with competitors, enabling the acquisition of new resources, capabilities, and opportunities through the implementation of a coopetition strategy [2, 85, 86]. However, the risks during the coopetition process should be considered additionally. For example, when we look at the financial value in coopetition entailing potential disadvantages in the literature, introducing risks to organizations such as opportunistic behaviour from involved parties, reduced efficiency in pursued activities and goals, a compromise in organizational and decision-making independence, a decline in market position, and an increase in operating costs (with the latter two outcomes identified as the most significant threats). Additionally, one should anticipate risks related to control over operations, an asymmetry in benefits compared to expenditures, conflictual relationships, and the potential threat of a decline in the overall value of the company during the coopetition process [87].

5 Limitations and Further Research

Although we identified significant findings through a comprehensive analysis, there are some limitations in our research. Firstly, we did not consider different coopetition types to be taken into consideration in CLD modelling such as the two primary types of coopetition: vertical versus horizontal coopetition. Prior studies have identified a multitude of factors that contribute to shaping the impact of coopetition, particularly concerning the various types of coopetition [88, 89]. On the other hand, the typology of coopetition has been defined by many authors [76, 90]. For example, our CLD modelling did not consider monoplayer, contender, partner or adapter coopetition types defined by Chin et al. (2008) [19]. However, the types of coopetition have not been involved in our modelling to further analyse. Our focus was on identifying the factors affecting co-opetitive performance so we did not include any specific actors.

Secondly, the adoption of a coopetition strategy is conducive to achieving low carbon emissions, thereby contributing to environmental sustainability [56]. Furthermore, Hafezi et al. (2022) indicates that the implementation of a coopetition strategy enables companies to produce more sustainable products, fostering the creation of environmentally friendly offerings [91] and Kumar et al. (2021) states that the interaction of coopetition between sustainability and manufacturers may generate tensions [13]. However, it is evident that coopetition can be a catalyst for the development of sustainability strategies in manufacturing. However, our study did not focus on environmental value. For further research, environmental aspect of coopetition should be considered in order to create a more inclusive causal loop diagram since they can be considered as coopetition dynamics.

Thirdly, the weights of the dynamics of coopetition have not been considered in this study. Even though the interactions between the variables can be found in the literature, the degree of interactions have not been analytically measured. Thus, a prioritization between the dynamics should be implemented to measure the weights of interaction by using a multi-criteria decision-making model using empirical data.

6 Conclusion

While coopetition seeks to leverage the benefits of both collaboration and competition, it inherently introduces a paradox by potentially conflicting with the simultaneous goals of creating and capturing value. In an effort to bridge this gap, this study strived to build a system dynamics model using causal loop diagrams (CLD) to elucidate the dynamics of coopetition in manufacturing and business ecosystems. The objective was to formulate a comprehensive framework outlining the factors that influence co-opetitive performance and relationships and, consequently, the dynamics inherent in coopetition integrated with manufacturing orientation and management concepts.

Our results show that business and manufacturing ecosystems embody both collaboration and competition, making the coopetition a highly relevant strategy to examine. We identified that the moderating role of mutual sharing in business and manufacturing ecosystem has been centred to create social, financial, and innovative value in coopetition. Furthermore, the experience in coopetition and the expertise of employees in the organisations and manufacturing companies serve as positive moderators in the relationship between social value and coopetition performance.

In our CLD modelling, we illustrated that co-opetitive performance depends on many factors, which can be identified as innovative, social or financial dynamics in the literature. Thus, co-opetitive performance is not easily measured given that its dynamics are related to many elements changing based on the firms' performance, experience, financial investments and innovation capability. Our examination challenges conventional viewpoints that prioritize trust, a trustworthy partnership, or technological advancement as singular determinants of coopetitive performance. Instead, we propose a nuanced and interconnected model. We argue that coopetitive performance operates cyclically, with trust acting as a foundational component that enhances the formation of trustworthy partnerships.

The chapter offers an insightful examination of coopetition dynamics within manufacturing and business ecosystems, utilizing causal loop modelling. While the practical implications and conclusions drawn from the study are significant, there remains an opportunity to delve deeper into both theoretical and practical implications. While the chapter effectively outlines the factors influencing co-opetitive performance and relationships, further elaboration on how these findings contribute to existing theoretical frameworks or extend current understanding would enhance the scholarly contribution. Providing insights into how the study's findings align with or challenge existing literature on coopetition dynamics, manufacturing orientation, and management concepts would enrich the theoretical discourse surrounding these topics. Moreover, the practical implications could be elucidated with more specificity. Concrete examples or case studies illustrating successful implementation of co-opetitive strategies in manufacturing and business ecosystems could strengthen the chapter's practical relevance.

Disclosure of Interest. No potential conflict of interest was reported by the author(s).

References

1. Nalebuff, B., Brandenburger, A.: Coopetition, p. 0. na (1996)
2. Bengtsson, M., Kock, S.: Coopetition—Quo vadis? Past accomplishments and future challenges. Ind. Mark. Manage. **43**(2), 180–188 (2014)
3. Bouncken, R.B., Fredrich, V.: Coopetition: performance implications and management antecedents. Int. J. Innov. Manag. **16**(05), 1250028 (2012)
4. Choi, P., Garcia, R., Friedrich, C.: The drivers for collective horizontal coopetition: a case study of screwcap initiatives in the international wine industry. Int. J. Strateg. Bus. Alliances **1**(3), 271–290 (2010)
5. Lacoste, S.: "Vertical coopetition": The key account perspective. Ind. Mark. Manage. **41**(4), 649–658 (2012)
6. Bengtsson, M., Kock, S.: "Coopetition" in business Networks—to cooperate and compete simultaneously. Ind. Mark. Manage. **29**(5), 411–426 (2000)
7. Bengtsson, M., Eriksson, J., Wincent, J.: Co-opetition dynamics–an outline for further inquiry. Competitiveness Rev. Int. Bus. J. **20**(2), 194–214 (2010)
8. Ricciardi, F., Zardini, A., Czakon, W., Rossignoli, C., Kraus, S.: Revisiting the cooperation–competition paradox: A configurational approach to short-and long-term coopetition performance in business networks. Eur. Manag. J. **40**(3), 320–331 (2022)
9. Morris, M.H., Kocak, A., Ozer, A.: Coopetition as a small business strategy: implications for performance. J. Small Bus. Strategy (archive only) **18**(1), 35–56 (2007)
10. Li, Y., Liu, Y., Liu, H.: Coopetition, distributor's entrepreneurial orientation and manufacturer's knowledge acquisition: evidence from China. J. Oper. Manag. **29**(1–2), 128–142 (2011)
11. Kim, Y., Choi, T.Y., Yan, T., Dooley, K.: Structural investigation of supply networks: a social network analysis approach. J. Oper. Manag. **29**(3), 194–211 (2011)
12. Rusko, R.: Exploring the concept of coopetition: a typology for the strategic moves of the Finnish forest industry. Ind. Mark. Manage. **40**(2), 311–320 (2011)
13. Kumar, A., Connell, J., Bhattacharyya, A.: Coopetition for corporate social responsibility and sustainability: drivers and success factors. Sustain. Acc. Manage. Policy J. **12**(6), 1208–1238 (2021)
14. Bengtsson, M., Raza-Ullah, T., Vanyushyn, V.: The coopetition paradox and tension: the moderating role of coopetition capability. Ind. Mark. Manage. **53**, 19–30 (2016)
15. Gnyawali, D.R., Madhavan, R., He, J., Bengtsson, M.: The competition–cooperation paradox in inter-firm relationships: a conceptual framework. Ind. Mark. Manage. **53**, 7–18 (2016)
16. Kim, J., Parkhe, A.: Competing and cooperating similarity in global strategic alliances: an exploratory examination. Br. J. Manag. **20**(3), 363–376 (2009)
17. Chen, X., Wang, X., Xia, Y.: Production coopetition strategies for competing manufacturers that produce partially substitutable products. Prod. Oper. Manag. **28**(6), 1446–1464 (2019)
18. Wilhelm, M.M., Kohlbacher, F.: Coopetition and knowledge co-creation in Japanese supplier-networks: the case of Toyota. Asian Bus. Manag. **10**, 66–86 (2011)
19. Chin, K.S., Chan, B.L., Lam, P.K.: Identifying and prioritizing critical success factors for coopetition strategy. Ind. Manag. Data Syst. **108**(4), 437–454 (2008)
20. de Resende, L.M.M., Volski, I., Betim, L.M., de Carvalho, G.D.G., De Barros, R., Senger, F.P.: Critical success factors in coopetition: evidence on a business network. Ind. Mark. Manage. **68**, 177–187 (2018)
21. Petter, R.R.H., Resende, L.M., de Andrade Júnior, P.P., Horst, D.J.: Systematic review: an analysis model for measuring the coopetitive performance in horizontal cooperation networks mapping the critical success factors and their variables. Ann. Reg. Sci. **53**, 157–178 (2014)

22. Cobben, D., Ooms, W., Roijakkers, N., Radziwon, A.: Ecosystem types: a systematic review on boundaries and goals. J. Bus. Res. **142**, 138–164 (2022)
23. Riquelme-Medina, M., Stevenson, M., Barrales-Molina, V., Llorens-Montes, F.J.: Coopetition in business ecosystems: the key role of absorptive capacity and supply chain agility. J. Bus. Res. **146**, 464–476 (2022)
24. Osarenkhoe, A.: A study of inter-firm dynamics between competition and cooperation–A coopetition strategy. J. Database Mark. Cust. Strategy Manag. **17**, 201–221 (2010)
25. Akpinar, M., Vincze, Z.: The dynamics of coopetition: a stakeholder view of the German automotive industry. Ind. Mark. Manage. **57**, 53–63 (2016)
26. McGrath, H., O'Toole, T., Canning, L.: Coopetition: a fundamental feature of entrepreneurial firms' collaborative dynamics. J. Bus. Ind. Mark. **34**(7), 1555–1569 (2019)
27. Forrester, J.W.: System dynamics, systems thinking, and soft OR. Syst. Dyn. Rev. **10**(2–3), 245–256 (1994)
28. Bala, B.K., Arshad, F.M., Noh, K.M.: Causal loop diagrams. In: System Dynamics: Modelling and Simulation, pp. 37–51. Springer, Singapore (2017). https://doi.org/10.1007/978-3-031-01919-7_4
29. Maani, K.: Systems Thinking. World Scientific Book Chapters, pp. 15–23 (2017)
30. Spector, J.M., Christensen, D.L., Sioutine, A.V., McCormack, D.: Models and simulations for learning in complex domains: using causal loop diagrams for assessment and evaluation. Comput. Hum. Behav. **17**(5–6), 517–545 (2001)
31. Richardson, G.P.: Problems with causal-loop diagrams. Syst. Dyn. Rev. **2**(2), 158–170 (1986)
32. Richardson, G.P.: Problems in causal loop diagrams revisited. Syst. Dyn. Rev. J. Syst. Dyn. Soc. **13**(3), 247–252 (1997)
33. Forrester, J.W.: Growth cycles. De Economist **125**(4), 525–543 (1977)
34. Sterman, J.D.: Learning from evidence in a complex world. Am. J. Public Health **96**(3), 505–514 (2006)
35. Akdoğan, A.A., Cingoz, A.: An empirical study on determining the attitudes of small and medium sized businesses (SMEs) related to coopetition. Procedia Soc. Behav. Sci. **58**, 252–258 (2012)
36. Lascaux, A.: Coopetition and trust: what we know, where to go next. Ind. Mark. Manage. **84**, 2–18 (2020)
37. Schmoltzi, C., Wallenburg, C.M.: Operational governance in horizontal cooperations of logistics service providers: performance effects and the moderating role of cooperation complexity. J. Supply Chain Manag. **48**(2), 53–74 (2012)
38. Gurnani, H., Erkoc, M., Luo, Y.: Impact of product pricing and timing of investment decisions on supply chain coopetition. Eur. J. Oper. Res. **180**(1), 228–248 (2007)
39. Hafezi, M., Zhao, X., Zolfagharinia, H.: Together we stand? Coopetition for the development of green products. Eur. J. Oper. Res. **306**, 1417–1438 (2022)
40. Jiang, W., Lu, W., Xu, Q.: Profit distribution model for construction supply chain with cap-and-trade policy. Sustainability **11**(4), 1215 (2019)
41. Luo, Y.: A coopetition perspective of global competition. J. World Bus. **42**(2), 129–144 (2007)
42. Kumar, A., Connell, J., Bhattacharyya, A.: Coopetition for corporate social responsibility and sustainability: drivers and success factors. Sustain. Acc. Manage. Policy J. **12**, 1208–1238 (2020)
43. Cohen, M.C., Zhang, R.: Competition and coopetition for two-sided platforms. Prod. Oper. Manag. **31**(5), 1997–2014 (2022)
44. Sun, S., Zhang, J., Lin, H.: Evolutionary game analysis on the effective coopetition mechanism of partners within high quality pork supply chain. In: 2008 IEEE International Conference on Service Operations and Logistics, and Informatics, vol. 2, pp. 2066–2071. IEEE (2008)
45. Mantena, R., Saha, R.L.: Coopetition between differentiated platforms in two-sided markets. J. Manag. Inf. Syst. **29**(2), 109–140 (2012)

46. Bakshi, N., Kleindorfer, P.: Co-opetition and investment for supply-chain resilience. Prod. Oper. Manag. **18**(6), 583–603 (2009)
47. Sethi, R., Smith, D.C., Park, C.W.: Cross-functional product development teams, creativity, and the innovativeness of new consumer products. J. Mark. Res. **38**(1), 73–85 (2001)
48. Dyer, B., Song, X.M.: Innovation strategy and sanctioned conflict: a new edge in innovation? J. Product Innov. Manage. Int. Publ. Product Dev. Manage. Assoc. **15**(6), 505–519 (1998)
49. Nakata, C., Im, S.: Spurring cross-functional integration for higher new product performance: a group effectiveness perspective. J. Prod. Innov. Manag. **27**(4), 554–571 (2010)
50. Levy, M., Loebbecke, C., Powell, P.: SMEs, coopetition and knowledge sharing: the role of information systems. Eur. J. Inf. Syst. **12**(1), 3–17 (2003)
51. Shih, M.H., Tsai, H.T., Wu, C.C., Lu, C.H.: A holistic knowledge sharing framework in high-tech firms: game and coopetition perspectives. Int. J. Technol. Manage. **36**(4), 354–367 (2006)
52. Zhang, X.B., Lin, C.H., Chung, K.C., Tsai, F.S., Wu, R.T.: Knowledge sharing and coopetition: turning absorptive capacity into effectiveness in consumer electronics industries. Sustainability **11**(17), 4694 (2019)
53. Pitelis, C.N., Desyllas, P., Panagopoulos, A.: Profiting from innovation through cross-border market co-creation and co-opetition: the case of global pharmaceuticals. Eur. Manag. Rev. **15**(4), 491–504 (2018)
54. Strese, S., Meuer, M.W., Flatten, T.C., Brettel, M.: Organizational antecedents of cross-functional coopetition: the impact of leadership and organizational structure on cross-functional coopetition. Ind. Mark. Manage. **53**, 42–55 (2016)
55. Saythongkeo, V., Le, V.D., Tran, L.T.T.: Leadership styles and innovation performance: the role of coopetition capability and in-learning in financial service firms. SAGE Open **12**(2), 21582440221079900 (2022)
56. Cassiman, B., Di Guardo, M.C., Valentini, G.: Organising R&D projects to profit from innovation: insights from coopetition. Long Range Plan. **42**(2), 216–233 (2009)
57. Clarke-Hill, C., Li, H., Davies, B.: The paradox of co-operation and competition in strategic alliances: towards a multi-paradigm approach. Manag. Res. News **26**(1), 1–20 (2003)
58. Gnyawali, D.R., Park, B.J.: Co-opetition and technological innovation in small and medium-sized enterprises: a multilevel conceptual model. J. Small Bus. Manage. **47**(3), 308–330 (2009)
59. BarNir, A., Smith, K.A.: Interfirm alliances in the small business: the role of social networks. J. Small Bus. Manage. **40**(3), 219–232 (2002)
60. Feranita, F., Kotlar, J., De Massis, A.: Collaborative innovation in family firms: past research, current debates and agenda for future research. J. Fam. Bus. Strat. **8**(3), 137–156 (2017)
61. Corbo, L., Kraus, S., Vlačić, B., Dabić, M., Caputo, A., Pellegrini, M.M.: Coopetition and innovation: a review and research agenda. Technovation **122**, 102624 (2023)
62. Shvindina, H.: Coopetition as an emerging trend in research: perspectives for safety & security. Safety **5**(3), 61 (2019)
63. Velu, C.: Evolutionary or revolutionary business model innovation through coopetition? The role of dominance in network markets. Ind. Mark. Manage. **53**, 124–135 (2016)
64. Parkman, I.D., Holloway, S.S., Sebastiao, H.: Creative industries: aligning entrepreneurial orientation and innovation capacity. J. Res. Mark. Entrep. **14**(1), 95–114 (2012)
65. Seepana, C., Paulraj, A., Huq, F.A.: The architecture of coopetition: strategic intent, ambidextrous managers, and knowledge sharing. Ind. Mark. Manage. **91**, 100–113 (2020)
66. Liu, G., Aroean, L., Ko, W.W.: Service innovation in business ecosystem: the roles of shared goals, coopetition, and interfirm power. Int. J. Prod. Econ. **255**, 108709 (2023)
67. Shamout, M.D.: A configural model of analytics capabilities, ambidexterity, co-opetition, and firm performance in the supply chain context. Bus. Strategy Dev. **6**, 128–139 (2022)

68. Xin, X., Miao, X., Cui, R.: Enhancing sustainable development: innovation ecosystem coopetition, environmental resource orchestration, and disruptive green innovation. Bus. Strateg. Environ. **32**(4), 1388–1402 (2023)
69. Nascimento, L.D.S., da Costa Júnior, J.C., Salazar, V.S., Chim-Miki, A.F.: Coopetition in social entrepreneurship: a strategy for social value devolution. Int. J. Emerg. Mark. **18**(9), 2176–2197 (2023)
70. Volschenk, J., Ungerer, M., Smit, E.: Creation and appropriation of socio-environmental value in coopetition. Ind. Mark. Manage. **57**, 109–118 (2016)
71. Maurer, I., Bartsch, V., Ebers, M.: The value of intra-organizational social capital: how it fosters knowledge transfer, innovation performance, and growth. Organ. Stud. **32**(2), 157–185 (2011)
72. Rai, R.K.: A coopetition-based approach to value creation in interfirm alliances: construction of a measure and examination of its psychometric properties. J. Manag. **42**(6), 1663–1699 (2016)
73. Bacon, E., Williams, M.D., Davies, G.: Coopetition in innovation ecosystems: a comparative analysis of knowledge transfer configurations. J. Bus. Res. **115**, 307–316 (2020)
74. Zineldin, M.: Co-opetition: the organisation of the future. Mark. Intell. Plan. **22**(7), 780–790 (2004)
75. Schiavone, F., Simoni, M.: Prior experience and coopetition in R&D programs. J. Knowl. Econ. **7**, 819–835 (2016)
76. Kumar, A., Connell, J., Bhattacharyya, A.: Coopetition, corporate responsibility and sustainability: why multi-dimensional constructs matter. Social Responsibility Journal **19**(1), 114–137 (2023)
77. Ziyae, B., Rezvani, M., Eynolghozat, M.: Coopetition in entrepreneurial city: an approach to complexity theory. Cities **121**, 103470 (2022)
78. Raza-Ullah, T., Bengtsson, M., Kock, S.: The coopetition paradox and tension in coopetition at multiple levels. Ind. Mark. Manage. **43**(2), 189–198 (2014)
79. Poulsen, M.B.J.: Competition and cooperation: what roles in scientific dynamics? Int. J. Technol. Manage. **22**(7–8), 782–793 (2001)
80. Tidström, A.: Managing tensions in coopetition. Ind. Mark. Manage. **43**(2), 261–271 (2014)
81. Christ, K.L., Burritt, R.L., Varsei, M.: Coopetition as a potential strategy for corporate sustainability. Bus. Strateg. Environ. **26**(7), 1029–1040 (2017)
82. Ritala, P.: Coopetition strategy–when is it successful? Empirical evidence on innovation and market performance. Br. J. Manag. **23**(3), 307–324 (2012)
83. Estrada, I., Dong, J.Q.: Learning from experience? Technological investments and the impact of coopetition experience on firm profitability. Long Range Plan. **53**(1), 101866 (2020)
84. Ritala, P., Hurmelinna-Laukkanen, P.: What's in it for me? Creating and appropriating value in innovation-related coopetition. Technovation **29**(12), 819–828 (2009)
85. Crick, J.M., Crick, D.: Coopetition and COVID-19: collaborative business-to-business marketing strategies in a pandemic crisis. Ind. Mark. Manage. **88**, 206–213 (2020)
86. Bouncken, R.B., Fredrich, V., Kraus, S.: Configurations of firm-level value capture in coopetition. Long Range Plan. **53**(1), 101869 (2020)
87. Cygler, J., Sroka, W.: Coopetition disadvantages: the case of the high tech companies. Eng. Econ. **28**(5), 494–504 (2017)
88. Le Roy, F., Robert, F., Hamouti, R.: Vertical vs horizontal coopetition and the market performance of product innovation: an empirical study of the video game industry. Technovation **112**, 102411 (2022)
89. Soppe, B., Lechner, C., Dowling, M.: Vertical coopetition in entrepreneurial firms: theory and practice. J. Small Bus. Enterp. Dev. **21**(4), 548–564 (2014)

90. Kotzab, H., Teller, C.: Value-adding partnerships and co-opetition models in the grocery industry. Int. J. Phys. Distrib. Logist. Manag. **33**(3), 268–281 (2003)
91. Zhang, H., Zhang, Y., Li, P., Zheng, H., Li, Z.: Low-carbon production or not? Coopetition supply chain manufacturers' production strategy under carbon cap-and-trade policy. Environ. Dev. Sustain. 1–21 (2022). https://doi.org/10.1007/s10668-022-02342-2

De-politicised in Court: The Interaction of Democracy with Innovation Projects

Atle M. Christiansen(✉), Kenn Steger-Jensen, and Kenneth Vidskjold

University of South-Eastern Norway, Post Office Box 4, 3199 Borre, Norway
{ac,kes,kv}@usn.no

Abstract. This study explores the interaction of politics and innovation. We discuss how entrepreneurs and innovators face policies, legislation, governmental rules and public opinion when compiling innovation projects. This paper highlights the power of politics in a democracy, emphasising the principle of checks and balances. We also touch upon the concept of deliberative democracy and how it should ideally function. Since the authors argue that excessive politicisation may hinder innovation, they seek to test this theory. The aim is to provide insight for overcoming undesired political influence, thereby helping those working on innovation projects to lower or mitigate risk. This study underlines the need to understand the implications of political leadership on innovation projects and statutory amendments. Further, it discusses the potential impact of political constraints on innovation, emphasising the need for strategies to influence policymakers effectively. The authors not only acknowledge the existence and need for political leadership in a democratic society, but also are interested in understanding the implications of such leadership on innovation. A two-sample case study is conducted with mixed methods research. Literature reviews and interviews are combined with theory in an iterative process to test the following hypothesis: If governments and departments become too politicised this may impede innovation. Results from our two cases vindicate our assumption. This study encompasses the maritime and marine sectors, mainly within ports, transport and logistics.

Keywords: Production management · Maritime transport · Innovation projects

1 Introduction

When entrepreneurs and industrial enterprises compile innovation projects they are confronted with a robust framework of legislation, governmental rules and public opinion. New, emerging technologies must not only verify technical feasibility and safe use, but also face diverging public and political interests that may be to their benefit, but quite often they face resistance and opposition. An innovation project predominantly strives for change, yet society or societal groups resist any such change in an already stable environment [1, 2].

In a functioning democracy, politics is essentially about the division of power, and then the executive branch of power. The judiciary and legislative branches of power shall

Published by Springer Nature Switzerland AG 2024
M. Thürer et al. (Eds.): APMS 2024, IFIP AICT 733, pp. 34–48, 2024.
https://doi.org/10.1007/978-3-031-71645-4_3

remain separated from the executive branch and, at least in theory, remain unpolitical and impartial. This division of the three powers upholds the principle of checks and balances. In Norway, through its negative parliamentarism, the executive branch, viz. the government, must gain support from the majority in the parliament. Failing this, the government and its ministers could be ousted by a vote of no confidence. Governments change when a general election occurs or due to a formal resignation. Prime ministers can, at their discretion, reshuffle members of cabinet, such as the minister of transport and minister of fisheries and industry, which leads to changes in their respective Departments [3].

Plato's critique of democracy is not universally adopted; hence democracy differs in form in Norway from that in the USA. However, quoting Philip Pettit: "It is now widely accepted as an ideal that democracy should be as deliberative as possible. Democracy should not involve a tussle between different interest groups or lobbies in which the numbers matter more than the arguments." [4] Further following Pettit: For a politician, politics should be the means, but not necessary the goal. When governments are formed, they should govern through a deliberative democracy. Theory and practice do not always converge. In the Norwegian guidelines specifying the seven essential duties of the department's civil servants, the second line in the foreword reads (our translation): "Departments are professional secretariats for political leadership, not political secretariats." [5] In essence, this means that the deputy general who wrote the foreword acknowledges that leadership will be politicised but warns of it potential perils. We decided to study innovation projects. The practical implications of this divergence from a deliberative democracy are numerous and diverse. One example is that a left-wing government could oppose automation and autonomous vessels to save national jobs, whereas a right-wing government would encourage such ventures to boost economic growth and international collaboration.

We are not disputing the existence and need of political leadership. In a democratic society there will probably always be this form of governance. Instead, we are trying to understand the implications of political leadership on innovation projects and amendments to statutory rules, regulations and support from politicians. *Our main hypothesis is that should governments and departments become too politicised this may impede innovation.* We will elaborate on entrepreneurs, project managers and others working with innovation who encounter political barriers that, in some cases, invoke all three powers of the state. Simply scrambling the press corps is not always enough. Such an action may shift public opinion, but in many cases does little to influence policymakers once they are elected and serving in government.

We contend that this hypothesis, when tested, provides insight into dealing with undesired political influence. Innovators can use this knowledge to lower or mitigate risk in innovation projects and start-ups. We will conduct a case study and sample one project that benefitted from political influence and another project that suffered from it. When reviewing our research work and previous studies done by others we recognised the need for more explanation about why decisions within transportation-and logistics projects are ruled by politics and not by knowledge. *Furthermore, this knowledge can be used to impose decisions based on data, information and algorithms.* Although we limit the focus of this study to the maritime transport sector, it may be relevant to other branches and industries.

2 Theory

Innovation, known as the act of introducing something new, is often accomplished through start-ups or entrepreneurs. However, innovation is not about the invention or the idea, rather it is the implementation in the society and the creation of a viable value chain. The increasing focus on climate change has shifted attention from purely monetary value towards sustainability. Sometimes this can result in innovation removing jobs, instead of creating them. Such a trend is often at odds with public and government policies, as this seemingly removes value. Global sustainability can be poorly valued in a local market. Autonomy and automation do not always remove value from the value-chain. For instance, many products are labelled "handmade", but we doubt that a salmon fillet labelled as "untouched by human hands" could possibly have any less value than one that has been handled. Thus, the value added by workers at Norwegian fish processing plants does not disappear with automation, it will materialise in the value chain.

From its inception until products and services are ready for market, an innovation loses adaptability, flexibility and it will be progressively resilient to any change or modification. In simple terms, ideas can be changed or modified, but finished products cannot, unless you spend a lot of resources modifying production plans and business models. Innovations need to be adapted to markets to materialise. Innovations and development in existing markets are designated *Blue ocean strategies and Red ocean strategies* correspondingly [14].

Innovators are wise at mapping applicable rules and regulations and meeting public and political policymakers early on in their projects. Such skills were practised by the innovators selected in this study. Unfortunately, the ministers, serving the role of politicians and not civil servants, changed their policies, rendering it impossible for the innovator in one of our studied cases to materialise the innovation.

2.1 Legal Mobilisation

Legal mobilisation is justified means to address law, justice and governance issues. It has mostly been discussed and used as a tool for the individual to enforce public policy and allocate power and authority. One may argue that this is a tool for political participation [6]. Traditionally, innovators use the legal system to protect their inventions by intellectual property rights and patents. The legal system can also be a framework for innovation, and hopefully, it is also being transformed and adapted by it. This relationship can foster a more efficient, accessible legal system and drive technological advancement. Innovators face restraints on the feasibility and development of their work, inhibiting them from achieving their aims and goals [7].

2.2 The Norway and Denmark Versus Sweden Conundrum

Norway has a ministerial government which vests considerable power in the politically installed minister. It could be stated that there is too much power in the hands of the politically installed minister. Some of this power should be transferred to the departmental civil servants. All in all, these civil servants' experts in their fields, politicians are, as a rule, not. Research shows that this could account for the greater success of Swedish over Norwegian post-war roadbuilding [8]. Sweden has three times more length of motorways than Norway has, planning is eight times greater and overall, the Swedish minor road networks are of a higher quality and standard of. A collaboration of three researchers from Denmark, Sweden and Norway describes it as follows:

> "In principle there is an important difference between the dual Swedish system and the monistic Danish/Norwegian one. In Sweden there is an organisational bifurcation between the ministries and the central agencies, which are subordinate to the cabinet as a collective body rather than to individual ministries. Denmark and Norway, on the other hand, have a purely ministerial government where the outer bands are subordinate to the minister himself, who can be held responsible for all decisions made in these secondary bodies." [9]

This confirms the fact that Norway and Sweden have different forms of democratic government, and it is argued that we have different systems of innovation which affect innovation paths differently [10]. Examining some of these differences can lead to a better understanding of the democratic, and not so democratic, leadership and government and the interaction between innovation and democracy.

2.3 The Norwegian Innovation System

The concept of a national innovation system was introduced in the late 1980s [10]. Historically, the innovation system in Norway has evolved in a resource-based small economy. In hindsight, it is evident that there has been a considerable triple helix or even a quadruple helix-based collaboration among industry, academia, government and the public [11, 12]. This is noticeable in large scale projects, such as the collaboration among the Norwegian Defence Research Establishment, Kongsberg and others, or the various projects emerging from The Foundation for Industrial and Technical Research. In addition, there is a considerable amount of innovation taking place in less formal innovation collaborations, often recognised and materialised in informal collaborations: "Perhaps the most important factor in Norway's innovative performance has been the ability of Norwegian entrepreneurs, firms, and public sector actors to recognise opportunities, mobilise resources, adapt existing capabilities and develop new ones, and develop appropriate institutions and policies." [10] Further, Fagerberg et. al. argue that political and institutional factors have been decisive in developing the Norwegian innovation system. There are incremental and disruptive innovations in Norway. Incremental innovations involve small improvements to existing products, services, or processes, enhancing efficiency and effectiveness. Disruptive innovations, however, originate in niche markets and eventually disrupt existing markets, displacing established Results systems or products. Both play crucial roles in business growth and industrial evolution [13]. Within the maritime industry, like ports and sea borne transport it is common to differentiate between new and existing industries.

3 Method

Since 2012 we have conducted miscellaneous projects concerning, inter alia, seaports and port development. In addition to traditional research projects funded by the university and different research bodies, we have conducted action research funded by EU Interreg projects. In many of the Interreg projects several ports have been partners and as a result of these years of collaboration we have amassed a vast amount of data and insight into port centred and maritime transport operations. Using this knowledge as a foundation for this project, we selected two cases from the maritime transport industry in Norway. These cases have received wide media coverage, both nationally and globally. The first case is an operation for collecting live salmon from breeding pens on the western coast of Norway and processing the fish at sea. It was the only current case were innovation interacted with litigation in the maritime transport sector we could identify. We will refer to this case as Hav Line [HL]. The other case is an autonomous container feeder vessel transiting the industry port at Herøya to the port of Grenland. Both ports lie on the eastern coast of Norway and only 10 nautical miles apart. We will refer to this case as Yara Birkeland [YB]. Both cases were selected through a process of screening potential maritime transport operations for innovation, technology, project start and primarily, duration.

Data were collected from numerous sources. Secondary data sources were websites, news articles, social media, press releases and court rulings. Primary data were sourced from interviews with representatives from industry, government, academia and individuals assigned to the projects. The sampling of relevant knowledge was conducted using either the naturalistic inquiry method or snowball sampling [15, 16]. All respondents remain anonymous and rigid ethical standards were applied for interviews data collection and analysis [17, 18]. Based on the retrieved data and derived information, we compiled a timeline for each case. Due to the sheer magnitude of input, we cannot include it in this paper, but make it is accessible on our university website for anyone interested.

4 Results

In all our data we could not find any indication that any politicians or political parties in Norway oppose innovation and development in general terms. However, one notable finding is that one political party, the Progress Party, which was in government from 2013 to 2020, denies in its manifesto that climate change can be attributed to human activity. However, they endorse CO_2 savings provided it is cost-efficient [19]. This is also in line with platform declarations made when entering government. Comparing the two cases on this level indicates that politicians *should* support both projects as they were both highly innovative and provided for large amounts of cost-efficient CO_2 savings and more sustainable transport and logistics operations overall. They were both co-funded by the same two bodies and *R&D and Innovation* programmes. Politically, the right-wing government remained in power from 2013 to October 2021, the last year without the Progress Party.

4.1 The Two Cases of Innovation

We studied two different cases of innovation in maritime transport and physical distribution, but they revealed many similarities and common ground. Both facilitated a significant shift of goods from land transport on trucks to sea transport. Both projects relied on a close cooperation with one or more seaports to materialise. Both could only be realised provided existing laws and/or other regulations were amended. There was one significant difference. HL was a disruptive market innovation to an already existing maritime market – Demanding a typical red ocean strategy. YB was a disrupting innovation in a market where no sea transport or trade was present – Demanding a typical blue ocean strategy.

4.2 Hav Line

Along most of the Norwegian coastline there are landing sites for fisheries and clusters of pens for fish farming. In addition to fishing vessels delivering their own catch to shore there is an extensive industry involved in collecting fish in the pens and landing them for further processing. This activity has mostly involved well-boats pumping the fish into big holding tanks (wells) on vessels, and then transporting them on shore when the boats arrived at harbours or ports. In 2008, a Norwegian shipowner, Napier, introduced a fish processing boat to the Norwegian market, whereby fish are sedated, bled, cooled, and stored in large tanks before being transported to shore. In 2018 Hav Line introduced the process ship, Norwegian Gannett, where the processing of the fish was even more comprehensive, since it performed almost all the steps towards rendering packaged and cooled fish fillets on board. Their business model was to collect the fish from sites mainly on the west coast of Norway and land their product at the port of Hirtshals situated in North-western Jutland, Denmark. Eventually, they could collect from other sites around the North sea and deliver to other ports closer to the European markets [20].

There is a lot to gain from this approach. The main advantages being a considerable shorter time to market, all transport is shifted from trucks to much more eco-friendly sea transport and improved quality. The downside could be a resultant loss of jobs in Norway. One can argue that these jobs are merely relocated from shoreside to on board ships. Further, these jobs would eventually phase out because of increased automation within the fish processing industry.

4.3 Hav Line Timeline

The idea behind the *Hav Line Method* arose in 2007 when an entrepreneur in the Norwegian seafood industry devised a concept for slaughtering and rapid cooling of fish onboard ship during transportation. This method was tested in the last week of 2007. When the quality of the fish was appraised, the results were promising. The plans were further discussed with one co-entrepreneur from the seafood industry, and together they started to plan a vessel.

Clarification from Relevant Authorities. From 2008 to 2016 the entrepreneurs, together with a consultant, actively sought clarification from the Norwegian Food Safety

Authority [NFSA] regarding certain regulations that could be relevant for building a vessel to operate according to the HL method. In a very short operation, the fish are loaded on board, then stunned, bled and slaughtered. Once the salmon is slaughtered, it is transferred to large tanks of saltwater [RSW], which are temperature regulated at −0.50 °C. The HL method was explained in meetings and documentation was provided to the Norwegian Food Safety Authority.

The Fish Quality Regulations and "Production Fish". The requirements relating to the sorting and grading of farmed salmon have played an important role in the "Hav Line case". The regulation, as it stood until April 29, 2019, had a provision on sorting in Sect. 17 with the following wording: *"Farmed fish shall be sorted so that fish with wounds, deformities, severe handling errors, or internal quality defects are not sold for human consumption.*" This sorted fish is termed *"production fish"* and can be sold to domestic companies who can rectify the errors before shipping the product to market as quality fish meant for human consumption. Historically, the purpose of this regulation was to ensure that all exported fish from Norway fulfilled the quality level required by the global fish market and to uphold Norway's reputation as an exporter of high-quality farmed fish. The development of the fish farming industry in recent decades has made the quality regulations less relevant and they have been revised several times. Notwithstanding the development of the HL method, there was already a discussion between the government and actors in the industry about replacing the regulation with an industry standard to ensure the quality of farmed fish. During the clarification process for building a vessel to operate according to the Hav Line method, *production fish* was mentioned, but the significance of this was not questioned in connection with the concept. According to HL and NFSA personnel, who also were witnesses in the trial between HL and the Norwegian government, the regulations regarding *production fish* were well past their expiry date.

The Development of the Concept Continued. Discussions were held with the authorities regarding different concepts and the work involved in developing a slaughter vessel utilising the HL method continued for some time. In March 2016 a meeting was held at the NFSA at which representatives from HL presented the final concept which later would be known as the Norwegian Gannet. After the meeting, HL drafted minutes which were sent to the Norwegian Food Safety Authority for reviewal and comments. The NFSA proposed some changes and additions, but in general, they agreed that nothing indicated that the vessel as presented would not qualify for the requisite approval. The NFSA was positive towards methods that can reduce risk and improve fish welfare.

Eventually, the HL case went to court in 2020, and we will cover this later. However, to better understand why HL initiated the shipbuilding, we include the following extract from the court proceedings: Several of the participants in the meeting in March 2016 provided testimony to the court. A witness from HL said that *the sorting of production fish was mentioned at the meeting, but that this issue was no longer relevant in 2016. The witness explained that nothing was said at the meeting indicating that the quality regulations could result in the project being terminated, or that there were any regulatory obstacles. The mood of the meeting was very encouraging; everyone from the NFSA seemed very positive about the project.* One witness from the NFSA explained that the *quality regulations and the provision regarding production fish were discussed at the meeting, and that the provision had long been debated within the Norwegian Food Safety Authority as the regulatory body preferred to focus on food safety and delegate quality assessment to the industry.* The witness also confirmed *that there was no discussion of sorting out production fish in Denmark and then filleting them in Norway. The witness explained that if the fish were delivered unsorted to Denmark, then Danish rules would apply to the sorting process.*

The Process of Building and Approving Norwegian Gannet. Based on the meeting in March 2016 and the feedback, HL had worked on the ship's drawings to such a detailed level that they could be sent to different shipyards to obtain bids for the vessel. It was the same drawings that were presented in the meeting with the NFSA. The company received bids from several shipyards and the company proceeded with the project. The company signed a contract for building Norwegian Gannet in July 2016. HL applied for approval of the vessel on July 27th, 2018. In August 2018, when a new Minister of Trade and Fisheries assumed office, a debate arose regarding the significance of Norwegian Gannet's planned operations. It concerned the interpretation of the regulations and the impact of Norwegian Gannet on the Norwegian fish farming industry. Consequently, HL wrote to the NFSA on September 4th, 2018, concerning the HL method and its status in accordance with quality regulations, as well as approval. Further, HL applied for an exemption from Sect. 17 of the quality regulations if necessary for the planned operations. The regional office rejected the application for both approval and exemption from the quality regulations on September 28th, 2018. HL appealed the decision to the NFSA's head Office.

The Department Acts Against Hav Line. On November 5th, 2018, the NFSA head office overturned the decision made by the regional office, and HL's planned use of Norwegian Gannet was legal under the existing quality regulations. HL took delivery of the vessel the day after, on November 6th. Later in November, the Ministry of trade and Fisheries (NFD) notified HL that they would overturn the decision made by NFSA head office. In May 2017 NFSA head office submitted, on behalf of NFD, a proposal for a revision of the quality regulation where Sect. 17 was proposed removed entirely. On November 26th, NFD sent an order to the FSA's head office requesting an amendment to only Sect. 17 to include domestic. The proposal for amending the fish quality regulations was sent for consultation on January 31st, 2019. The regulation amendment was adopted on April 24th, 2019, and announced on April 29th, 2019.

The Court Order and the Lawsuits. The amended regulation was, in reality, a stoporder for HL. They had already filed for a temporary permission to operate on the 6th of January 2019, and they brought a claim to challenge the validity of the Norwegian Food Safety Authority's decision before the court by filing a lawsuit with the Bergen District Court on April 11th, 2019. HL filed a claim in court on the validity of the regulation amendment on August 5th, 2019. Both lawsuits were consolidated for joint consideration by the court's decision on November 7th. In June 2020 the Bergen District Court delivered its judgment largely in favour of the HL. The court concluded that the Norwegian government had abused its power over HL. The court has nevertheless concluded that the decision is invalid due to gross unreasonableness. The highlights from the court rulings read (our translation):

"…HL had legitimate expectations that the Norwegian Food Safety Authority and political leadership would approve the use of Norwegian Gannet for its intended purpose. In reaching this assessment, the court considered that HL could not have done anything more or different to mitigate the risk associated with the investments made by the company and its subsidiaries in Norwegian Gannet and in Denmark. The sudden and unexpected change in the political evaluation of the company's operations was also considered by the court."

4.4 Yara Birkeland

When reviewing the costs involved with small-scale barge-, short sea- and feeder operations, wages account for a considerable portion of the total costs. The situation is most challenging when vessels are in operation 24/7/365, and particularly, when shipowners are not sailing under flags of convenience. Earlier, many Norwegian shipowners were sailing under flags of convenience to benefit from the favourable tax and tariff conditions at that time. With the launch of the Norwegian International Ship Register [NIS] in 1987, this led to the return of many shipowners to the Norwegian flag. When transiting between Norwegian ports there are limitations regarding the benefits of NIS. To overcome this caveat, autonomous ships can be a solution to minimise the impact of labour costs.

Another, and probably more important, reason for investigating and developing an autonomous ship was to provide a different way of conducting logistics operations altogether. Drawing on experiences gained from fully automated and robotised warehouses, it might be feasible to instal similar facilities in an out-of-warehouse logistics operation. In short, to shift cargo carriers like containers on to feeder ships, from ships to port and back on to trucks again, without any human operators. Creating a wholly new mode of logistics. Lowering costs not only on ships, but also in port and other terminals. Automated ships should be considered as more than just a ship without a crew [7, 21]. In general, most of our respondents shared this line of thought and supported the notion that this would be one of the reasons for building YB.

4.5 Yara Birkeland Timelines

Discussions about autonomous vessels and drones have been held for many years in Norway. The Norwegian Defence Research Establishment and its partner, Kongsberg, commenced work on an autonomous underwater vessel in 1991, and their first commercial survey was in 1997. Kongsberg Maritime has worked on dynamic positioning systems since the late 1980s. The combination of these technologies provided for much of the systems needed for autonomous operations of larger size vessels. The idea behind YB arose in 2016 when a Norwegian industrial company was seeking for a new and more sustainable method for transportation of their product between the factory and the export port.

Developing the Concept and Meeting with Authorities. Staff members of the industrial company and technology company presented their concept to the Norwegian Maritime authority. During the following months, several meetings were held at the authority's office where the idea was further developed. In the spring of 2017, the industrial company and the technology company formed a joint venture, announcing that their goal was to build a full-scale autonomous container ship and have it sailing without any crew onboard by the end of 2020. cf. Fig. 1, planned timeline. During the announcement several representatives of the Norwegian government made their supporting statements regarding the project. The prime minister of Norway said this was an important step in promoting Norway as a world leader in the development of autonomous technology. The head of the Norwegian Maritime authority said that there *may be some challenges with regulations*, but they were willing to investigate the matter and that their goal was to actively work for amending the regulations towards making the autonomous ship feasible. The statement from the maritime authority was also supported by the minister of transport.

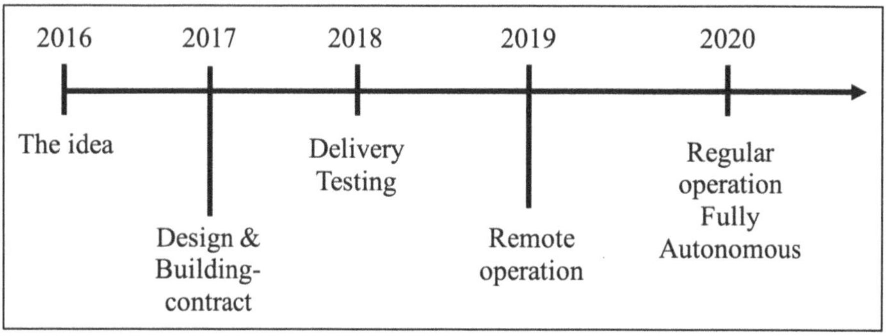

Fig. 1. The planned project timeline.

After the announcement of the project, the two companies received considerable attention both nationally and internationally. The initiative had evolved into an important prestige project not only for the companies, but also for other stakeholders like the maritime authorities, departments, and the parliament. According to our interviewees this attention caused a lot of stress with different actors. Staff at the Maritime authority's office were told to do whatever necessary to facilitate progress of the project and the minister of transportation visited their office several times.

Early in the project phase, the complexity involving a full-scale autonomous ship was starting to appear for the actors involved in the project. The main challenge was the existing rules and regulations which are based on manned ship operations. The plan, initially, was that the ship could be operated according to applicable rules regulating the inland waterways, monitored by control centre personnel on shore. However, due to the magnitude of the planned concept, the ship had to comply with the existing rules and regulations governing manned ship operations. The lack of pertinent regulations resulted in the concept having to be revised in accordance with existing regulations that, among other things, involved using a crew on board the ship. Hence, the evolving complexity caused several delays to the project. These were of a diverse nature involving technological, engineering, building and regulatory issues. See Fig. 2 below, which illustrates the actual progress. The development of autonomous technology was a milestone for the government. The minister of transport liaised regularly with the technology company, trying to facilitate the process. In addition, the Prime Minister's Office was regularly in contact with the technology company, requesting status reports.

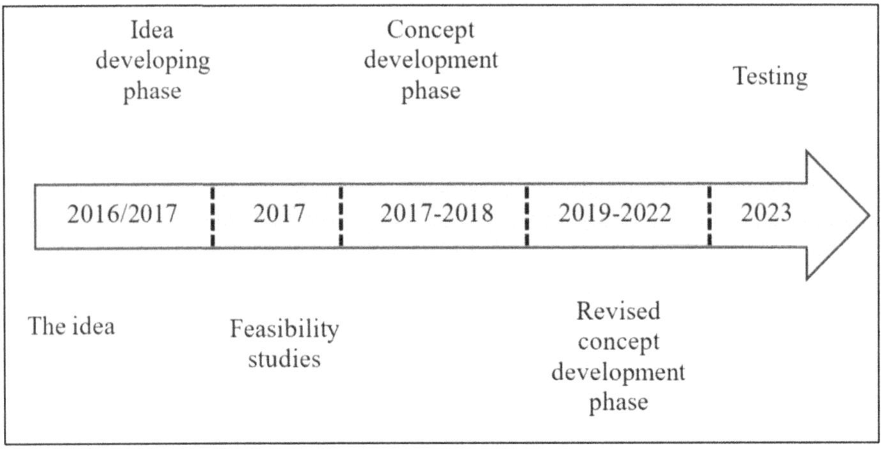

Fig. 2. The actual project timeline.

5 Discussion

We sampled two different innovation cases from the maritime transport sector. The support from the government through its ministers was positive and productive for the YB case, but negative and restrictive for the HL case. We decided to analyse the cases without scrutinising their motivation and focusing on their political intent, which would go beyond the scope of our research. Our objective was to show that two almost identical innovation projects seem to gain support and trigger opposition in a seemingly random manner. The District Court of Bergen concluded that the HL case was in line with current political policies yet, despite this, they gained no support when the development phase was completed, and the ship was ready to operate. The YB, on the other hand, gained widespread political support and was backed by both the minister and the government. There were probably ulterior motives beyond the official political agenda present. We were able to ask the incumbent minister of transport at that time about the HL case and why it was not supported, and we can quote him on this because it was in a public debate; «It was due to diverging political interests». This implies: We need to differentiate between political leadership and political interests. Political interest may not always be of a democratic nature if endorsed. Initially, we discussed the need for a deliberative democracy and how this was necessary to depoliticise political leadership and democracy [4]. In a project report comparing the operational potential of a ship like YB we wrote in 2017: "The new ferry-free E39 motorway between Bergen and Stavanger, scheduled for 2025, will shorten the travelling time between the two cities from 4.5 h to 2 h and moreover, compromise the competitiveness of any short sea feeder." [22] In the

National Transport Plan presented to Parliament on March 22nd 2024 the ferry-free E39 project was excluded in full [23], rendering a scheduled autonomous electric vessel route linking Bergen to Stavanger entirely viable. Scenarios of the future whenever politics and politicians are involved are almost impossible to make with any degree of certainty. We could investigate more cases underlining the need for depoliticisation and more deliberation. Aristotle was probably right when he called for *phronesis – practical wisdom* as the most important of all virtues. This is most true for politicians [24].

The Relevance of Democracy to an Innovation Project. It is easier to adapt to the existing political system than it is to change it. We argue that this is crucial for innovators and entrepreneurs. Adaptation often invokes the use of tools, inter alia, public relations, lobbying, press, public opinion and litigation. Innovators and others alike need momentum and progress to materialise their innovation. Time is of essence. HL used almost every possible tool to ensure feasible and viable operations. Both projects benefitted from a quadruple helix collaboration and gained huge support from their sponsors and most of their stakeholders. However, the HL project had little involvement from the Norwegian Research Council. Forging a closer collaboration or funding through one of their programmes could lead to augmenting political support for the project.

Norway is a democracy and governed by a principle of openness and checks and balances. This facilitates litigators and the judge finding and assessing pertinent information to the HL case. Such information was available to the ministers of trade, industry and fisheries and transport. A process of deliberation, together with an arguably broader use of this knowledge, *could* lead to a more supportive stance towards HL from the ministers and their respective departments. The court ruling was quite clear that it *should*.

Use of the Knowledge Gained to Reduce Risk. There was no obvious reason for HL to scrutinise the support they initially got from the NFSA and its parent department. Had the authorities presented HL with barriers to the feasibility and viability of their project, then HL would have dealt with them. If potential issues were presented and dealt with early in the project, they could be much more easily resolved than when presented later. The economic consequences would also be diminished. However, much evidence suggests our hypothesis to be true within this scope of study, namely, *if governments and departments become too politicised this may impede innovation.* Further, sufficient evidence suggests that governments and departments are politicised. They choose between which one of diverging political interests to support. Seemingly similar projects like HL and YB are either supported or not based on political interests and external influences. In the HL case, the government went from a rational and deliberative stance to a politicised stance towards the project. The YB project was supported all the way. *Ministers, governments and policies all undergo change. People working with innovation should take this into account, using different scenarios and political options in their risk analysis to develop strategies and actions to remove or mitigate such risk.*

Innovation Spawns New Innovations or New Industry. Currently, several projects in Norway and other countries must be considered as spin-offs from the YB case, inter alia, ASKO drones in operation between the ports of Moss and Horten in the Oslo Fjord, and several planned autonomous feeder vessels. There are no signs yet of any similar spin-offs from the HL case. We did not study similar cases in Sweden, but regarding objective performance indicators as we addressed initially, the Swedish system with its "organisational bifurcation between the ministries and the central agencies" [9], might be better at diluting political agendas. This again could facilitate a more deliberative democracy, and coherent support for innovators and entrepreneurs across different projects.

5.1 Lessons Learned

It can be a challenge to generalise from a two-case study, but because of their inherently similar nature some lessons can be learned. We have designated these *(i)* to *(vi)*. Our informants from the industry and government told us that politics is about trade-offs. *(i)* Innovators and entrepreneurs do wise in learning about the whole map relevant to their project. Politicians change their opinion and direction. Sometimes dormant interests awake with the potential to derail or stop a project.

Politicians and civil servants alike have an obligation towards the system of openness and checks and balances. This will leave a trail to follow, and things said cannot easily be unsaid. *(ii)* Innovators and entrepreneurs will benefit from open and visible dialogue with everybody involved in their project. People working with inventions and new technology sometimes shroud their work in secrecy. Not always to their benefit. Some oceans are blue, and some are red. *(iii)* Innovators and entrepreneurs should realise that disruptions to already existing markets require a red ocean strategy. Projects that are in line with current policies, visions and political guidelines are probably more prone to support by politicians, and projects that are not will probably be hindered. *(iv)* Innovators and entrepreneurs should wrap their project in the current and correct sheet of wrapping paper. If other things fail: *(v)* Innovators and entrepreneurs can de-politicise their innovation in court. *(vi)* Politicians should govern through a deliberative democracy.

6 Conclusions and Further Research

The two cases investigated very much confirm our main hypothesis that politicisation can impede innovation. Politicisation can also promote innovation. In democratic societies, innovators and entrepreneurs can apply this knowledge to mitigate risk and deal with unforeseen factors influencing the progress and materialisation of their innovation. In short: Democracy interacts with innovation projects. The implication of this is presented in the lessons learned section. Since our research is restricted to the maritime transport sector, any further or additional research should be conducted to determine its relevance to other areas of innovation. We could only identify one current project like the Hav Line case, but projects with some similarity exist. These could be investigated to establish more general results than those presented here. The associated seaports played pivotal roles in the development of both projects, further research should be conducted to investigate seaports and other stakeholders in the maritime transport domain with regard to innovation and democracy.

Acknowledgements. Finally, we would like to express our gratitude to the EU Interreg programme and the project "Value4sea", granted June 2023, and the MARKOM II project "New

Maritime Competencies: NyMK", granted December 2022, for funding our research. We would also like to thank Richard W. Nelson of Neltec Language for valuable input to this manuscript.

References

1. Lewin, K.: Action research and minority problems. J. Soc. Issues **2**, 34–46 (1946)
2. Burnes, B.: The origins of lewin's three-step model of change. J. Appl. Behav. Sci. **56**, 32–59 (2020). https://doi.org/10.1177/0021886319892685
3. Nordby, T.: Parlamentarisme. Store norske leksikon (2023)
4. Pettit, P.: Depoliticizing democracy. Ratio Juris **17**, 52–65 (2004). https://doi.org/10.1111/j.0952-1917.2004.00254.x
5. Kommunal og moderniseringsdepartementet. Om forholdet mellom politisk ledelse og embetsverk. Regjeringen.no 2019. https://www.regjeringen.no/no/dokumenter/om-forholdet-mellom-politisk-ledelse-og-embetsverk/id2626841/. Accessed 13 Mar 2024
6. Zemans, F.K.: Legal mobilization: the neglected role of the law in the political system. Am. Polit. Sci. Rev. **77**, 690–703 (1983)
7. Christiansen, A.M., Vidskjold, K.: Management and emerging technology in maritime logistics: a lewin force field analysis. In: Alfnes, E., Romsdal, A., Strandhagen, J.O., von Cieminski, G., Romero, D. (eds.) Advances in Production Management Systems. Production Management Systems for Responsible Manufacturing, Service, and Logistics Futures, p. 34–48. Springer Nature Switzerland, Cham (2023). https://doi.org/10.1007/978-3-031-43670-3_3
8. Boge, K.: Votes count but the number of seats decides: a comparative historical case study of 20th century Danish, Swedish and Norwegian road policy. Doctoral thesis, Sandvika (2006)
9. Lægreid, P., Pedersen, O.K.: Transforming states: continuity and change in the europeanisation of the nordic central governments. In: Scancor Workshop Transnational Regulation and the Transformation of States, pp. 22–3. Stanford (2001)
10. Fagerberg, J., Mowery, D.C., Verspagen, B.: The evolution of Norway's national innovation system. Sci. Public Policy **36**, 431–444 (2009)
11. Leydesdorff, L., Etzkowitz, H.: Can 'the public' be considered as a fourth helix in university-industry-government relations? Report on the fourth triple helix conference, 2002. Sci. Public Policy **30**, 55–61 (2003)
12. Leydesdorff, L., Etzkowitz, H.: Emergence of a triple helix of university—industry—government relations. Sci. Public Policy **23**, 279–286 (1996)
13. Christensen, C.M.: The Innovator's Dilemma: When New Technologies Cause Great Firms to Fail. Harvard Business Review Press (2013)
14. Kim, W.C., Mauborgne, R.A.: Blue ocean Strategy, Expanded Edition: How to Create Uncontested Market Space and Make the Competition Irrelevant. Harvard Business Review Press (2014)
15. Lincoln, Y.S., Guba, E.G.: Naturalistic Inquiry, vol. 75. Sage (1985)
16. Bowen, G.A.: Naturalistic inquiry and the saturation concept: a research note. Qual. Res. (2008). https://doi.org/10.1177/1468794107085301
17. Flick, U., von Kardorff, E., Steinke, I. (eds.): A Companion to Qualitative Research. Sage Publications, London, Thousand Oaks, Calif (2004)
18. Dubois, A., Gadde, L.-E.: Systematic combining: an abductive approach to case research. J. Bus. Res. **55**, 553–560 (2002). https://doi.org/10.1016/S0148-2963(00)00195-8
19. Klima, n.d.: https://www.frp.no/var-politikk/energi-og-miljo/klima. Accessed 17 Mar 2024
20. Bli med om bord i «Norwegian Gannet» (2018). https://www.kyst.no/fartoy-hav-line-havbruk-og-fiskeri/bli-med-om-bord-i-norwegian-gannet/497599. Accessed 16 Mar 2024

21. Rødseth ØJ, Tjora A. A system architecture for an unmanned ship. Proceedings of the 13th International Conference on Computer and IT Applications in the Maritime Industries (COMPIT), 2014
22. Christiansen, A.M.: Autonomous Coastal Container Feeder: Opportunities in Logistics. HSN (2017)
23. Regjeringen. Nasjonal Transportplan 2025–2036 (2024)
24. Surprenant, C.W.: Politics and practical wisdom: rethinking aristotle's account of phronesis. Topoi **31**, 221–227 (2012). https://doi.org/10.1007/s11245-012-9117-z

Dynamic Dispatching of DDMRP Replenishment Orders

Lucas Alves(✉), Maha Benali, and Pierre Baptiste

Department of Mathematics and Industrial Engineering, Polytechnique Montréal, 2500 Chemin de Polytechnique, Montréal, QC H3T 1J4, Canada
lucas.alves-da-silva-santos@polymtl.ca

Abstract. The Demand Driven Material Requirements Planning (DDMRP) is a production control system originally tailored for the VUCA world. Despite research increasingly focusing on the dynamic aspects of this production control system, it has failed to address important execution aspects. This study bridges the literature gap by focusing on Demand Driven Planning and Execution, particularly the dispatching rules. The use of static dispatching rules, such as the one proposed in DDMRP, does not respond to the volatility and uncertainty of the VUCA world. Through the simulation of a six-machine flow shop, we investigate several dispatching rules to assess their impact on the system stability and service level. Our findings reveal that, when compared to a static dispatching rule, the use of a dynamic dispatching rule in the DDMRP execution significantly improves the service level and reduces the number of stockouts and the time to fulfill backorders.

Keywords: Dynamic Order Dispatching · Demand Driven Planning and Execution · DDMRP · VUCA environments

1 Introduction

Nowadays, organizations face accelerated change with increased levels of unpredictability and complexity, resulting in unforeseen problems and vulnerability which require the development of new skills and the adoption of new strategies [1]. Traditional push or pull Production Control Systems (PCS) fail to respond to organizations' needs in this new environment. Thus, managers increasingly adopt hybrid PCSs [2], such as the Demand Driven Material Requirements Planning (DDMRP) which is proposed as a responsive PCS adapted to VUCA environments and as an alternative to traditional PCSs designed for more stable and predictable environments.

Considering its operationalization, DDMRP has five sequential phases: Strategic Inventory Positioning; Buffer Profiles and Levels; Dynamic Adjustment; Demand Driven Planning (DDP), and Demand Driven Execution (DDE). The first three phases define the configuration of DDMRP, such as where the decoupling points are placed, the amount of protection at those decoupling points, and how that level of protection increases or decreases based on operating parameters. The fourth and fifth phases define the actual operational aspects of a DDMRP system [3].

Published by Springer Nature Switzerland AG 2024
M. Thürer et al. (Eds.): APMS 2024, IFIP AICT 733, pp. 49–64, 2024.
https://doi.org/10.1007/978-3-031-71645-4_4

There is a clear distinction between the Planning and Execution phases of DDMRP operationalization. The planning of replenishment orders based on the Net Flow Positions (NFPs) of the buffered items is part of the DDP, while the management of open replenishment orders based on the On-Hand quantities is part of the DDE [3]. While some studies related to DDMRP (e.g., [4–7]) considered mainly the planning phase parameters (i.e. the NFP) to build the dispatching rules, others (e.g., [8–11]) consider mainly the execution phase parameters (i.e. the On-Hand). The variety of approaches results in research not acknowledging how order dispatching can affect DDMRP performance facing dynamic changes characterized by VUCA environments.

According to [12], dispatching rules centered around the buffer status reflected by the NFPs surpass due-date based rules in DDMRP because orders of the same product in the same queue can have different priorities according to their NFP. However, static dispatching rules relying only on NFPs are not adapted to a dynamic production system. In the modern manufacturing industry, methods to support real-time decision-making, such as dynamic evaluation of dispatching rules, are an urgent requirement to respond to the uncertainty and complexity [13]. From a practical point of view, managing a queue with different orders of the same product with different priorities incurs excessive execution complexity and issues with lot management and changeover times. This paper contributes to the DDMRP literature by investigating the utility of dispatching replenishment orders using both DDMRP planning and execution parameters (i.e., the NFPs and the On-Hand quantities). Considering a six-machine flow shop managed by DDMRP, three different dispatching rules are evaluated by simulation at various loading rates and compared in terms of their impact on the system stability and service level.

This article is structured as follows. Section 2 brings the current literature state regarding the use of dispatching rules in DDMRP applications. In Sect. 3, we introduce the case study and the proposed dispatching rules. Sections 4 and 5 present respectively the experiments and the result analysis. Finally, in Sect. 6, a conclusion is made and some perspectives for future research work are depicted.

2 DDMRP Concepts

Conceived as a rethinking of Material Requirements Planning (MRP), the DDMRP draws elements of Lean Systems and the Theory of Constraints to improve the overall performance and effectiveness of the MRP logic. It challenged some MRP practices, such as defining safety stocks based on demand forecasts [8].

For DDMRP practice, a buffer is a level of stock that is sized and maintained to protect a decoupling point. It serves to absorb demand variability, lead time compression, and replenishment order generation in terms of the NFP (see Fig. 1). A buffer is divided into three zones: (1) Green Zone – determining the frequency of order generation and minimum size of each order; (2) Yellow Zone – the demand coverage in the buffer while replenishment orders are fulfilled; (3) Red Zone – the safety embedded in the buffer position [3].

The buffer zones are proportional to the Average Daily Usage (ADU) and the Decoupled Lead Time (DLT) – or the Lead Time (LT) for non-decoupled product structures. The variability of both the ADU and the LT are considered in terms of the Lead Time

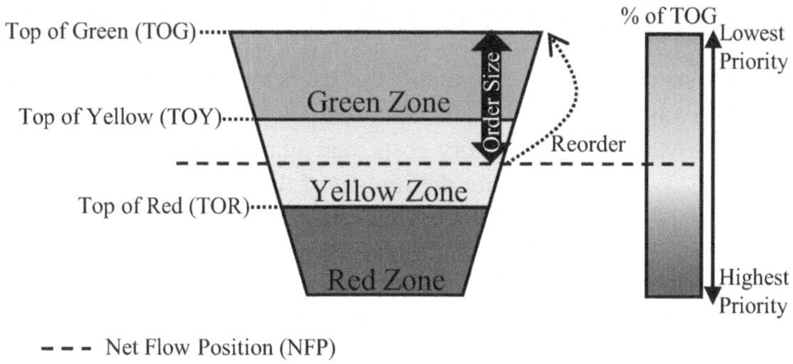

Fig. 1. DDMRP buffer zones and Net Flow Position (NFP)

Factor (LTF) and Variability Factor (VF). Refer to [3] for suggested ranges. The sizing of each zone is demonstrated in the following equations:

$$\text{Red Base} = \text{ADU} \times \text{LT} \times \text{LTF} \tag{1}$$

$$\text{Red Safety} = \text{ADU} \times \text{VF} \tag{2}$$

$$\text{Red Zone} = \text{Red Base} + \text{Red Safety} \tag{3}$$

$$\text{Yellow Zone} = \text{ADU} \times \text{LT} \tag{4}$$

$$\text{Green Zone} = \text{ADU} \times \text{LT} \times \text{LTF} \tag{5}$$

After the three zones are sized, the Top Of each zone can be calculated. The following equations demonstrate the complete calculations of the buffers.

$$\text{Top Of Red(TOR)} = \text{Red Zone} \tag{6}$$

$$\text{Top Of Yellow(TOY)} = \text{TOR} + \text{Yellow Zone} \tag{7}$$

$$\text{Top Of Green(TOG)} = \text{TOY} + \text{Green Zone} \tag{8}$$

As stated by [3], replenishment order generation is a function of the buffers. A replenishment order is created if the NFP is below the TOY and the requested quantity corresponds to a replenishment to the TOG. Expressed by Eq. 9, the NFP corresponds to the sum of the On-Hand (i.e. the current stock level of a particular item within a buffer) and the On-Order (i.e. the quantity of stock that has been ordered but not received) minus the Qualified Demand (QD). The QD is the past-due demand, plus today's demand, plus qualified order spikes in the future horizon [3]. The NFP is used to create orders which then need to be scheduled or prioritized at the production control level.

$$\text{NFP} = \text{On-Hand} + \text{On-Order} - \text{Qualified Demand} \tag{9}$$

2.1 Dispatching Rules in DDMRP Research

Dispatching consists of a priority rule that is used to select the next job to be processed from a set of jobs awaiting service [14]. Efficient dispatching is critical for on-time deliveries, minimizing overdue and inventory costs, and adapting to dynamic and stochastic environments [15]. When considering the impact of dispatching on the flow shop performance, it is known that the change in the dispatching rules does not have an impact on the average Flow Time (FT) and the average Waiting Time (WT), but on the FT and the WT variability [16]. In fact, some replenishment orders can be accelerated at the expense of others, increasing the distribution width of the FTs and the distribution width of the WTs, even if the expected means remain unchanged [17].

When analyzing the dispatching rule in DDMRP systems, it is crucial to take into account that one of the main objectives of DDMRP is to protect and promote the flow by using NFPs [3]. Initial DDMRP authors [3] propose to use the NFP as a priority rule at the control production level. They note that the NFP characterizes the urgency for the order's creation, and therefore its priority. This priority (Eq. 10) depends on the inverse of the value of the NFP over the TOG (i.e. smaller the value, the greater the priority) [3]. In this paper, we propose using different parameters to calculate the priority and consequently dispatch replenishment orders in both planning and execution.

$$\text{Priority} = \text{NFP}/\text{TOG} \tag{10}$$

Initial DDMRP authors propose that the buffered item with the smallest value of the NFP over TOG at the moment of the order creation (i.e., the most 'endangered' buffer) has priority over the other buffered items in the planning stage [3]. In a static approach, priority is given on an order basis, while in a dynamic approach, priority is given on a product basis and in both cases they are dependent on the NFP (see Fig. 2).

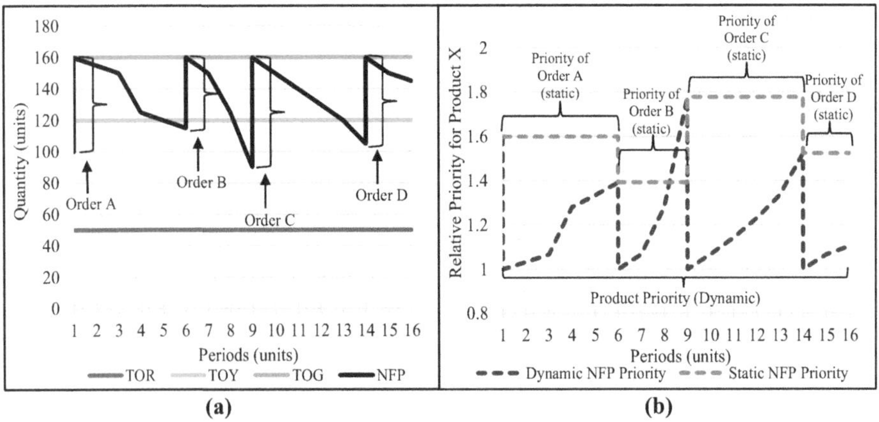

Fig. 2. Product X's NFP evolution over time (a) and relative priority evolution over time (b)

When using the static NFP as proposed by [3], different orders of the same product on the same flow may have different priorities (i.e., the priority of order A is greater than that of order B and less than that of order C), while in a dynamic approach, different orders of the same product always have the same priority (i.e., all orders have the same priority that changes over time). For orders of the same product, the NFP at the moment of order generation determines the priority, with a smaller NFP resulting in a higher priority. According to [3], the main function of the NFP is to dictate whether a replenishment order should be created for a buffer position while maintaining its integrity. The NFP is required for the planning phase, while the buffer integrity has to be considered in the execution to ensure that the buffer effectively performs its decoupling role.

While most research on DDMRP [4–7] used the combination of NFP and First In First Out (FIFO) to dispatch replenishment orders, based only on DDP parameters, a few studies proposed different approaches. In [8, 9], single dispatching rules are proposed for both DDP and DDE phases based only on DDE parameters. In [8], the dispatching rule is a result of the percentage of the On-Hand over the Red Zone plus the Green Zone. For [9], the dispatching rule depends on the percentage of the On-Hand over the TOR. In [10], the proposed dispatching rule is based on both DDP and DDE parameters. The authors used a single dispatching rule for DDP and DDE phases giving priority to replenishment orders with an NFP in the Red Safety Zone with a stockout alert, if none of the conditions apply, priority followed the FIFO rule. [11] proceeds differently and uses the expected production start date as the dispatching rule (i.e. the earliest expected production start date has the priority) at both DDP and DDE phases.

Considering the buffer status to dispatch replenishment orders, using NFPs as recommended by [3] seems to outperform due-date oriented dispatching rules and FIFO rules [12]. However, the performance of NFP is highly dependent on two factors: the reorder quantity associated with the replenishment orders and the fact that orders of the same product may have different priorities [12]. This suggests that the static NFP rule proposed by [3] is not adapted to all situations, as using only NFPs fails to cover some issues in the DDE phase. To manage open replenishment orders in the execution phase, [3] proposes to prioritize replenishment orders with issued On-Hand Alerts. An On-Hand Alert is generated whenever the On-Hand Position is below half of the TOR [3]. However, it is unclear in theory and practice how to use both On-Hand alerts and NFPs for dispatching considering dynamic production systems.

Despite the increase in research to evaluate dynamic aspects of DDMRP such as the decouple lead time (DLT) [4], buffer sizing [18, 19], and parametrization [20, 21], none of the previously mentioned applications considered a dynamic dispatching rule for DDMRP systems. Since [12] findings demonstrate that dynamically adjusting the NFP may not be the best option when dispatching replenishment orders, it is interesting to further investigate DDMRP dynamic dispatching rules. In contrast to the previous literature, this paper proposes to use real-time buffer data for order dispatching in a way that better connects the DDP and the DDE phases of the DDMRP methodology.

3 Dispatching Rule Models

This paper aims to investigate the utility of dynamic dispatching in DDMRP applications in terms of system stability and service level. To this end, first, we consider a Discrete Event Simulation (DES) model of a DDMRP flow shop, using ARENA 16.10. The model is inspired by [19] and uses the static NFP rule as the most commonly used dispatching rule in DDMRP applications [4–7]. Then, we add two dynamic dispatching rules that will be compared to the static NFP rule (which serves as a benchmark for evaluation). Finally, we experiment with the model with the three dispatching rules in scenarios considering different loading rates and compare the two proposed dynamic dispatching rules to the static rule.

3.1 Static Dispatching Rule Model (Static)

The DES model represents a six-machine flow shop where final products managed with DDMRP buffers must pass through all the machines. We consider 5 operating days a week with 1 shift of 8 h per day. Replenishment orders are generated every time the NFP is below TOY to restore the TOG quantity (Sect. 2.1). We consider 30 products manufactured in the flow shop. The customer demands have a similar interarrival rate. The quantities requested by customer demands are calculated to meet the Average Daily Usage (ADU) presented in Table 1 with a 20% variability. To simplify, the visibility of customer demands is restricted to a 1-day simulation horizon. Thus, the model does not consider qualified spikes in the QD. If a customer demand cannot be met from the On-Hand inventory, it would be back-ordered to be fulfilled as soon as possible.

Table 1. Average Daily Usage (ADU) in units

Product	P1	P2	P3	P4	P5	P6	P7	P8	P9	P10
ADU	65	23	55	63	19	38	29	33	61	34
Product	P11	P12	P13	P14	P15	P16	P17	P18	P19	P20
ADU	61	62	57	32	52	21	61	42	48	44
Product	P21	P22	P23	P24	P25	P26	P27	P28	P29	P30
ADU	35	52	23	56	55	37	32	51	55	52

The Processing Times (PT) (see Table 2) are randomly adjusted with a coefficient following a triangular law TRI (0.8; 1; 1.2). Transfer times between different processes are not considered neither changeover times between products. The variability factor is 0.50 for every product and the lead time factor is 0.75, 0.5, and 0.25 respectively for small, medium, and long PTs, following the factor ranges suggested by Ptak and Smith [3]. The Lead Time (LT) is 9.75 h for all experiments. Buffers are randomly initialized between 50% and 100% of the TOG.

The Static dispatching rule for replenishment orders considers the static value of the NFP over the TOG. The replenishment order with the smallest value has the highest priority. In the case of two replenishment orders having the same priority, the FIFO rule is used. This dispatching rule is used only for the first machine, all the subsequent machines use the FIFO dispatching rule (see Fig. 3). For a detailed description of DDMRP functioning refer to Ptak and Smith [3].

Table 2. Processing Times in Seconds

Products / Machines	M1	M2	M3	M4	M5	M6
P1 – P10 (small PT)	5	5	5	10	15	5
P11 – P20 (med. PT)	5	10	10	10	10	10
P21 – P30 (high PT)	5	15	15	10	5	15

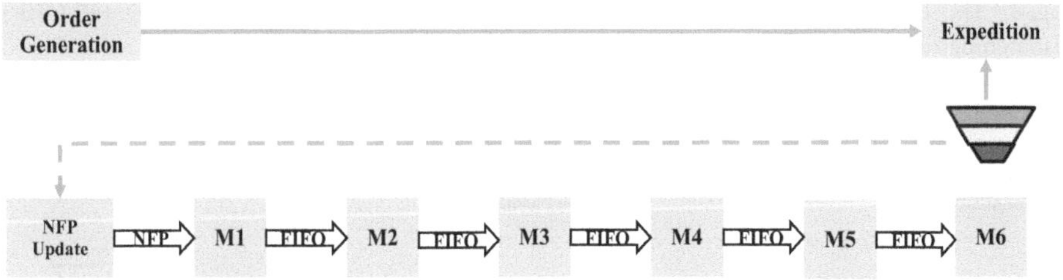

Fig. 3. The DDMRP flow shop managed by the Static dispatching rule.

3.2 Dynamic Dispatching Rule Model

To assess the performance of the dynamic dispatching rules, we adapted the static dispatching model to function with the two new dispatching rules. The first dynamic rule uses the **Net Flow On-Hand Positioning (NFOHP)**. We propose to use the static NFP to define the priority in the first machine queue, in such a way that the lowest value of the NFP over the TOG has the highest priority. For the subsequent machines, the On-Hand Position (expressed by Eq. 11 as the On-Hand stock divided by the TOR), is dynamically adjusted each time a decision is made about the next replenishment order to process. The replenishment order with the smallest On-Hand Position is placed first in the queue for processing (see Fig. 4). In contrast to Ptak and Smith's [3] which uses the On-Hand Position only for managing orders with an active On-Hand Alert, we propose to consider the On-Hand Position for all replenishment orders.

$$\text{On-Hand Position(OHP)} = \text{On-Hand}/\text{TOR} \tag{11}$$

The second dynamic rule that we propose is the **Dynamic Available On-Hand Positioning (DAOHP)**. With this rule, we dynamically adjust what we call the Available On-Hand Position (AOHP) for the replenishment orders in all machines' queues each time a decision is made about the next replenishment order to process. As presented in Eq. 12, we define the AOHP as the units in inventory minus the backorders and the demand due today. The replenishment order with the smallest AOHP is placed first in the queue for processing (see Fig. 5).

$$\text{AOHP} = (\text{On-Hand} - \text{Backorders} - \text{Demand})/\text{TOR} \tag{12}$$

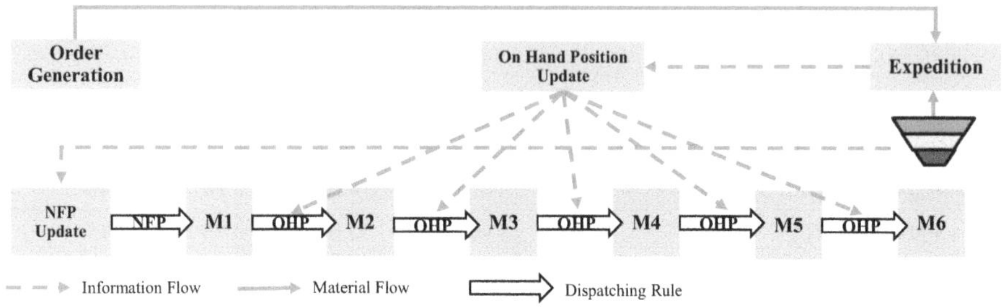

Fig. 4. The DDMRP flow shop managed by the Net Flow On-Hand Positioning (NFOHP) dispatching rule

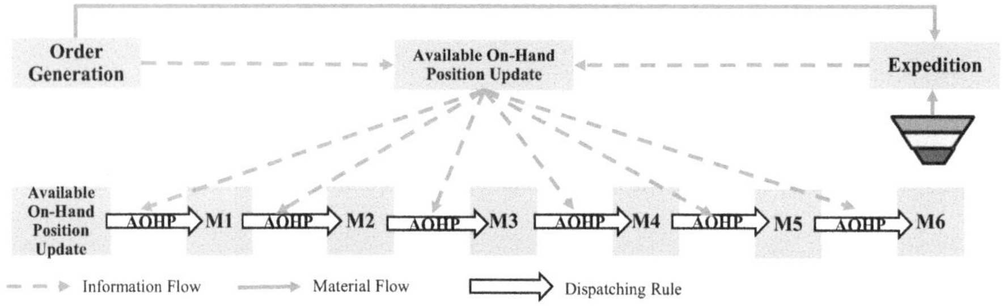

Fig. 5. The DDMRP flow shop managed by the Dynamic Available On-Hand Positioning (DAOHP) dispatching rule

4 Experiments

To achieve the objective of assessing the dispatching rules' impact on the system stability and service level of DDMRP applications, we propose to experiment with a full factorial design with two factors characterizing the DDMRP system: the Loading Rate (LR) and the dispatching rule (Static, NFOHP, DAOHP). For each dispatching rule, 20 experiments (20 levels of LR from 80% to 99% in 1% increments) are simulated. A total of 60 experiments are conducted and the performance is measured using six performance indicators:

- Service Level (SL): the number of customer demands filled in time over the total number of customer demands;
- Average Inventory (I): the average inventory (Work In Process + Finished Products) during the experiment;
- Stockouts (SO): the number of times a buffer position reaches zero;
- Backorders (BO): the number of units delivered after the expected time;
- On-Hand Alerts (OHA) percentage: the percentage of On-Hand Alerts over the number of deliveries. An On-Hand Alert is generated whenever the On-Hand Position is below half of the TOR.
- Mean Backorder Fulfillment Time (MBOT): the mean time to fulfill a backorder.

Considering 50 replications for each experiment and 95% confidence intervals (CI) on estimates, the results were collected during 550 simulation days, after 50 warmup days.

5 Results and Discussion

The simulation results obtained for the three proposed dispatching rules are summarized in the Annex A. We can see that the three different rules had the same performance in highly loaded systems (i.e., LR greater than 95%). Figure 6 focuses on the results up to 95% of LR and illustrates the Service Level SL and the average Inventory I with the three dispatching rules. First, as expected, as the LR increases, the system fails to maintain its stability, as demonstrated by I increasing exponentially when the LR approaches 95%. Second, we notice that both static and dynamic rules are equally affected by increasing the LR. We can conclude then that the effectivity of the dispatching rules is not connected to the LR.

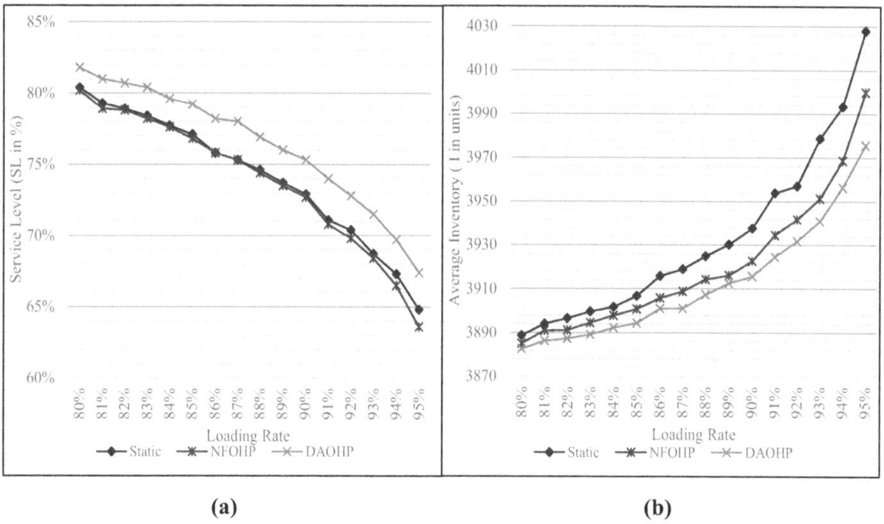

Fig. 6. Impact of the Loading Rate on the Service Level and Average Inventory achieved with the three dispatching rules

While the Static rule and the NFOHP have mostly the same results in terms of SL, the DAOHP rule achieves a higher SL up to 95% of LR. This demonstrates that even if the NFOHP's parameters are dynamically adjusted, it fails to improve the DDMRP performance. This can be explained by the On-Hand not considering the backorders and the current demand, and then failing to prioritize the most endangered buffers. In contrast, the DAOHP rule increases the SL compared to the static rule and the NFOHP. This is because the DAOHP rule prioritizes the most urgent replenishment orders for processing. In contrast to Ptak and Smith's [3] On-Hand Position proposition which does not consider backorders and current demand, the DAOHP rule proposed in this paper prioritizes replenishment orders with the smallest availability. We can say that the DAOHP rule gives the DDMRP system the visibility of the most endangered buffer in real time and reorganizes it accordingly to define the next replenishment order.

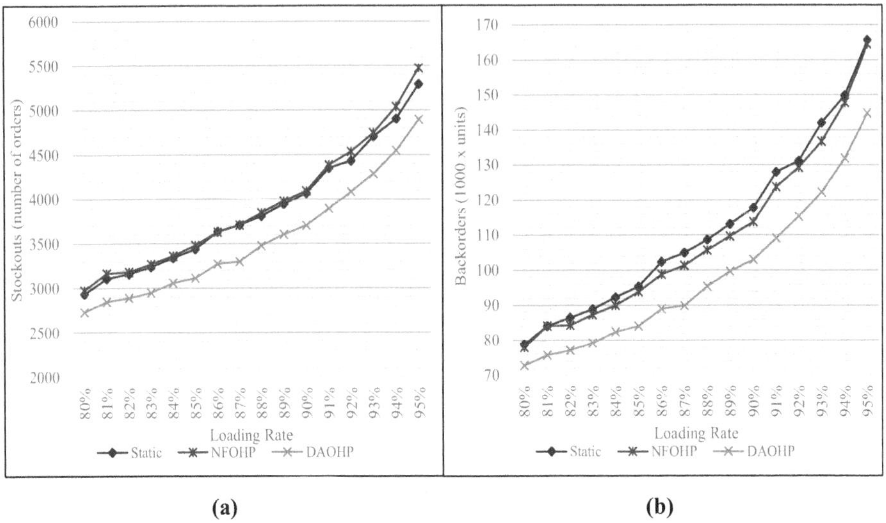

Fig. 7. Impact of the Loading Rate on Stockouts and Backorders with three dispatching rules

Figure 7 shows the results of Stockouts and the Products Backordered. As expected, when the LR approaches 95% the number of Stockouts and Products Backordered starts to grow exponentially. It is interesting to note that up to 95% of LR, the DAOHP rule increases the buffer protection. When compared to the Static rule, the DAOHP rule consistently reduced the percentage of Stockouts (e.g., at 90% of LR, 9% reduction) and the number of products backordered (e.g., at 90% of LR, 12.6% of reduction).

Last but not least, with regards to the On-Hand Alerts percentage and the MBOT (see Fig. 8), we can see that the effectivity of a dispatching rule is connected to its ability to rapidly fulfill a backorder while reducing the number of On-Hand Alerts. Compared to the Static rule, the NFOHP rule has a smaller MBOT. Thus, this dynamic rule can reduce the time to fulfill backorders, avoiding stockouts. However, the NFOHP resulted in a higher number of On-Hand Alerts due to its lack of visibility of the most endangered buffers which resulted in the NFOHP rule and the Static rule having the same performance. In contrast, the DAOHP rule, reduced both the On-Hand Alerts percentage and the MBOT, resulting in a higher performance compared to the other rules.

Finally, a comparison of the confidence intervals confirms the above results which indicate that, up to an LR of 95%, the DAOHP rule outperforms the Static and NFOHP rules, with the two latter performing similarly except for the MBOT. Both the Static and the NFOHP dispatching rules produced similar results mainly because of the routing characteristics of the flow shop (i.e., both rules treat the first machine queue similarly). The results between the Static and NFOHP rules may differ when examining other shop arrangements. For the DAOHP rule, since the same rule is applied to every queue, there

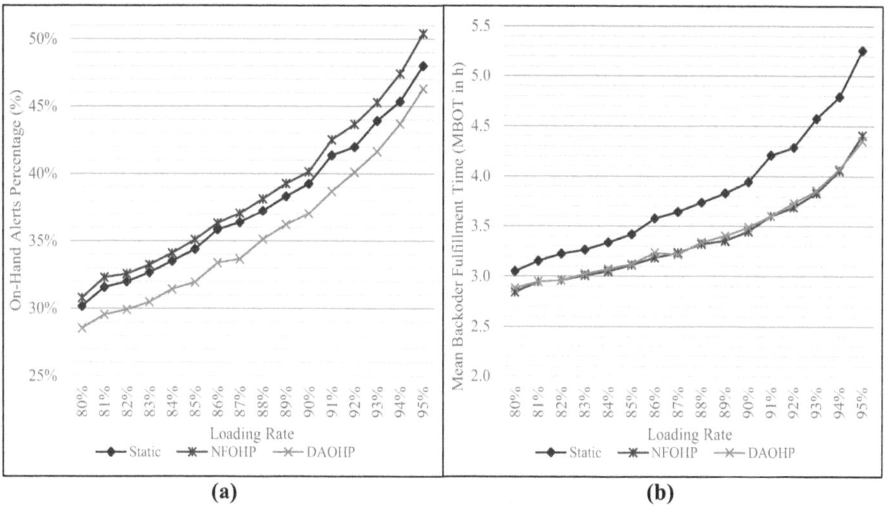

Fig. 8. Impact of the Loading Rate on On-Hand Alerts Percent and Mean Backorder Fulfillment Time with the three dispatching rules

is no reason for the results to differ in another production arrangement such as a job shop.

6 Managerial Insights

Dynamic dispatching rules that consider the DDMRP execution parameters can significantly improve service levels when compared to static rules. The DAOHP rule proposed in the study achieved higher service levels by prioritizing the most urgent replenishment orders for processing. The implementation of dynamic dispatching rules can reduce stockouts, backorders, and the time required to fulfill backorders. This improved responsiveness to demand fluctuations and disruptions is critical in today's volatile and uncertain business environment.

While traditional static dispatching rules in DDMRP cannot quickly adjust the production to changes in buffers, dynamic rules such as DAOHP can recognize the varying urgency levels of different buffers and prioritize accordingly. This tailored prioritization is consistent with the DDMRP philosophy of promoting flow and protecting decoupling points. The results of this study challenge the assumption of using the NFP to dispatch replenishment orders in the planning phase of DDMRP. The use of a dispatching rule based on real-time buffer data that incorporates backorder and future demand information can lead to better performance in DDMRP applications.

The implementation of dynamic rules such as the DAOHP requires the integration of real-time execution parameters, such as On-Hand inventory, backorders, and current demand with planning parameters such as buffer levels. This level of integration may require efforts to enable seamless data exchange between DDP and DDE modules. Another important consideration is the need to ensure high data quality and real-time visibility across the supply chain, which can be challenging, especially in multi-site, multi-echelon environments with disparate systems.

7 Conclusion

This paper investigates the problem of dynamically adjusting the dispatching rule in DDMRP applications. We contribute to the theory by extending the research on dynamic aspects of DDMRP that have not been addressed in the literature so far. Two new dynamic dispatching rules are proposed and compared to the conventional static dispatching rule under different loading rates. The results suggest that the dynamic rule DAOHP based on the Available On-Hand Positioning can significantly improve the service level, reduce the number of stockouts, and shorten the time to fulfill backorders.

Protecting the buffer position is one of the main objectives of DDMRP. However, buffer protection is different from inventory protection. At the execution level, taking into account the actual position of the inventory is key to improving the performance of the system. In this paper, using the Available On-Hand Position as the dispatching rule increased the SL due to its accuracy to represent real inventory status. The NFP does not represent the real inventory position because it takes into account orders that are already in the system. At the planning level, the NFP is an important parameter to determine the quantity to be replenished, but it should not be used at the execution level.

Despite the consistency of the results, there is a limitation related to the simulated flow shop. Further research should focus on uncovering the effects of using the static NFP as a prioritization parameter in DDMRP applications. Given the improvements resulting from using execution parameters for dispatching, it seems practical to consider other execution aspects to dispatch replenishment orders in DDMRP applications.

Acknowledgments. This work was supported by the Natural Sciences and Engineering Research Council of Canada (NSERC).

Disclosure of Interests. The authors have no competing interests to declare that are relevant to the content of this article.

Annex A

See Tables A.1, A.2 and A.3.

Table A.1. Simulation Results for STATIC dispatching rule (95% CI)

LR (%)	SL (%)	I (units)	SO (units)	BO (units)	OHA (units)	MBOT (hours)
80%	80.4 ± 0.18	3888.74 ± 12.92	2936 ± 31.46	78844 ± 1030	4526 ± 38.94	3.05 ± 0.03
81%	79.3 ± 0.2	3894.02 ± 15.32	3107 ± 35.02	84105 ± 1184	4744 ± 43.76	3.15 ± 0.03
82%	78.9 ± 0.23	3896.49 ± 14.38	3161 ± 38.99	86498 ± 1359	4800 ± 45.24	3.22 ± 0.03
83%	78.4 ± 0.23	3899.67 ± 15.68	3239 ± 39.94	88863 ± 1412	4902 ± 46.68	3.26 ± 0.03
84%	77.7 ± 0.27	3901.67 ± 18.89	3341 ± 46.59	92308 ± 1718	5027 ± 55.26	3.34 ± 0.03
85%	77.1 ± 0.2	3906.73 ± 14.54	3438 ± 33.78	95335 ± 1290	5153 ± 41.5	3.42 ± 0.03
86%	75.8 ± 0.33	3915.87 ± 24.31	3636 ± 56.86	102365 ± 2117	5382 ± 67.17	3.58 ± 0.04
87%	75.3 ± 0.33	3918.91 ± 23.611	3709 ± 55.81	104930 ± 2220	5467 ± 66.25	3.64 ± 0.05
88%	74.6 ± 0.29	3924.87 ± 22.97	3811 ± 49.24	108635 ± 1858	5591 ± 59.86	3.74 ± 0.05
89%	73.7 ± 0.36	3930.25 ± 26.65	3945 ± 62.13	113057 ± 2286	5752 ± 71.16	3.83 ± 0.05
90%	72.9 ± 0.3	3937.57 ± 21.88	4068 ± 50.59	117762 ± 1973	5898 ± 57.73	3.94 ± 0.04
91%	71.1 ± 0.45	3953.88 ± 33.36	4350 ± 74.42	128000 ± 2950	6221 ± 82.52	4.21 ± 0.08
92%	70.4 ± 0.47	3957.07 ± 34.42	4432 ± 78.9	131153 ± 3052	6294 ± 89.19	4.29 ± 0.07
93%	68.7 ± 0.59	3978.69 ± 44.11	4705 ± 97.69	142000 ± 3795	6597 ± 104.1	4.58 ± 0.11
94%	67.3 ± 0.55	3993.27 ± 46.22	4904 ± 90.23	149833 ± 3785	6809 ± 97.6	4.79 ± 0.11
95%	64.8 ± 0.72	4027.93 ± 58.26	5294 ± 116.8	165607 ± 4987	7209 ± 120.7	5.26 ± 0.16
96%	62.4 ± 0.86	4064.95 ± 71.39	5644 ± 139.7	180045 ± 6115	7562 ± 138.7	5.71 ± 0.19
97%	58.5 ± 1.35	4140.49 ± 116.03	6236 ± 214.8	205157 ± 9468	8141 ± 206.7	6.57 ± 0.35
98%	52.7 ± 2.13	4306.49 ± 215.52	7107 ± 333.2	245474 ± 15900	8942 ± 295.7	8.1 ± 0.71
99%	37.9 ± 3.71	5107.14 ± 569.4	9334 ± 581	355141 ± 30090	10838 ± 475.5	13.82 ± 2.2

Table A.2. Simulation Results for NFOHP dispatching rule (95% CI)

LR (%)	SL (%)	I (units)	SO (units)	BO (units)	OHA (units)	MBOT (hours)
80%	80.2 ± 0.2	3885.08 ± 14.46	2977 ± 34.64	78044 ± 1187	4620 ± 43.35	2.84 ± 0.02
81%	78.9 ± 0.31	3890.96 ± 21	3165 ± 53.61	84108 ± 1899	4856 ± 66.29	2.94 ± 0.03
82%	78.8 ± 0.23	3891.08 ± 16.97	3181 ± 39.36	84269 ± 1360	4883 ± 48.18	2.96 ± 0.03
83%	78.2 ± 0.23	3894.47 ± 17.4	3268 ± 41.42	87297 ± 1379	4984 ± 50.75	3.01 ± 0.03
84%	77.6 ± 0.29	3897.84 ± 21.09	3361 ± 50.32	89990 ± 1706	5115 ± 61.21	3.05 ± 0.03
85%	76.8 ± 0.27	3900.74 ± 20.1	3481 ± 48.14	93848 ± 1666	5262 ± 61.51	3.11 ± 0.03
86%	75.8 ± 0.31	3905.75 ± 20.49	3631 ± 52.23	98775 ± 1733	5448 ± 60.98	3.18 ± 0.04
87%	75.3 ± 0.31	3908.78 ± 23.53	3709 ± 52.93	101320 ± 1861	5560 ± 67.65	3.23 ± 0.03
88%	74.4 ± 0.29	3914.32 ± 21.39	3846 ± 49.8	105636 ± 1735	5722 ± 58.58	3.32 ± 0.03

(*continued*)

Table A.2. (*continued*)

LR (%)	SL (%)	I (units)	SO (units)	BO (units)	OHA (units)	MBOT (hours)
89%	73.5 ± 0.39	3916.14 ± 28.53	3981 ± 65.57	109625 ± 2360	5890 ± 77.57	3.35 ± 0.04
90%	72.7 ± 0.34	3922.62 ± 24.6	4092 ± 56.64	113720 ± 2005	6019 ± 65.49	3.45 ± 0.04
91%	70.8 ± 0.48	3934.49 ± 34.03	4385 ± 79.61	123749 ± 2993	6383 ± 88.08	3.6 ± 0.06
92%	69.8 ± 0.49	3941.69 ± 36.2	4535 ± 80.12	129247 ± 3001	6553 ± 91.57	3.69 ± 0.06
93%	68.4 ± 0.52	3951.21 ± 40.52	4744 ± 86.89	136678 ± 3255	6791 ± 97.19	3.83 ± 0.06
94%	66.5 ± 0.64	3968.55 ± 47.63	5038 ± 105.5	147799 ± 4024	7118 ± 114.9	4.06 ± 0.09
95%	63.6 ± 0.99	3999.74 ± 79.98	5474 ± 158.7	164444 ± 6712	7575 ± 158	4.41 ± 0.2
96%	61.1 ± 1.36	4029.68 ± 106.07	5842 ± 215.6	179340 ± 9147	7947 ± 211.1	4.78 ± 0.26
97%	55.5 ± 1.49	4113.73 ± 125.98	6694 ± 234.1	214139 ± 10400	8796 ± 217.7	5.62 ± 0.44
98%	51.7 ± 1.63	4182.68 ± 136.88	7241 ± 255.1	238022 ± 11490	9300 ± 233.6	6.36 ± 0.39
99%	35.6 ± 3.78	4974.96 ± 519.1	9679 ± 589.6	359376 ± 30800	11311 ± 465.7	12.14 ± 2.03

Table A.3. Simulation Results for DAOHP dispatching rule (95% CI)

LR (%)	SL (%)	I (units)	SO (units)	BO (units)	OHA (units)	MBOT (hours)
80%	81.8 ± 0.22	3882.62 ± 16.38	2733 ± 37.66	72786 ± 1209	4285 ± 46.2	2.88 ± 0.02
81%	81 ± 0.23	3886.3 ± 19.52	2848 ± 40.66	75772 ± 1273	4439 ± 55.43	2.94 ± 0.03
82%	80.7 ± 0.21	3887.32 ± 17.2	2891 ± 35.55	77123 ± 1218	4485 ± 48.25	2.96 ± 0.02
83%	80.4 ± 0.23	3889.14 ± 17.34	2948 ± 39.13	79200 ± 1363	4575 ± 49.73	3.02 ± 0.03
84%	79.6 ± 0.2	3892.24 ± 18.85	3058 ± 34.5	82368 ± 1158	4721 ± 45.5	3.07 ± 0.02
85%	79.2 ± 0.24	3894.35 ± 19.27	3114 ± 40	84063 ± 1382	4793 ± 50.04	3.12 ± 0.03
86%	78.2 ± 0.31	3900.94 ± 21.85	3274 ± 51.51	88951 ± 1795	5014 ± 61.94	3.23 ± 0.04
87%	78 ± 0.31	3901.03 ± 24.23	3299 ± 53.12	89840 ± 1821	5043 ± 66.62	3.22 ± 0.03
88%	76.9 ± 0.3	3907.32 ± 23.14	3476 ± 50.67	95364 ± 1768	5274 ± 61.56	3.34 ± 0.03
89%	76 ± 0.43	3912.54 ± 32.92	3603 ± 71.73	99617 ± 2437	5438 ± 86.92	3.4 ± 0.04
90%	75.3 ± 0.4	3915.62 ± 29.59	3703 ± 65.31	102945 ± 2365	5554 ± 76.3	3.49 ± 0.05
91%	74 ± 0.44	3924.55 ± 32.77	3893 ± 72.29	109112 ± 2557	5803 ± 82.51	3.6 ± 0.05
92%	72.8 ± 0.43	3931.62 ± 34.18	4084 ± 70.86	115302 ± 2668	6021 ± 82.35	3.73 ± 0.07
93%	71.5 ± 0.66	3940.86 ± 47.57	4284 ± 106.7	122236 ± 4027	6247 ± 117.7	3.86 ± 0.1
94%	69.7 ± 0.73	3956.32 ± 56.27	4546 ± 118.2	131928 ± 4522	6557 ± 128.9	4.07 ± 0.12
95%	67.4 ± 0.96	3975.58 ± 71.16	4897 ± 152.9	144702 ± 6049	6945 ± 158.5	4.35 ± 0.15
96%	63.6 ± 1.47	4016.61 ± 111.43	5473 ± 233.6	166345 ± 9533	7552 ± 231.3	4.81 ± 0.27
97%	57.9 ± 1.91	4103.17 ± 153.73	6319 ± 300.5	201786 ± 13530	8369 ± 273.5	5.75 ± 0.44
98%	53.7 ± 2.44	4219.27 ± 248.59	6956 ± 377.5	230211 ± 18240	8951 ± 322	6.76 ± 0.87
99%	36 ± 3.66	4763.71 ± 384.1	9616 ± 569.8	356796 ± 29170	11201 ± 453.5	10.74 ± 1.22

References

1. Taskan, B., Junça-Silva, A., Caetano, A.: Clarifying the conceptual map of VUCA: a systematic review. Int. J. Organ. Anal. **30**(7), 196–217 (2022)
2. Miguel, H., González, G., Antonio, L., Moreno, G., Luisa, A., Corralejo, G.: Hybrid push-pull systems: classification, optimal methods and uncertainty. In: Gurrola, J. (ed.) International Conference on Computational Science and Computational Intelligence 2021. CSCI, vol. 1, pp. 633–641. IEEE, Las Vegas (2022)
3. Ptak, C., Smith, C.: DDMRP Demand Driven Material Requirements Planning: Version 3, 3rd edn. Industrial Press Inc., New York (2019)
4. Dessevre, G., Martin, G., Baptiste, P., Lamothe, J., Pellerin, R., Lauras, M.: Decoupled lead time in finite capacity flowshop: a feedback loop approach. In: Zheng, F., Chu, F., Liu, M. (eds.) International Conference on Industrial Engineering and Systems Management 2019. IESM, vol. 1, pp. 1–6. IEEE, Shangai (2020)
5. Shofa, M., Moeis, A., Restiana, N.: Effective production planning for purchased part under long lead time and uncertain demand: MRP Vs demand-driven MRP. IOP Conf. Ser. Mater. Sci. Eng. **337**(1), 1–7 (2018)
6. Kortabarria, A., Apaolaza, U., Lizarralde, A., Amorrortu, I.: Material management without forecasting: from MRP to demand driven MRP. J. Ind. Eng. Manage. **11**(4), 632–650 (2018)
7. Dessevre, G., Baptiste, P., Lamothe, J., Pellerin, R.: Visual charts produced by simulation to correlate service rate, resource utilization and DDMRP parameters. Int. J. Prod. Res. **61**(3), 741–753 (2021)
8. Miclo, R., Lauras, M., Fontanili, F., Lamothe, J., Melnyk, S.: Demand driven MRP: assessment of a new approach to materials management. Int. J. Prod. Res. **57**(1), 166–181 (2018)
9. Pekarcıkova, M., Trebuna, P., Kliment, M., Trojan, J.: Demand driven material requirements planning. Some methodical and practical comments. Manage. Prod. Eng. Rev. **10**(2), 50–59 (2019)
10. Velasco Acosta, A., Mascle, C., Baptiste, P.: Applicability of demand-driven MRP in a complex manufacturing environment. Int. J. Prod. Res. **58**(14), 4233–4245 (2019)
11. Thürer, M., Fernandes, N., Stevenson, M.: Production planning and control in multi-stage assembly systems: an assessment of Kanban, MRP, OPT (DBR) and DDMRP by simulation. Int. J. Prod. Res. **60**(3), 1036–1050 (2020)
12. Fernandes, N., Guedes, N., Thürer, M., Ferreira, L., Ávila, P.: How to prioritize replenishment orders in demand driven MRP: a simulation study. In: Silva, F., Ferreira, L., Sá, J., Pereira, M., Pinto, C. (eds.) Flexible Automation and Intelligent Manufacturing 2023. FAIM LNME, vol. 2, pp. 95–102. Springer, Cham (2024). https://doi.org/10.1007/978-3-031-38165-2_12
13. Zhang, L., Hu, Y., Tang, Q., Li, J., Li, Z.: Data-driven dispatching rules mining and real-time decision-making methodology in intelligent manufacturing shop floor with uncertainty. Sensors **21**(14), 4863 (2021)
14. Blackstone, J.H., Phillips, D.T., Hogg, G.L.: A state-of-the-art survey of dispatching rules for manufacturing job shop operations. Int. J. Prod. Res. **20**(1), 27–45 (1981)
15. Zheng, S., Gupta, C., Serita, S.: Manufacturing dispatching using reinforcement and transfer learning. In: Brefeld, U., Fromont, E., Hotho, A., Knobbe, A., Maathuis, M., Robardet, C. (eds.) Machine Learning and Knowledge Discovery in Databases 2019. ECML PKDD, LNCS, vol. 11908, pp. 655–671. Springer, Cham (2020). https://doi.org/10.1007/978-3-030-46133-1_39
16. Hubl, A., Jodlbauer, H., Altendorfer, K.: Influence of dispatching rules on average production lead time for multi-stage production systems. Int. J. Prod. Econ. **144**(2), 479–484 (2013)

17. Jäger, Y., Roser, C.: Effect of prioritization on the waiting time. In: Moon, I., Lee, G., Park, J., Kiritsis, D., von Cieminski, G. (eds.) Advances in Production Management Systems 2018. APMS, IFIPAICT, vol. 535, pp. 21–26. Springer, Cham (2018). https://doi.org/10.1007/978-3-319-99704-9_3
18. Martin, G., Lauras, M., Baptiste, P.: Dynamical multi-parameter sizing of DDMRP buffers in finite capacity flow-shops. Comput. Ind. Eng. **175**, 1–10 (2023)
19. Azzamouri, A., Baptiste, P., Pellerin, R., Dessevre, G.: Impact of the continuous and periodic assessment of a buffer replenishment on the DDMRP method. Int. J. Prod. Res. **61**(16), 5637–5658 (2023)
20. Duhem, L., Benali, M., Martin, G.: Parametrization of a demand-driven operating model using reinforcement learning. Comput. Ind. **147**, 1–12 (2023)
21. Damand, D., Lahrichi, Y., Barth, M.: Parameterisation of demand-driven material requirements planning: a multi-objective genetic algorithm. Int. J. Prod. Res. **61**(15), 5134–5155 (2023)

Assembly Line Design for Industrialized Electrolyser Production

Sven-Vegard Buer[1](✉) and Lars Skjelstad[2]

[1] Department of Technology Management, SINTEF Digital, Oslo, Norway
sven-vegard.buer@sintef.no
[2] Department of Technology Management, SINTEF Digital, Trondheim, Norway

Abstract. The electrolyser industry is under rapid development, where manufacturing companies seek opportunities to develop industrialized solutions for electrolyser production. However, electrolyser production implies major logistical challenges related to for instance product characteristics (expensive and fragile components, rare materials), the stack assembly process (precise alignment of cells and pressure), the supply chain (immature, limited supplier base, quality issues, delivery constraints), and the production equipment (immature). Designing a high efficiency electrolyser assembly line is thus a challenging task. This paper presents results of a study with the aim to assess design alternatives for electrolyser stack assembly with focus on logistics and flow efficiency. Design elements such as the number and locations of buffers and quality inspection are considered. A discrete-event simulation model is developed and tested based on scenarios reflecting situations related to for instance poor quality yield of incoming components and equipment downtime. Data from an empirical case involving the design of a new assembly line, are used to set up the model. Based on the findings from the simulations, recommendations for the design of industrialized electrolyser assembly lines are formulated. This study contributes with insight to critical aspects and guidance to design decisions for efficient electrolyser assembly lines. Moreover, it shows how simulation can support assessment of different design alternatives in the development of efficient stack assembly lines.

Keywords: factory planning · electrolyser stack assembly · flow efficiency · modelling and simulation

1 Introduction

Hydrogen is expected to play a key role in the transition to renewable energy sources and in reaching the goals of climate neutrality [1]. The European Union sees hydrogen as an intrinsic part of the future energy system, and has a strategic objective of installing at least 40 GW of renewable hydrogen electrolysers by 2030 [2]. Hydrogen is seen as a promising energy carrier and can be generated from renewable energy sources through electrolysis [3]. However, the production of hydrogen electrolysers is expensive, both due to the design of the electrolyser and the immaturity of the electrolyser industry.

Published by Springer Nature Switzerland AG 2024
M. Thürer et al. (Eds.): APMS 2024, IFIP AICT 733, pp. 65–77, 2024.
https://doi.org/10.1007/978-3-031-71645-4_5

Electrolysers are still often made by hand in small volumes [1]. To reach the projected future demand of hydrogen, there is a need to find solutions for efficient and sustainable production of electrolysers to realize high-volume production. Krishnan et al. [4] expect that the reduction of electrolyser stack cost will be driven by more efficient electrolyser technology, the replacement of expensive materials with cheaper alternatives and the reduction of manufacturing and labour costs enabled by mass production.

Start-ups in young industries need to focus on developing logistics and manufacturing solutions alongside their product development since there is no proven way how to set up the production for the new product. Often the outset is limited to product information and a plan for volume and speed in operations in the short term. The design of operations must start early when even some details of the product still are not known. Balancing resources across product development, logistics planning, and operational setup can be complex, since time, capital and manpower need to be exploited effectively. Changes or delays in one aspect influence the other domains, i.e., changes in product design can influence packaging. Early product components from suppliers are prototypes and might be adjusted due to technical advancements. In concurrent development projects, there is a risk of overlooking critical aspects that have the potential to influence costs, quality and product yield, but despite the complexities involved, successfully synchronizing product, logistics, and operations development can provide startups with a competitive advantage and set the foundation for sustainable growth.

Product characteristics constitute an important input to the design of any logistics system. This is grounded in the notion that there is no "one-size-fits-all" solution to logistics, and that the superior performers have systems that are adapted to their unique context. The characteristics of an electrolyser stack (from a logistical point of view) can be highlighted as follows:

- High value components
- "Brittle" and "floppy" components
- Few different components, but many components per product (makes assembly important)
- Simple assembly process, the complexity is in high-precision and low-tolerance processes
- Components require operations in a cleanroom environment
- Long lead times for delivery of components from suppliers
- Both the product, components and production technology are under development, and a new electrolyser market is being developed, with currently limited availability of mature solutions

Although manufacturing process optimization is seen as a key factor in reducing the unit production cost of electrolysers [5], there is still a limited number of studies that focus on the production logistics challenges of electrolyser manufacturing. In order to transition to high-efficiency and mass production hydrogen electrolysers, there is a need to find appropriate solutions to automated production including material handling, parts feeding systems, quality inspection, inventory management, integrated production steps etc. A zero-defect strategy should be sought, given the high cost of components and rework and repairs.

In this study, the aim was to assess operations design alternatives with respect to efficiency and robustness towards quality and to compare them using simulation at an early stage. When comparing different manufacturing setups, using the same assumptions and input data in the simulation models, aspects such as the need for buffers and operators, and the robustness of the manufacturing process with respect to achieved production volume under incremental alternations in input parameters can be tested and studied. Equipment failures, supply chain disruptions or changes in demand can be provoked in the simulation models.

The paper is organized as follows: Sect. 2 gives a brief introduction to some of the relevant earlier studies on this topic, while Sect. 3 presents the methodological approach used in this study. Section 4 describes the case, the stack assembly process, and the relevant assembly line design decisions. Section 5 describes the simulation scenarios, while Sect. 6 summarizes the findings. Finally, Sect. 7 discusses the results and concludes the paper.

2 Theoretical Background

The large-scale industry of production of electrolysers is only starting to emerge and thus research literature on the industrial production of electrolysers is scarce. To contribute with input to this new industry, it is possible to take into consideration recent developments within the fuel cell industry. There are several similarities between fuel cells and electrolysers and fuel cell manufacturing has developed rapidly over several years.

Similar to the electrolyser industry, high-volume manufacturing of fuel cell stacks is limited. Still there are mostly manual or semi-automated assembly, e.g., processes and procedures that have been proven in a pilot line environment and are ready for low-rate production [6]. Some manual and semi-automated approaches are available in the literature [7]. Current literature mostly focuses on a single assembly work cell/station, while few studies include the production/assembly line. An exception is the mass manufacturing machine described by Porstmann et al. [7] that includes feeder, stacking and compression.

There is a need for the industrialization of stack assembly in order to meet the rapidly increasing demand [7]. Through an expert assessment study, Kampker et al. [6] found that experts in the fuel cell industry see the production-related challenges as considerably more hindering than the product-related challenges. Furthermore, they identified a great uncertainty regarding the production technologies needed for high-volume production. More research is needed within the development of automated stacking and automation challenges need to be solved.

Porstmann et al. [7] and Song et al. [8] point out some of the requirements that need to be satisfied to achieve industrialized stack assembly. This includes:

- Repeating accuracy and consistent precision in all processes
- Automated handling of fragile components
- Automated feeding of components
- High technological flexibility to be able to handle different materials and component designs, while fulfilling future scalability demands requirements

- Possibility of tracking of products and batches (full traceability)

Moreover, earlier research has pointed to three critical concerns in designing a fuel cell manufacturing line [9]. The first is *flow*, which relates to how components move within the manufacturing system. Another concern is *balancing* the line, as different fuel cell operations may differ in cycle times and cycle time variations, leading to imbalances resulting in waiting time in the line. The third concern is designing a suitable *buffer* system. This is especially relevant in this industry due to immature processes and products, and the anticipated machine downtimes and failure rates are mostly unknown. This paper will focus on the aspect of buffers in particular.

Previous research has looked into the relationship between buffer allocations and production line performance. McNamara et al. [10] present a simulation study investigating how different configurations in buffer capacity affect the overall performance of unreliable production lines. They find that a buffer capacity that is allocated as evenly as possible between workstations generates the highest throughput rates. Kłos & Trebuna [11] use discrete-event simulation to evaluate different allocations of buffer capacities and different numbers of employees in an automotive industry context.

3 Methodological Approach

The research presented in this paper is based on a mixed-methods approach combining quantitative and qualitative methods. The research involved collecting data from a case company setting which was used to create a discrete-event simulation model of the planned production line.

The empirical data was collected at the case company which is currently in the process of designing an electrolyser assembly line. The researchers have followed this process and collected data regarding the objectives of the production system, the observed challenges and issues, and specific process data on the different assembly line designs of the assembly line. The data have been collected through workshops where the different concepts from assembly line suppliers were discussed. In addition, underlying documentation for the concepts provided the necessary operational data required to build the simulation model.

To investigate and test the different design decisions, a simulation approach was chosen. A simulation model allows for thorough testing of multiple scenarios in the virtual world, without the need to make physical adjustments. It can help anticipate future performance and potential issues, making it a valuable tool for insight into the uncertainties and complexities inherent in emerging industries. While other approaches such as system dynamics, agent-based modelling or mathematical optimization also may have offered valuable insights, discrete-event simulation was chosen as the method in this study. Discrete-event simulation provides a balanced approach that combines detailed modelling of individual events with the ability to handle variability and stochasticity [12]. Furthermore, it gives a precise measurement and overview of different operational metrics. This makes it suitable for operational-level analysis and optimization, especially in emerging industries with processes that are dynamic, complex and subject to frequent changes.

The discrete-event simulation model was built using the simulation software FlexSim. To model the production system, relevant data were collected from the case company. This included data on the relevant components of the system, such as the machines, conveyors and inventories/buffers, as well as operational data such as cycle times, machine reliability and quality yield of components. The workflow was then modelled to represent the sequence of the operations and the interactions and material flows between the model components.

Model verification and validation were done to ensure the accuracy of the simulation results. Verification was done by checking that the model worked as intended, meaning that the model components and the various material flows followed the intended logic. This was done through a joint review of the model together with representatives from the case company. Validation involves comparing the output of the simulation model with real-world data. As this production system is yet to be established, the simulated output of the theoretical capacity was compared with the performance data from the supplier of the assembly line. The simulation model was validated showing similar results to the designed concept.

Next up was to define the experimental design, which involved defining different design decisions, simulation scenarios and key performance indicators (KPIs). The simulation model was used to test the different design alternatives through different scenarios, and the performance and robustness of the different designs were evaluated based on the chosen KPI. The case and the different design decisions are further discussed in Sect. 4, the simulated scenarios are presented in Sect. 5, and the chosen KPI is discussed in Sect. 6.

4 Case Description and Assembly Line Design Decisions

This study focuses on the setup of an assembly line for an anonymous case company currently establishing its production of electrolyser stacks. The case company is still in a start-up phase and is anticipating a rapid increase in production volume in the coming years. This implies that they need to establish a production system that is efficient, scalable and resilient. The system must be able to detect and handle any incoming quality issues while ensuring zero defects in the outgoing products.

The case company are currently in the process of designing its electrolyser stack assembly line and is cooperating with different suppliers of production equipment, each proposing different solutions and adhering to various production philosophies. This study focuses on evaluating one of these and experimenting with various design decisions to increase the production output.

The assembly of electrolysers can be described as repetitive precision work. A high number of cells must accurately be placed on top of each other forming what is known as an electrolyser stack. Existing literature points to two different approaches to stack assembly: 1) assembling single cells which are later assembled into a final stack, and 2) assembling the entire stack directly by stacking the individual components [12]. The first approach will be the focus of this study.

Each cell contains the same components, that also need alignment and precision in their assembly. The stack is then compressed to engage all seals and gaskets between cells, and even between the components within each cell. When fully compressed, solid end plates are placed at each end of the stack before it is secured with bolts. Identified defects most likely result in rework, because of the costs of the components. Opening a compressed stack for disassembly, failure detection, and reassembly is time-consuming and expensive. Hence, to secure quality on each component and the stacking process itself is as crucial as obtaining industrial speed in operations.

The assembly system for electrolysers at the case company is still in the design phase and has several uncertainties. Although the designed production system has sufficient theoretical capacity, these uncertainties can affect the performance of the production systems and reduce the practical capacity. Related logistics challenges identified through the case study include:

- New types of production equipment with uncertain uptime and reliability
- New and immature suppliers result in an uncertain quality yield of incoming components
- Uncertainties regarding the required quality specification of the product make the requirements for quality control even more challenging
- Uncertainties regarding the quality of finished stacks, there may be a need for rework
- New quality standards might arise as the product design matures and the product is tested more

These aspects can seriously hamper the performance of the assembly line, for instance affecting the takt time and production output. This highlights the importance of designing a robust production system. Different resilience measures can be taken to increase the robustness of the assembly line. Some of these measures/design elements are discussed and tested in this paper. The tested design decisions were:

1. Location of Component Quality Inspection – In-Line vs. Decoupled: Quality inspection is another crucial aspect of the assembly line design. Ensuring that defects are detected and handled promptly can prevent faulty products from progressing through the production process. We examined two approaches to quality inspection. The in-line inspection approach (Fig. 1) integrates quality checks directly into the production line, allowing for immediate detection and exchange of defective components. In contrast, the decoupled inspection approach (Fig. 2) separates quality inspection from the main production flow, allowing the quality inspection to run at a different pace and enabling the possibility of building a buffer of quality-assured components.

2. Location of Buffer Inventories: The use and placement of buffer inventories in a production system is a critical design decision that can significantly impact overall efficiency and resilience. Buffers decouple different stages of the production process, mitigating the effects of variability and disruptions.

In this study, 3 types of buffers strategically placed at key points in the assembly line were tested:

I. **Buffers of OK cell components.** The first type of buffer tested was buffers of quality-assured cell components. This is enabled by having a decoupled QI, which allows the QI process to run at a different takt than the cell manufacturing. Detecting components with insufficient quality in the cell manufacturing stage will reduce the available machine time for cell manufacturing, possibly reducing the overall production capacity.

II. **Buffers of cell sub-assemblies.** The second type of buffer tested were buffers of sub-assemblies, i.e., cells currently under assembly and between the different process steps in cell manufacturing. Quality issues or machine downtime might disturb the flow of cells in the line, and these buffers may help to maintain the flow during such fluctuations.

III. **Buffer of finished cells.** The final type of buffer tested was a buffer of finished cells ready to be stacked. This buffer decouples the cell manufacturing from the stacking, meaning that one process can continue producing for some time despite issues in the other.

Different configurations of the various buffers were tested through the simulation. Table 1 presents the 8 different configurations tested. Figures 1 and 2 illustrate the assembly line and present the different design decisions, marked in grey.

Table 1. Evaluated buffer configurations

	In-line component quality inspection				Decoupled component quality inspection			
	Buffer test 1	Buffer test 2	Buffer test 3	Buffer test 4	Buffer test 5	Buffer test 6	Buffer test 7	Buffer test 8
Buffers of OK cell components					✓	✓	✓	✓
Buffers of cell sub-assemblies		✓		✓		✓		✓
Buffer of finished cells			✓	✓			✓	✓

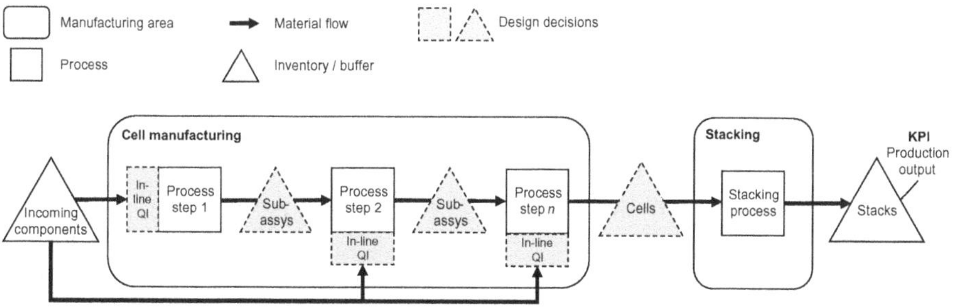

Fig. 1. In-line quality inspection and the different buffers evaluated using simulation

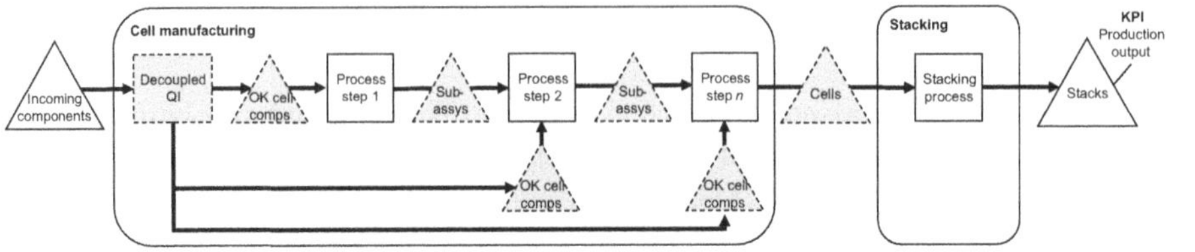

Fig. 2. Decoupled quality inspection and the different buffers evaluated using simulation

5 Simulation Scenarios

In order to evaluate the different design alternatives, a discrete-event simulation model of the proposed solutions was created. This allowed for virtual testing of different alternatives regarding the design decisions. To test the robustness of the different design alternatives, we ran 4 different scenarios to simulate how the solution was able to handle the deviation(s). The different scenarios are presented in Table 2.

To avoid initialization bias [13], the model had a warmup period to reach steady state. In this model, the warmup period was 3 work shifts of 8 h. After the warmup period, the statistics were reset, and the model simulated a period equal to 90 work shifts of 8 h. Each combination of design alternative and scenario was replicated 5 times. 5 replications were deemed as sufficient using the confidence interval method and graphical method described by Robinson [13], as there was little variability between the runs. Running the 8 eight different tests in 4 different scenarios with 5 replications resulted in a total of 160 simulation runs.

Table 2. Simulated scenarios

	Quality yield	Reliability
1. Ideal case (theoretical capacity)	No quality issues	No downtime
2. Reference case	98% quality yield on incoming components	MTBF ~ $N(40,16^2)^*$ MTTR ~ $U(0.167,1)^\dagger$
3. Low quality yield	70% quality yield on one of the incoming cell components, 98% on the other incoming components	MTBF ~ $N(40,16^2)^*$ MTTR ~ $U(0.167,1)^\dagger$
4. Extensive downtime	98% quality yield of incoming components	One period of 24 h downtime in one of the cell manufacturing process steps For the rest of the equipment: MTBF ~ $N(40,16^2)^*$ MTTR ~ $U(0.167,1)^\dagger$

Notes: MTBF = Mean time between failures; MTTR = Mean time to repair; *Normal distribution with mean 40 h and standard deviation 16 h; †Uniform distribution between 10 and 60 min

6 Findings

A total of 8 different variations/tests of the simulation model were tested, each with different combinations of the different types of buffers and locations of component quality inspection that we wanted to test. The KPI used to compare the solutions was the total production output, as this is considered a key indicator of the overall performance of the assembly line. The production output is suited as an aggregated measure, as it is influenced by other aspects such as takt time and quality yield. The actual production output in number of units is anonymized as this is considered confidential information. Instead, we use relative numbers in %, where the theoretical capacity measured by simulating the ideal case is considered as the benchmark of 100%. As we can see in Table 3, the ideal case maintains a production output of 100% through all the different tests. This is expected since the ideal case does not include downtime and quality issues, and there is therefore no need for buffers to balance out any unexpected events or variations in processing time. The max capacity of the added buffers is equal to the number of components needed for producing one electrolyser stack. The evaluation focused on design alternatives regarding the use of buffers and the placement of the component quality inspection, and the results from the simulations are presented in Fig. 3 and Table 3.

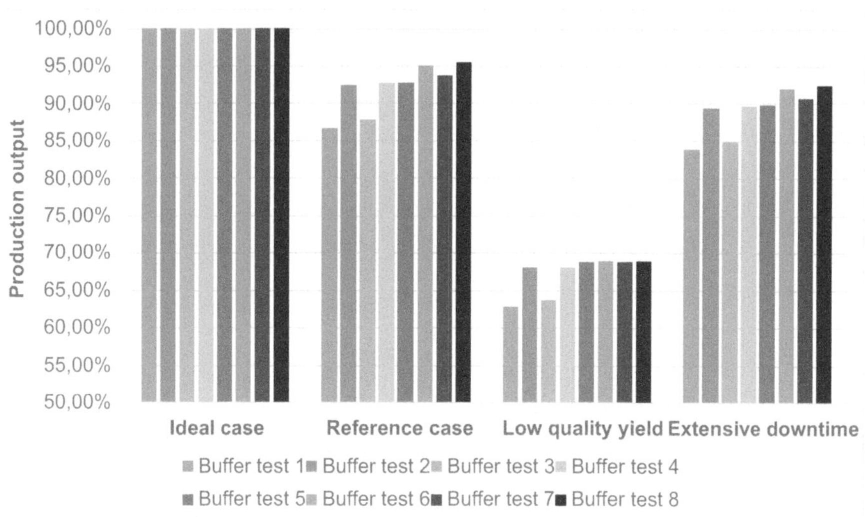

Fig. 3. Comparison of the different buffer tests in different simulation scenarios

The simulated tests showed that all the different types of buffers had positive effects on the total production output when different types of uncertainty were added to the model. This was as expected given the uncertainties resulting from the context of this case, such as the immaturity of this kind of production and the start-up environment of the case company.

Table 3. The results of the different buffer tests (Mean is estimated with a confidence interval of 95%)

	In-line component quality inspection								Decoupled component quality inspection							
	Buffer test 1		Buffer test 2		Buffer test 3		Buffer test 4		Buffer test 5		Buffer test 6		Buffer test 7		Buffer test 8	
Buffers of OK cell components									✓		✓		✓		✓	
Buffers of cell sub-assemblies			✓				✓				✓				✓	
Buffer of finished cells					✓		✓						✓		✓	
	Production output		Production output		Production output		Production output		Production output		Production output		Production output		Production output	
	Mean	±	Mean	±	Mean	±	Mean	±	Mean	±	Mean	±	Mean	±	Mean	±
Ideal case	100%	0.00%	100%	0.00%	100%	0.00%	100%	0.00%	100%	0.00%	100%	0.00%	100%	0.00%	100%	0.00%
Reference case	86.72%	0.72%	92.43%	0.54%	87.78%	0.65%	92.72%	0.46%	92.81%	0.61%	95.09%	0.44%	93.72%	0.39%	95.54%	0.26%
Low quality yield	62.83%	0.57%	68.06%	0.27%	63.69%	0.51%	68.06%	0.27%	68.81%	0.22%	68.98%	0.23%	68.89%	0.23%	68.98%	0.23%
Extensive downtime	83.93%	0.73%	89.44%	0.49%	84.93%	0.57%	89.69%	0.42%	89.80%	0.64%	92.07%	0.44%	90.69%	0.39%	92.48%	0.31%

Across the different scenarios, the single buffer with the largest positive impact on production output was the buffer of OK component parts. This also suggests that, for this case, having a decoupled quality inspection will have a favourable effect on the total production output compared to in-line quality inspection. The relatively low expected quality yield of incoming parts means that valuable machine time will be lost if a defective component reaches cell manufacturing, and it is favourable to detect and remove a component with insufficient quality as early as possible. Feeding the cell manufacturing with OK parts means less machine time lost, and a higher production output can be obtained. However, it should be noted that a decoupled quality inspection also has some drawbacks which are not necessarily reflected in this model, which only focuses on production output. There will most likely be some extra handling associated with a decoupled quality inspection, such as feeding in and out of the quality inspection system and transporting quality assured items to the cell manufacturing.

Combining two types of buffers, having buffers for both OK components and cell sub-assemblies had the largest improvement. Nevertheless, it can be noted that the total improvement of this buffer combination is less than the sum of the individual improvements associated with each of these buffer types. That these two types of buffers have the largest improvement could be understood that it is important to keep the cell manufacturing running steadily, as this part of the electrolyser production system easily can become a bottleneck.

As expected, the largest improvement was seen with having all three types of buffers in the process. It should however be noted that there is already a large improvement in only adding the buffer of OK components and that additional improvements of adding the other types of buffers are relatively minor.

7 Discussions and Concluding Remarks

Simulation of manufacturing set-ups to verify potential or test ideas of change is well described as a viable method in the literature. So is the need for accurate and reliable data when modelling. In the early phases of a logistics development project, a lot of the information necessary to build a representative model is not confirmed, and it may seem counterintuitive to invest resources in simulation when information is uncertain. On the other hand, doing so can actually help identify and mitigate potential risks before they manifest in the physical manufacturing process. By simulating various scenarios, decision-makers can better understand the impact of different logistical strategies and make more informed choices to minimize risk. In this research, an early model was built, and testing with it gave valuable insight into some of the strengths and weaknesses of the manufacturing design at hand. Moving forward, the simulation model can be the foundation of a digital twin of the future electrolyser stack factory, integrating real-time data from the factory into the model. This allows continuous monitoring and facilitates optimization of production, logistics, and planning and control processes.

In our case, the simulation model was used to explore different what-if scenarios, evaluating potential outcomes of various decisions. This enabled decision-makers to make more confident and data-driven choices, and fostered new ideas that also could be tested. There is only one product in this start-up at the time of designing the factory.

The need for flexibility is not prioritized. The case contained expensive components that should be assembled in a zero-defect system, and hence quality control of components is crucial to fit into the system. Four scenarios were described for the model to test. This illustrated the impact defective parts and downtime has on overall performance of the manufacturing system. The findings indicate that adding buffers to an electrolyser assembly line can significantly increase the production output. Especially adding a buffer of quality-assured cell components enabled by a decoupled quality inspection showed promising results.

From a lean manufacturing perspective, inventory is seen as a source of waste. This also includes buffer inventories as lean manufacturing philosophy strives towards one-piece flow. Building up inventories is a great way to hide problems. Instead of building inventories and buffers in anticipation of machine or component issues, lean philosophy emphasizes that rather the root cause of the problem should be addressed [14]. Moving forwards, the electrolyser manufacturing industry should work towards increasing the stability of the production and supply chains, reducing, or potentially eliminating, the need for buffers in the first place.

A few limitations of this research study should be noted. Depending on the picking solution, it can be extra time associated with picking from a buffer rather than from a line. However, in this model, additional picking time of picking from a buffer is not modelled. Furthermore, this research focuses on a single case. Although we expect that these findings and recommendations are relevant also for other electrolyser manufacturers, we cannot claim that this is the case. Finally, this simulation study only focuses on the KPI of production output. Investigating other KPIs, such as overall equipment efficiency (OEE), takt time, or inventory turnover might provide additional insights into the design of such assembly lines.

There is a scarcity of literature on research studies discussing the operational and logistics aspects of electrolyser stack production. This is a growing industry and will be important in the transition to more environmentally friendly energy solutions. There is thus a need to understand the contextual conditions of this industry, and how efficient industrialized solutions can be designed. Lessons can be learned from operational excellence in other industries but must be adapted to the context of electrolyser manufacturing.

Numerous open issues in designing and implementing industrialized solutions for electrolyser manufacturing are relevant for further research. This includes aspects related to production and supply chain planning, quality control, material handling and inventory management. These areas can be especially challenging to handle in an electrolyser production environment characterized by immature supply chains, a product under continuous development, very high-quality standards, expensive and fragile components, and few proven industrial demonstrators. An issue for further work can be to expand the simulation model to investigate other issues in the electrolyser value chain, for instance, related to supply chain planning, incoming logistics and inventory management.

Acknowledgments. The authors would like to acknowledge the support from the anonymous case company for providing valuable input data for this study.

Disclosure of Interests. The authors have no competing interests to declare that are relevant to the content of this article.

References

1. Brandstetter, A., Himmelstoss, H., Risling, M., Schel, D., Oberle, M.: Towards a design of a software-defined manufacturing system based on a systematic literature review for enabling a decentralised high-rate electrolyser production. In: Herberger, D., Hübner, M., Stich, V. (eds.) Proceedings of the Conference on Production Systems and Logistics: CPSL 2023 - 1, pp. 479–488. Hannover publishing (2023). https://doi.org/10.15488/13466
2. European Commission: A hydrogen strategy for a climate-neutral Europe, COM/2020/301 (2020)
3. Arsad, A.Z., et al.: Hydrogen electrolyser for sustainable energy production: a bibliometric analysis and future directions. Int. J. Hydrog. Energy **48**, 4960–4983 (2023). https://doi.org/10.1016/j.ijhydene.2022.11.023
4. Krishnan, S., et al.: Present and future cost of alkaline and PEM electrolyser stacks. Int. J. Hydrogen Energy **48**(83), 32313–32330 (2023). https://doi.org/10.1016/j.ijhydene.2023.05.031
5. Pastore, L.M., Lo Basso, G., Sforzini, M., de Santoli, L.: Technical, economic and environmental issues related to electrolysers capacity targets according to the Italian hydrogen strategy: a critical analysis. Renew. Sustain. Energy Rev. **166**, 112685 (2022). https://doi.org/10.1016/j.rser.2022.112685
6. Kampker, A., Ayvaz, P., Schön, C., Karstedt, J., Förstmann, R., Welker, F.: Challenges towards large-scale fuel cell production: results of an expert assessment study. Int. J. Hydrog. Energy **45**, 29288–29296 (2020). https://doi.org/10.1016/j.ijhydene.2020.07.180
7. Porstmann, S., Wannemacher, T., Richter, T.: Overcoming the challenges for a mass manufacturing machine for the assembly of PEMFC stacks. Machines. **7**, 66 (2019). https://doi.org/10.3390/machines7040066
8. Song, K., et al.: Assembly techniques for proton exchange membrane fuel cell stack: a literature review. Renew. Sustain. Energy Rev. **153**, 111777 (2022). https://doi.org/10.1016/j.rser.2021.111777
9. Laskowski, C., Derby, S.: Fuel cell ASAP: two iterations of an automated stack assembly process and ramifications for fuel cell design-for-manufacture considerations. J. Fuel Cell Sci. Technol. **8**(3), 031004 (2011). https://doi.org/10.1115/1.4000684
10. McNamara, T., Shaaban, S., Hudson, S.: Simulation of unbalanced buffer allocation in unreliable unpaced production lines. Int. J. Prod. Res. **51**, 1922–1936 (2013). https://doi.org/10.1080/00207543.2012.720726
11. Kłos, S., Trebuna, P.: Using computer simulation method to improve throughput of production systems by buffers and workers allocation. Manag. Prod. Eng. Rev. **6**, 60–69 (2015). https://doi.org/10.1515/mper-2015-0037
12. Robinson, S.: Simulation: The Practice of Model Development and Use. Bloomsbury Publishing (2014)
13. Gurau, V., Fowler, D., Cox, D.: Robotic technologies for proton exchange membrane fuel cell assembly. In: Taner, T. (ed.) Proton Exchange Membrane Fuel Cell. InTech (2018). https://doi.org/10.5772/intechopen.71470
14. Ohno, T.: Toyota Production System: Beyond Large-Scale Production. CRC Press, Boca Raton, FL (1988)

Empirical Grounded Simulation Models for Make-To-Order (MTO) Supply Chains: An Application in the Furniture Industry

Alexandra Lagorio[1], Giulio Mangano[2], Roberto Pinto[1], Carlo Rafele[2], and Giovanni Zenezini[2](✉)

[1] Department of Management, Information and Production Engineering, University of Bergamo, Viale Marconi 5, 24044 Dalmine, BG, Italy
{alexandra.lagorio,roberto.pinto}@unibg.it

[2] Department of Management and Production Engineering, Politecnico di Torino, Corso Duca degli Abruzzi 24, 10129 Torino, Italy
{giulio.mangano,carlo.rafele,giovanni.zenezini}@polito.it
http://www.unibg.it , https://reslog.polito.it

Abstract. The Make-to-Order (MTO) supply chain seeks to balance cost reduction with satisfactory customer service, especially concerning order lead times. Simulation plays a vital role in this balance, identifying risks, analyzing scenarios, and evaluating key performance indicators. However, existing simulation models often overlook suppliers and inventory management, focusing more on production, sales, and distribution. To address this, a simulation model tailored for the furniture industry integrates supplier selection with inventory management strategies, considering geographical complexities. Through a case study, various scenarios are assessed, revealing a trade-off between lead times and costs. Close MTO suppliers decrease lead times but increase costs due to transportation expenses, while distant sourcing minimizes costs but extends lead times, challenging customer expectations. This simulation model offers insights for MTO companies navigating supplier selection and inventory management, enhancing decision-making and customer satisfaction. Future research aims to expand the model into a supply chain Digital Twin, incorporating resilience and risk management to tackle broader MTO supply chain challenges.

Keywords: Make-to-order (MTO) supply chains · Inventory Management · Distant sourcing · Suppliers selection · Simulation · Furniture

1 Introduction

In Make-to-Order (MTO) supply chains, similar to other supply chains, the aim is to strike a balance between reducing costs and maintaining satisfactory customer service levels, typically assessed by order lead time [8,17]. Customers are,

Published by Springer Nature Switzerland AG 2024
M. Thürer et al. (Eds.): APMS 2024, IFIP AICT 733, pp. 78–91, 2024.
https://doi.org/10.1007/978-3-031-71645-4_6

in fact, very sensitive to order lead times in MTO supply chains, emphasizing both the timeliness and reliability of promised due-dates [19]. Consequently, determining the appropriate order lead time stands out as a primary objective in MTO research [23]. To expedite lead times and ensure consistent order fulfillment, companies often resort to maintaining strategic inventories of standardized products or semi-finished components [16]. To address these strategies, extant studies on MTO supply chains focus on the integration between production and distribution processes [8], as well as the impact of inventory management decisions on manufacturing lead time [14]. For the latter issue MTO industries encounter various challenges associated with sourcing materials from distant locations. [17]. The availability of data unlocks the potential for data-driven, real-time decision making capabilities [12]. Data-centric decision-making methods are essential to mitigate risks and navigate disruptions [15]. However, these data-driven real-time methodologies have presented a formidable challenge to both scholars and industry professionals. For instance, they require real-time coupling with data from the field as well as rigorous validation of underlying models [7,11]. In this context, simulation can be viewed as a very useful and important complement to these recent developments [21]. Simulation can help identify potential supply chain risks, including disruptions in the supply chain, unexpected changes in demand, or delays in production. Finally, simulation models enable scenario analysis, allowing decision-makers to evaluate the impact of different scenarios on key performance indicators (KPIs) such as lead time, on-time delivery, and inventory levels. However, while simulation models are widely available for traditional Make-to-stock (MTS) industries like retailing or food supply chains, most simulation models in MTO contexts focus on optimizing production and order planning policies [13,22,28] only few treat other issues of MTO supply chains such as inventory management with distant sourcing [17] and the impact of uncertainties and risks. A generic modelling and solution technique for MTO supply chains that include various sources of uncertainties (i.e. unforeseen customer order arrivals, varying production processing times and stochastic transportation times) is available in [20]. However, this paper does not explicitly consider suppliers selection and inventory management practices. The research objective of this study is to evaluate the impact on inventory management costs and order lead time in a MTO supply chain of different strategies that combine suppliers selection with inventory management policies. To this end, we develop and validate a simulation model with data from a case study application in the furniture sector. The paper is structured as follows. A literature review of simulation models in MTO contexts is presented in Sect. 2. Then, Sect. 3 highlights the main steps undertaken for the model development, including parameters calibration and experimental design for scenario analyses. Section 4 is then dedicated to the analyses of results, while a final section underlines the main implications of this study as well as its limitations and future research.

2 Theoretical Background

A range of simulation studies have explored the complexities of make-to-order (MTO) and build-to-order (BTO) supply chains, offering valuable insights for decision-making. [13,17] both propose simulation-based optimization approaches for production planning and sales and operations planning, respectively. The aim of these models is to find an optimal or maximum production lot size to achieve the best trade-off between logistics cost and customer satisfaction. To this end, both [22,28] explore the impact of different order promising policies to the above-mentioned trade-off, considering customer satisfaction levels and production capacity constraint. A similar objective is shared by [10] who instead focus on optimizing product configuration in order to sell products with available components. Lastly, [20] propose another simulation-optimization approach that instead focuses on production and distribution planning policies. [3] focus on the impact of performance determinants in MTO settings. The authors show that workload and supply chain complexity negatively affect delivery time and total cost. On the contrary, outsourcing yields positive impacts on both variables.

The choice of suppliers can also be a cause of uncertainties, especially when suppliers are located far from the company. The issue of distant sourcing is fairly overlooked in the literature. For instance, [17] treat distant sourcing only as part of the context rather than a dynamic variable. In terms of suppliers-related policies (e.g., inventory management practices) the authors only accounts for the possibility of emergency supplies if needed, with a negligible lead time and not considering procurement costs. To the extent of the authors' knowledge, inventory management policies are more thoroughly considered in very few papers, among which the one by [29]. The authors indicate that, compared with a purely MTO or Make-to-stock (MTS) production strategy, the flexibility to choose between MTS (i.e., purchasing parts to stock) and MTO (i.e., purchasing parts to order) could reduce the amount of stock owned but also caused more production delays.

3 Model Development

The methodology applied to the research developed in this paper is shown in Fig. 1. Several steps were followed in the design and testing of the model before proceeding to various simulation scenarios. The goal was to evaluate the impact on inventory management costs and order lead time in a Make-To-Order (MTO) supply chain. This was done using different strategies that combine supplier selection with inventory management policies.

3.1 Case Study

The proposed simulation model is applied to a company operating in the furniture sector at an international level and struggling with its service level in terms of promised order lead time to customers. The company under study sells

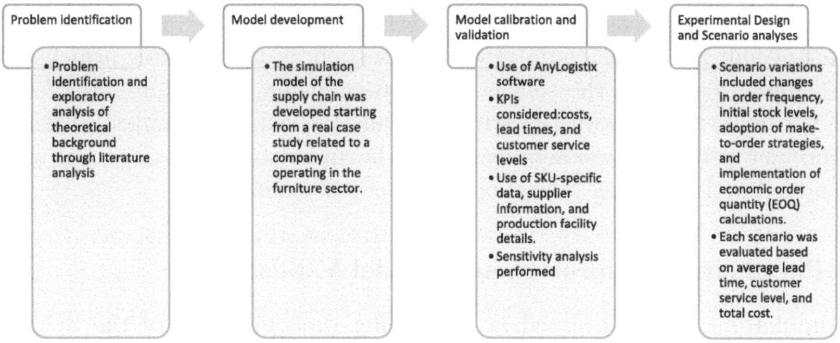

Fig. 1. Research workflow

sofas and armchairs and its headquartered in Italy, where automated and manual processes are carried out to ensure the highest possible quality. The production process is integrated among different plants in Italy, China, Brazil and Romania. In order to understand the complexity of the logistics management of the different components, it is necessary to describe also the main activities required to be carried out to produce a sofa. First, the coating is prepared, the material is inspected, and then there is the cutting phase. After the cutting phase, there are the enrichment, sewing, and final assembly phases. This production process results in a quite complex SC involving components having different nature and characteristics. For this case, we have decided to focus the attention on the most sold nine models of sofas. These models could be chosen by customers in 11 leather types and 107 colours. These different possibilities bring thousands of possible Stock Keeping units (SKUs). The distribution of the finished products is in charge of wholesalers and retailers' networks. Most products are managed according to the Make-to-Order approach in order to have lower inventory levels. The procurement activity is classified into two different groups. The first group includes component having extended order lead times. These components are stored in the warehouse and managed through a reorder policy. The other group includes components that are purchased based on a Just in Time approach. The developed Information System can show the progress of every placed order (i.e., in transit, delivered, produced, and blocked). Once customer orders are placed, the IT system reads the related bill of materials and sends the request to the warehouse. If any of the required components is out of stock, the systems immediately place a new order to the suppliers, to be aligned with the expected lead time equal to three weeks (production lead time), backlog and transit time (that also includes temporary storage). Frame assembly only starts when all final assembly materials are ready, minimising waste. Cutting happens independently a week earlier as it is organised by colour, unlike subsequent phases in the following week on contiguous workstations. In-line operators perform sub-tasks progressively in one of three assembly cells. Daily sequences organise sofas by model within each cell, so operators repeat identical operations multiple times.

The time needed for assembly can range between an hour for fixed sofas and two and a half hours for complex products such as the Chester armchair (its most defining feature of the sofa is the craftsmanship called *capitonné lozenge*, which is carried out in both the armrests and the backrest seamlessly) with the predominant part of the assembly being the upholstery. As the total sum of value-added activity times, the average time is 300 min

3.2 Parameters Calibration and Model Validation

The simulation model is aimed at evaluating the behaviour of the SC of the company under study focusing on the trade-off between costs (namely order, transport and warehouse holding costs) and customer order lead time. The model has been developed on AnyLogistix software, which is used to create and analyse supply chain models and perform network optimization and simulation [27]. It is largely exploited for solving problems including supply chain design, risk assessment, inventory and sourcing, transportation and production planning, and bullwhip effect quantification [9]. It is based on several tables that need to be fed in order to more appropriately simulate the system. A depiction of the supply chain structure in Anylogistix is shown in Fig. 2.

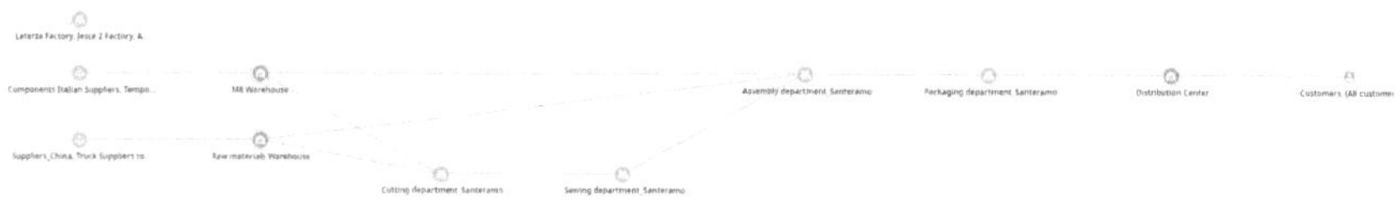

Fig. 2. Supply Chain structure in Anylogistix

In addition several Key Performance Indicators (KPIs) have been taken into account. In particular the Customer Service level expressed as the percentage of orders delivered to customers on time is included since it is a crucial measure to assess the quality of logistics services [18]. Then, the order lead time, that measures the time elapsed between the order placement and the final deliver date [4] is assessed, and the total costs that are the sum of the inventory holding cost, the unitary price for every component for every supplier and the transportation. Knowing these costs is a vital part of any industrial logistics systems management [2]. To this end, the calibration of parameters has been carried out in accordance with real data provided by the company. First, the inventory holding cost has been evaluated to be, together with the company, equal to the 15% of the annual unit cost. The unitary price is provided by company and it depends on the cost corresponded to suppliers and to the quantities that are ordered. Finally, the cost of transport is defined based on the distance travelled. The model has been developed on an initial subset of SKUs in agreement with the case study company. Each of the sofa models can be ordered on a single colour. The bill of materials at issue includes: - the coating (leather or fabric);

- the frame kit; - stuffing; -strap; -feet (metal or wood); - mechanisms kit (if needed); - electric engine (if needed). 5 of the model considered are made of leather and the remaining 4 ones can be ordered with fabric. Each sofa requires 15 or 20 squared meters of leather of fabric depending on whether is a two-seat or three-seat configuration. The total demand for each sofa has been estimated based on the procurement of upholstery during 2023. Each sofa model is associated with a type of upholstery. A reordering policy is used for each of them, of which we have the following data: lead time (LT), safety stock (SS) and reorder point (R). The demand has been considered according last year volume (2023) on a weekly basis. The average demand is computed as follows:

$$AVG(D) = \frac{R - SS}{LT} \tag{1}$$

Assuming the random variable of distributed demand according to a normal of mean AVG and standard deviation STD, it is demonstrable that ideal R corresponds to:

$$R = LT * AVG(D) + z * STD(D) * \sqrt{LT} \tag{2}$$

Fluctuations in demand will be mitigated by the presence of an additional stock with a probability equal to the desired Service Level, whose standardized equivalent value is the z of the formula.

$$STD(D) = \frac{R - LT * AVG(D)}{z * \sqrt{LT}} \tag{3}$$

In the model, demand is considered to follow a Poisson distribution because customers' orders, by sofa model and by customer, are unitary and arrive at each time interval. In fact, to formalize a number of random events that occur in a unit of time, a Poisson process is usually used, which corresponds to independent and stationary inter-arrival times [26]. The variance of a Poisson's distribution is equal to its mean.

Three customers from northern, central and southern Italy are taken into account. The level of demand is assumed to be equal across the three areas. An order for a single sofa is generated by the individual customer at constant time intervals on average. A customer orders a model on a day of the interval that may be different from that of another model, as well as another customer. The time interval is computed as time span between two orders as to have a resulting annual demand equal to the one expected. In Table 1, for every SKU under study, the lead time, the safety stock, reorder point, standard deviation of the demand, and average demand expressed in squared meters of fabric and leather, are shown. Seventeen suppliers are considered in the model. They are located in Italy, Romania, Turkey, Germany and China. For every supplier the purchased SKU and related purchase costs are shown in Table 2.

Table 1. SKU data.

PRODUCT	LT(d)	SS(m2)	R(m2)	STD D(m2/d)	AVG D(m2/d)
Beige and pink Leather	35	10,357	24,157	2,594	0,394
Orange Leather	42	3,675	10,371	0,840	0,159
Beige and yellow Leather	42	25,349	144,431	5,795	2,835
Red Leather	35	34,129	160,484	8,546	3,610
Gray Leather	118	90,936	658,073	9,966	4,806
Medium pink Fabric	35	14,449	56,694	4,652	1,207
Orange Fabric	63	46,766	281,189	8,729	3,721
White Fabric	42	10,631	27,881	2,430	0,411
Antique pink Fabric	42	29,263	94,765	6,689	1,560

Table 2. Products, Suppliers and purchase costs.

SUPPLIER	PRODUCT	UNIT COST (€/pcs)
Italy1	Beige and pink Leather	15,28
Italy1	Beige and yellow Leather	15,28
Italy1	Red Leather	15,28
Italy2	Orange Leather	16
Italy2	Beige and yellow Leather	16
Italy2	Red Leather	20,58
India	Gray Leather	8,5
Turkey	Medium pink Fabric	7,9
Turkey	Orange Fabric	7,6
Italy3	White Fabric	26,90
Italy3	Antique pink Fabric	27,80
Romania	Orange Leather	17,5
Romania	Beige and yellow Leather	16,5
Romania	Red Leather	22
Romania	Gray Leather	10,35
Romania	Medium pink Fabric	8,12
Romania	Orange Fabric	8,3
China1	Straps	0,3
Italy4	Straps	0,6
Germany	Straps	0,35
Italy5	Frame kit	120
China2	Mechanism kit	12,38
Romania	Mechanism kit	15
Italy6	Metal feet	13
China3	Black metal feet	11,7
Romania	Wooden feet	0,5
Italy7	Cherry wooden feet D40/D90	0,125
Italy8	Motor	10
Italy9	Stuffing	10

European suppliers ship by road with vans, while extra European ones deliver goods by sea in containers in ships. Taking into account the difference in transportation cost for shipping between various suppliers, different vehicles with different costs have been identified. The product costs will also be different by type of supplier and material, for which the real prices have been defined, based on the data provided by the company. For the remainder, average selling prices for the product type were used. In the SC there are five production plants (that have been geolocated), a distribution center and a raw material warehouse wherein coatings and other components are stored for longer periods. All the production lines are divided into 4 activity steps: cut, sewing, assembly and packaging. Through the anyLogistix "Events" table, weekend and evening closures have been configured with total operating time of 16 h × 5 d a week. Shipments are also scheduled, from any supplier or site, from Monday to Friday. The 'production time' of all the phases phases from cutting to packaging has been calculated according to the condition that the plant is able to produce, in the manner conceived by anyLogistix, the number of units that corresponds to its maximum production capacity, defined as cycle time. A dummy time was later added so that the average transit time within the system, barring any waits for slowdowns, was about 300 min for a three-seater sofa and 250 for a two-seater sofa. In fact, from the packaging departments to the DC each shipment requires a certain processing time. Due to the wait, a time is generated that is added to the actual processing time to total a total time of 250/300 min on average. A sensitivity analysis has been also carried out in order to understand how the model behaves under different conditions, by varying the input parameters [5]. The sensitivity analysis has been conducting considering the number of trucks and their related capacity. The result is that the transportation cost tends to increase as fleets and vehicle capacity decrease. Finally, these results were tested with company managers who verified that the model baseline behavior reflected the reality of their operations.

3.3 Experimental Design

The simulations have started with a baseline scenario whereby the values of the parameters are created according to the real company data, as mentioned in the previous section. Based on that scenario, all the other ones were implemented. For every component, the initial level of stock and the quantity ordered was considered. The definition of every scenario and the related value of R (i.e., Reorder point) and Q (i.e., quantity ordered) has been agreed with the company operations manager. In Scenario 1, the order frequency for item purchased in China is no more equal to 30, but it becomes equal to 115 d, with a consequent increase of the quantity ordered. In Scenario 2, the initial level of stocks have been changed. Scenario 3 is intended to further improve Scenario 2 in term of initial inventory levels. In Scenario 4 a Make-to-order strategy is adopted for a subset of components, while other components are still replenished to stock adopting the company policy. Scenario 5 builds on scenario 4 by assigning a reorder point to all the make-to-stock components. In Scenario 6 a more objective approach has been applied. In particular, the Economic Order Quantity (EOQ) has been used for minimizing the cost of order [1] or the transportation cost in the case at issue. Given a certain the cost of the single order (K), the optimal quantity to be shipped in each cycle is calculated. If this exceeds the capacity of the vehicle, this K is updated and K* will be the new cost of issuing the order, i.e. single cost*number of vehicles needed. The procedure is repeated n times until the EOQ calculated at iteration n converges with the EOQ calculated at iteration n-1. By dividing the resulting EOQs by the average annual demand, it is checked whether the established simulation length is suitable. For instance, orders for black metal feet equal to 10767 can satisfy the demand for 78 years. The last scenario (Scenario 7) is based on an improvement of Scenario 6 in terms of Reordered points. In Table 3 the value of R and Q for every above mentioned scenario are shown.

Table 3. Description of scenarios according to the inventory management policy

		Scenario							
Product	Parameter	Baseline	1	2	3	4	5	6	7
Mechanism Kit	R	345	245	245	245	345	345	345	245
	Q	250	1,988	250	250	250	250	1,988	1,988
Black Metal Feet	R	60	10	40	40	60	40	60	40
	Q	100	10,767	100	100	100	100	10,767	10,767
Beige & Pink Leather	R	24	60	69	69	24	69	24	60
	Q	300	318	300	300	300	300	318	318
Orange Leather	R	10	10	10	10	10	10	10	10
	Q	300	130	300	300	300	300	130	130
Beige & Yellow Leather	R	144	84	84	84	144	144	114	84
	Q	300	2,217	300	300	300	300	2,217	2,217
Red Leather	R	160	120	120	120	160	160	160	120
	Q	300	2,812	300	300	300	300	2,812	2,812
Grey Leather	R	658	508	508	508	658	658	568	508
	Q	1,000	3,571	1,000	1,000	1,000	1,000	3,571	3,571
Pink Fabric	R	57	97	97	97	57	57	57	97
	Q	100	1,079	100	100	100	100	1,069	1,079
Orange Fabric	R	281	262	262	262	281	281	281	262
	Q	100	3,269	100	100	100	100	3,269	3,269
White Fabric	R	28	28	28	28	28	28	28	28
	Q	70	331	70	70	70	70	331	331
Antique Pink Fabric	R	95	72	72	72	95	72	95	72
	Q	48	1,092	48	47	48	48	1,092	1,092
Straps	R	7,617	7,617	7,617	7,617	7,617	7,617	7,617	7,617
	Q	48,962	48,962	10,000	10,000	48,962	48,962	48,962	48,962

4 Scenario Analyses

All the above mentioned scenarios have been compared by means of a dashboard of indicators as follows: - Average Lead Time: it represents the average value of lead times for the products considered computed as the difference between the time when an order to the manufacturer is place and when that order is delivered to the final customer; - Customer Service Level: it is the ration between the orders delivered on time and the total number of order shipped; -Total Cost: it is the sum of the inventory carry costs (measured as the ratio between €and cubic meter per year), order cost, and transport cost measured as the ratio between €and km time the total distance covered. The analyses of the different scenarios (Fig. 3) indicate that, not surprisingly, higher delivery lead times are associated with lower total costs and vice versa. This is particularly striking in scenarios 6 and 7, where an EOQ policy with distant sourcing is exploited thus leading to lower costs but higher lead times due to supply issues. On the contrary, Scenario 4 exploits closer suppliers' ability to fulfill MTO approaches by the company. This strategy leads to lower lead times but higher costs, due to increasing purchase costs charged by closer suppliers and higher transport costs due to shorter requested lead times.

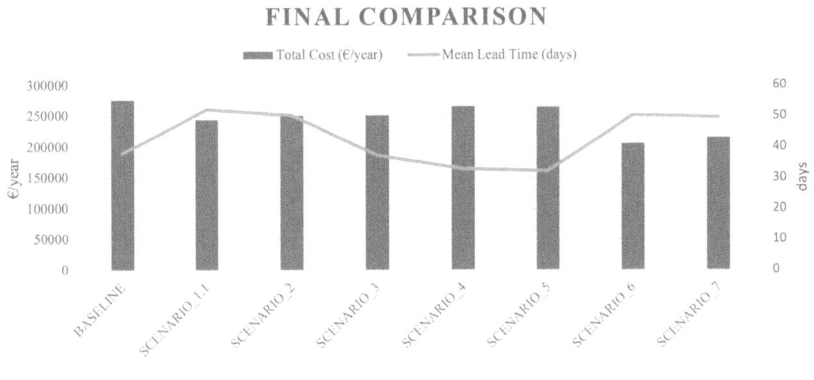

Fig. 3. Final comparison between scenarios

5 Discussions and Conclusions

In summary, this paper presents a simulation model tailored for Make-to-Order (MTO) supply chains, focusing specifically on inventory management strategies involving suppliers situated across diverse geographical regions. Each supplier is characterized by unique parameters, including transportation costs, purchase costs, and delivery lead times. The application of this simulation model is demonstrated through a case study in the furniture industry and its baseline behavior is tested against real operations together with company managers.

The next two sub-sections aim at discussing the results of the proposed simulation model, also in relation with previous literature, and highlighting its main implications, limitations and future research.

5.1 Discussions of Results

A trade-off becomes evident between lead times and total costs as a result of the chosen inventory management policies. Although this trade-off is not new in the literature, this study moves beyond the consolidated approach of production and sales planning to study the supply and procurement side of the MTO enterprise. In fact, various inventory management policies, each associated with different suppliers, are evaluated by considering both lead time and total costs, commencing from a Business-as-Usual (BAU) scenario. Notably, the impact of the choice between MTO or Make-to-Stock (MTS) suppliers is thoroughly investigated. Opting for nearby MTO suppliers leads to decreased lead times but increased costs due to higher expenses associated with purchase and transportation. These results are in line with previous work on inventory management strategies in MTO supply chains by [29], who however only provided quantification of the number of orders delayed. Furthermore, the proposed simulation model delves deeper into the MTS suppliers management by addressing parameters such as order quantity and reorder points, finding that an EOQ policy minimizes total cost. In this regard, we differentiate from [24] who did not explore transportation costs and found that through EOQ an MTO company can achieve higher profits due to lower stockout rather than lower inventory cost.

5.2 Study Implications, Limitations and Future Research

In terms of theoretical implications, this study may pave the way for further research on flexible, hybrid MTS/MTO sourcing considering other factors beyond costs and lead times such as resilience and sustainability. This is an understudied topic considering that most papers in this field focus on flexible production rather than flexible suppliers' selection [6,25].

This model also offers practical utility for MTO companies grappling with supplier selection challenges and seeking insights into how inventory management decisions influence delivery lead times. Presently, it aids the decision-making process of the company under study, enhancing visibility into lead times and positively impacting reputation and customer satisfaction. The model enables the company to concurrently pursue two objectives: selecting the most suitable strategy based on desired outcomes and accurately forecasting promised delivery dates to customers.

This study has some limitations. The model's development encompasses only a subset of Stock Keeping Units (SKUs), and a limited range of inventory management policies are examined. Future research endeavors will address these shortcomings by expanding the model's scale to encompass a greater number of products and a wider array of inventory policies. In addition, further research will be aimed at broadening the scope of the model towards a supply chain Digital Twin. A Digital Twin emulates the physical flows and interactions among supply chain actors to predict unexpected events in advances and test alternative courses of action [15]. This evolution will facilitate the incorporation of additional critical elements of MTO supply chains, such as resilience and supplier risk, into the model.

Acknowledgments. Project funded under the National Recovery and Resilience Plan (NRRP), Mission 4 Component 2 Investment 1.3 - Call for tender No. 341 of 15/03/2022 of Italian Ministry of University and Research funded by the European Union - NextGenerationEU. Award Number: PE00000004, Concession Decree No. 1551 of 11/10/2022 adopted by the Italian Ministry of University and Research, CUP B83C22004890007. MICS (Made in Italy - Circular and Sustainable). Spoke 7: New and consumer-driven business models for resilient and circular supply chains. The authors would like to extend their sincere gratitude to Alessia Noviello and Camilla Ricciardi, whose MSc thesis was integral to the development of the simulation model discussed in this paper.

Disclosure of Interests. The authors have no competing interests.

References

1. Agarwal, S.: Economic order quantity model: a review. VSRD Int. J. Mech. Civil Autom. Prod. Eng. **4**(12), 233–236 (2014)
2. Azzi, A., Battini, D., Faccio, M., Persona, A., Sgarbossa, F.: Inventory holding costs measurement: a multi-case study. Int. J. Logistics Manage. **25**(1), 109–132 (2014)
3. Barbosa, C., Azevedo, A.: Assessing the impact of performance determinants in complex MTO/ETO supply chains through an extended hybrid modelling approach. Int. J. Prod. Res. **57**(11), 3577–3597 (2019). https://doi.org/10.1080/00207543.2018.1543970
4. Ben-Ammar, O., Dolgui, A., Hnaien, F., Ould-Louly, M.A.: Supply planning and inventory control under lead time uncertainty: a literature review and future directions. IFAC-PapersOnLine **55**(10), 2749–2754 (2022)
5. Cagliano, A.C., Carlin, A., Mangano, G., Rafele, C.: Analyzing the diffusion of eco-friendly vans for urban freight distribution. Int. J. Logistics Manage. **28**(4), 1218–1242 (2017)
6. Calle, M., González-R, P.L., Leon, J.M., Pierreval, H., Canca, D.: Integrated management of inventory and production systems based on floating decoupling point and real-time information: a simulation based analysis. Int. J. Prod. Econ. **181**, 48–57 (2016)
7. Defraeye, T., et al.: Digital twins are coming: will we need them in supply chains of fresh horticultural produce? Trends Food Sci. Technol. **109**, 245–258 (2021). https://doi.org/10.1016/j.tifs.2021.01.025
8. Demirli, K., Yimer, A.D.: Fuzzy scheduling of a build-to-order supply chain. Int. J. Prod. Res. **46**(14), 3931–3958 (2008). https://doi.org/10.1080/00207540601113273

9. Echefaj, K., Charkaoui, A., Cherrafi, A., Ivanov, D.: Design of resilient and viable sourcing strategies in intertwined circular supply networks. Ann. Oper. Res. **29**, 1–40 (2024)
10. Ervolina, T., Ettl, M., Lee, Y., Peters, D.: Simulating order fulfillment with product substitutions in an assemble-to-order supply chain. In: Proceedings of the 2006 Winter Simulation Conference, pp. 2012–2020. IEEE, Monterey, CA, USA (2006). https://doi.org/10.1109/WSC.2006.322988, http://ieeexplore.ieee.org/document/4117846/
11. Ferrari, A., Carlin, A., Rafele, C., Zenezini, G.: A method for developing and validating simulation models for automated storage and retrieval system digital twins. Int. J. Adv. Manuf. Technol. **131**, 5369–5382 (2023). https://doi.org/10.1007/s00170-023-12660-y
12. Frazzon, E.M., Freitag, M., Ivanov, D.: Intelligent methods and systems for decision-making support: toward digital supply chain twins. Int. J. Inf. Manage. **57**, 102281 (2021). https://doi.org/10.1016/j.ijinfomgt.2020.102281
13. Hachicha, W., Ammeri, A., Masmoudi, F., Chabchoub, H.: A multi-product lot size in make-to-order supply chain using discrete event simulation and response surface methodology. Int. J. Services Econ. Manage. **2**(3/4), 246 (2010). https://doi.org/10.1504/IJSEM.2010.033366
14. Ioannou, G., Dimitriou, S.: Lead time estimation in MRP/ERP for make-to-order manufacturing systems. Int. J. Prod. Econ. **139**(2), 551–563 (2012). https://doi.org/10.1016/j.ijpe.2012.05.029
15. Ivanov, D., Dolgui, A., Das, A., Sokolov, B.: Digital supply chain twins: managing the ripple effect, resilience, and disruption risks by data-driven optimization, simulation, and visibility. In: Ivanov, D., Dolgui, A., Sokolov, B. (eds.) Handbook of Ripple Effects in the Supply Chain. ISORMS, vol. 276, pp. 309–332. Springer, Cham (2019). https://doi.org/10.1007/978-3-030-14302-2_15
16. Kuroda, M., Mihira, H.: Strategic inventory holding to allow the estimation of earlier due dates in make-to-order production. Int. J. Prod. Res. **46**(2), 495–508 (2008)
17. Lim, L.L., Alpan, G., Penz, B.: A simulation-optimization approach for sales and operations planning in build-to-order industries with distant sourcing. Comput. Ind. Eng. **112**, 469–482 (2017). https://doi.org/10.1016/J.CIE.2016.12.002
18. Liu, H., Zhao, J., Zhou, L., Yang, J., Liang, K.: Intelligent performance evaluation of e-commerce express services using machine learning: a case study with quantitative analysis. Expert Syst. Appl. **240**, 122511 (2024)
19. Marino, G., Zotteri, G., Montagna, F.: Consumer sensitivity to delivery lead time: a furniture retail case. Int. J. Phys. Distrib. Logistics Manage. **48**(6), 610–629 (2018)
20. Meisel, F., Bierwirth, C.: The design of make-to-order supply networks under uncertainties using simulation and optimisation. Int. J. Prod. Res. **52**(22), 6590–6607 (2014)
21. Melnyk, S.A., Thürer, M., Blome, C., Schoenherr, T., Gold, S.: (re)-discovering simulation as a critical element of OM/SCM research: call for research. Int. J. Oper. Prod. Manage. **44**, 1376–1389 (2023)
22. Montreuil, B., Labarthe, O., Cloutier, C.: Modeling client profiles for order promising and delivery. Simul. Model. Pract. Theory **35**, 1–25 (2013). https://doi.org/10.1016/j.simpat.2013.03.003
23. Moses, S., Grant, H., Gruenwald, L., Pulat, S.: Real-time due-date promising by build-to-order environments. Int. J. Prod. Res. **42**(20), 4353–4375 (2004). https://doi.org/10.1080/00207540410001716462

24. Parasian, A., Sakti, H.R., Novarianti, I.K., Nurcahyo, R.: Application of economic order quantity (EOQ) for make-to-order (MTO) business. In: Proceedings of the International Conference on Industrial Engineering and Operations Management. IEOM Society International, Bangalore, India (2021). https://doi.org/10.46254/IN01.20210074, https://index.ieomsociety.org/index.cfm/article/view/ID/8046
25. Perona, M., Saccani, N., Zanoni, S.: Combining make-to-order and make-to-stock inventory policies: an empirical application to a manufacturing SME. Prod. Plann. Control **20**(7), 559–575 (2009). https://doi.org/10.1080/09537280903034271
26. Schulte, B., Sachs, A.L.: The price-setting newsvendor with poisson demand. Eur. J. Oper. Res. **283**(1), 125–137 (2020)
27. Stewart, M., Ivanov, D.: Design redundancy in agile and resilient humanitarian supply chains. Ann. Oper. Res. **319**(1), 633–659 (2022)
28. Volling, T., Spengler, T.S.: Modeling and simulation of order-driven planning policies in build-to-order automobile production. Int. J. Prod. Econ. **131**(1), 183–193 (2011). https://doi.org/10.1016/j.ijpe.2011.01.008
29. Yamaba, H., Tomita, S.: On an object-oriented modeling of supply chain and its operational strategy. Comput. Aided Chem. Eng. **15**, 660–665 (2003). https://doi.org/10.1016/S1570-7946(03)80620-X

Modeling of Logistics Objectives at Rail-Road Inland Transshipment Terminals

Volodymyr Alieksieiev(✉), Rafat Beigpoor Shahrivar, Carlos Jahn, and Hermann Lödding

Hamburg University of Technology, Am Schwarzenberg-Campus 1, 21073 Hamburg, Germany
{volodymyr.alieksieiev,rafat.beigpoor.shahrivar,carlos.jahn, loedding}@tuhh.de

Abstract. The logistics performance of inland transshipment terminals, such as rail-road container terminals, is crucial for establishing a reliable combined transport and for accomplishing sustainability goals. Main logistics objectives, such as high output rate and short throughput times, cannot be achieved simultaneously, creating an objective conflict. Modeling the interdependencies between the objectives can support terminals in determining an optimal operation point depending on their goals. Previously, these interdependencies have been analyzed primarily using complex simulation or optimization models. This paper presents an approach to model these interdependencies using Logistic Operating Curves (LOCs) and validates it using simulation.

Keywords: Combined Transport · Inland Container Terminal · Logistics Objectives · Logistic Operating Curves

1 Introduction

Combined transport, i.e. the combination of environmentally friendly transport modes, such as railways or waterways, for long-distance transportation and flexible road transport for short distances is essential for achieving the sustainability goals of European Union [1]. Inland transshipment terminals, such as rail-road container terminals, are important nodes of combined transport and influence with their logistics performance an overall combined transport capability. One of the main logistics objectives of transshipment terminals is a short throughput time of vehicles. On the other hand, a high output rate of the cost-intensive main working systems of the terminals, the Rail-Mounted Gantry Cranes (RMGC), is important to reduce their unproductive idle times and transshipment costs. Both logistics objectives depend on terminal's work in progress (WIP) and are opposed to each other: while a low level of WIP enables short throughput times of trains and trucks, it goes along with a lower crane output rate. Conversely, an increased output rate results in with increased WIP levels and throughput times. Therefore, terminals have to find an optimal operating position in this objective conflict, depending on their individual management strategies (cf. [2] and [3]).

Published by Springer Nature Switzerland AG 2024
M. Thürer et al. (Eds.): APMS 2024, IFIP AICT 733, pp. 92–107, 2024.
https://doi.org/10.1007/978-3-031-71645-4_7

Previously, these interdependencies have been primarily studied using simulation or mathematical optimization models. The advantage of simulation models is that they usually enable a highly detailed modeling of terminal operations. However, they are time-intensive, typically require many input parameters and need to be adapted for each use case [3]. Optimization models can find a nearly optimal solution for many problems. However, they are often very complex to implement and usually restrict decision flexibility (cf. [4]).

Models of production logistics, such as throughput diagram and Logistic Operating Curves (LOCs), enable relatively simple, but sufficiently precise modeling of the terminal objectives and dependencies between them. This paper presents hence some fundamentals of modeling the logistics objectives of rail-road inland transshipment terminals, one of the most common inland terminal types, using throughput diagrams and LOCs. The paper is structured as follows: Sect. 2 describes the theoretical background and state of research on inland rail-road transshipment terminals and the theory of LOCs. Section 3 presents the modeling fundamentals. Section 4 evaluates the models using simulation. Section 5 discusses the evaluation results. Finally, Sect. 6 concludes the paper and provides an outlook of further research.

2 Theoretical Background and State of Research

This Section describes firstly some fundamentals of inland rail-road transshipment terminals (Subsect. 2.1). Afterwards, Subsect. 2.2 briefly introduces the modeling of LOCs.

2.1 Inland Rail-Road Transshipment Terminals

Inland transshipment terminals serve as pivotal nodes within both national and international logistics networks, facilitating the transfer of loading units (LUs) across road, rail and waterway. These terminals encompass diverse functional areas, including storage facilities for LUs such as containers, swap bodies and semi-trailers, alongside dedicated zones for the integration of different transport modes (see Fig. 1). This paper focuses specifically on bimodal rail-road terminals equipped with gantry cranes.

Inland intermodal terminal logistics has been extensively studied, employing mathematical optimization, queueing theory and simulation techniques to improve efficiency. Previous research, such as [5] on minimizing transshipment duration and costs or [6] on gantry crane scheduling in rail-road container terminals, explore optimization strategies. Clausen et al. [2] highlight the objective conflict in container terminal operations, which involves balancing inventory, utilization, cycle time, and schedule adherence. Basallo-Triana et al. [7] propose an analytical method to evaluate rail-road transshipment yard efficiency, focusing on crane cycle times, truck waiting times, and yard throughput. Simulation results underscore the significance of train-truck synchronization for optimizing yard performance and reducing congestion. Roy et al. [8] analyze the throughput times of trucks for different handling equipment and priority strategies. They show in a numerical experiment, that the throughput time of trucks increases with their growing arrival rate. Zając et al. [9] explore the container handling time depending on the terminal occupancy for two different handling strategies. They emphasize waiting time and operation time as the components of container handling time. The authors mention, that the handling time increases with growing terminal occupancy. Möller [10] and Roy et al. [8] highlight the potential of queuing theory and Little's Law [11] for modeling the trucks queue and their corresponding waiting times at inland terminals depending on their arrival frequency.

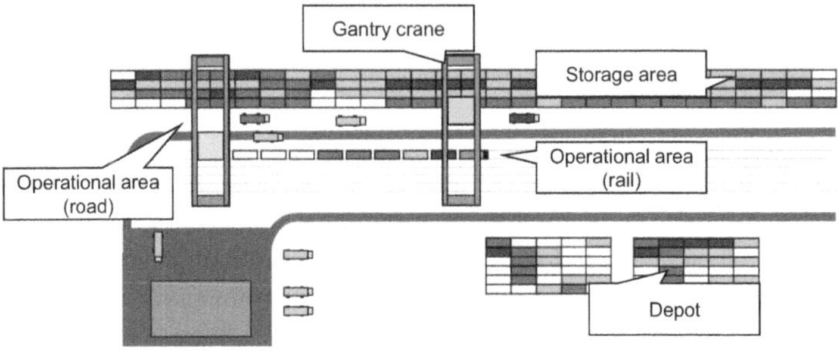

Fig. 1. Layout of inland rail-road transshipment terminal (cf. [12])

However, to the authors knowledge, existing research notably lacks a comprehensive approach to model conflicting terminal objectives. The Theory of Logistic Operating Curves promises a high potential for this modeling and is introduced in the next Subsection.

2.2 Modeling of Logistic Operating Curves

One of the basic models for deriving the Logistic Operating Curves is the Funnel model [13]. It describes an arbitrary working system as a funnel. Figure 2a illustrates the Funnel model of a sample working system. Orders, that are coming into the funnel, along with the waiting orders, represent a queue or work in progress (WIPO) of a system. After completion, orders leave the funnel. Output rate (ROUTO) defines the number of orders completed per time unit [3, 14].

A throughput diagram (Fig. 2b) can be derived from the Funnel model by cumulatively plotting the input and output times of orders over time. Using the throughput diagram, it is possible to derive the main logistics Key Performance Indicators of a working system, such as mean load and mean output rate. The horizontal distance between the input and output curves represents the mean throughput time. The vertical distance is equal to the mean work in progress of the system and describes the mean number of orders in system's queue. The mean work in progress can be calculated directly from the throughput diagram by dividing the WIP area (grey area in Fig. 2b) by the reference period [3, 14].

Similar to Little's Law [11], the Funnel formula describes the interdependency between mean values of WIP, output rate and throughput time [3, 13]:

$$TTP_m = \frac{WIPO_m}{ROUTO_m} \tag{1}$$

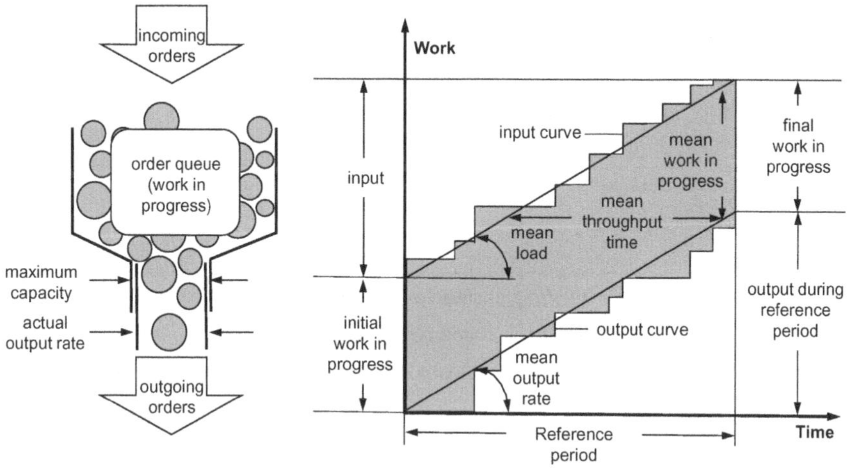

a) Funnel model **b) Throughput diagram**

Fig. 2. Funnel model and throughput diagram for a sample working system [3, 13, 14]

A throughput diagram represents only one operation state of a system. Plotting the obtained mean values of output rate and throughput time in dependence of mean WIP derives real LOCs for output rate and throughput time correspondingly [3].

For modeling LOC, it is important to describe an ideal process. In the ideal process, neither the working system waits for an order (no idle time), nor an order waits for the system (no queue, full utilization) [3]. A working system operating under ideal conditions is characterized by a minimum ideal work in progress ($WIPIO_{min}$), a maximum possible output rate ($ROUTO_{max}$) and a minimum throughput time (TTP_{min}).

In the ideal process, the mean output rate grows first proportionally to the WIP (so-called underload operating zone) until full utilization is realized, while the throughput time remains constant (see Fig. 3 and Eq. (1)). After the maximum output rate has been reached, further increasing the WIP does not increase the output rate anymore, but results in building up an orders queue and growing throughput times (overload operating zone, see Fig. 3) [3, 14].

Real LOCs do not have a clear angulation, but a fluid transition to the overload (transitional operating zone). The deviation from the ideal LOC is caused by fluctuations in work content of orders and interarrival times. In the real processes, a so-called buffer WIP is needed to compensate the load variance (see Fig. 3) [3].

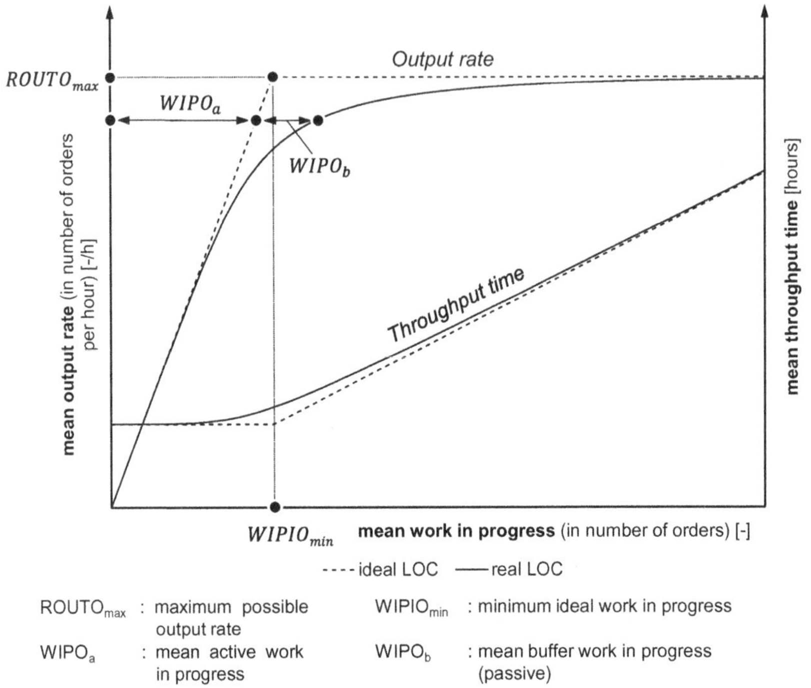

Fig. 3. Ideal and real LOCs for a sample working system [3]

Modeling of both WIP components and an approximation of the real LOC enables e.g. the Kingman Eq. (2), which models the mean work in progress depending on the mean utilization of the working system, variation of order processing time and variation of interarrival time (a mean time between two order arrivals) ([15] in notation of [3, 16]). The first summand in this equation represents the active WIP, while the second corresponds to the buffer WIP (see Fig. 3).

$$WIPO_m = U_m + \frac{U_m^2}{1 - U_m} \cdot \left(\frac{t_{p,v}^2 + IAT_v^2}{2} \right) \tag{2}$$

where

$WIPO_m$: mean work in progress (in number of orders) [-]
U_m: mean utilization of a working system [-]
$t_{p,v}$: coefficient of variation of order processing time [-]
IAT_v: coefficient of variation of interarrival time [-]

Utilization in the formula (2) represents an output rate in relation to its maximum possible value, which is defined by the system's capacity [3].

The Kingman equation enables an approximation of the mean utilization, systematically considering load and processing fluctuations in a working system. Important assumptions are the independency of processing and interarrival times as well as rigid system capacities (cf. [16]). Egli [17] and Seitz [18] model the LOCs of transport processes using an approximation equation of Nyhuis and Wiendahl [3]. Grafelmann et al. [19] construct a simulation model for deriving simulated LOCs for unloading processes at inland terminals and demonstrate the potential of LOCs for managing inland terminals operations. The goal of this paper is to model such LOCs using the LOC theory and the Kingman equation.

3 Modeling of Logistics Objectives at Inland Intermodal Terminals

This Section presents some fundamentals of modeling logistics objectives at inland intermodal terminals, using an example of a rail-road container terminal with one RMGC. Section 3.1 first describes an order structure, Funnel model and throughput diagram of such terminals. Further, the derivation of ideal and real LOCs for inland intermodal terminals is presented (Sect. 3.2).

3.1 Funnel Model and Throughput Diagram

One of the features, that distinguish a terminal system from a production system is a hierarchical order structure. A crane order (CO) is analogous to a transport order (cf. [17, 18]) and corresponds to each handling of a loading unit, such as its loading to or unloading from train or truck (see e.g. [10, 19]). Its processing time $t_{p,CO}$ consists analogous to [10, 17, 18] of an empty run time to a start position (t_{empty}), time for picking the container ($t_{picking}$), transport time to the end position ($t_{transport}$) and time for setting the container down (t_{down}):

$$t_{p,CO} = t_{empty} + t_{picking} + t_{transport} + t_{down} \tag{3}$$

A dispatch order (DO) includes all crane orders, which are necessary to handle all required containers from and onto the vehicle, thereby preparing it for dispatching. Its processing time $t_{p,DO}$ can be derived similar to a work content of a production order [3] from the sum of processing times of corresponding crane orders:

$$t_{p,DO} = \sum\nolimits_{i=1}^{n} t_{p,CO,i} \tag{4}$$

where

$t_{p,DO}$: processing time of a dispatch order [seconds]
$t_{p,CO,i}$: processing time of crane order i [seconds]
n: number of crane orders in the dispatch order [-]

Figure 4 illustrates the processing of dispatch and crane orders at the inland intermodal terminal using the Funnel model. It shows three dispatch orders, representing a transshipment of two trucks and one train. The bigger size of the train order indicates the higher number of handlings or crane orders, which have to be processed to fully unload and load the train (in this example – 12 crane orders). Correspondingly, the truck orders have smaller number of crane orders (two crane orders for each truck order in this example).

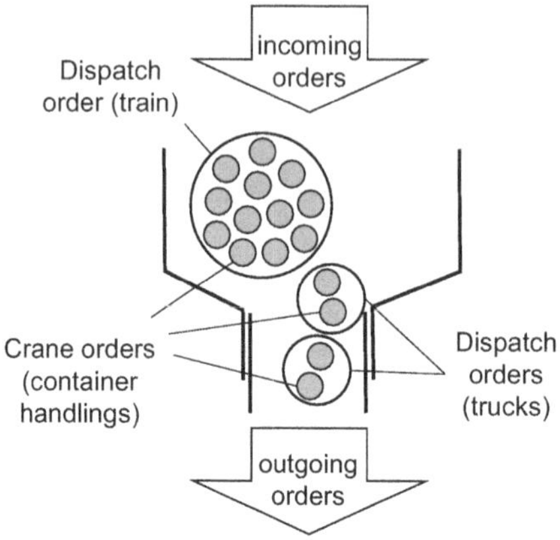

Fig. 4. Funnel model of transshipment process at inland rail-road container terminal

Figure 5 (above) plots a throughput diagram for both order types, corresponding to the Funnel model presented in Fig. 4. Assuming that all crane orders of a dispatch order are being released immediately after its arrival in the terminal, all crane orders of one train or truck arrive at the same time and input curves of both order types are identical (see input curves in Fig. 5 above). A crane can normally simultaneously handle one container only. Therefore, crane orders are being completed sequentially (see output curve of crane orders in Fig. 5 above). A dispatch order is being completed with the completion of its last crane order (see output curve of dispatch orders in Fig. 5 above). The vertical distance between the input and output curves corresponds to the WIP level. Similarly, the horizontal distance equals the range of the crane and the mean throughput time of the orders (see Fig. 2). The WIP of dispatch orders describes the number of trains and trucks being processed in the operational area of a terminal at the time of measurement. The WIP values of dispatch orders can be used e.g. to analyze the capacity utilization of tracks. The WIP of crane orders corresponds to the number of crane jobs in the queue. Its value at a time moment depends on the number of trains and trucks at the terminal and the number of containers being unloaded from and loaded onto them, i.e. number of crane orders of corresponding dispatch orders. Figure 5 (below) depicts exemplary the WIP trend of the crane orders to the throughput diagram.

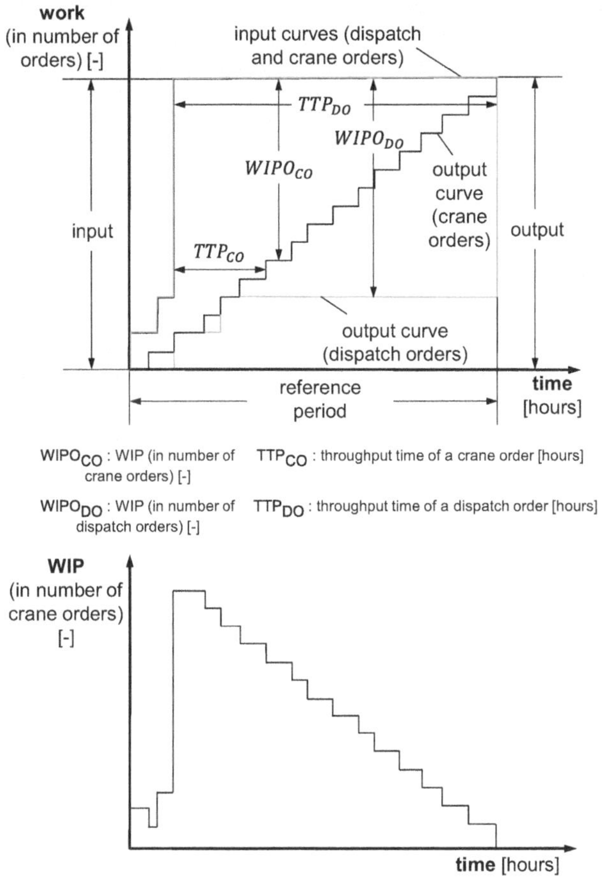

Fig. 5. Throughput diagram for dispatch and crane orders and WIP trend of crane orders

3.2 Modeling of Logistic Operating Curves

This Section covers modeling Logistic Operating Curves of transshipment process at inland rail-road terminals. It firstly describes ideal LOCs. Afterwards it presents an approximation of real LOCs using the Kingman equation.

Ideal Logistic Operating Curves

Transferring the conditions of the ideal process (see Sect. 2.2) to the dispatch orders at inland terminals means, that there is only one train or truck at the terminal at any point in time. The minimum ideal WIP then corresponds to one dispatch order. Since one dispatch order can contain many crane orders, which arrive as a batch (see Sect. 3.1), the minimum ideal work in progress for crane orders is greater than one and can be calculated according to (5) as ratio of WIP area to the length of a reference period (see also Fig. 5).

$$WIPIO_{min,CO} = \frac{\sum_{j=1}^{n}\left(\frac{NCO_{DO,j}+1}{2} \cdot NCO_{DO,j}\right)}{\sum_{j=1}^{n} NCO_{DO,j}} \tag{5}$$

where

$WIPIO_{min,CO}$: minimum ideal work in progress (in number of crane orders) [-]
$NCO_{DO,j}$: number of crane orders per dispatch order j (in number of crane orders) [-]
n: number of dispatch orders [-]

The maximum possible output rate of a gantry crane in dispatch orders per hour can be calculated by dividing the handling capacity of a crane in minutes per hour (i.e. the crane working time per hour) by the mean processing time of a dispatch order in minutes (see Eq. (4), [3], cf. also [10]). Similarly, the maximum possible output rate of a gantry crane in crane orders per hour can be determined using the processing time of crane orders (3) instead of that of dispatch orders. Figure 6 shows the resulting ideal LOCs for a minimum ideal WIP of 5.25 crane orders, a maximum possible crane output rate of 25 crane orders per hour and consequently a minimum throughput time of a crane order of 0.2 h (12 min).

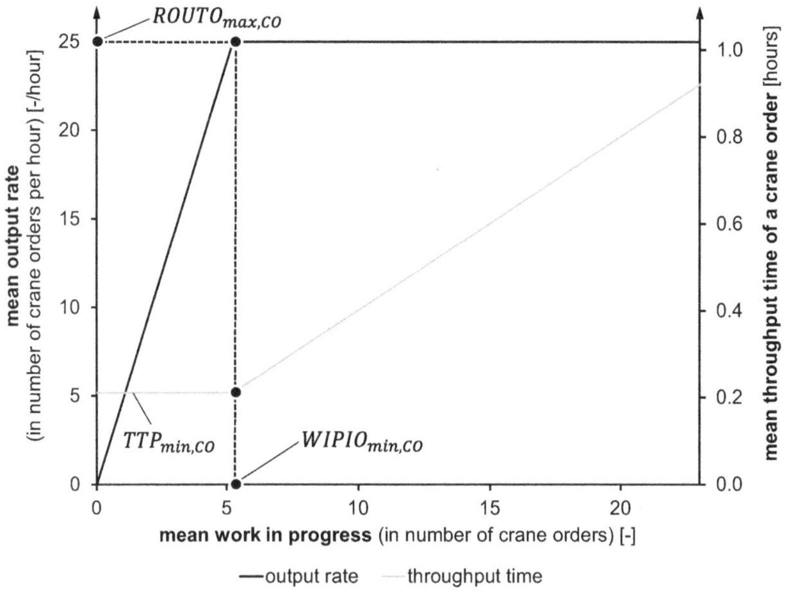

$ROUTO_{max,CO}$	: maximum possible output rate of a gantry crane (in number of crane orders per hour [-/hour]
$WIPIO_{min,CO}$	: minimum ideal work in progress (in number of crane orders) [-]
$TTP_{min,CO}$	: minimum throughput time of a crane order [hours]

Fig. 6. Sample ideal LOCs for output rate and throughput time of crane orders

Approximate Logistic Operating Curves

As already mentioned in Sect. 2, the conditions of the ideal process do not correspond to practical applications. In real terminal operations, trains and trucks throughput times overlap (see Fig. 5). The variance of the load defines the deviation of real LOC from the ideal curve (see Sect. 2). Therefore, additional WIP is needed to compensate the fluctuations and to achieve a certain crane output rate, the buffer WIP (see Sect. 2 and [3]). The basic equations for approximation of the mean WIP level for a certain output rate or utilization level are hence according to [3] for dispatch orders (6) and for crane orders (7) (see also Fig. 3).

$$WIPO_{m,DO} = WIPO_{m,active,DO} + WIPO_{m,buffer,DO} \tag{6}$$

$$WIPO_{m,CO} = WIPO_{m,active,CO} + WIPO_{m,buffer,CO} \tag{7}$$

The Kingman equation, presented in Sect. 2, encompasses the most crucial variations of the transshipment process at inland intermodal terminals – the variations of interarrival and processing times. The variation of processing times comprises in this case variations of transport and empty run times as well as the variation of number of crane orders of train and truck orders. If trains and trucks arrive independently of each other, it is possible to directly apply the Kingman equation (see also Sect. 2.2):

$$WIPO_{m,DO} = U_m + \frac{U_m^2}{1 - U_m} \cdot \left(\frac{t_{p,DO,v}^2 + IAT_{DO,v}^2}{2} \right) \tag{8}$$

where

$WIPO_{m,DO}$: mean work in progress (in number of dispatch orders) [-]
U_m: gantry crane utilization [-]
${t_{p,DO,v}}^2$: coefficient of variation of dispatch order processing time [-]
${IAT_{DO,v}}^2$: coefficient of variation of dispatch order interarrival time [-]

The first summand in the Eq. (8) models the active WIP, which is limited by minimum ideal WIP, i.e. by one in case of dispatch orders processing (see descriptions of ideal LOCs in Sects. 2 and 3). The second part of the equation approximates the buffer WIP, required to compensate the arrival and processing fluctuations, which are described by corresponding coefficients of variations. The fluctuations of processing times of dispatch orders at inland terminals are extremely high, alone due to the large difference in mean number of crane orders per train and truck orders (see also Figs. 4 and 5). Hence, the order structure at such terminals requires high buffer WIP levels, if operating under standard conditions of queuing theory.

In contrast, a direct application of the Kingman equation for crane orders processing is not possible due to their batch arrival behavior (see Sect. 3.1). Under the previously mentioned assumptions of Kingman approximation (see Sect. 2.2), the mean utilization dependent WIP in crane orders can be derived from (8) and interdependencies between crane and dispatch orders. As the minimum ideal WIP in crane orders describes the mean number of crane orders per dispatch order in the ideal process (see derivation of ideal LOCs), its multiplication with utilization provides the active WIP in crane orders (cf.

active WIP in approximation of Nyhuis and Wiendahl in [3]). To approximate the buffer WIP in crane orders, the buffer WIP in dispatch orders has to be multiplied with the mean number of crane orders per dispatch order. The approximated equation for mean work in progress in number of crane orders is presented in (9).

$$WIPO_{m,CO} = U_m \cdot WIPIO_{m,CO} + \left(\frac{U_m^2}{1 - U_m} \cdot \left(\frac{t_{p,DO,v}^2 + IAT_{DO,v}^2}{2} \right) \right) \cdot NCO_{DO,m} \tag{9}$$

where

$WIPO_{m,CO}$: mean work in progress (in number of crane orders) [-]
U_m: gantry crane utilization [-]
$t_{p,DO,v}$: coefficient of variation of dispatch order processing time [-]
$IAT_{DO,v}$: coefficient of variation of dispatch order interarrival time [-]
$NCO_{DO,m}$: mean number of crane orders per dispatch order (in number of crane orders) [-]

After obtaining the approximated values for mean WIP in dispatch and crane orders, the corresponding mean values of throughput time result from the Funnel formula (1):

$$TTP_{m,DO} = \frac{WIPO_{m,DO}}{ROUTO_{m,DO}} \tag{10}$$

$$TTP_{m,DO} = \frac{WIPO_{m,DO}}{ROUTO_{m,DO}} \tag{11}$$

Variation of parameters utilization and output rate in Eqs. (8)–(11) derives approximative LOCs for dispatch and crane orders. For obtaining the modeled output rate curve, corresponding values of mean output rate can be derived from the mean utilization (see description of interdependency between utilization and output rate after (2)).

4 Model Evaluation

Evaluation of the model has been conducted using discrete-event simulation in Microsoft Excel VBA. For this purpose, a simulation model of inland rail-road container transshipment terminal with one gantry crane, four train tracks, one load track and one drive track for trucks as well as three storage tracks has been created (cf. [20]). Figure 7 describes the simulation parameters.

As bridge and trolley of a gantry crane are usually being moved simultaneously, empty and transport times of crane orders result from the maximum value [20]. Acceleration and deceleration of the bridge and trolley of the gantry crane are not considered. It is assumed, that the number of crane orders per each truck and train order is approximately equally divided for loading and unloading operations. Direct transshipments between trains and trucks are not considered. Further, it is assumed, that the number of containers being loaded onto and unloaded from the trains is approximately equal to the number of containers being loaded onto and unloaded from the trucks within the simulation period.

Parameter	Value	Parameter	Value	Parameter	Value
Width (train track)	*4.5 m*	Crane trolley velocity	*1.33 m/s*	Distribution of number of crane orders per dispatch order	*rectangular*
Width (storage area track)	*3.3 m*	Crane bridge velocity	*2 m/s*	Distribution of dispatch orders arrival	*exponential*
Width (truck drive and load track)	*4.75 m*	Container picking and taking down time (sum)	*15 s*	Number of simulated dispatch orders	*10000*
Length (container places at all tracks)	*17.6 m*	Number of crane orders per truck dispatch order	*1-4 (3-4 as trucks with trailers)*	Dispatch orders sequencing	*FIFO*
Length (track)	*370 m/ 21 places*	Number of crane orders per train dispatch order	*20-40*	Crane orders sequencing	*First unloading, then loading*

Fig. 7. Layout and process parameters of a simulated terminal (cf. [20])

Consequently, the number of trucks is significantly higher than the number of trains. Other assumptions are a random distribution of container start and end positions in the storage area and no restacking operations.

The simulation procedure is as follows: firstly, the simulation program creates dispatch orders with corresponding numbers of crane orders per each dispatch order. Then it assigns trains to a random train track position and a truck to a random truck track position. The track capacity is not restricted. Afterwards, the crane orders of corresponding dispatch orders are created and their processing times are calculated (see Eq. (3)). From the processing times of crane orders, the program calculates corresponding processing times of dispatch orders (see Eq. (4)). Using this value, it determines afterwards the required mean interarrival times of dispatch orders to ensure the mean crane utilization from 1% to 99%. In the next step, simulation experiments have been conducted for each mean interarrival time of dispatch orders, creating different utilization and output rate levels in crane and dispatch orders. For the distribution of dispatch order arrival times, common random numbers apply.

Figures 8 shows approximated and simulated LOCs for dispatch orders (Fig. 8a) and crane orders (Fig. 8b) based on two simulation scenarios with different mean processing times of dispatch orders (scenario 1, blue – 13 min; scenario 2, green – 11 min) and hence, different maximum output rates (scenario 1, blue – 4.6 dispatch orders and 21.2 crane orders per hour; scenario 2, green – 5.4 dispatch orders and 24.7 crane orders per hour). The approximations of mean WIP and throughput time values have been conducted using Eqs. (8)–(11). The coefficients of variation of the processing times are 1.7 (green curves) and 1.4 (yellow curves). The coefficient of variation of the interarrival times equals to 1 for both scenarios (exponential distribution).

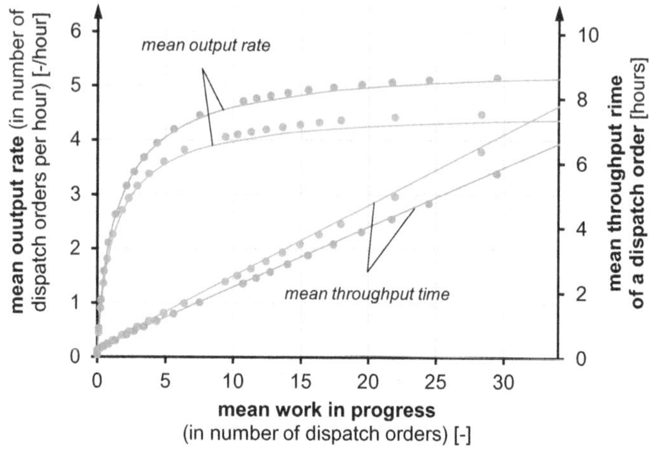

a) Approximated and simulated LOCs for dispatch orders

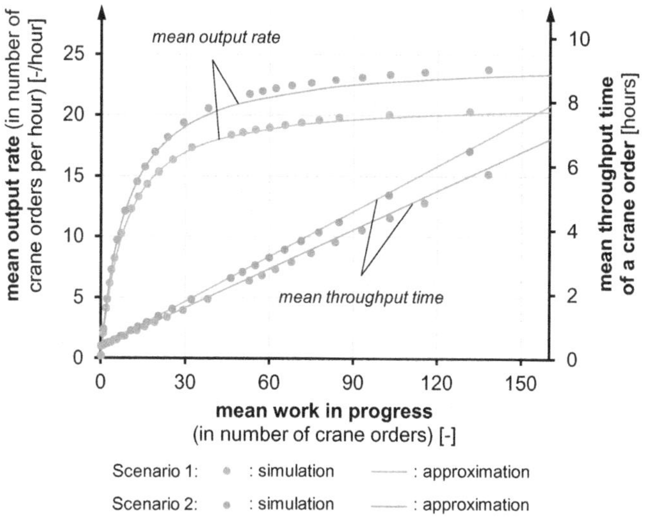

b) Approximated and simulated LOCs for crane orders

Fig. 8. Approximated and simulated LOCs for dispatch (a) and crane orders (b) and two scenarios (Color figure online)

5 Discussion

Modeled values of mean output rate and throughput time for both order types show slight deviations from simulated values (see Fig. 8). For the mean output rate in dispatch orders, the mean deviations of the model results from the simulation results are −1% (green curve) and −2% (blue curve). The corresponding mean absolute deviation (MAD) is 2% for both curves. Modeled mean throughput times of a dispatch order deviate from the corresponding simulation results to +2% (mean deviation) and 2% (MAD). The modeled mean output rate in crane orders differs from the simulated values to −1% (mean deviation, green curve) and to −2% (mean deviation, blue curve), whereas the MAD equals to 2% for both scenarios. The corresponding deviation of mean throughput time amounts in both scenarios to +2% (mean deviation), while the MAD for green curve equals to 2% and for blue curve to 3%. In addition, it can be seen, that although the processing time of crane orders is smaller, the values of throughput times for both order types are close to each other. It can be referred to the batch arrival behavior of crane orders. Their throughput time depends hence not only on the dispatch order queue, but also on the number of crane orders of corresponding dispatch order.

Analysis of model's accuracy concludes, that the Kingman equation and Funnel model can be applied for sufficient approximations of mean output rate and throughput time. At the same time, from the practical perspective the modeled LOCs do not provide competitive values of both terminal logistics objectives. Particularly, due to extremely high load variation, very high WIP levels are needed for reaching high output rates. For instance, for ensuring the maximum possible output rate in both scenarios, the WIP of more than 30 dispatch orders or more than 150 crane orders is needed, resulting in extremely long mean throughput times. In addition, assumed maximum capacity of a truck track of 21 truck places, and four train tracks enables the maximum WIP of 12 dispatch orders (by assumed 2.5 crane orders per each truck order in average). It makes it impossible to reach high output rate values due to the conflict between order WIP and track capacity.

This demonstrates, that operating the inland terminals under the standard conditions of queuing theory may not be practical. Specifically, interarrival and processing times should not be independent from each other. Rather, it is the task of terminal management systems to schedule train arrivals and to synchronize truck arrivals accordingly. Additional measures to control the WIP level can be capacity flexibility (cf. [3, 14]) and efficient operational strategies. However, the Kingman equation will not be able to consider these factors, that highlights the need for new modeling approaches.

6 Conclusion and Outlook

This paper introduces some fundamentals of modeling of logistics objectives at inland intermodal terminals. The conducted literature review has revealed a lack of models for the logistics objectives at inland terminals, such as output rate, WIP and throughput time as well as interdependencies between them. Based on the theory of LOC, Sect. 3 firstly introduced throughput diagrams for two terminal order types – dispatch and crane orders. The throughput diagram can support the visualization and analysis of the interdependencies between throughput time, output rate and WIP of both order types. Afterwards, ideal LOCs have been derived, enabling to model the active WIP in dispatch and crane orders. In the next step, approximative LOCs for mean output rate and mean throughput time of crane and dispatch orders have been derived using the Kingman equation and the Funnel formula. The models were evaluated using simulation. A comparison of simulation and modeling results shows, that the Kingman equation achieves sufficient modeling accuracy. However, the model clearly suggests, that under the standard premises of queuing theory (independency of processing and interarrival times, rigid capacities), inland terminals cannot achieve competitive levels of throughput time and output rate. Particularly, the model demonstrates the need for high WIP levels for achieving sufficiently high output rates. As high WIP levels go along with long throughput times and the need for high track capacities, load control is very important at inland terminals. In summary, our findings thus also support the need for new models for approximating the WIP of inland terminals under real conditions, e.g. considering restacking operations and different sequencing rules. Further research could also focus on increasing the productivity of gantry cranes, e.g. with new operational strategies.

References

1. United Nations: Terminology of Combined Transport, New York and Geneva (2001)
2. Clausen, U., Kaffka, J., Meier, F.: CONTSIM–container terminal management with simulation. Procedia Soc. Behav. Sci. (2012). https://doi.org/10.1016/j.sbspro.2012.09.752
3. Nyhuis, P., Wiendahl, H.-P.: Fundamentals of Production Logistics. Springer, Heidelberg (2009)
4. Hopp, W.J., Spearman, M.L.: Factory Physics, 3rd edn. The McGraw-Hill/Irwin Series. McGraw-Hill/Irwin, Boston (2008)
5. Kuhlmann, A.S.: Konstruktion und Implementierung eines Optimierungsmodells für den Kombinierten Güterverkehr. Dissertation. Information - Organisation - Produktion. Springer Fachmedien Wiesbaden, Wiesbaden (2013)
6. Guo, P., Cheng, W., Wang, Y., Boysen, N.: Gantry crane scheduling in intermodal rail-road container terminals. Int. J. Prod. Res. **56**, 5419–5436 (2018)
7. Basallo-Triana, M.J., Bravo-Bastidas, J.J., Vidal-Holguín, C.J.: A rail-road transshipment yard picture. Transp. Res. Part E Logist. Transp. Rev. (2022). https://doi.org/10.1016/j.tre.2022.102629
8. Roy, D., van Ommeren, J.-K., de Koster, R., Gharehgozli, A.: Modeling landside container terminal queues: exact analysis and approximations. Transp. Res. Part B Methodol. (2022). https://doi.org/10.1016/j.trb.2022.05.012
9. Zając, M., Rozic, T., Bajor, I.: Model for evaluating the effectiveness of cargo operation strategy in an inland container terminal. Appl. Sci. (2023). https://doi.org/10.3390/app13127127
10. Möller, H.C.: Beitrag zur Potenzialanalyse im kombinierten Verkehr für Güterverkehrszentren. Zugl.: Dortmund, Univ., Diss., 2001. Logistik für die Praxis. Verl. Praxiswissen, Dortmund (2002)
11. Little, J.D.C.: A proof for the queuing formula: L = (lambda) W. Oper. Res. 383–387 (1961)
12. Krüger, S., Grafelmann, M., Jahn, C.: Digital transformation of inland terminals. In: Advances in Intelligent Traffic and Transportation Systems, pp. 93–103. IOS Press (2023)
13. Bechte, W.: Steuerung der Durchlaufzeit durch Belastungsorientierte Auftragsfreigabe bei Werkstattfertigung. Dissertation, Universität Hannover, 1980. 70, vol. 2. VDI-Z, Düsseldorf (1984)
14. Lödding, H.: Handbook of Manufacturing Control. Springer, Heidelberg (2013)
15. Kingman, J.F.C.: The single server queue in heavy traffic. Math. Proc. Camb. Phil. Soc. (1961). https://doi.org/10.1017/S0305004100036094
16. Winter, M., Schoop, B., Lödding, H.: Produktionskennlinien mit Belastungsflexibilität. Zeitschrift für wirtschaftlichen Fabrikbetrieb (2021). https://doi.org/10.1515/zwf-2021-0187
17. Egli, J.: Transportkennlinien: ein Ansatz zur Analyse von Materialflusssystemen. Zugl.: Dortmund, Univ., Diss., 2001. Unternehmenslogistik. Verl. Praxiswissen, Dortmund (2002)

18. Seitz, K.-F.: Modellierung von Transportprozessen in Lieferketten. Dissertation. Berichte aus dem IFA, 2016, Band 4. PZH-Verlag, [Garbsen] (2016)
19. Grafelmann, M., Alieksieiev, V., Lödding, H., Jahn, C.: Logistics objective conflicts at intermodal terminals. 20. ASIM Fachtagung Simulation in Produktion und Logistik 2023, p. 313 (2023). https://doi.org/10.22032/dbt.57778
20. Lampe, H.: Untersuchung von Dispositionsentscheidungen in Umschlagterminals des kombinierten Verkehrs Schiene-Straße. Zugl.: Dortmund, Univ., Diss., 2005. Logistik, Verkehr und Umwelt. Verl. Praxiswissen, Dortmund (2006)

Resilience Management in Supply Chains

Forging Resilience Through Supply Chain Collaboration: Insights from the Chinese Automotive Industry

Weiyuan Li(✉), Ming K. Lim, and James Wilson

Adam Smith Business School, University of Glasgow, Glasgow, UK
weiyuan.li@glasgow.ac.uk

Abstract. This paper explores the pivotal role of supply chain collaboration in building resilience within the Chinese automotive industry amidst the new kind of disruption catalysed by the COVID-19 pandemic. Resource Dependency Theory (RDT) provides the theoretical foundation for this research, which used case studies and interviews to examine resilience strategies. The research found information sharing, resource sharing (supporting smaller suppliers), and adjustments (seeking external help) played key roles. This research highlights collaboration's importance in building supply chain resilience and offers practical guidance for navigating disruptions.

Keywords: Supply Chain Resilience · Supply Chain Collaboration · Resource Dependency Theory · Automotive Supply Chain

1 Introduction

In the dynamic landscape of global commerce, the resilience of supply chains has emerged as a critical determinant of organizational success, particularly in industries marked by complexity and interdependence. The automotive sector stands at the forefront of these challenges, characterized by intricate global supply networks, just-in-time manufacturing practices, and a relentless pursuit of efficiency. However, the advent of the COVID-19 pandemic ushered in a new era of disruption, presenting unprecedented challenges to the automotive supply chain and underscoring the imperative for enhanced resilience strategies.

Existing research emphasises the positive effect of collaboration on supply chain resilience [1, 2], but there is a lack of detailed research on specific implementation strategies. Supply chain collaboration was important in enhancing supply chain resilience and navigating the challenges posed by disruptions [3, 4]. In order to fully understand how these strategies affect supply chain resilience, this paper embarks on a comprehensive exploration of resilience within the Chinese automotive supply chain against the backdrop of the post COVID-19 era. Defined by the seismic shifts induced by the pandemic, this period tested the resilience of industry stakeholders and prompted a re-evaluation of traditional supply chain paradigms.

Published by Springer Nature Switzerland AG 2024
M. Thürer et al. (Eds.): APMS 2024, IFIP AICT 733, pp. 111–122, 2024.
https://doi.org/10.1007/978-3-031-71645-4_8

The aim of this research is twofold: to illuminate the experiences of how organisations in the automotive industry collaborated amidst the disruptions from COVID-19 and to distil valuable insights for fortifying supply chain resilience in anticipation of future uncertainties.

The remaining contents of this paper are structured as follows: Sect. 2 reviews the current research on supply chain resilience and supply chain collaboration, and pointed out the gap; Sect. 3 demonstrates the theoretical lens of this research; Sect. 3 illustrates the methodology used for data collection and data analysis; Sect. 4 presents the findings from the collected data; Sect. 5 discusses how the results could fill the gap; Sect. 6 gives a conclusion of this paper.

2 Literature Review

2.1 Supply Chain Resilience

Supply chain resilience (SCRES) refers to the ability of a supply chain to anticipate, prepare for, respond to, and recover from disruptions to ensure continuity and sustainability of operations [5]. The concept has gained widespread attention due to the increasing frequency and severity of disruptions such as natural disasters, pandemics, and technological changes. SCRES involves a variety of strategies and capabilities, including agility, flexibility, redundancy, and adaptability, that enable a supply chain to withstand unfavourable events and recover quickly. The interplay between supply chain agility and resilience is critical, with agility positively affecting resilience, which in turn affects supply chain performance, especially in the presence of environmental risks [6]. Dynamic capabilities and relational perspectives are also crucial as they help to understand the antecedents and consequences of SCRES, thereby improving sustainable operational performance [7].

In terms of research methodology, in addition to empirical studies, modelling is also a widely adopted approach in SCRES research. For example, a supply chain modelling study was conducted for the Portuguese automotive supply chain to evaluate alternatives to improve the resilience of the supply chain, aiming to understand how mitigation strategies affect the performance of each entity in the supply chain [8]. Deterministic modelling approaches and graph theory have also been applied to measure SCRES, by understanding the main contributors to resilience and their interrelationships to support organisational decision-making processes [9]. There are also studies that emphasise the importance of management processes and capabilities in building resilience by developing specific supply chain frameworks [10]. Alternatively, SCRES has been further enhanced by integrating smart technologies with decision-making frameworks such as the Plan-Do-Check-Act cycle through improved demand forecasting and risk management [11].

Collaboration is also a key factor in enhancing SCRES, as described in [1]. Collaborative activities can go beyond the impact of individual organisations on resilience, so as to demonstrate the need for co-operation between stakeholders. A contingent resource view can help to understand the relationship between specific resources, capabilities and performance in collaboration. And in specific collaboration methods, information sharing, connectivity and visibility play a crucial role in establishing SCRES

[12]. In conclusion, SCRES is a multifaceted concept that requires a comprehensive approach integrating strategic, technological, and collaborative efforts to ensure supply chain robustness and sustainability in the face of disruption. Future research directions include further exploration of principles and strategies to enhance overall SCRES [13].

2.2 Supply Chain Collaboration

In today's globalised business environment, Supply Chain Collaboration (SCC) is essential for achieving sustainable development, innovation and competitive advantage. SCC enhances sustainable operational performance of supply chains through improved capacity building and resource utilisation, which is critical to the achievement of broader sustainable development goals [14]. The nature and characteristics of collaboration, and its positive impact on organisational performance, including planning, decision-making and execution, have been confirmed by existing literature [15, 16]. Effective SCC requires a framework that encompasses internal and external knowledge sharing, collaborative culture, and problem-solving capabilities, which are essential for readiness and relationship strengthening [17]. Inter-organisational team members in collaboration activities have significantly improved business performance by optimising network connections and enhancing process innovation to achieve collaborative advantage in uncertain environments [18]. Co-operation with external partners would strengthen internal collaboration to improve logistics service performance, which is an effective way to achieve SCC. And the relationship between external and internal collaboration is crucial for behavioural change and performance improvement [19]. On the other hand, information technology plays a pivotal role in facilitating SCC by enabling knowledge transfer and promoting the behaviour of product innovation [20]. Empirical evidence suggests that industries such as pharmaceutical, automotive and electronics tend to have higher levels of collaboration, which is important for achieving performance excellence [21]. In conclusion, supply chain collaboration is indispensable for organisations wishing to thrive in competitive markets and can be achieved through a strategic framework, effective use of information technology, strong knowledge management practices and secure information sharing mechanisms.

Collaboration plays a key role in enhancing supply chain resilience by facilitating better communication, coordination, and integration among supply chain partners, which is essential for mitigating disruptions and ensuring continuity. Both horizontal and vertical forms of collaboration have been shown to improve cost, sustainability, and responsive indicators, and strengthening these collaborations can improve the overall resilience of supply chains when facing natural disasters [4, 22]. Effective collaboration can be achieved through a variety of subcomponents such as information sharing, connectivity, coordination, integration, and visibility, all of which have been proved positive to SCRES [1]. Integrated logistics capabilities, such as digitisation, innovation and modularity, also play a crucial role in improving supply chain resilience, with collaboration and trust between supply chain members positively moderating these effects [2]. The positive impact of IT capabilities on SCRES is further confirmed by the experience during the COVID-19 pandemic [23], which highlighted the importance of collaboration as it enables supply chains to return to their original or more optimal state after a disruption [24]. Collaborative emergency adaptation, which involves the coordination of re-purposed capacities and shared stocks, can mitigate the ripple effect of disruptions in intertwined supply networks, thereby improving recovery performance and contributing to the development of a resilient "network-of-networks" [25]. Furthermore, investment in flexibility and innovation, which are closely linked to collaborative efforts, can significantly enhance an organization's response to disruptions, as these elements positively impact resilience by decreasing the impact of disruptions [26].

In summary, these studies highlight the importance of collaboration in enhancing SCRES and emphasise the need for further research to explore the mechanisms by which collaboration improve resilience in various supply chain environments.

3 Methodology

3.1 Theoretical Lens

This paper takes Resource Dependency Theory (RDT) as the support of the research. The literature on RDT in the context of supply chain management offers valuable insights into the dynamics of inter-enterprise relationships. [27] provide a collaborative perspective on resource dependency theory, suggesting that engaging in supply chain relationships with key retail accounts can enhance supplier performance, particularly based on market share dynamics. [28] identify resource dependency theory as one of the predominant theories used to analyse information sharing in supply chains, alongside transaction cost economics, contingency theory, and relational governance theories. Furthermore, [29] introduce a supply chain risk management process grounded in Bayesian Belief Networks and resource dependency theory to prioritize interdependent risks and mitigation strategies. [30] explore the influence of green warehousing, logistics optimization, and social values on supply chain sustainability and economic performance through the lens of resource dependency theory in manufacturing firms. [31] incorporate resource dependency theory to assess how trust, satisfaction, and commitment impact firms' decisions on logistics integration, highlighting the role of strategic relationships in improving business and operational performance in the supply chain. Lastly, [32] use resource dependency theory to examine the challenges of implementing community benefits in public procurement, emphasizing the impact of resource dependence on the successful implementation of sustainable procurement policies. Overall, the literature review demonstrates the significance of resource dependency theory in understanding and enhancing supply chain relationships, risk management, sustainability, and performance across various industries and contexts.

3.2 Data Collection

This research utilises semi-structured interviews to delve into recovery strategies amid disruptions in the automotive supply chain (ASC). The aim of the interviews is to validate the effectiveness of existing strategies and explore adaptations by ASC practitioners. Considering the unprecedented nature of the current disruption, the study seeks to assess the adequacy of existing strategies and identify areas for improvement. In order to gather diverse perspectives on the impact of the COVID-19 Disruption, this research engaged with mid to senior-level professionals from various departments, regions, and supply chain roles. Participants were from production, logistics, and procurement departments, to obtain strategies in dealing with different aspects. It started from 6 organisations and ended while data saturation occurred, totally with 15 respondents in 9 organisations (Table 1).

Table 1. Profile of participants in Interviews

Interviewee	Organisation	Position	Years of experience
1	Organisation 1	Production manager	7
2	Organisation 1	Supply chain manager	10
3	Organisation 2	Logistics manager	5
4	Organisation 2	Production manager	8
5	Organisation 3	Production manager	9
6	Organisation 3	International logistics manager	9
7	Organisation 4	Procurement manager	6
8	Organisation 4	Logistics planning specialist	7
9	Organisation 4	Production manager	11
10	Organisation 5	Export business manager	5
11	Organisation 6	Supply chain manager	8
12	Organisation 7	Chief operation officer	18
13	Organisation 7	Production planning manager	6
14	Organisation 8	General manager	10
15	Organisation 9	General manager	14

3.3 Sampling

For the selection of the interview samples, this study established specific practice selection criteria. Two of the most important criteria were that the sample organisations should be as large as possible and have as high a market share as possible to ensure that the results obtained were representative of the Chinese automotive industry; and that the sample organisations' operations were as affected as possible by the COVID-19 Disruption to allow for a more diverse set of data to be collected. In addition to this, in order to better reflect the overall situation of ASCs in China, other criteria included the selection of organisations from different regions of China, as the policies adopted in response to COVID-19 may differ from region to region. In terms of product type, the sample organisations should be selected to cover the majority of the automotive product market, including both traditional internal combustion engine vehicles and electric vehicles. Finally, there should be supply chain linkages between the selected sample organisations, i.e., they are located in the same supply chain, so that the researcher is able to observe from a supply chain perspective. After all the above conditions were met, secondary factors such as data accessibility, ease of conducting interviews and social reputation of the organisations were also considered.

3.4 Data Analysis

All interview data was analysed by thematic analysis. Thematic analysis is a widely used method of analysing qualitative data. It provides a systematic and flexible approach to analysing and interpreting complex information and plays a vital role in understanding qualitative data. Thematic analysis is applicable to this study as it is effective in exploring different perspectives and experiences [33]. The semi-structured interviews were based on participants' different perspectives on disruption and their experiences of coping with it. These formed the core of the data collected for this study. The use of thematic analysis allows coping strategies to be summarised into major themes and demonstrates the diversity of coping strategies within sub-themes to address different types of challenges. Its flexible and rich analytical capabilities make thematic analysis a valuable tool for discovering insights and generating meaningful interpretations from the qualitative data in this study.

4 Findings

Thematic analysis categorised the interview data into five themes, yielding a total of 29 recovery strategies. Among them, there were 11 strategies involving supply chain collaboration. Based on the combined analysis of the data and literature, they were classified into three categories: information sharing, resource sharing, and alignment (see Table 2). Among them, the information sharing methods were used throughout the emergence of COVID-19 to the post-COVID era, while the other two categories basically appeared in the period of the outbreak.

The interview data suggest that collaboration played an important role in helping the automotive supply chain to cope with this new type of disruption, both in the initial unprecedented period and during the period when the situation was under control and the economy was gradually recovering. In the initial phase of the COVID-19 disruption, collaboration was reflected in the sharing of information and resources up and down the supply chain. These sharing strategies enabled organisations to minimise losses through timely and effective communication of information under constraints (e.g. travel restrictions) and to face challenges efficiently using appropriate methods. At the same time, adjusting appropriately to the conditions along the supply chain effectively mitigated the ripple effect i.e. the continuous diffusion of the disruption, and prevented problems at certain nodes from creating a bullwhip effect throughout the supply chain. Collaboration methods generally existed within supply chains in which firms are located, and there are few methods of collaboration across supply chains. An exception is the use of cluster effects to share logistics resources with competitors, which existed among some OEMs to help them receive timely replenishment of parts and components from common suppliers. Subsequently, during the recovery period, collaboration between organisations was significantly reduced compared to the previous period, gradually returning to the

Table 2. Recovery strategies from different ways of collaboration

Collaboration	Strategies
Information Sharing	Form an information exchange group
	Expand new collaboration opportunities
	Develop an overall stockpiling plan
Resource Sharing	Release some passes for suppliers
	Help SME suppliers communicate with local governments
	Enhance supplier production
	Utilise cluster effect for transportation
Adjustment	Require overseas suppliers to disinfect parts in advance
	Ask suppliers to increase inventory levels
	Convert certain components to VMI (Vendor Managed Inventory)
	Match supplier capacities

pre-COVID-19 situation. During this phase, most of the organisations only retained the necessary channels of information exchange without further collaboration with each other.

The findings from the interviews also corroborate the positive impact of supply chain collaboration on resilience in the literature. However, the qualitative data in this study provides more detailed ways of realising supply chain collaboration than the prevalent use of simulation or decision-making models in the literature to argue for the effectiveness of collaboration. It is interesting to note that although the positive effects of collaboration on SCRES have long been recognised in the literature, it has not been widely applied in practice. The recovery strategies mentioned by the interviewees in relation to collaboration all occurred after the hit of the disruption, and there was a lack of forethought about such problems and solutions. According to their descriptions, prior to this, most supply chain partners basically only maintained a contractual employment relationship. Therefore, in this disruption organisations demonstrated more temporary trust in the face of a common problem than in the literature which emphasises trust as an essential foundation for collaboration and the importance of building long-term trust. However, the effects of these collaborative strategies were still strong and were effective in response to this incident.

5 Discussion

5.1 Information Sharing

Information sharing, although not part of the original RDT definition, essentially reflects an organisation's dependence on other partners. This reliance was not directly on a particular resource from the partner, but rather on the need to have timely access to the partner's information. This sharing of information helped the organisation to adjust its capacity in time to match the upstream and downstream partners and the overall situation of the supply chain. For example, the supply chain learned to share important information such as production capacity and inventory levels during shutdowns. One successful strategy was to establish information exchange groups. These groups had members from all levels of the supply chain, allowing everyone to stay informed and to skilfully focus limited resources to overcome challenges. In addition to sharing basic details, these groups improved communication. This allowed for quicker coordination when problems arose anywhere, minimising the ripple effect and stopping disruptions from spreading further down the supply chain.

As well as mitigating the impact of disruptions during their occurrence, information sharing also involved collaborating on recovery after a pandemic. Organisations in the supply chain had found this collaborative approach to be effective and therefore continued to share and rely on each other in the post-COVID era. For example, the company worked with its supply chain partners to develop an annual backup plan for parts. They built redundancy by coordinating inventory levels for critical parts, sharing production capacity and sales data, and planning for larger parts reserves. This included discussing issues such as funding, transport and storage space with partners. Collaboration also extended to exploring new opportunities, such as original equipment manufacturers (OEMs) finding new suppliers or establishing additional distribution channels. This enhances the supply chain's ability to respond to future surprises. The information sharing highlights the enhancement of inter-organisational collaboration for supply chain resilience.

5.2 Resource Sharing

Resource sharing is when an organisation shares resources with other entities in the supply chain, both upstream and downstream, to meet challenges and keep operations running smoothly. The content shared is more physical than information sharing. According to the RDT, the sharing of resources by organisations in the ASC reflects their interdependence and demonstrates that they collaborate to overcome problems caused by disruptions. According to the interview data, shared resources are usually related to production and transport capacity, and the main problems they overcome usually occur during disruptions.

Many resources in the ASC are concentrated in OEMs and suppliers are always centred around these OEMs, which is due to the characteristics of the automotive supply chain. Therefore, the sharing of their resources by OEMs with their co-operating suppliers plays a very important role in this. Although these strategies are rarely discussed in the literature, they are specifically designed to assist small and medium-sized suppliers. This assistance can take many forms, such as providing passes to suppliers during blockades and travel restrictions, and facilitating communication between suppliers and local authorities to speed up the resumption of production, especially when faced with shutdowns due to government policies. Such collaborative efforts have not been historically common, but have proved invaluable in mitigating the knock-on effects of disruptions and minimizing the overall negative impact on the supply chain. At the same time, contrary to reports of pandemics leading to permanent closure of businesses, this study found that many of the affected upstream suppliers, particularly smaller secondary suppliers, were able to maintain at least some of their production as a result of resource-sharing initiatives. Larger organisations, often OEMs, played a key role in securing favourable conditions for their smaller partners, highlighting the importance of resource sharing as a collaborative approach to building supply chain resilience.

Another noteworthy strategy deployed during disruptive periods was to increase the production capacity of suppliers. This approach often entails moving equipment or moulds to less affected facilities or finding alternative suppliers for critical components. However, due to the sensitivity of the proprietary technology involved, these efforts require a high level of trust between organisations, making it impractical to implement this approach after a disruption has occurred. Its needs to be a long-term trust already established by the partners to be an effective means of improving supply chain resilience.

In addition, some organisations have used the 'cluster effect' within automotive industrial parks to address logistics challenges. By co-locating and collaborating with other industry players, they optimise resource utilisation and minimise logistics bottlenecks. While effective, this approach is limited by geographic dispersion, particularly among OEMs in China, suggesting that its applicability may be more feasible at lower levels of the supply chain.

5.3 Adjustment

Unlike resource sharing, which involves organisations sharing resources with partners, alignment strategies involve reaching out to partners when organisations are unable to achieve certain goals or overcome challenges on their own. These adjustments involve the entire supply chain, from suppliers to OEMs, both upstream and downstream. Feedback from interviewees suggests that these strategies arose from the contingency stage, where an emergency exceeded the organisation's resilience and external assistance was required. While these strategies reflect the organisation's dependence on partners or the external environment, they are not directly related to the need for a particular resource or information. For example, one strategy is to require upstream suppliers to sterilise goods prior to shipment to prevent delays caused by customs inspections. Suppliers may also be required to maintain higher inventory levels due to reduced logistics capacity. In addition, suppliers may seek help from downstream partners, such as asking original equipment manufacturers to match their reduced production capacity to meet current challenges. These strategies help mitigate the impact of disruptions in the short term, highlighting the positive impact of supply chain collaboration on overall resilience.

In addition to short-term strategies, transitioning certain components to Vendor Managed Inventory (VMI) was also a strategy adopted by one of the OEMs surveyed after the outbreak phase. They initially implemented this strategy with one supplier during the outbreak, and then recognised that VMI was a more effective way of improving the long-term resilience of their supply chain compared to the pre-outbreak approach, and adopted it as standard practice. These adapted strategies face more complex situations than the first two forms of collaboration, so rather than relying on one type of resource, they are more like turning to supply chain partners when problems arise and relying on the collective power of the organisation to solve them together, thus improving overall resilience.

6 Conclusion

In conclusion, this study integrated case study analysis and interviews to delve deep into the resilience strategies deployed within the Chinese automotive supply chain. Grounded in the theoretical framework of Resource Dependency Theory, this research examined how organisations relied on each other and collaborated for resources, and how this interdependence influences decision-making and helped organisations to adapt to exogenous shocks such as the COVID-19 disruption, offering a nuanced understanding of resilience dynamics.

The findings of this study shed light on the multifaceted response of industry participants, spanning from diversification of sourcing and strengthened collaboration with suppliers to the integration of digital technologies for agile decision-making. By elucidating the strategies employed to mitigate disruptions and enhance resilience, this research contributes to the growing body of knowledge on supply chain resilience, particularly within the context of the manufacturing industry. Through its empirical insights and practical recommendations, this paper suggests practitioners and policymakers on effective strategies to navigate turbulent environments and fortify supply chain resilience.

This study limits on supply chain collaboration in the context of large-scale disruptions. Future research opportunities lie in exploring supply chain collaboration patterns for other disruption types (e.g., small-scale, more frequent disruptions) and their impact on supply chain resilience. In addition, further quantification of the extent to which these forms of collaboration enhance supply chain resilience could be considered.

References

1. Scholten, K., Schilder, S.: The role of collaboration in supply chain resilience. Supply Chain Manag. Int. J. **20**(4), 471–484 (2015)
2. Mandal, S., Sarathy, R., Korasiga, V.R., et al.: Achieving supply chain resilience: the contribution of logistics and supply chain capabilities. Int. J. Disaster Resilience Built Environ. **7**(5), 544–562 (2016)
3. Gružauskas, V., Vilkas, M.: Managing capabilities for supply chain resilience through IT integration. Econ. Bus. **31**(1), 30–43 (2017)
4. Umar, M., Wilson, M.: Supply chain resilience: Unleashing the power of collaboration in disaster management. Sustainability **13**(19), 10573 (2021)
5. Ponomarov, S.Y., Holcomb, M.C.: Understanding the concept of supply chain resilience. Int. J. Logist. Manag. **20**(1), 124–143 (2009)
6. Wieland, A., Wallenburg, C.M.: The influence of relational competencies on supply chain resilience: a relational view. Int. J. Phys. Distrib. Logist. Manag. **43**(4), 300–320 (2013)
7. Pu, G., Qiao, W., Feng, Z.: Antecedents and outcomes of supply chain resilience: integrating dynamic capabilities and relational perspective. J. Conting. Crisis Manag. **31**(4), 706–726 (2023)
8. Carvalho, H., Barroso, A.P., Machado, V.H., et al.: Supply chain redesign for resilience using simulation. Comput. Ind. Eng. **62**(1), 329–341 (2012)
9. Soni, U., Jain, V., Kumar, S.: Measuring supply chain resilience using a deterministic modeling approach. Comput. Ind. Eng. **74**, 11–25 (2014)
10. Scholten, K., Sharkey Scott, P., Fynes, B.: Mitigation processes–antecedents for building supply chain resilience. Supply Chain Manag. Int. J. **19**(2), 211–228 (2014)

11. Fu, W., Jing, S., Liu, Q., et al.: Resilient supply chain framework for semiconductor distribution and an empirical study of demand risk inference. Sustainability **15**(9), 7382 (2023)
12. Brandon-Jones, E., Squire, B., Autry, C.W., et al.: A contingent resource-based perspective of supply chain resilience and robustness. J. Supply Chain Manag. **50**(3), 55–73 (2014)
13. Pettit, T.J., Croxton, K.L., Fiksel, J.: The evolution of resilience in supply chain management: a retrospective on ensuring supply chain resilience. J. Bus. Logist. **40**(1), 56–65 (2019)
14. Chauhan, C., Kaur, P., Arrawatia, R., et al.: Supply chain collaboration and sustainable development goals (SDGs). Teamwork makes achieving SDGs dream work. J. Bus. Res. **147**, 290–307 (2022)
15. Cao, M., Vonderembse, M.A., Zhang, Q., et al.: Supply chain collaboration: conceptualisation and instrument development. Int. J. Prod. Res. **48**(22), 6613–6635 (2010)
16. Ramanathan, U., Gunasekaran, A.: Supply chain collaboration: impact of success in long-term partnerships. Int. J. Prod. Econ. **147**, 252–259 (2014)
17. Holweg, M., Disney, S., Holmström, J., et al.: Supply chain collaboration: making sense of the strategy continuum. Eur. Manag. J. **23**(2), 170–181 (2005)
18. Cao, M., Zhang, Q.: Supply chain collaboration: Impact on collaborative advantage and firm performance. J. Oper. Manag. **29**(3), 163–180 (2011)
19. Stank, T.P., Keller, S.B., Daugherty, P.J.: Supply chain collaboration and logistical service performance. J. Bus. Logist. **22**(1), 29–48 (2001)
20. Jimenez-Jimenez, D., Martínez-Costa, M., Sanchez Rodriguez, C.: The mediating role of supply chain collaboration on the relationship between information technology and innovation. J. Knowl. Manag. **23**(3), 548–567 (2019)
21. Kumar, G.: Collaboration between supply chain partners: when does it matter? Evidence from collaborative profiles. J. Bus. Ind. Mark. **36**(6), 1042–1057 (2020)
22. Lotfi, M., Larmour, A.: Supply chain resilience in the face of uncertainty: how horizontal and vertical collaboration can help? Continuity Resilience Rev. **4**(1), 37–53 (2022)
23. Zhou, J., Hu, L., Yu, Y., et al.: Impacts of IT capability and supply chain collaboration on supply chain resilience: empirical evidence from China in COVID-19 pandemic. J. Enterp. Inf. Manag. (2022)
24. Kim, B., Kim, J.G., Lee, S.: A multi-agent reinforcement learning model for inventory transshipments under supply chain disruption. IISE Trans. **56**(7), 715–728 (2024)
25. Ivanov, D.: Collaborative emergency adaptation for ripple effect mitigation in intertwined supply networks. Ann. Oper. Res. **2023**, 1–17 (2023)
26. Ghomi, V., Nooraei, S.V.R., Shekarian, N., et al.: Improving supply chain resilience through investment in flexibility and innovation. Int. J. Syst. Sci. Oper. Logist. **10**(1), 2221068 (2023)
27. Hofer, C., Jin, H., Swanson, R.D., et al.: The impact of key retail accounts on supplier performance: a collaborative perspective of resource dependency theory. J. Retail. **88**(3), 412–420 (2012)
28. Kembro, J., Selviaridis, K., Näslund, D.: Theoretical perspectives on information sharing in supply chains: a systematic literature review and conceptual framework. Supply Chain Manag. Int. J. **19**(5/6), 609–625 (2014)
29. Qazi, A., Quigley, J., Dickson, A., et al.: Exploring dependency based probabilistic supply chain risk measures for prioritising interdependent risks and strategies. Eur. J. Oper. Res. **259**(1), 189–204 (2017)
30. Agyabeng-Mensah, Y., Ahenkorah, E., Afum, E., et al.: Green warehousing, logistics optimization, social values and ethics and economic performance: the role of supply chain sustainability. Int. J. Logist. Manag. **31**(3), 549–574 (2020)
31. Kim, S.T., Lee, H.H., Hwang, T.: Logistics integration in the supply chain: a resource dependence theory perspective. Int. J. Qual. Innov. **6**(1), 5 (2020)

32. Wontner, K.L., Walker, H., Harris, I., et al.: Maximising "community benefits" in public procurement: tensions and trade-offs. Int. J. Oper. Prod. Manag. **40**(12), 1909–1939 (2020)
33. Nowell, L.S., Norris, J.M., White, D.E., et al.: Thematic analysis: striving to meet the trustworthiness criteria. Int. J. Qual. Methods **16**(1) (2017)

Anticipating VUCA by Utilizing the Potential of Technological and Logistical Degrees of Freedom

Julian B. Maier[1(✉)] and Hans-Hermann Wiendahl[1,2]

[1] Fraunhofer Institute for Manufacturing Engineering and Automation IPA, 70569 Stuttgart, Germany
Julian.Maier@ipa.fraunhofer.de

[2] Institute of Industrial Manufacturing and Management IFF, University of Stuttgart, 70569 Stuttgart, Germany

Abstract. Facing increased market volatility necessitates robustness and flexibility to mitigate short-term disruptions and transformability to adjust to long-term changes. The degrees of freedom available for (re-)acting in this environment are determined by product requirements, production system characteristics, and production control system flexibility. The often-found sequential and decoupled consideration of technological and logistical decisions provides a narrow action space, resulting in limited flexibility.

Production control with integrated consideration of logistical and technological degrees of freedom has the potential to increase flexibility by opening up the decision space.

So far, research in this context is focused on job shop environments. Aiming to extend this integrated consideration to production systems with other characteristics, this article proposes a method to identify available degrees of freedom and introduces an approach to quantify their potential to mitigate the effects of company-specific turbulence patterns through a daydreaming-enhanced digital twin.

Keywords: Production Control · Process Planning · Digital Twin · Daydreaming · Resilience

1 Volatility and the Demand for Flexibility

To be successful, production companies must strive in volatile market environments characterized by an overlay of changes on different time scales [1]. On the one hand, they must adapt to long-term changes, e.g., caused by shorter product lifecycles, megatrends, and economic shifts [2]. On the other hand, the sudden occurrence of unexpected adverse events (i.e., turbulence germs [3]) requires robustness to absorb their effects (e.g., through stock), as well as agility and flexibility to react to them [4, 5]. While agility describes the scope of action at the level of a production network, this article focuses on the use of

Published by Springer Nature Switzerland AG 2024
M. Thürer et al. (Eds.): APMS 2024, IFIP AICT 733, pp. 123–137, 2024.
https://doi.org/10.1007/978-3-031-71645-4_9

flexibility, referring to the scope of action within a production system [6, 7]. This scope is determined by the structural flexibility of the physical production system, combined with the logical flexibility of the technological and logistical decision-making system [8]. Together, they determine the achievable flexibility of the production system and, consequently, its degrees of freedom (DoF) to (re-)act [8].

However, the sequential decision pattern of computer-aided process planning (CAPP) and production planning and control (PPC), commonly observed in today's production companies, decouples the technological and logistical considerations. This, in turn, limits the DoF to respond to unexpected events to the logistical scope of PPC. Existing approaches to extend these DoF by integrating production control with process planning promise additional potential for increased flexibility in job shop environments (Sect. 3.1). Despite these efforts, current research is focused on this narrow application scope, either concentrating on solving the resulting optimization problem [9] or the application of approaches in production companies [10]. This results in a lack of guidance for practitioners on how to identify their specific DoF and how to decide under which conditions this additional flexibility has a noticeable impact on performance.

To address this gap, this article expands the focus of integrated process planning and production control to a broader spectrum of production systems. Building on previous work to qualitatively identify use-case-specific DoF [11], this article aims to quantitatively assess the potential DoF in the face of characteristic turbulence patterns the company is faced with. This is achieved through a daydreaming-enhanced digital twin, advocating for a shift from problem-specific simulation experimentation to a continuous evaluation of virtual alternatives.

The goal is to enable production managers faced with complex interrelationships between decisions and their outcomes to assess the effectiveness of the current flexibility and the potential for further flexibility through redesign of the production system in the face of potential perturbations. This facilitates risk-based thinking and contingency planning in the context of production management [12].

The design science research method, selected to bridge the gap between relevance and rigor, guides this work [13]: First, to ensure relevance, the problems and opportunities of the application domain have been described in this section. Next, rigor is ensured by describing the knowledge base in Sects. 2 and 3. Based on this, the developed two-phase method is introduced in Sect. 4. The evaluation of the developed method and its artifacts is described in Sect. 5.

2 Decision Functions in CAPP and PPC

The decision functions of process planning, focused on production technology, and production control, focused on logistics, are closely intertwined by the information required in the decision-making process and the objects to be planned [14]. For example, the results of the sales and operations planning (logistical perspective) determine the design of a production system, its facilities, and resources (technological perspective). Downstream, the bill of materials and the routing sheet (technological) form the basis for material and capacity planning (logistical). The resulting production order (logistical), in turn, forms the basis for the realization of the production process (technological). These decisions ensure the product quality and shape the achievable logistical performance, and consequently, the cost structure of a production system [15].

To this point, technological and logistical decisions in industrial practice are often made sequentially and decoupled from each other (Sect. 2.1). However, as shown in Fig. 1, increased flexibility and transformability of production systems change the decision structure, requiring an integrated technological and logistical perspective for decision-making (Sect. 2.2).

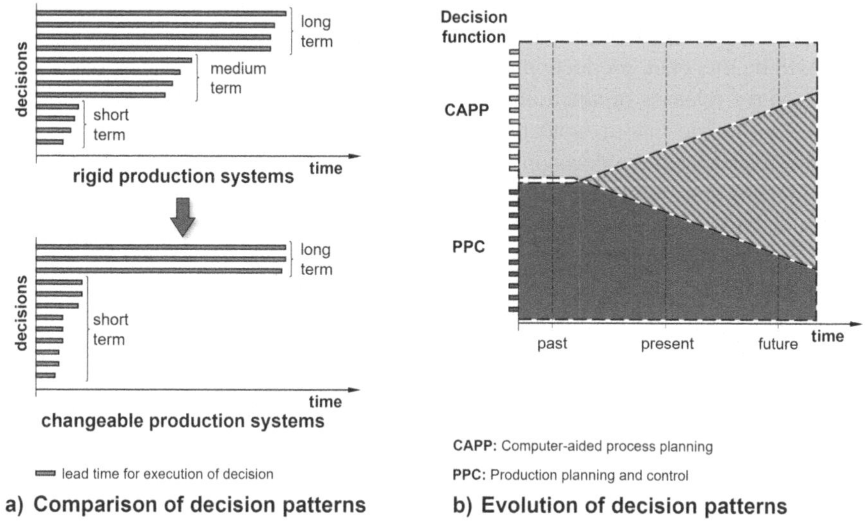

Fig. 1. Decision patterns of rigid and changeable production systems.

2.1 Decision Patterns in Rigid Production Systems

The traditional decision process from customer or market requirements to product engineering and production is characterized by sequential decision-making steps, each limiting the DoF for the subsequent steps. CAPP and PPC decide the fundamental questions of how, when, and by which resource products are manufactured.

Based on functional requirements and design features, process planning determines the required production features, translates them into operations, resources, and parameterizes them from a technological point of view [16]. ELMARAGHY divides the planning activities into a hierarchy with four levels of granularity [16]: On the coarsest level, the generic planning determines suitable technologies and processes. Next, the macro planning level determines the operational routing and selects corresponding resources. The detailed planning level further determines the step sequence within a resource, along with the required tools and fixtures. Finally, the micro planning level determines the parameters and develops control programs.

Previously, the term computer-aided process planning (CAPP) has been used to distinguish between manual planning and automated computer-aided planning approaches.

However, advancements in the digitalization of process planning now require a more nuanced understanding. Therefore, expert-based process planning using computers (CAPP) and automated computer-aided process planning (ACAPP), based on decision models, are distinguished [17]. Despite the digitalization of the design and planning processes, the availability of digital part models, and many existing research approaches, the challenge of fully automating process planning remains unresolved [17].

Utilizing the routing information, along with the bill of materials, PPC transforms market demands into production orders and procurement orders to ensure their fulfillment at the required time and in the required quantity [18]. To fulfill these requirements, production planning determines demands based on which material and capacity are planned [19]. Based on this plan, production orders and procurement orders are generated. Production control releases production orders, monitors order progress, and ensures the plans are fulfilled despite unavoidable perturbations [15]. Besides the release strategy, this is ensured through the decision functions of capacity control, order and operation sequencing, and scheduling [15].

Turbulent market environments drive the demand for changeability in production systems [8]. Increased technological and logistical changeability lead to a shift of traditional decision patterns of CAPP and PPC, as described in the next section.

2.2 Decision Patterns in Changeable Production Systems

Originating in American military literature, authors in supply chain management are adopting the term VUCA to describe the environment for global supply networks in the context of the disruptions experienced within recent years [20].

Increasing production companies' resilience capabilities has gained recognition in the literature as an antidote to this volatility [21]. Resilient production systems are characterized by: 1. Anticipation to prepare for events before they occur, 2. Robustness to respond to minor perturbations, 3. Agility and flexibility to face major perturbations, and 4. Adaptability to recover and adjust to shifted conditions, also described as transformability [1]. Transformability and flexibility, as two enablers for resilience [1, 22], are further detailed below.

Changeability describes the ability of a production system to perform "[…] *foresighted adjustments of the factory's structures and processes on all levels to change impulses economically*" [23]. Structured by five product levels and five production levels, the following five classes of changeability are distinguished: Changeoverability, reconfigurability, flexibility, transformability, and agility [23]. Flexibility and transformability characterize tactical changes on the segment level (former) and the site level (latter):

- **Flexibility** describes the change of production processes, material flows, and logistical functions to adapt to similar product families with little time and effort.
- **Transformability** describes medium to long-term decisions for structural changes of the production and logistics systems, as well as the facilities and buildings, from the factory planning perspective.

Increased flexibility enables higher DoF to be used for short-term technological decisions in CAPP and logistical decisions in PPC. Increased transformability, up to changeable production systems (e.g., fluid production systems [24]), reduces the effort for structural changes, and subsequently, the lead time to implement factory planning decisions. This enables short-term structural changes and has the potential to change the decision pattern from sequential and decoupled decisions of a more rigid production system to short-term transformations, as seen in Fig. 1a). Thus, the strict separation between the design phase, dominated by factory planning and CAPP decisions, and the operational phase, dominated by production planning and control decisions, will blend in changeable production systems, as illustrated in Fig. 1b). This opens up the possibility of extending the DoF through transformations. At the same time, the increased possibilities to (re-)act make decision processes more complex owing to the high number of alternatives. In addition, decision processes require more interdisciplinary competence through an evaluation from a technological perspective (factory planning and CAPP) and a logistical perspective (PPC). The following section provides an overview of the existing approaches used for integrating CAPP and PPC.

3 Related Work

This section first provides an overview of existing approaches for integrated CAPP and PPC (Sect. 3.1). In addition, considering that decision-making relies on timely and accurate information, Sect. 3.2 highlights current developments in the context of digital twins, promising a near-to-real-time image of the production system.

3.1 Existing Approaches for Integrated Process Planning and Production Control

Early works on integrated process planning and production control date back to 1985 [25]. Since then, the problem has been addressed from the domain perspective of production research (e.g., [10, 26]), which focused on the application aspects, as well as computer science (e.g., [9, 27]), which focused on the algorithmic aspects of the optimization problem. Three main categories of methodical approaches can be differentiated, which are characterized by the point in time of the process plan creation, the temporal relation of the process planning and production control, and the distinction between multi-stage planning and single-stage planning [28]: 1. Distributed process planning (e.g., [29]) is characterized by a two-phase approach of coarse technological planning and subsequent detailed planning. 2. Closed-loop process planning (e.g., [30]) develops the process plan incrementally in parallel to production. Technological and logistical aspects are jointly considered. 3. Non-linear process planning (e.g., [31]) describes the process plan as a network to model potential technological alternatives and then selects the best path based on logistical criteria.

Various methods instantiated with different characteristics have been developed for each of these categories. A more in-depth introduction is provided by the books of Li [28] and Phanden [32], as well as several literature reviews (e.g., [33, 34]).

Apart from the planning method, different combinations of technological and logistical DoF are considered. Commonly considered are job shop environments with flexible production routes, enabling the use of alternative resources and alternative technologies [10, 26]. Additionally, some authors consider alternative parameters (e.g., [35]).

3.2 Digital Twins and Daydreaming in Production Management

Decision-making requires the availability of accurate and timely information. This poses a challenge, as technical, engineering-related data and logistical data are often fragmented across different IT systems. Furthermore, data quality is often insufficient, either due to a lack of timeliness or consistency.

The broader availability of networked sensors for large-scale data acquisition in the context of Industry 4.0 and the demand for timely information fostered the idea of a real-time digital image of a production system, a digital shadow, by linking this fragmented data [36]. Digital twins offer a further integration between the physical object (production system) and its digital representation, with a bidirectional data flow between them and the possibility for simulation [37]. Initially being focused on data availability and modeling, the current research focus is on extending the cognitive capabilities of digital twins through a combination with artificial intelligence to derive actions (e.g., [38]) and obtain insight into a potential future through foresighted [39] and resilient digital twins [40].

The concept of daydreaming factories, introduced by NASSEHI, expands on these ideas by providing a framework to explore and learn from possible alternative futures through large-scale scenario generation based on domain randomization, thus achieving antifragility in production systems [41]. Addressing the trade-off between investigating relevant scenarios and the effort for their definition, MAIER introduces an approach for the semi-automatic scenario evaluation of turbulences in production systems [42].

To summarize, the methods developed for integrated CAPP and PPC have been extensively researched. However, while existing approaches offer solutions on how to solve the integrated planning problem, they lack a detailed analysis of the DoF for other types of production systems, as well as a method for deciding under which circumstances (e.g., unexpected events) they might be beneficial.

At the same time, the latest technological developments in the context of digital twins offer the potential to virtually explore alternative scenarios without the risk of investment or interruption of production. However, more concrete use cases are required to establish their role as an integral part of decision-making in the context of production management.

Through a combination of both approaches, this work aims to address these limitations by systematically identifying the use-case-specific degrees of freedom and building on digital twins for a quantitative evaluation and, thereby, the possibility of a continuous evaluation of technological and logistical decisions.

4 Evaluation of Technological and Logistical DoF

Figure 2 shows the proposed method for identifying potential technological and logistical DoF, consisting of two phases: In the first phase, experts qualitatively assess the potential of available DoF. The second phase builds on this assessment and uses the concept of daydreaming to quantitatively evaluate the DoF in the face of use-case-specific turbulence patterns. The following sections describe the relevant technological and logistical degrees of freedom and introduce the individual steps (A, B, and C) of the two phases in detail.

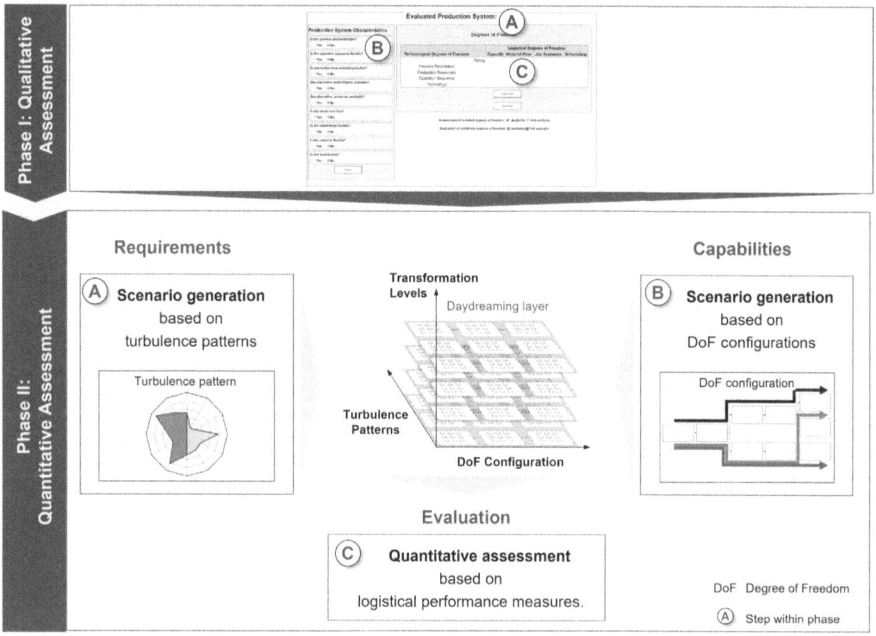

Fig. 2. Method for evaluating the potential of technological and logistical DoF.

4.1 Technological and Logistical Degrees of Freedom

First, customer requirements and regulatory restrictions limit the degrees of freedom. The characteristics of the production system, regarding the technological and organizational capabilities, further shape the available technological and logistical DoF and, thereby, the resulting flexibility. Figure 3 provides an overview of how nine characteristics of a production system shape the flexibility through an n:n relationship with the DoF [11, 15, 16, 24]:

- Technological flexibility for process planning is achieved by the following four DoF: Process parameter, operation sequence, technology, and production resource.
- Logistical flexibility for production control is achieved by the following four DoF: Scheduling, job sequence, material flow, and capacity.

The DoF 'production resource' and 'scheduling', as well as the DoF 'operation sequence' and 'material flow', show a high overlap in terms of the considered objects. However, they are distinct based on the perspectives used in decision-making. The selection of the production resource is driven by technological considerations (e.g., process requirements). In contrast, logistical criteria (e.g., shortest queue length) drive the scheduling decision. Similarly, the operation sequence is determined by constraint relationships between operations from the process perspective, and the material flow evaluates the material handling flexibility, enabling alternative operation sequences [7]. These DoF build the foundation of the qualitative assessment, described in the next section.

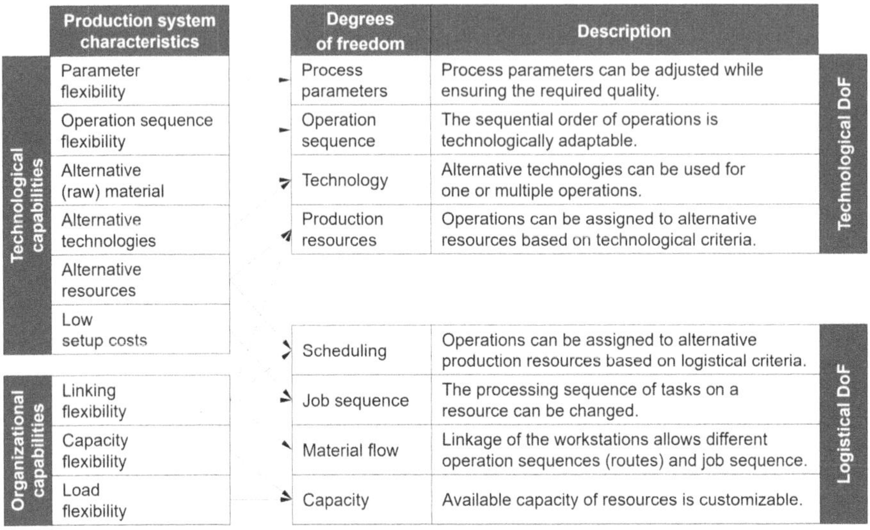

Fig. 3. Influence of the production system characteristics on technological and logistical DoF.

4.2 Phase I: Qualitative DoF Assessment

The initial qualitative assessment extends the evaluation scheme introduced in [11] by deducing the DoF from the characteristics of the production system acquired through a questionnaire, which is implemented in a software tool (Fig. 4). Besides visualizing the resulting DoF as a cross table, this also provides the basis for automatically configuring the model to perform the subsequent analysis in Phase II (Sect. 4.3).

The user interface contains two sections: First, an assessment section to capture the production system characteristics, consisting of the technological and organizational capabilities in Fig. 3. Second, a results section to visualize the determined DoF, providing orientation regarding which DoF can be used individually or in an integrated manner.

The assessment of the characteristics is binary: Each characteristic is either available (A) or unavailable (-). If all the characteristics required for a DoF are available, the DoF can be utilized. Similarly, if one DoF from the technological perspective and one from the logistical perspective are available, they can be considered in an integrated manner.

Technological domain experts from the company conduct the assessment, which is carried out according to the following three steps:

A. Define the object of consideration from a product and production perspective. Typically, a part or product family is considered (e.g., "shaft production").
B. Fill out the questionnaire regarding the characteristics of the production system.
C. Review the qualitative results regarding the available DoF.

Based on these results, the quantitative evaluation described in the next section provides further insight.

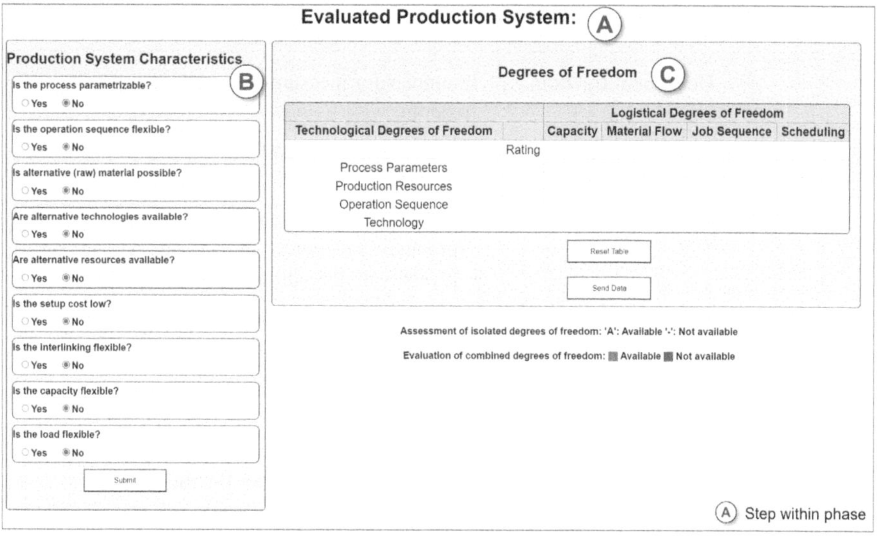

Fig. 4. Evaluation scheme for the qualitative assessment of DoF (software prototype).

4.3 Phase II: Quantitative DoF Assessment

The scenario generation and quantitative assessment requires a digital twin or simulation model of the production system that can be parameterized and reconfigured. If this is unavailable, automated modeling techniques can reduce the required manual effort (e.g., [42, 43]). Reconfiguration of the simulation model can be achieved through a modular modeling strategy with reusable logical and structural components. The model entities within the production system, but also the entities to model supply and demand must be parameterizable for the scenario generation. To assess the available DoF, the model has to support a reconfiguration of the control logic to support the different combinations of integrated CAPP and PPC. To assess extended DoF, the model has to support a structural reconfiguration, e.g., by changing the layout, which is often accompanied by a reconfiguration of the control logic.

The first step (A) of the second phase in Fig. 2 involves the generation of relevant scenarios that the production system may encounter. An overview of possible unplanned events for production systems is provided by the morphology of turbulence germs in [3]. Based on their turbulence profile and the potential impact, companies can prioritize the scenarios. An initial parameterization for the scenarios, e.g., regarding the frequency of machine breakdowns, must be provided by experts, as this is company-specific. Based on this, a large number of potentially disruptive scenarios can be generated either automatically using domain randomization [41] or semi-automatically by the experts [42].

For each of these scenarios, different configurations of the available DoF are tested in the second step (B). This is achieved by connecting the output of the identified DoF in Phase I to alternative model building blocks (e.g., assignment strategies or process plans) to implement these different configurations for DoF. Thereby, a plane of possible

Table 1. Measures to extend the available DoF.

	Degrees of freedom	Examples for measures to extend degrees of freedom
Technological	Process parameters	• Adjustments to enable different working speed parameters
	Production resources	• Consideration of technologically similar production resources • Investment into additional production resources
	Operation sequence	• Use of machines with increased flexibility • Use of flexible clamping devices • Adjustments of product design
	Technology	• Consideration of alternative production technologies
Logistical	Capacity	• Shift model with higher flexibility • Additional resources • Customer management for load balancing
	Job sequence	• Adjustments for increased flexibility of buffers between production resources
	Scheduling	• Additional resources • Adjustments for increased flexibility of linkage between resources
	Material flow	• Adjustments for increased flexibility of linkage between resources

experiments is created for potential turbulences in one dimension and configurations of integrated CAPP and PPC in the other dimension. To evaluate the DoF (C), each experiment is assessed by the weighted product of the four normalized logistical objectives: short throughput time, high delivery reliability, low WIP, and high utilization.

Based on these results, the extension of the available DoF through logical reconfiguration of the PPC system, structural or technological reconfiguration of the production system, or product redesign can provide further potential. Consequently, a third dimension in the scenario evaluation opens up. Available measures for reconfiguration or redesign are also company-specific. The PPC design matrix describes the three levels of reconfiguration of production planning and control systems [44]. Table 1. provides examples of measures to extend the DoF by structural or technological reconfiguration. These structural reconfigurations often require subsequent reconfigurations of the control system.

5 Application of the Potential Analysis

The job shop environment of a hypothetical individual and small series production company is used to demonstrate the practical application of the qualitative and quantitative assessment of potentials. The available machines, products, and process steps shown in Fig. 5 are inspired by the example described in [35] and extended using additional data from an industrial client with a job shop production system. The data consists of production orders extracted from a manufacturing execution system, to model the duration of operations and downtimes of machines tracked using tablets. The qualitative DoF evaluation and the automated scenario generation are implemented in Python. The multi-method simulation tool AnyLogic is used to execute the daydreaming approach. Based on a Python interface, the models are configured and parametrized.

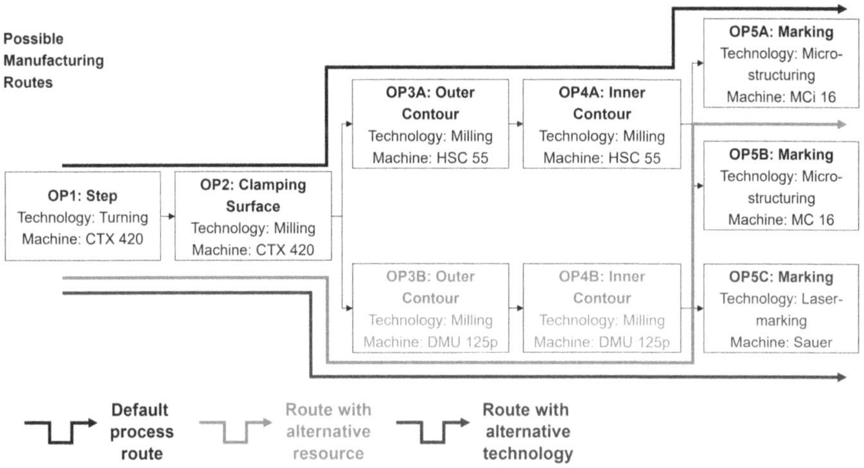

Fig. 5. Manufacturing routes in the application scenario.

From a technological perspective, alternative processes are available for individual operations. For example, the milling of the outer and inner contours can be performed on the HSC 55 and the DMU 125p. Two technologies are available for the marking step: Laser marking and micro structuring. The operations for the workpiece are technologically defined as a strict sequence.

Due to the flexible linking in combination with alternative resources, the DoF 'job sequences', 'material flow', and 'scheduling' are flexible from a logistics perspective. In addition, the company can adjust capacities by adding or removing shifts.

Based on these characteristics, a qualitative assessment is performed. The result is shown in Fig. 6. Except for the operation sequence, all DoF are available individually, and thereby also in combination. However, this high flexibility also generates high combinatorial complexity for decision-making.

The second phase validates the initial assessment and quantifies the effects. Figure 7 describes the three steps. First, the relevant turbulence germs are quantified (A). In this practical example, these are predominantly determined by demand fluctuations, unreliable processes, and heterogeneous replenishment times. After this expert-based modeling step, domain randomization is used to explore the search space further.

Twenty different parameter combinations are created. Additionally, three different DoF configurations of the production system are implemented in the model. The first uses a fixed process flow, the second uses alternative resources, and the third uses alternative

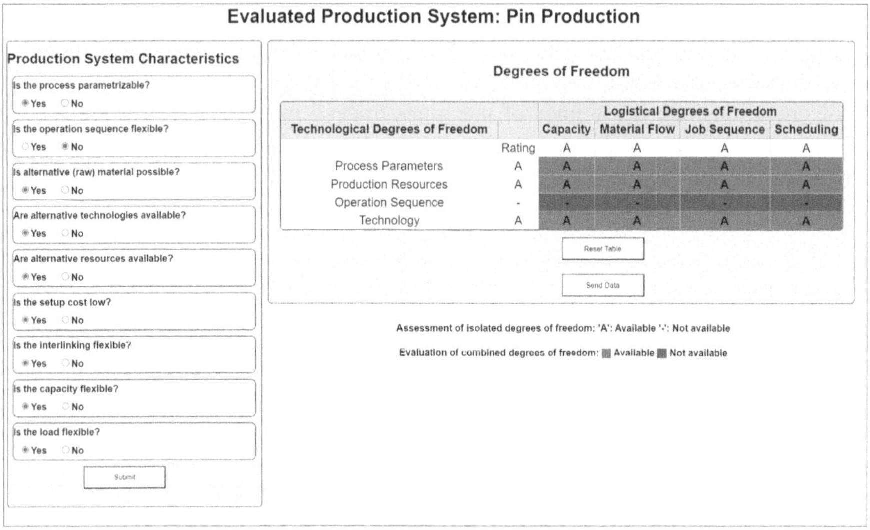

Fig. 6. Qualitative evaluation of the DoF (results of the practical example).

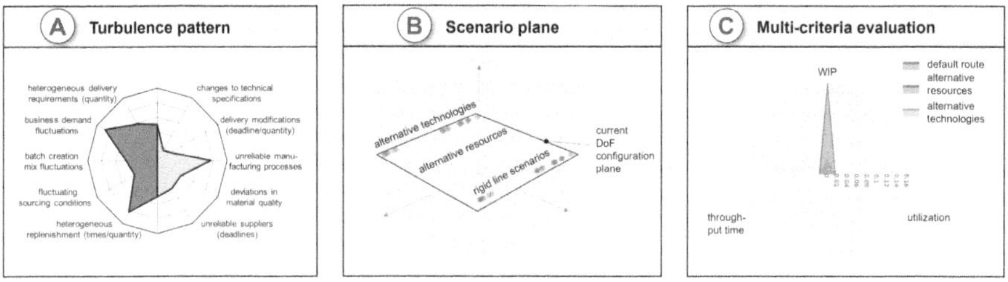

Fig. 7. Pipeline for quantitative evaluation (results of practical example).

technologies (see Fig. 5). Parameters and configurations span the scenario plane, as shown in Fig. 7 (B). For visualization purposes, the dimensionality of the parameters is reduced using principal component analysis. The simulation experiments are then executed automatically. The last step (C) involves the detailed evaluation of the results, to gather insight regarding weak spots and to develop mitigation strategies.

6 Summary and Outlook

The proposed method aims to help production companies increase their flexibility by identifying their available technological and logistical degrees of freedom and evaluating their potential. In the first phase, a qualitative assessment is performed. Building on this, the second phase assesses the potential quantitatively based on the flexibility requirements for the system. A short use case description exemplifies the approach. Based on the described characteristics enabling these DoF, production companies can be systematically transformed to extend their DoF. In a broader context, this enables production managers and planners to systematically analyze potential risks by evaluating the effect of flexibility measures under different turbulence patterns. Additionally, the potential of reconfigurations to extend the existing DoF can be evaluated using the same structure. This method serves as a building block to enhance resilience.

Further empirical research into the DoF used in production companies may uncover patterns of DoF that are commonly used for different production systems. Further practicalities of using this method should be investigated through an implementation study in an industrial company. Additionally, several steps of the quantitative evaluation still require manual effort. While the automation of the parameter generation is readily available, the automatic generation of production system configurations would further automate the approach and thereby extend the possibilities for scenario spaces.

Acknowledgments. The research presented in this paper was supported by the European Commission through the H2020 project EPIC (https://www.centre-epic.eu/) under grant No. 739592. This project has received funding from the European Union's Horizon 2020 research and innovation programme under grant agreement No 873086. The authors are responsible for the content of this publication.

References

1. Lanza, G., et al.: Global production networks: design and operation. CIRP Ann. **68**(2), 823–841 (2019)
2. Joglekar, N., Parker, G., Srai, J.: Winning the race for survival: how advanced manufacturing technologies are driving business-model innovation (2020). https://doi.org/10.2139/ssrn.3604242
3. Wiendahl, H.H.: Turbulence germs and their impact on planning and control – root causes and solutions for PPC design. CIRP Ann. **56**(1), 443–446 (2007)
4. Wiendahl, H.H.: Auftragsmanagement im turbulenten Umfeld: Teil 2 - Lösungsansätze. wt Werkstatttechnik online **96**(5), 325–330 (2006)
5. Heinicke, M.: Implementation of resilient production systems by production control. Procedia CIRP **19**, 105–110 (2014)
6. Fayezi, S., Zutshi, A., O'Loughlin, A.: Understanding and development of supply chain agility and flexibility: a structured literature review. Int. J. Manag. Rev. **19**(4), 379–407 (2016)
7. Wiendahl, H.-H.; Reichardt, J.; Nyhuis, P.: Handbuch Fabrikplanung. Konzept, Gestaltung und Umsetzung wandlungsfähiger Produktionsstätten. Hanser, München (2024)
8. ElMaraghy, H.A., Wiendahl, H.-P.: Changeability - an introduction. In: ElMaraghy, H.A. (ed.) Changeable and Reconfigurable Manufacturing Systems, pp. 3–24. Springer, London (2009). https://doi.org/10.1007/978-1-84882-067-8_1
9. Jin, L., Zhang, C., Wen, X., Christopher, G.G.: A neutrosophic number-based memetic algorithm for the integrated process planning and scheduling problem with uncertain processing times. IEEE Access **8**, 96628–96648 (2020)
10. Schmidt, B.C.: Integration von Arbeitsplanung und Fertigungssteuerung mit Netzarbeitsplänen. Fortschritt-Berichte VDI, Reihe 20, Rechnerunterstüzte Verfahren, vol. 226. VDI Verlag, Düsseldorf (1996)
11. Maier, J.B., Wiendahl, H.-H.: Potenzialanalyse technologischer und logistischer Freiheitsgrade der Produktionssteuerung. Integrierte Arbeitsplanung und Produktionssteuerung. wt Werkstattstechnik online **114**(6), 351–359 (2024)

12. ISO 9001: Quality Management Systems (2015)
13. Hevner, A.R., March, S.T., Park, J.: Design science in information systems research. MIS Q. **28**(1), 75 (2004)
14. Scheer, A.-W.: Wirtschaftsinformatik. Referenzmodelle für industrielle Geschäftsprozesse. Springer, Heidelberg (1997)
15. Wiendahl, H.P., Wiendahl, H.H.: Betriebsorganisation für Ingenieure. Hanser eLibrary, Hanser, München, 9., (2019)
16. ElMaraghy, H., Nassehi, A.: Computer-aided process planning. In: Laperrière, L., Reinhart, G. (eds.) CIRP Encyclopedia of Production Engineering, pp. 266–271. Springer, Heidelberg (2014). https://doi.org/10.1007/978-3-642-20617-7_6551
17. Al-wswasi, M., Ivanov, A., Makatsoris, H.: A survey on smart automated computer-aided process planning (ACAPP) techniques. Int. J. Adv. Manufact. Technol. **97**(1–4), 809–832 (2018)
18. Vollmann, T.E., Berry, W.L., Whybark, C.D., et al.: Manufacturing Planning and Control for Supply Chain Management, 5th edn. McGraw-Hill Irwin, Boston (2005)
19. Wight, O.W.: Manufacturing Resource Planning: MRP II. Unlocking America's Productivity Potential. Wight, VT: Essex Junction (1984)
20. Becker, T.: Supply chain strategies in the VUCA world. J. Supply Chain Manag. Logist. Procure. **5**(4), 372–389 (2023)
21. Benfer, M., Verhaelen, B., Peukert, S., Lanza, G.: Resilience measures in global production networks: a literature review and conceptual framework. Die Unternehmung **75**(4), 491–520 (2021)
22. Ihlenfeldt, S., et al.: Increasing resilience of production systems by integrated design. Appl. Sci. **11**(18), 8457 (2021)
23. Wiendahl, H.P., et al.: Changeable manufacturing – classification, design and operation. CIRP Ann. **56**(2), 783–809 (2007)
24. Fries, C., et al.: Fluid manufacturing systems (FLMS): a novel approach for versatility in production. In: Weißgraeber, P., Heieck, F., Ackermann, C. (eds.) Advances in Automotive Production Technology – Theory and Application, ARENA 2036, pp. 37–44. Springer, Heidelberg (2021). https://doi.org/10.1007/978-3-662-62962-8_5
25. Chryssolouris, G., Chan, S., Suh, N.P.: An integrated approach to process planning and scheduling. CIRP Ann. **34**(1), 413–417 (1985)
26. Sielemann, M.: Integration von Arbeitsplanung und Fertigungssteuerung bei autonomer, kooperativer Fertigung. Fortschritts-Berichte, Reihe 2. Nr. 535. VDI-Verlag, Düsseldorf (2000)
27. Wen, X., Li, X., Gao, L., Wang, K., Li, H.: Modified honey bees mating optimization algorithm for multi-objective uncertain integrated process planning and scheduling problem. Int. J. Adv. Rob. Syst. **17**(3), 172988142092523 (2020)
28. Li, X., Gao, L.: Effective Methods for Integrated Process Planning and Scheduling, vol. 2. Springer, Heidelberg (2020). https://doi.org/10.1007/978-3-662-55305-3
29. Huang, S.H., Zhang, H.C., Smith, M.L.: A progressive approach for the integration of process planning and scheduling. IIE Trans. **27**(4), 456–464 (1995)
30. Zhang, H.C., Merchant, M.E.: IPPM – a prototype to integrate process planning and job shop scheduling functions. CIRP Ann. **42**(1), 513–518 (1993)
31. Kempenaers, J., Pinte, J., Detand, J., Kruth, J.P.: A collaborative process planning and scheduling system. Adv. Eng. Softw. **25**(1), 3–8 (1996)
32. Phanden, R.K., Jain, A., Davim, J.P. (eds.): Integration of Process Planning and Scheduling: Approaches and Algorithms. CRC Press, Boca Raton (2019)
33. Varela, M.L.R., Putnik, G.D., Manupati, V.K., Rajyalakshmi, G., Trojanowska, J., Machado, J.: Integrated process planning and scheduling in networked manufacturing systems for I4.0: a review and framework proposal. Wirel. Netw. **27**(3), 1587–1599 (2021)

34. Shen, W., Wang, L., Hao, Q.: Agent-based distributed manufacturing process planning and scheduling: a state-of-the-art survey. IEEE Trans. Syst. Man Cybern. Part C (Appl. Rev.) **36**(4), 563–577 (2006)
35. Schmidt, J.: Integrierte Arbeitsplanung und Fertigungssteuerung auf Basis von Zustandsinformationen. Berichte aus dem IFW, vol. 2015, 8 PZH Verlag (2015)
36. Bauernhansl, T., Hartleif, S., Felix, T.: The digital shadow of production – a concept for the effective and efficient information supply in dynamic industrial environments. Procedia CIRP **72**, 69–74 (2018)
37. Kritzinger, W., Karner, M., Traar, G., Henjes, J., Sihn, W.: Digital twin in manufacturing: a categorical literature review and classification. IFAC-PapersOnLine **51**(11), 1016–1022 (2018)
38. Rožanec, J.M., et al.: Actionable cognitive twins for decision making in manufacturing. Int. J. Prod. Res. **60**(2), 452–478 (2021)
39. May, M.C., Overbeck, L., Wurster, M., Kuhnle, A., Lanza, G.: Foresighted digital twin for situational agent selection in production control. Procedia CIRP **99**, 27–32 (2021)
40. van der Aalst, W.M.P., Hinz, O., Weinhardt, C.: Resilient digital twins: organizations need to prepare for the unexpected. Bus. Inf. Syst. Eng. **63**(6), 615–619 (2021)
41. Nassehi, A., Colledani, M., Kádár, B., Lutters, E.: Daydreaming factories. CIRP Ann. **71**(2), 671–692 (2022)
42. Maier, J.B., Colangelo, E., Hinrichsen, T.-F., Tran, D.K., Wiendahl, H.-H., Huber, M.F.: Simulation discovery and semi-automatic scenario generation for evaluation of turbulence in production systems. Procedia CIRP (2024, in press)
43. Schlecht, M., de Guio, R., Köbler, J.: Automated generation of simulation model in context of industry 4.0. Int. J. Model. Simul. 1–13 (2023)
44. Wiendahl, H.H.: Adaptive production planning and control – elements and enablers of changeability. In: ElMaraghy H.A. (ed.) Changeable and Reconfigurable Manufacturing Systems, pp. 197–212. Springer, London (2009). https://doi.org/10.1007/978-1-84882-067-8_11

Material Shortages Propagation: Using Network Science to Evaluate Inventory Efficacy

Michele Martignago(✉), Martina Calzavara, and Daria Battini

Department of Management and Engineering, University of Padua, Stradella San Nicola 3, 36100 Vicenza, Italy
michele.martignago@phd.unipd.it, {martina.calzavara, daria.battini}@unipd.it

Abstract. The ability to manage the effects of supply chain (SC) disruptions has played a major role in industrial environments in recent years. Pandemics, natural disasters, and political tensions increased difficulties in procuring components, resulting in delays in delivering finished products and thus in lost sales, along with other relevant effects that strained companies' survivability (e.g. delayed financial flows). Companies have tried to avoid material shortages by increasing inventory levels, with increased costs and further risks (e.g., inventory devaluation, obsolescence), but with hard-to-evaluate benefits. Efficient resilience requires these cost-benefit analyses: their computation is even harder in today's complex and intertwined supply networks (SN), where it is also hard to align SC players with contrasting goals. In this paper, we aim to understand the effect of increasing inventory levels in two different network types. Hence, we simulate the shortages of materials propagating through the SN. For the first time in this area of study, we extend the common SIR (susceptible, infectious, recovered) model to a SEIR (susceptible, exposed, infectious, recovered) model, where the state 'E' helps in modelling the time-dependent effect of different inventory levels to preserve operations. This paper's practical value lies in offering a way for professionals to estimate the benefit coming from the investment in higher inventory levels, towards a choice that balances resilience and efficiency. On the theoretical side, we show that increasing inventory levels is more effective in SNs corresponding to scale-free network type (closer to automotive industry), compared to SNs corresponding to small-world type (typical of electronics industry).

Keywords: supply chain network · viability · ripple effect · stockpiling · simulation · risk

1 Introduction

Supply chains (SCs) strive towards being profitable and being able to overcome disruptions, that is to be both efficient and resilient [1]. This is easier said than done, since the disruptions to overcome are often unprecedented and unforeseeable, meaning that they may arrive anytime soon. Dealing with at least one of the unknown characteristics of

Published by Springer Nature Switzerland AG 2024
M. Thürer et al. (Eds.): APMS 2024, IFIP AICT 733, pp. 138–152, 2024.
https://doi.org/10.1007/978-3-031-71645-4_10

disruptions (unknown cause, unknown timing, and unknown effects, [2]) is a complex task, but the huge economic and social relevance of those phenomena push companies and researchers towards this area of study. Supply network (SN) is a term used to stress the relevant number of players of some SCs, with wide geographical extension and the intertwined links among some of them. In SNs, relationships between players are multiple, diverse, and mutable, in contrast to simpler and linear SCs. The term SCN (supply chain network) is often used to encompass these two definitions [3].

In times of disruption risks, as a first reaction many companies enhance inventory levels [4], even more for products subject to panic buying [5]. This is a curious case where a solution from one problem is used to alleviate the symptoms of another. In fact, warehouses and higher inventory levels are a way to cope typically with operational risks [6], that are high-frequency and low-effects SC risks, such as periodic oscillations of demand, transportation times, production mixes, besides being a way to aggregate goods by convenient lot sizes and reduce distribution costs. Disruption risks are different because they come from unpredictable low-frequency and high-impact events [7]. After a disruption, the system may even find a new equilibrium, different from the starting one: on this point, it is important to consider SC viability [8, 9]. While enhancing inventory levels may help against instantaneous and short-time disruptions, it is clear that it is expensive and of little help against long-lasting disruptions [10].

Increasing inventory levels has been extensively used even against disruption risks, and the reasons are easy to understand: first, a higher inventory level increases robustness (at least temporarily), that is the ability to withstand a disruption to maintain performance as planned. Resilience is more than this because it considers also the adaptation and recovery aspects [2]. Second, disruptions come with little or no notice, and typically managers do not have time or resources to change SN configuration, find and validate new suppliers, or redesign processes. This comes from the fact that some industries were not prepared for the many disruptions that hit the world in early 2010s and even less to the SC crisis represented by COVID-19 [11], and even after having experienced their harsh effects, they are often reluctant in investing in SN redesign with disruption consideration. Many managers are still attracted by an approach closer to the just-in-time (JIT) philosophy because it promises savings on a daily basis, compared to a just-in-case approach, that typically requires changes and investments that may payoff only years later, if ever [12, 13]. On the relationship between JIT and resilience, academic literature is still divided [14].

The goal of this paper is to understand at a network level how effective it is to increase inventory levels. Our research is in fact guided by the following question:

> Does the supply chain network type play a role in the effectiveness of improving resilience through increased inventory levels?

The basic idea of our study is that a higher inbound inventory level is correlated to longer temporal coverage for the single player to continue its operations, in case its supplier is disrupted and is not delivering components. We conduct this study using the network science approach also because some relevant SNs have so many nodes (representing locations) that can hardly be tackled exactly with a classical optimization approach [15, 16], especially if stochastic considerations or scenarios come into play.

We compare the effect of enhancing inventory levels against the material shortages on different network types: in fact, previous study found out that different industries' SNs can be represented by very different graph models (see Sect. 2). We apply for the first time the SEIR model (susceptible, exposed, infectious, recovered) to SCN resilience with focus on inventory level. SEIR is a well-known simulation model that has extensively been used in epidemiology and other fields. It represents a common extension of the SIR model: this latter has already been used in SN studies to study risks propagation [15, 17, 18], sometimes even allowing reinfection.

The results from our simulations show that the efficacy of increasing inventory level against the propagation of shortages of materials is influenced by the graph type. Besides general results, practitioners can use the same approach to implement their own simulations customized on their specific data for more detailed estimations. Managers can then collect data from their SN and consider the cost of increasing inventory levels and of opening or renting new distribution centers (DC), running costs, risks (such as goods perishability and inventory depreciation). This will help them in making a more informed and conscious choice towards efficient resilience on a tactical and operational level, where literature lacks of quantitative contributions at the tactical-operational level [11]. In fact, practitioners have little possibility to redesign different industries' SNs: what they can do is to act on some parameters such as inventory levels, especially in the short time. Even in long time-horizon, it is hard to change structural properties "without significant redesign and cooperation from major partners" [17].

The rest of the paper is organized as follows: Sect. 2 is for literature review. Section 3 is for model description and setting the parameters. Section 4 is for results and discussion. Section 5 is for the conclusion and for discussing the future research opportunities opened by this study.

2 Literature Review

Many studies in SCN design aim at finding the optimal solution against known threats, or against a set of possible disruptions. While giving interesting managerial insights and indicate the direction of the redesign process [19, 20], the implementation of these optimal solutions has drawbacks that are related to time. For example, changing sites' location, hiring and training workforce, setting up efficient processes cannot be done overnight, and they all represent sources of new risks. On the other hand, disruptions typically come with little notice, rarely allowing for the theorical best solution to be implemented. These time-dependent dimensions explain the attractiveness of increasing the average inventory level in disruption times. Higher inventory levels can thus increase the "time to survive", but even the "proactive component" of the "time to adapt" [21]. This once again motivates towards studies where the time-dependent behavior of material shortages is considered. A supplier's ability to recover and thus deliver components is function of time; in fact, the consequences of a disruption with fast recovery are negligible in mid-term perspective [22]. In short, disruption is also a function of time and adaptability [22].

This article is based on many assumptions made in the literature by previous studies that introduced network science in the supply network [15, 17, 23–26]. In fact, the usual operational research approach has difficulties in analyzing complex networks with hundreds of nodes, while the network science provides convenient tools to analyze the macroscopic behavior of large-sized graphs, considering the complexity of the interaction between nodes [27]. For example, Basole and Bellamy [17] summarized how SNs with small-world characteristics decelerate risk propagation, while SNs closer to a scale-free network typically enhance risk propagation: this latter threatens the network's health more. For a simple example of what is meant by small-world and scale-free, see Fig. 1. Small-world networks are a good representation of what is found in electronics SCNs, while the automotive SCNs are more like a scale-free network [15, 17], and we will refer to them due to their importance. The correlation between industry type and network type comes from the processes required to transform and assemble the products. In small-world type, the variability of number of links across nodes is ideally null. For example, different networks can have approximately the same average degree *d*, that is the average number of links per node, and be classified as small-world type if almost all the nodes have the same number of links. By contrast, a network with the same average degree *d* where few nodes (hubs) have many connections, and many nodes (hubs) have few, will not be classified as small-world type; it will be classified as scale-free network if the relationship between the number of nodes and the number of connections can be approximated by an exponential law.

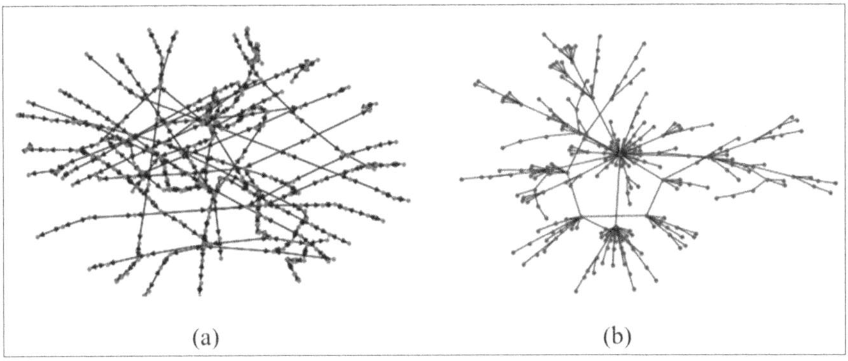

Fig. 1. Considered network types. **a** Small-world network type. **b** Scale-free network type

Li et al. [15] noted that real-life SCs are not always close to a particular type of theorical network. To close this gap with reality, they proposed to focus not on network type, but on network characteristics, such as degree centrality, betweenness centrality, eigenvector centrality, and more. Then they put those in correspondence with network SC resilience measures, such as size of the largest connected component (at initial and full impact) and minimum number of healthy nodes during the considered horizon, or recovery time. Other network science studies show that a scale-free network is in general more robust than small-world one against disruptions on random nodes only [25], but this property deteriorates against targeted disruption on key nodes (attacks) [24]. Like many other, Li et al. [15] used the susceptible – infectious – recovered (SIR) model, where the three letters represent the three consecutive states in which a node can be.

This comes from epidemiology studies, but has found a wide variety of applications where the interest is in how the phenomena spread from node to node, such as new product adoption, marketing [28], cybersecurity [29]. Sometimes slight modifications in application field or in the spreading logic led to different acronyms [28].

A simple extension of SIR is SEIR model, where the additional *E* stands for the exposed state. This has been widely used for example in recent COVID-19 pandemic analyses [30], where there is a window of time in which the infected person can transmit the disease, without noticing to be affected (pre-symptomatic, asymptomatic [31, 32]). This is crucial in epidemics modelling because the lack of symptoms makes it harder to isolate spreaders and, thus, to contain the spread of the disease.

3 Model Description

3.1 Problem Definition

In epidemiology, the transformation from SIR to SEIR model follows the relevant delay that some diseases present between infection and symptoms: the state *E* plays a crucial role in modelling a more realistic situation. Using the analogy with this delay, we aim at extending the SIR model to SEIR model in SCN studies, where the additional *E* is a state put into correspondence with the time for which the node is able to continue its internal operations after being affected by the disease, and thus with the available level of stocks. In the case we study, the disease is the material shortage that propagates through the SCN, and the ability of the node to continue its own operations is ensured by the inventory. The higher the inventory levels, the longer the nodes can continue their own operations while the suppliers are not delivering new material. After this time expires, the symptoms appear also on the node. In the epidemic case, a long permanence in state *E* is dangerous, because it helps the spread of the disease by hiding the spreaders, whereas in SCN a long stay in state *E* helps the operations of the nodes to continue.

Nonetheless, we decided to keep the same notation to better communicate that this parameter helps to better model some time-dependent behavior in terms of delay of symptoms and because node in *E* state may continue its operations when suppliers fail to deliver materials, but it is already exposed to the risk of shortages that will become clear once the inventory is depleted. The higher the inventory level (represented by a higher the value of the parameter λ), the longer it takes to run out of material. In the following, we describe the different states and how different parameters influence the process, as summarized in Fig. 2.

Susceptible (*S*): SC players in the *S* state have not yet been affected by the material shortage and thus operate normally. All nodes are in the *S* state before the initial disruption event happens. When the disruption happens, a subset of the players in *S* transits to the *E* state as in Fig. 2. Then the material shortages propagation process continues with the infection process: we labelled infection probabilities as α.

Exposed (*E*): players in this state experience a latent impact of the disruption: manufacturers in this state have not yet been forced to stop production, but their suppliers are not able to deliver new material in a short time. The longer a player can stay in the *E* state without falling into the infected state, the more time it has: to continue operations without damage, to allow its suppliers to recover, and to adapt. This underlines the importance of the parameter λ, that is in relation with the maximum possible number of periods in which a player can stay in the *E* state, because it describes how long the inventory allows operations to run normally. From this state, nodes can recover with probability β_1 at each period. If they fail to recover before running out of material to process, they transit to state *I*.

Infectious (*I*): players in this state are actively affected by the material shortage. For example, in a factory the lack of one component is enough to stop production of a product, or in a distribution center that runs out of stock of some material. An infected player also spreads the impact of its production halt downstream in the SCN to its customers with a certain probability. The probability is influenced by many factors, such as the diversity of components in the product portfolio. While in *I* state, a node propagates the material shortage to its direct customers that are in state *S* with probability α. From state *I*, nodes recover with probability β_2.

Recovered (*R*): players in this state have recovered from the exposition to the disruption, some avoiding the effects of it (by recovering through β_1 in Fig. 2), others after having fully experienced the material shortage (by recovering through β_2 in Fig. 2). The model of this paper follows two assumptions: uniform inventory policy across players, and average inventory level. In the following sections, these assumptions are discussed with reference to the results and future research opportunities.

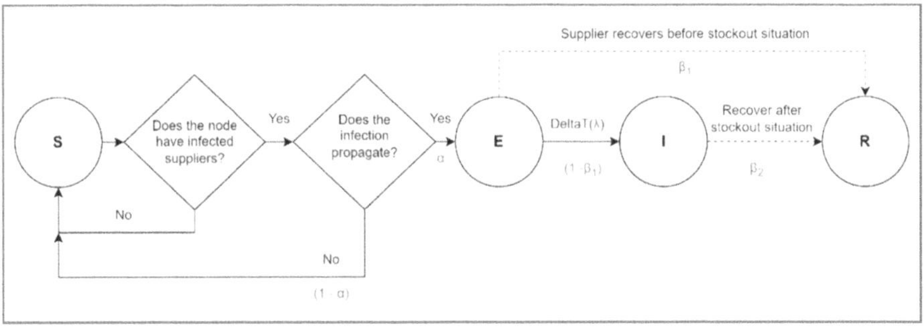

Fig. 2. Scheme of the SEIR model used in this paper

3.2 Parameters: Definition and Settings

The networks are generated from NetworkX Python package [33]. With reference to parameters description in Table 1, for small-world networks we used Watts-Strogatz model [33], that is f(*n*, *d*, *p*). For scale-free networks we used the Barabási-Albert model [33], that is f(*n*,*m*), where *m* is chosen depending on the desired *d* (see columns Notes in Table 1). The settings used lead to 96,000 total simulation runs.

Table 1. Parameters used for generation of graphs and for simulations.

Item	Description	Values	Notes
n	Number of nodes in a graph	$n = 250$	As in [15]
n_0	Number of initially infected nodes	$n_0 = 25$	As in [15]
d	Average degree of a graph	$d = \{2, 4, 6, 8\}$	As in [15]
m	Input for Barabási-Albert model [33]	$m = d/2$	$d = 2m(n-1)/n$ $\lim_{n \to \infty} m = d/2$
p	Input for Watts-Strogatz model [33]	$p = 0.25$	As in [15]
λ	Number of periods covered by inventory in case of disruptions	$\lambda = \{0, 2, 5, 10, 15, 20\}$	
α	Infection rate	$\alpha = 0.2$	As in [15]
β_1, β_2	Recovery rate from state E and from state I	$\beta_1, \beta_2 = 0.1$	As in [15]
G	Number of different networks tested per each of the 2 considered network types	$G = 50$	As in [15]
runs	Simulation runs on the same generated network with different set of disrupted nodes	$runs = 40$	[15] limited to 30 simulation runs

4 Results and Discussion

Metrics to evaluate SN resilience usually consider the largest connected component (LCC) after the disruption event, specifically its size, its average path and largest path. The use of these assumes that different nodes play similar roles: this is very interesting for general considerations and for some kind of networks (e.g., powergrids) but it is far from the scope of this paper and from the study of most industrial applications, as explained in [25]. Zhao et al. [25] used the largest functional subnetwork that considers the different roles of different nodes in a military SN. In contrast to [25], we focus on industries where typically one missing component in the bill of materials is enough to stop the whole production. Thus, the size of the sub-network the company is connected to is not relevant. In fact, operations may be halted even if just one of its key suppliers is not delivering components. Accordingly, we display results highlighting the number of uninfected nodes per each loop (Fig. 3, Fig. 4), the maximum number of infected nodes in a single period (Table 2), and the total number of nodes that were never interested by such shortages (Fig. 6). This latter reflects also how the network type combined with the increase warehouse levels impacted the shortage propagation. Given the fact that a recovered node cannot be re-infected, the risk can propagate in one direction only, that is following the supplier-customer relationships.

Table 2 is useful to analyze the different ranges of maximum infected nodes for equal values of d and λ in different network types. Differences between different network types are more relevant for smaller values of λ. This is because the effect of very high inventory levels successfully hinders shortages propagation in every case. Table 2 describes only a facet of the results but shows how different graphs with the same properties perform similarly against disruptions starting from sets of nodes that are different in each repetition. Showing that data from different repetition with the same parameters are contained within a very limited range, allows us to focus on the comparison between the average behavior across 40 runs showing the effects of changing the parameter λ on only one small-world network and only one scale-free network. This allows for a simpler presentation and better analysis. To support this, consider that in Fig. 3 and Fig. 4 we only report the average values across 40 repetitions for 1 of the 50 graphs that lead to the results from Table 2, per each of the 2 network types, and consider that Fig. 5 reports average values, the ranges of the standard deviations, and the ranges of minimum and maximum values across repetitions.

Figure 3 and Fig. 4 highlight how the performance in terms of maximum impacted nodes in one period is affected by the network type. In fact, this shows that small-word networks can contain the contemporaneous infection of nodes to lower values, compared to scale-free networks, for equal inventory levels.

In scale-free network type (Fig. 4), the material shortages quickly spread due to the presence of hubs (i.e. nodes with many connections) that affect many nodes, while small world network type (Fig. 3) allow for slower propagation due to lower variability of number of links per node, that is the absence of hubs. To avoid the maximum impact of the shortages, it is thus more important to raise the inventory level in scale-free networks compared to small-world networks. This suggests also to tend to change networks towards scale-free networks, in the re-design processes. From Table 2, we notice that the efficacy of higher inventory level affects the maximum number of infected nodes especially for networks with higher values of average degree, due to faster propagation through well-connected nodes. Small-world and scale-free networks tend to have smaller differences in terms of maximum nodes impacted at the same time when the average degree *d* increases.

For the scope of our investigation, Fig. 6 is even more important: it summarizes the total number of nodes that have never been affected by shortages propagations. From Fig. 6, we can see that in small-world type less nodes are impacted by disruptions, compared to scale-free type. A reason for the worst performance of the scale-free type is that in it the material shortages propagate quickly. In the range where λ is between 2 and 10, the small-world network type performs as well as a scale-free network that has an inventory that allows it to cover about 2 extra periods of time.

In other words, scale-free networks with the ability to preserve the same number of nodes as a small-world network are more expensive by the amount required to have an inventory able to cover 2 extra periods of time. This holds until the value of λ increases so much that it overrules the influence of network type, as in Fig. 6 for $\lambda = 15$ and for $\lambda = 20$.

Even if on average the values of never infected nodes are similar in different types of networks for higher values of λ, it is worth noting that the variability is higher for scale-free networks, and in particular the worst-case results make us prefer small-word networks. In fact, higher variability is related to higher uncertainty and unpredictability, and if other conditions hold equal, the solution that minimizes the effects of the worst-case scenario is preferred [23]. This once again comes from the fact that small-world networks have more regular distribution of links across nodes, while in scale-free networks even high levels of inventory make it less likely but not impossible the unfortunate event where many hubs are hit by material shortages and infect a lot of nodes.

Table 2. Statistics of θ = maximum number of nodes impacted by disruptions in one period (I state only), per each of the 40 runs with different starting disruptions. Δ is the difference between Mean(θ) in the case with small-world network types and with scale-free network types.

d	λ	Small-world network				Scale-free network				Δ
		Mean(θ)	*Min(θ)*	*Max(θ)*	*SD(θ)*	*Mean(θ)*	*Min(θ)*	*Max(θ)*	*SD(θ)*	
2	0	50.76	43	57	3.13	85.24	68	101	6.07	34.48
	2	39.84	35	47	2.71	64.98	54	77	5.82	25.14
	5	22.78	20	31	2.43	33.12	25	44	3.99	10.34
	10	14.14	12	17	1.17	15.04	12	22	2.41	0.90
	15	9.98	8	14	1.19	9.92	8	13	1.26	-0.06
	20	7.08	5	10	1.13	7.04	6	9	0.89	-0.04
4	0	112.00	104	123	4.06	134.98	126	149	5.39	22.98
	2	83.50	73	96	4.72	106.3	95	120	4.98	22.80
	5	37.38	32	48	3.46	53.86	42	66	4.81	16.48
	10	15.04	12	20	1.40	20.36	15	28	3.04	5.32
	15	9.88	8	13	1.23	10.80	8	18	1.72	0.92
	20	7.00	6	9	0.85	6.80	5	9	0.87	-0.20
6	0	144.96	135	155	4.21	155.72	147	167	4.71	10.76
	2	111.82	103	121	4.51	124.92	116	140	4.58	13.10
	5	53.94	47	62	3.64	67.50	60	76	3.97	13.56
	10	18.66	15	23	1.99	26.48	22	35	3.23	7.82
	15	10.28	8	14	1.27	12.14	9	20	1.96	1.86
	20	6.88	5	9	1.03	7.48	6	11	1.12	0.60
8	0	161.58	154	172	4.37	170.00	161	179	4.13	8.42
	2	129.80	120	146	5.46	139.76	133	152	4.43	9.96
	5	65.50	59	81	3.88	77.20	69	93	4.96	11.70
	10	23.92	19	30	2.42	31.72	25	40	3.08	7.80
	15	11.18	9	14	1.24	14.38	11	23	2.15	3.20
	20	7.06	5	11	1.16	7.84	6	10	1.03	0.78

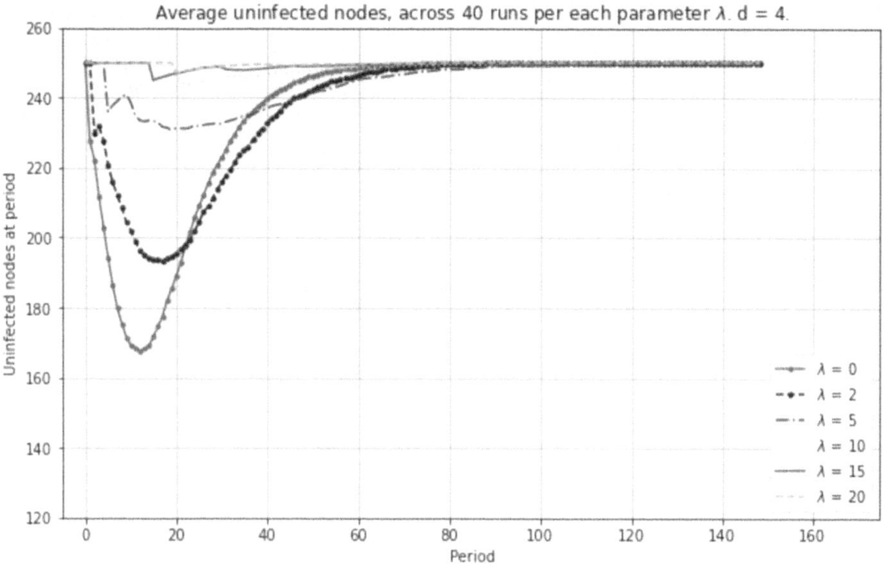

Fig. 3. Results for small-world network type

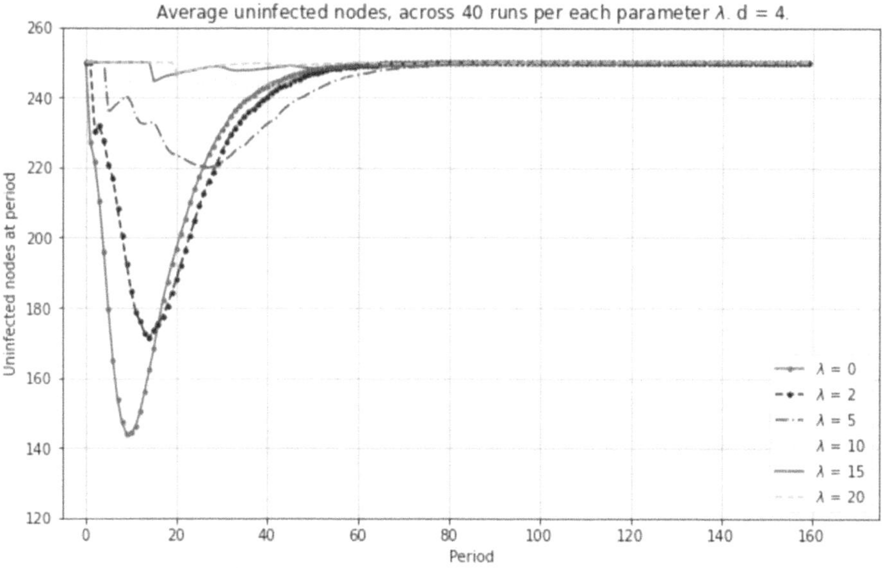

Fig. 4. Results for scale-free network type

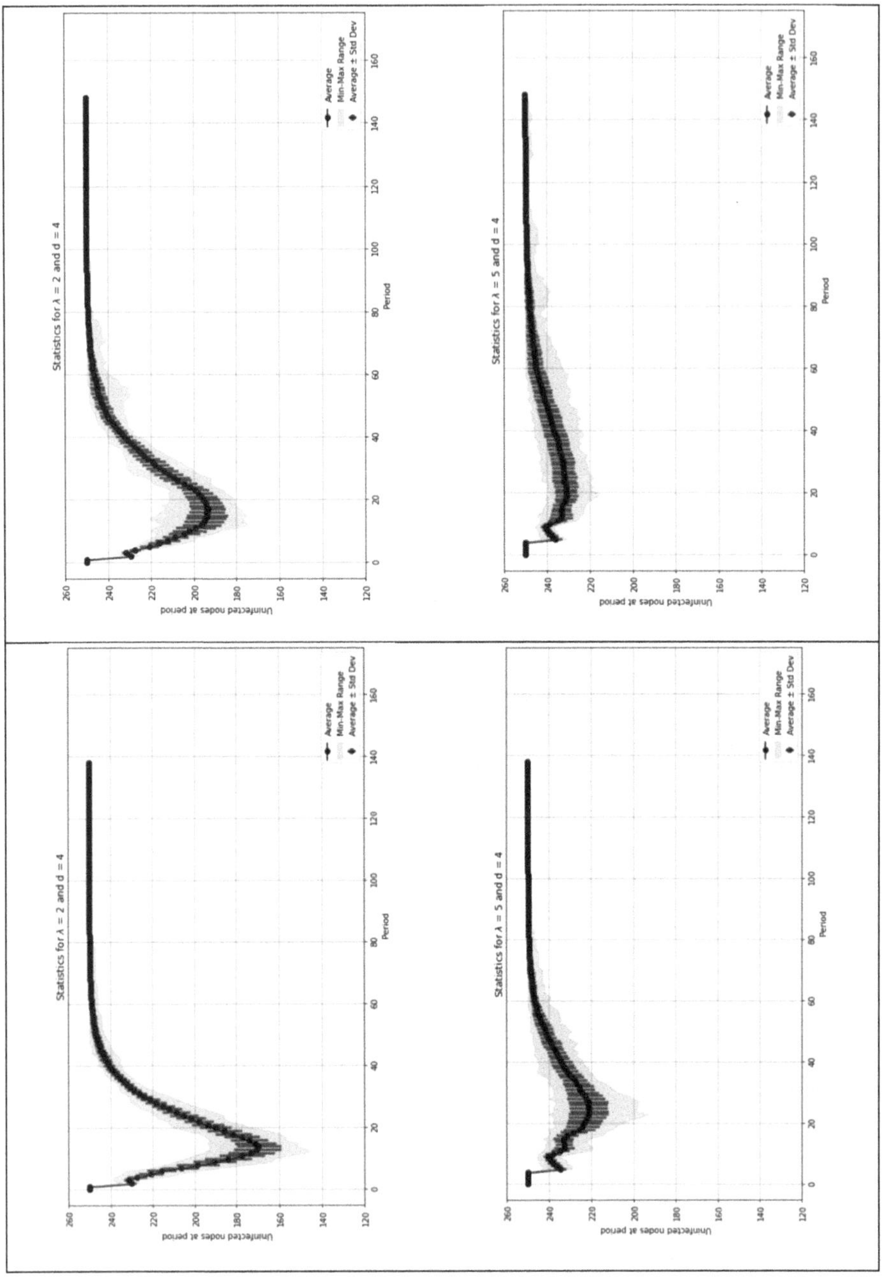

Fig. 5. Statistical dispersion of two selected results from small-world network type (top) and for scale-free type (bottom). 40 runs with disruption starting from different nodes.

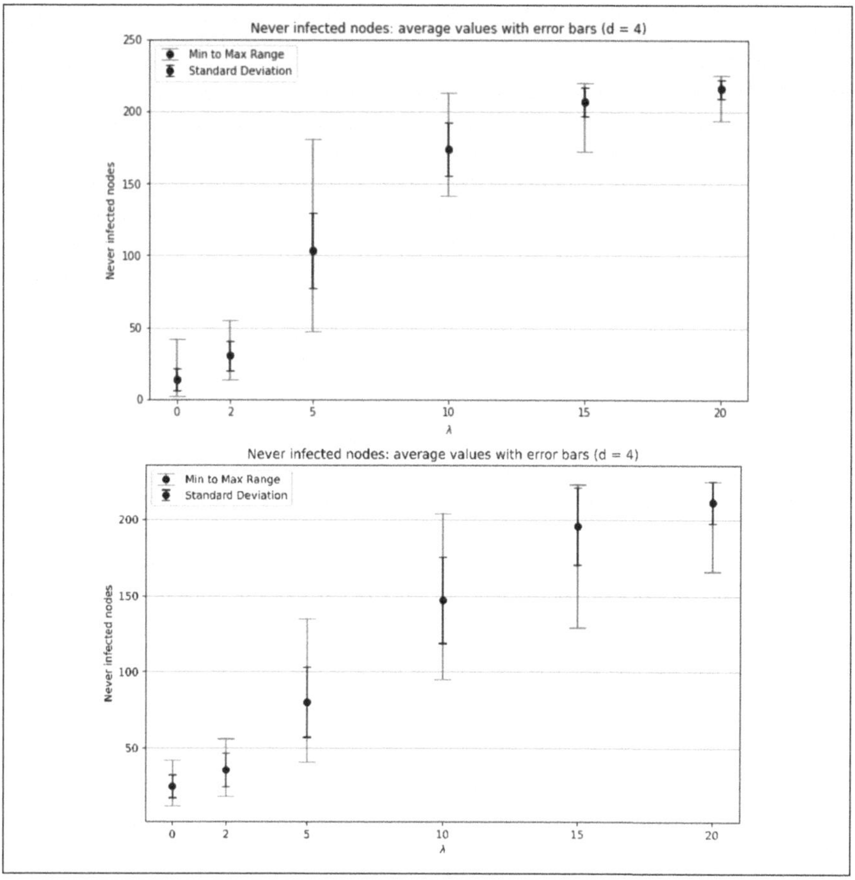

Fig. 6. Number of nodes that are never infected, as function of parameter λ. Top: small-world network type. Bottom: scale-free network type

5 Conclusions

In this study, we used network science and SEIR simulation model to investigate the influence of increasing inventory levels to mitigate the effects of material shortages propagations. The results can help in understanding to which level of inventory investments are efficient, and when further investment does not provide sensible improvement in terms of resilience. From a practical point of view, this can help managers in assessing the need for further investments in inventory levels. Managers must tailor these considerations to the specific case, keeping in mind the negative effects of using inventory as a buffer, especially for perishable or short-life products.

From a theoretical point of view, by repeating experiments on different kind of networks, we showed how small-world networks are in general better performing against material disruptions and that practitioners should put attention in increasing inventory in SN showing scale-free characteristics due to the ease of propagation of disruptions in this kind of network. We introduced measurements of performance in line with the specific interest of this study: we focused on the number of nodes that were not operating during the different periods, and on the number of nodes that, due to the network type and inventory level, never faced material shortages. Again, this suggests that managers redesign SNs to more closely resemble a small-world network, when possible.

Limitations of this study lay in the tools used and in the underlying hypotheses: we assumed average inventory level, not considering that a disruption can occur just shortly before the material arrival, where the inventory level is at its minimum, thus amplifying the negative effect of the material shortages. Moreover, we assumed that every player follows the same inventory policy (uniform inventory policy), while different companies could have in general different inventory management strategies. For further research, some of these facets could be investigated, understanding thus what is the impact of few players stockpiling: is it less effective than doing it in a coordinated way with all other players? Is it necessary to reach a certain threshold percentage of aligned players to have substantial increase against disruptions?

While roles are diverse in complex SNs, further studies can investigate the role that different nodes have in their cost-benefit analysis of raising inventory levels: inventory is cheaper for players far from final customers, due to the lower added value of products that underwent fewer operations. This could also help in defining more in detail propagations logic among different kinds of players. Moreover, further studies could understand the efficacy of stockpiling in terms of network characteristics instead of network types, in analogy with the work done by [15] for studying risks.

References

1. Aldrighetti, R., Calzavara, M., Martignago, M., Zennaro, I., Battini, D., Ivanov, D.: A methodological framework for the design of efficient resilience in supply networks. Int. J. Prod. Res. **62**, 271–290 (2024). https://doi.org/10.1080/00207543.2023.2285424
2. Ivanov, D.: Introduction to Supply Chain Resilience: Management, Modelling, Technology. Springer, Cham (2021). https://doi.org/10.1007/978-3-030-70490-2
3. MacCarthy, B.L., Ahmed, W.A.H., Demirel, G.: Mapping the supply chain: why, what and how? Int. J. Prod. Econ. **250**, 108688 (2022). https://doi.org/10.1016/j.ijpe.2022.108688
4. Karakatsoulis, G., Skouri, K., Lagodimos, A.G.: EOQ with supply disruptions under different advance information regimes. Appl. Math. Model. **125**, 772–788 (2024). https://doi.org/10.1016/j.apm.2023.08.012
5. Xu, Q., He, Y., Shao, Z.: Retailer's ordering decisions with consumer panic buying under unexpected events. Int. J. Prod. Econ. **266**, 109032 (2023). https://doi.org/10.1016/j.ijpe.2023.109032
6. Tang, C.S.: Perspectives in supply chain risk management. Int. J. Prod. Econ. **103**, 451–488 (2006). https://doi.org/10.1016/j.ijpe.2005.12.006
7. Akkermans, H., Van Wassenhove, L.N.: Supply chain tsunamis: research on low-probability, high-impact disruptions. J. Supply Chain Manag. **54**, 64–76 (2018). https://doi.org/10.1111/jscm.12162
8. Ivanov, D., Dolgui, A.: Viability of intertwined supply networks: extending the supply chain resilience angles towards survivability. A position paper motivated by COVID-19 outbreak. Int. J. Prod. Res. **58**, 2904–2915 (2020). https://doi.org/10.1080/00207543.2020.1750727
9. Ivanov, D., Keskin, B.B.: Post-pandemic adaptation and development of supply chain viability theory. Omega **116**, 102806 (2023). https://doi.org/10.1016/j.omega.2022.102806

10. Aldrighetti, R., Battini, D., Ivanov, D.: Efficient resilience portfolio design in the supply chain with consideration of preparedness and recovery investments. Omega **117**, 102841 (2023). https://doi.org/10.1016/j.omega.2023.102841
11. Batista, M., Ribeiro, J.P., Barbosa-Póvoa, A.: Supply chain resilience: tactical-operational models, a literature review. In: Almeida, J.P., Alvelos, F.P.E., Cerdeira, J.O., Moniz, S., Requejo, C. (eds.) Operational Research, pp. 157–177. Springer, Cham (2023). https://doi.org/10.1007/978-3-031-46439-3_12
12. Sodhi, M.S., Choi, T.Y.: Don't Abandon Your Just-In-Time Supply Chain, Revamp It (2022). https://hbr.org/2022/10/dont-abandon-your-just-in-time-supply-chain-revamp-it
13. Ye, Y., Suleiman, M.A., Huo, B.: Impact of just-in-time (JIT) on supply chain disruption risk: the moderating role of supply chain centralization. Ind. Manag. Data Syst. **122**, 1665–1685 (2022). https://doi.org/10.1108/IMDS-09-2021-0552
14. Choi, T.Y., Netland, T.H., Sanders, N., Sodhi, M.S., Wagner, S.M.: Just-in-time for supply chains in turbulent times. Prod. Oper. Manag. **32**, 2331–2340 (2023). https://doi.org/10.1111/poms.13979
15. Li, Y., Zobel, C.W., Seref, O., Chatfield, D.: Network characteristics and supply chain resilience under conditions of risk propagation. Int. J. Prod. Econ. **223**, 107529 (2020). https://doi.org/10.1016/j.ijpe.2019.107529
16. Lodi, A., Mossina, L., Rachelson, E.: Learning to handle parameter perturbations in Combinatorial Optimization: an application to facility location. EURO J. Transp. Logist. **9**, 100023 (2020). https://doi.org/10.1016/j.ejtl.2020.100023
17. Basole, R.C., Bellamy, M.A.: Supply network structure, visibility, and risk diffusion: a computational approach. Decis. Sci. **45**, 753–789 (2014). https://doi.org/10.1111/deci.12099
18. Kim, Y., Chen, Y.-S., Linderman, K.: Supply network disruption and resilience: a network structural perspective. J. Oper. Manag. **33–34**, 43–59 (2015). https://doi.org/10.1016/j.jom.2014.10.006
19. Aldrighetti, R., Battini, D., Ivanov, D.: Increasing supply chain resilience through efficient redundancy allocation: a risk-averse mathematical model. IFAC-PapersOnLine **54**, 1011–1016 (2021). https://doi.org/10.1016/j.ifacol.2021.08.120
20. Aldrighetti, R., Battini, D., Ivanov, D., Zennaro, I.: Costs of resilience and disruptions in supply chain network design models: a review and future research directions. Int. J. Prod. Econ. **235**, 108103 (2021). https://doi.org/10.1016/j.ijpe.2021.108103
21. Ivanov, D.: Two views of supply chain resilience. Int. J. Prod. Res. 1–15 (2023). https://doi.org/10.1080/00207543.2023.2253328
22. Ivanov, D.: Probability, adaptability, and time: some research-practice paradoxes in supply chain resilience and viability modeling. IJISM **15**, 1 (2022). https://doi.org/10.1504/IJISM.2022.10048806
23. Dixit, V., Verma, P., Tiwari, M.K.: Assessment of pre and post-disaster supply chain resilience based on network structural parameters with CVaR as a risk measure. Int. J. Prod. Econ. **227**, 107655 (2020). https://doi.org/10.1016/j.ijpe.2020.107655
24. Nair, A., Vidal, J.M.: Supply network topology and robustness against disruptions – an investigation using multi-agent model. Int. J. Prod. Res. **49**, 1391–1404 (2011). https://doi.org/10.1080/00207543.2010.518744
25. Zhao, K., Kumar, A., Harrison, T.P., Yen, J.: Analyzing the resilience of complex supply network topologies against random and targeted disruptions. IEEE Syst. J. **5**, 28–39 (2011). https://doi.org/10.1109/JSYST.2010.2100192
26. Albert, R., Jeong, H., Barabási, A.-L.: Error and attack tolerance of complex networks. Nature **406**, 378–382 (2000). https://doi.org/10.1038/35019019
27. Bellamy, M.A., Basole, R.C.: Network analysis of supply chain systems: a systematic review and future research. Syst. Eng. **16**, 235–249 (2013). https://doi.org/10.1002/sys.21238

28. Saadatniaki, F., Khan, U.A.: Product adoption in heterogeneous networks: an epidemiological perspective. Presented at the Conference Record - Asilomar Conference on Signals, Systems and Computers (2018). https://doi.org/10.1109/ACSSC.2018.8645221
29. López, M., Peinado, A., Ortiz, A.: An extensive validation of a SIR epidemic model to study the propagation of jamming attacks against IoT wireless networks. Comput. Netw. **165** (2019). https://doi.org/10.1016/j.comnet.2019.106945
30. Rahimi, I., Chen, F., Gandomi, A.H.: A review on COVID-19 forecasting models. Neural Comput. Appl. **35**, 23671–23681 (2023). https://doi.org/10.1007/s00521-020-05626-8
31. Ding, X., Huang, S., Leung, A., Rabbany, R.: Incorporating dynamic flight network in SEIR to model mobility between populations. Appl. Netw. Sci. **6** (2021). https://doi.org/10.1007/s41109-021-00378-3
32. Vo, M.V., Feng, Z., Glasser, J.W., Clarke, K.E.N., Jones, J.N.: Analysis of metapopulation models of the transmission of SARS-CoV-2 in the United States. J. Math. Biol. **87** (2023). https://doi.org/10.1007/s00285-023-01948-y
33. Hagberg, A., Schult, D., Swart, P.J.: Exploring network structure, dynamics, and function using NetworkX. In: Proceedings of the 7th Python in Science Conference (SciPy 2008), pp. 11–15 (2008)

Digital Twin Concepts in Production and Services

A Digital Twin Framework for Flexible Manufacturing System

Mokhtar Nizar Sid-Lakhdar[1,2](✉), Hichem Haddou Benderbal[3], and Mehdi Souier[1]

[1] Manufacturing Engineering Laboratory of Tlemcen (MELT), University of Tlemcen, Tlemcen, Algeria
mokhtarnizar.sidlakhdar@essa-tlemcen.dz, mehdi.souier@univ-tlemcen.dz
[2] Higher School in Applied Sciences, Tlemcen, Algeria
[3] Aix-Marseille University, University of Toulon, CNRS, Marseille, France
hicham.HADDOU-BEN-DERBAL@univ-amu.fr

Abstract. In the era of digitalization and automatization, several technologies and manufacturing paradigms emerged and became popular and attracted the interest of both researchers and industrials. Flexible Manufacturing System that is one of these paradigms is a production system able to switch between tasks easily making it adaptable to varying production needs. Furthermore, the concept of Digital Twin—a virtual counterpart of a tangible object—emerged and gained widespread popularity across diverse domains. Despite the focus of several research studies to develop these concepts, there is still a lack of work on an integrated and comprehensive framework that encompass a Digital Twin for Flexible Manufacturing System with a focus on smart manufacturing. Thus, this paper provides an attempt to propose a detailed Digital Twin framework for Flexible Manufacturing System with a focus on smart manufacturing and outlining various components and information. This is done by integrating different paradigms namely Machine learning; Simulation based Optimization, and Acquisition technologies. The proposed framework is composed of Physical part, Virtual part, and Stakeholders.

Keywords: Flexible Manufacturing System · Digital Twin · Smart manufacturing · Agile · Flexible and Reconfigurable Manufacturing

1 Introduction

The emergence of Industrie 4.0, a concept introduced in 2011 by German Professor Wolfgang Wahlster, has revolutionized manufacturing processes. This paradigm shift towards the digital transformation of the manufacturing sector is characterized by the seamless integration of advanced technologies such as the Internet of Things (IoT), Artificial Intelligence (AI), Cyber-Physical Systems (CPS), Cloud Computing, and Digital Twin (DT), among others. These technologies enable enhanced automation, interconnectivity, and real-time optimization of processes, fostering seamless communication and control within the manufacturing ecosystem [1].

Published by Springer Nature Switzerland AG 2024
M. Thürer et al. (Eds.): APMS 2024, IFIP AICT 733, pp. 155–167, 2024.
https://doi.org/10.1007/978-3-031-71645-4_11

Another important element of the Industrie 4.0 paradigm is the Flexible Manufacturing System (FMS), which represents an advanced production system linking machines, workstations, and logistics equipment [2]. In the literature, there are several definitions of the FMS. The majority of them, as Byrkett *et al.* [3] and O'Keefe and Kasirajan [4] focused on the hardware used in the system. Meanwhile Kaltwasser *et al.* [5] and Sarin and Chen [6] directed on the performance of the system [7]. This system integrates the entire manufacturing process under computer coordination [2]. FMS is designed to handle manufacturing tasks characterized by a wide range of types, high complexity, strict time requirements, and minimal costs [2]. It accommodates unpredictable production demands by organizing manufacturing into small batches and facilitating frequent changes as needed [2]. Therefore, the FMS decreases inventories, lead times, labor times, equipment cost, and floor space [8]. In addition, this system increases machine utilization, product quality, and financial benefits [8].

Additionally, the DT concept, serving as a dynamic virtual replica of a physical system, continuously updated with real-time data, represents a cornerstone of this technological revolution. This model utilizes simulation, machine learning, and reasoning to improve decision-making processes. The concept of the DT was introduced and developed by John Vickers in collaboration with Dr. Michael Grieves in 2002. It offers numerous advantages, including the ability to predict and optimize system behavior, perform continuous monitoring through real-time data acquisition, and enhance the security, resilience, performance, and efficiency of a system. Furthermore, employing DTs can lead to reduced costs and risks while significantly improving decision-making capabilities [9–11].

In the context of integrating the concepts DT in FMS, studies have been published, each proposing distinct frameworks aimed at achieving specific objectives. For instance, Magalhães *et al.* [12] introduced a reference architecture designed to examine the accuracy of information and results shared between a FMS and its corresponding DT. Similarly, Coito *et al.* [13] developed a framework that simulates their workshop environment, thereby furnishing stakeholders with insightful data to inform optimal decision-making. Furthermore, Neto *et al.* [14] proposed a framework aimed at addressing challenges related to manufacturing flexibility, while Li and Chen [15] focused their framework on optimizing key performance indicators such as Makespan, Total carbon emissions, Total production costs, and the stability of Product quality.

Despite these advancements, there remains a notable gap which is, a universal and adaptable framework that provides a comprehensive overview of a DT for FMS, particularly one that emphasizes the intelligent aspects of manufacturing tailored to the specific needs of businesses or laboratories employing FMS and seeking to design and implement a fitting DT. This paper attempts to fill this gap by presenting a detailed DT framework for FMS. It offers a comprehensive description of both the DT and FMS components, while incorporating the perspective of stakeholders, based on the works of Magalhães *et al.* [12], Coito *et al.* [13], and Haddou-Benderbal *et al.* [18]. This approach aims to provide a holistic and actionable framework that can be adapted to the unique requirements of various manufacturing environments, thereby advancing the integration of smart manufacturing practices. Also, this approach puts in value supply chain elements that are the manufacturing environment and its virtual twin. As well as Stakeholders which are composed of suppliers and customers based on the work of Ivanov and Dolgui [28].

The rest of the paper is organized as follows; in Sect. 2, we highlight the related works. In Sect. 3, we explain our framework in detail. Section 4 is reserved to discuss the framework. The conclusions and future works are given in Sect. 5.

2 Related Works

In the multifaceted domain of smart and changeable manufacturing, concomitant with the evolving field of DT technology, research endeavors are extensive and diverse. Therefore, in this section, our emphasis is based on two key points: (i) FMS incorporating smart manufacturing aspects, and (ii) the nuanced development of DT frameworks tailored for smart and changeable manufacturing systems.

2.1 Smart and Changeable Manufacturing: A Focus on Flexible Manufacturing Systems (FMS)

During this decade, several works have been carried out in combining DT with smart and changeable Manufacturing Systems whether they are Flexible or Reconfigurable. For the FMS paradigm, Magalhães *et al.* [12] made a DT Framework for an FMS composed of a Computer Numerical Control, a Machine and two Robot arms based on ISA 95 reference model for industrial automation. This framework covers DT, FMS, and Encapsulation Interface. The DT part covers several layers like Model View, Data Storage, Data Analysis, Simulation etc. Also; the FMS part includes Cyber Physical System with Computer Numerical Control (CNC), Conveyor, and Robotic Arm, Physical controllers and Management system (ERP). Interface for Clients and Users and Analysis and Management System are also considered. However, they did not provide details about Process and Product Data which are important for the FMS, as well as information about Model Data and Knowledge Base for DT.

In addition, Coito *et al.* [13] made the DT, Real world and Stakeholders framework. The latter contains DT with Simulation and Scheduling, and then Stakeholders like Operator and Clients, and finally Real system. These three-parts share information with each other such as Scheduling decisions, and rules, Sensor and Current Data. Yet, they did not provide a wide description of the Real system as well as the DT and the Stakeholders.

Also, Neto *et al.* [14] made a DT general architecture of an FMS where they highlight Flexibility challenges and DT-based solutions in addition of modeling the production system composed of a Shop floor, in addition of Industrial engineering and Production planning modules and its digital model composed of Shop floor's process behavior model as well as information about machine and buffer. Still, the model does not bring FMS components such as CNC, Robots, Material handling and Storage system etc. as well as information about Product and Process Data in the FMS part. Also, the model does not provide information about simulation and optimization part as well as Knowledge base and Model data.

Furthermore, Li and Chen [15] made a DT-driven multi-memory process flexible job shop scheduling problem (MPFJSP). This framework covers two parts i.e. Physical workshop, and Virtual workshop. The two-parts swap information with each other via the Cyber physical systems (CPS). The Physical workshop includes Production data (Human, Machines, Task information…), Scheduling execution (Execution of Machines, Worker, Workpiece etc.). In addition, the virtual workshop comprises Virtual model (Model Data, Simulation, Updating…) and Virtual Scheduling (Production data, Optimization Execution, Scheduling scheme, and Simulation verification.). Nevertheless, they did not provide a detailed description of virtual and real workshops; they did not outline the FMS components in addition of Product and Process data.

Besides, Fan *et al.* [16] made a DT framework for machine tools based on Geometric information, Historical samples, and Object Attributes, Snapshot Collection, and Topological constraint. Still, the physical part of this model covers machine tools only; it does not cover the other FMS components such as robots, material handling and storage systems, control and communication system. In addition, the virtual part does not include information about model data.

Finally, Fan *et al.* [17] described through an architecture the information flow for Production planning and control of FMS based on its Asset Administration Shell, this architecture encompasses Stakeholders' elements i.e. Enterprise Resource Planning (ERP), Customer and other elements such as Manufacturing Execution System.

2.2 Digital Twin Frameworks for Smart and Changeable Manufacturing Paradigm

In the context of smart and changeable manufacturing system, Haddou-Benderbal *et al.* [18] made a modular framework of a Reconfigurable Manufacturing System (RMS) and its DT. In this framework, the physical part covers CNC machines, Robots etc. The virtual part encompasses different layers namely Raw data archive, Processing layer, and Optimization based Simulation. However they centered their framework on virtual part only, they did not offer a detailed description about the physical part.

Also, Tang *et al.* [19] proposed a four-layer Cyber physical system cooperative framework for DT-enabled RMS. This framework is based on ISA 95 and DT parts as well as RMS features. These layers are Physical Execution and Sensing Layer (PESL), Autonomous Control Layer (ACL), Monitoring and Management Layer (MML), and Knowledge Service Layer (KSL). Still, they did not schematize the description of the RMS.

Moreover, Yang *et al.* [20] offered a DT-based application that can quantitatively assess several aspects of reconfiguration for the RMS. This suggested application is made to realize responsiveness to quick Market changes through the Planning and operation stages for Manufacturing systems. Yet, the model does not provide a description about Optimization, Knowledge base and Model data in the DT part in addition of Layout configuration, Process and Product data in their manufacturing system.

Furthermore, Mo *et al.* [21] proposed a framework for managing manufacturing system reconfiguration optimization that affords a general methodology to achieve reconfiguration in various cases and uses the benefit of DTs to enhance the performance and the accuracy of the reconfiguration. This framework consists of three stages of reconfiguration in addition of flows and agents. But it does not provide a description about RMS's components as well as DT's components.

In addition, Leng *et al.* [22] made architecture of DT Monitoring and Simulation Integrated Platform (DTMSIP) for RMS. This architecture is composed of CPS Bus that connects each device and application, Physical shop floor containing machine tools with sensors; Shop-floor Service System that encompasses Monitoring Service and Manufacturing Execution System (MES) and affords many services to support the management and control of the shop-floor. And finally the DTMSIP as Cyber-shop floor which joins the function of production monitoring and simulation. Nevertheless, they focus only on Machine tools as component for RMS; they did not put into the value the process and product data. In addition there is no optimization module as well as key performance indicators, knowledge base in the DT part.

Besides, Caesar *et al.* [23] proposed DT Framework for reconfiguration management that covers all reconfiguration planning tasks and highlights how the operator will be supported depending on the components included in a solution. This framework is founded on the classic architecture of a DT covers Reconfiguration Trigger, Reconfiguration Planner, Decision Maker, Eastbound Interface (ERP/MES), Data Layer, and Reconfigurable Manufacturing System. Still, they did not provide Simulation based Optimization module in the virtual space, as well as Layout configuration, Process and Product Data in the RMS part.

Finally, Flores-García *et al.* [24] suggested a framework that highlights the application of CPS comprising Simulation-based multi-objective optimization (SMO) to tackle the dynamic allocation of resource of Material handling in the Configuration analysis (CA) of RMS and enhance operational performance in order to realize flexible Decision-making. This framework is divided into three layers: Digital Services that covers SMO for CA of RMS, Material Handling execution and monitoring, Cyber layer that includes Capturing Data, Standardizing, Processing, and Synchronizing Information, and finally a Physical layer with Machines AGVs etc. Nevertheless, they focus only on material handling without taking Machines' production features into consideration.

The literature indicated that there is lack of a detailed description of the FMS part in terms of configuration (by showing FMS components), and information and data flow (process and product data) as well as the DT part in the Smart manufacturing environment. In addition, the external environment covering the Stakeholders which is necessary for the survival of every business is often taken into consideration. Based on this information, we provide a detailed DT framework for FMS. The proposed framework integrates the principle components and data flow of the FMS. In addition; we provide a detailed description of the DT outlining key performance indicators and Product data. Furthermore, a Stakeholders part which interacts with the FMS's DT is added. This part covers important supply chain elements that are Distribution and Forecasting. This framework gives a general view about FMS component and information, important modules needed in its virtual twin, and Stakeholders with its relevant elements. In addition of how these three parts interact with each other.

3 The Proposed FMS-DT Framework

In this section, we are going to describe the proposed FMS-DT framework which is inspired from those presented in [12, 13, 18, and 28]. It is divided into three parts: The Physical part, The Stakeholders, and The Virtual part. The Framework is shown in Fig. 1. The Physical part is in green color, the Virtual part is in blue, and Stakeholders are in red. In addition, the bidirectional information flow between modules is represented in a brown color. While unidirectional information flow is in violet.

3.1 The Physical Part

The considered Physical part is an FMS. It receives instructions from the Virtual part. These instructions cover the optimal configuration of FMS resources in the shop floor, as well as the parts to be produced and their routing. The FMS is an automated system that contains three separated modules: Product Data, Process Data, and Resources. These modules concern information about products to be manufactured, production process, and the machines needed respectively.

3.1.1 Product Data

Product Data module covers information about finished products to be realized. Products might be used in various sectors such as heavy equipment (Commercial or Military), Aerospace (Commercial or Military), and Automotive industry [25]. This module collects information from Stakeholders about the forecasted demand through real-time data flow module. This module encompasses the Type which is types of manufacturing parts as well as units, prices and costs of each part. In addition of part's Bill of Materials, Assembly, and finally Components needed to design parts. Furthermore, this module transfers the information about the manufactured parts to Process Data module. These information concern part types, bill of materials and components.

3.1.2 Process Data

Process Data module includes information collected before and through FMS execution. It includes Routing and Scheduling of each part type, Planning of each part's operation into each resource. As well as Processing times of each FMS resources and Setup times. In addition of Machine workload and state, Reliability data, and Capacity data, those are features of each FMS resource. Additionally, Transportation task of pieces or parts through Material Handling System. Furthermore, Repair and Preventive Maintenance tasks must be considered and carried out by operators. The former is carried out in case of machine breakdown and the latter periodically before the FMS execution. Furthermore, this module sends information about routing, scheduling and planning of each part type to Resources module. As well as transportation, repair and preventive maintenance tasks.

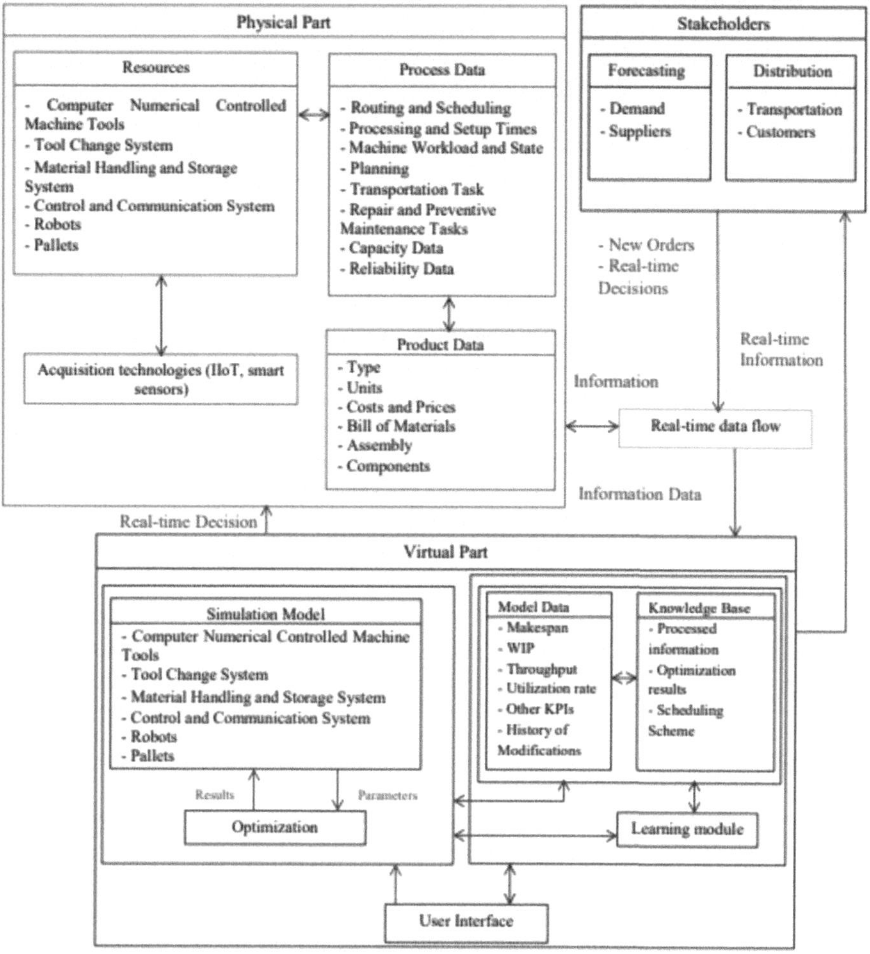

Fig. 1. The FMS – DT framework.

3.1.3 Resources

The Resources module covers the main components of the FMS according to MacCarthy and Liu [26]. These components cover at first Computer Numerical Controlled (CNC) machine tools which designed for diverse operations (milling, drilling, etc.). Also, Tool change system for changing tools during operations. Additionally, there are Material Handling and Storage system which is a set of Conveyors, Automated Guided Vehicles (AGVs) as well as Automated Storage and Retrieval System (AS/RS) for carrying out and storing parts during manufacturing. Moreover, Control and Communication System for synchronizing machine tools as well as work pieces [27]. Finally, there are Robots and Pallets for carrying pieces. In most FMSsetups, raw material workpieces are fixtured onto pallets at a load station separate from the machine tools [27]. These workpieces and pallets are then transported via a material handling system to queues at production machines for processing, with the flow directed by a central control computer [27]. Well-designed systems ensure minimal idle time for machines, as holding queues are typically occupied, and with short pallet exchange times, idle time is further reduced. Such systems usually comprise between two to twenty machines [27]. Furthermore, this module highlights the FMS resources as well as their disposition in the shop floor. We did not specify how many units of each FMS component do we need but in general FMS covers these components. This module receives information from Process Data module about routing and scheduling operations. As well as it receives instructions from the Virtual part through Acquisition technologies. These instructions concern the processes of scheduling or rescheduling. We should note that the scheduling in FMS pertain to efficiently managing the execution of production orders [29]. These challenges encompass sequencing the input of raw parts, scheduling machine and vehicle operations, monitoring system performance, and implementing corrective actions as needed [29]. Moreover, Rescheduling becomes necessary to integrate new orders into the existing production schedule with minimal disruptions and to maintain the efficiency of the current schedule [30]. This involves adjusting the schedule in a way that accommodates the new orders while minimizing the number of changes required [30]. In addition, this module sends back information to Process Data module about Reliability, Capacity, Workload and State. Further, it sends real-time information such as Manufacturing Lead Time of each part type to The Virtual part. It also sends other information about Workload and Reliability and Capacity of each Machine in the FMS. This is enabled through acquisition technologies such as Smart sensors and real-time data flow module.

3.2 Stakeholders

Stakeholders represent the external environment of the manufacturing system. Many researches have done about manufacturing systems. However, few of them take into consideration the external environment. This latter is important for each business's survival. Since each business manufacture products in order to sell them to Customers. In addition, it needs raw materials from suppliers to produce their products. In our study, we focus on the manufacturing process. The Stakeholders part contains two essential elements: Forecasting and Distribution. Forecasting involves estimated customers' demand during a period of time (In months) through a set of methods. It encompasses information about Demand and Suppliers [28]. On the one hand, information about Demand encompasses its type (deterministic or stochastic), the Quantity of products, as well as Order intervals [28]. On the other hand, information about Suppliers concerns their name and location [28]. Meanwhile, Distribution refers to the process of transporting the manufactured products to Customers through trucks in general. Additionally, Distribution covers information concerning Transportation of products and Customers [28]. Information about product transportation includes the routes from the Factory to Customers, the type of Vehicle and its Capacity, and Priority rules (FIFO, urgent etc.) [28]. Moreover, information about Customers involves their location, Delivery time, and Payment terms [28]. This module sends real-time decisions to both The Physical part and the Virtual part. These decisions involve the production of forecasted demand, urgent demand, or demand to be remanufactured from one or more Customers. Therefore, this part receives the batch of finished products from the Physical part in order to get distributed to Customers via vehicles. And it receives information from The Virtual part that the products have been manufactured.

3.3 The Virtual Part

The Virtual part here is the virtual twin of the FMS. It receives information from both Physical part and Stakeholders. This information is about current shop floor's state customer's demand, rescheduling process for an urgent demand from certain customers, or disruption data [28]. This part covers simulation model; optimization module, Learning module, Model Data, Knowledge Base, and User Interface. These modules are detailed hereafter.

3.3.1 Simulation Model

The Simulation model is the virtual representation of FMS resources' configuration using simulation software. This representation is done in order to test various scenarios of the FMS configuration virtually. It receives information from Stakeholders about customer's demand, urgent demand, and disruptive data [29]. Then, it shares this information as parameters with the Optimization module. This latter seeks to find the best solution to a mathematical problem for realizing the most favorable solution [18]. The solution concerns the best configuration, the best routing possible etc. Further, the Optimization module returns back the resulted configuration and resulted data [18] to the Simulation model.

3.3.2 Model Data

Model Data covers key performance indicators collected from current virtual representation of FMS configuration acquired from Simulation model. The key performance indicators include Makespan, Work in process (WIP), Throughput, Utilization rate and other key performance indicators (KPIs) like Overall equipment effectiveness and so forth; in addition of History of modifications concerning these indicators. This module sends the different information based key performance indicators values to the Knowledge Base module to be stored.

3.3.3 Knowledge Base

Knowledge Base module encompasses processed information, Optimization results, and Scheduling Scheme. Processed information contains various data and information gathered from both the physical part and the Stakeholders as well as the results learned from the learning module. Also, the results of Optimization of each KPI are stored in Optimization results. Then, the resulting configurations of FMS resources are stored in scheduling scheme. This module collects data from the Physical part (Number of finished products, number of products in storage…) the Virtual part (Processed Information, Optimization results, scheduling scheme…) and the Stakeholders (Production Quantity, Customer demand…). In addition, the bloc consisting of Knowledge Base and Model

Data (KMB) sends Input Data to the Simulation and Optimization Bloc (SOB) consisting of the Simulation Model and The Optimization module This Input data is information about the Resources' configuration. The SOB sends updated values of optimized Key Performance Indicators.

3.3.4 Learning Module

The learning module is a module that learns all information provided by the KMB as well as results provided by the SOB through Machine Learning techniques. This information encompasses the different scenarios of the virtual part of FMS and their outcome (key performance indicators, history of modifications etc.) as well as their comparison with data related to the real FMS. This learning enables better configurations and better key performance indicators like Outputs, machine utilization etc. Thus, this module enables adapting to shifts in market demand. As well as minimizing the costs associated with inefficient production, enhancing economic returns, and improving sustainable development capabilities [31].

3.3.5 User Interface

User Interface is an interface that enables the user (an Operator, or a Decision-maker) to get access to all data and information related to the Physical, the Virtual and the Stakeholders parts. It also enables the decision maker the ability to study new situations and their results and modify or introduce new parameters to the Virtual part [18]. It also enables the user to get précised data about the Physical part and Stakeholders. Furthermore, the user can get the results about the optimized configuration of the FMS.

3.4 Real-Time Data Flow

Real-time data flow module is the communication bridge between the Physical part, the Virtual part and Stakeholders. It can be protocols such as Modbus TCP/IP, Message Queuing Telemetry Transport (MQTT), or Open Platform Communications Unified Architecture (OPC UA). This module enables data and information transfer between the abovementioned three parts. This information covers the optimal configuration of the resources and part routing and scheduling sent by the Virtual part to the Physical part. It is enabled through acquisition technologies that transmit instructions to the Control and Communication System. This latter manages workstations individually [8]. It distributes control instructions efficiently [8]. It oversees production control for smooth workflow [8]. It coordinates traffic to optimize material flow. It controls shuttle movements for effective transfer [8]. It manages work handling and monitoring [8]. It monitors and reports system performance for optimization [8]. In addition, Real-time data flow module sends results' feedback about manufacturing lead time and machines' workload and state sent by the Physical part to the Virtual part. As well as Customers requests from Stakeholders to the Virtual part. Additionally, information about manufactured products is transferred from Physical part to Stakeholders.

4 Discussion

Since its appearance in 1960s, FMS is still becoming an important research subject. Over the years, there have been several works about FMSs and DT, but few of them focus on implementing DT in a real FMS. Furthermore, several frameworks were proposed concerning DT for FMS, still they do not put in value the major systems that make the FMS. Our framework fills this gap by highlighting the FMS components and the information we can get from them. This information involves Machine Workload and State, Routing and Scheduling, Capacity Data, and Reliability Data. In addition, there is a lack of consideration about enabling technologies of the DT. Consequently, the suggested framework covers several modules composing the DT namely Simulation, Optimization, Machine Learning, Real-time data flow, and Human Machine Interface through User Interface module. Moreover, there is a lack of consideration of the external environment which is necessary for an FMS. Our framework fills this gap by integrating the Stakeholders part. This part covers important information about Supply chain elements which are suppliers and customers. Furthermore, the framework encompasses branches of Production Management such as Planning, Scheduling, Maintenance, Optimization, Simulation, Forecasting, Distribution, and Machine Learning; illustrating data flow between each module and part in real-time.

This framework gives to FMS users, researchers as well as practitioners a general view about The FMS, its virtual counterpart, and the External Environment as well as the real-time data flow between these three environments. The advantage of implementing DT in FMS is to enhance its efficiency, increase its utilization rate, and to foster its ability to react efficiently to sudden market changes. This is enabled through the simulation and optimization modules as well as the learning module.

The challenge we face with this framework is implementing it in an industrial use case. In fact, there are use cases of implementing DT in a FMS. However, the majority of these cases were either simple case of FMS or laboratory case. While our framework is manufacturing-oriented, outlining Forecasting demand, Part type manufacturing, and Product distribution, the framework has yet to undergo real-world testing in a real FMS in order to test its efficiency as well as its responsiveness to market demands.

Although the framework puts forward the aspect of data gathering, learning, and storing in Knowledge Base module. It does not put forward data protection through the Cybersecurity paradigm. Another relevant limitation of the framework is the absence of human resources. According to literature, the role of human in an FMS is to undergo repair and predictive maintenance tasks as well as loading machines and adjusting tool change systems as well as programming the control system.

This framework gives a general view about the FMS, its virtual twin, and Stakeholders as well as the data flow between them. However, further studies are suggested concerning the integration of Human Resources as well as Cybersecurity in our framework.

5 Conclusion

This article is an attempt to integrate the Digital twin, Flexible manufacturing system, and supply chain elements in the Smart manufacturing environment. It offers a new industrial perspective of the FMS outlining its virtual counterpart and the external environment.

In this work, we have presented a DT framework for FMS, highlighting the crucial resources and information about the Physical part which is FMS. As well as the Virtual part where simulation model is put into value in addition of optimization based on key performance indicators, history of modifications, optimization results, and processed information. Moreover, the Stakeholders part which shares information with the Physical part and the Virtual part.

This framework can serve as an important decision-making tool for FMS users. It gives a detailed view about how DT can help an FMS in various operations. These operations involve Planning, Scheduling, Real-time monitoring, Forecasting demand, and Distribution to Customers. Also, this framework shows how DT can improve FMS's efficiency and productivity, making it more resilient thanks to Optimization module as well as Learning module. Finally, this framework displays how FMS can interact with its virtual twin as well as the external environment to the manufacturing system.

Further studies are needed in this work. First of all, a use case is necessary to validate our proposed framework. Additionally, we will take into consideration Human resources as well as Cybersecurity of data and information gathered in our framework and use case.

References

1. Santhi, A.R., Muthuswamy, P.: Industry 5.0 or industry 4.0S? Introduction to industry 4.0 and a peek into the prospective industry 5.0 technologies. Int. J. Interact. Des. Manufact. (IJIDeM) **17**, 947–979 (2023)
2. Florescu, A., Barabas, S.A.: Modeling and simulation of a flexible manufacturing system—a basic component of industry 4.0. Appl. Sci. **10**, 1–20 (2020)
3. Byrkett, D.L., Ozden, M.H., Patton, J.M.: Integrating flexible manufacturing systems with traditional manufacturing, planning, and control. J. Prod. Inventory Manag. **29**, 15–21 (1988)
4. O'Keefe, R.M., Kasirajan, T.: Interaction between dispatching and next station selection rules in a dedicated flexible manufacturing system. Int. J. Prod. Res. **30**(8), 1753–1772 (1992)
5. Kaltwasser, J., Hercht, A., Lang, R.: Hierarchical control of flexible manufacturing systems. In: IFAC Information Control Problems in Manufacturing Technology, pp. 37–44. Suzdal USSR (1986)
6. Sarin, S.C., Chen, C.S.: The machine loading and toll allocation problem in a flexible manufacturing system. Int. J. Prod. Res. **25**, 1081–1094 (1987)
7. Chan, F.T.S., Chan, H.K., Lau, H.C.W.: The state of the art in simulation study on FMS scheduling: a comprehensive survey. Int. J. Adv. Manuf. Technol. **19**, 830–849 (2002)
8. Kaushal, A., Vardhan, A., Rajput, R.S.: Flexible manufacturing system a modern approach to manufacturing technology. Int. Refereed J. Eng. Sci. (IRJES) **5**(4), 16–23 (2016)
9. Bhattacharya, M., Penica, M., O'Connell, E., Southern, M., Hayes, M.: Human-in-loop: a review of smart manufacturing deployments. Systems **11**, 1–25 (2023)
10. Mazumder, A., et al.: Towards next generation digital twin in robotics: trends, scopes, challenges, and future. Heliyon **9**, 1–25 (2023)
11. Segovia, M., Garcia-Alfaro, J.: Design, modeling and implementation of digital twins. Sensors **22**, 1–30 (2022)
12. Magalhães, L.C., et al.: Conceiving a digital twin for a flexible manufacturing system. Appl. Sci. **12**, 1–18 (2022)
13. Coito, T., et al.: Digital twin of a flexible manufacturing system for solutions preparation. Automation **3**, 153–175 (2022)

14. Neto, A.A., Ribeiro da Silva, E., Deschamps, F., Alves do Nascimento Junior, L., Pinheiro de Lima, E.: Modeling production disorder: procedures for digital twins of flexibility-driven manufacturing systems. Int. J. Prod. Econ. **260**, 1–18 (2023)
15. Li, Z., Chen, Y.: Dynamic scheduling of multi-memory process flexible job shop problem based on digital twin. Comput. Ind. Eng. **183**, 1–14 (2023)
16. Fan, Y., et al.: A digital-twin visualized architecture for flexible manufacturing system. J. Manuf. Syst. **60**, 176–201 (2021)
17. Fan, Z., Shi, D., Meyer, O., Seidelmann, J., Wang, H.: Asset administration shell-based flexible manufacturing system. In: IEEE 6th International Conference on Industrial Cyber-Physical Systems (ICPS), Wuhan, China, pp. 1–6 (2023)
18. Haddou-Benderbal, H., Dolgui, A., Yelles-Chaouche, A.: A digital twin modular framework for reconfigurable manufacturing systems. In: Lalic, B., Majstorovic, V., Marjanovic, U., von Cieminski, G., Romero, D. (eds.) APMS 2020, pp. 493–500. Springer, Cham (2020)
19. Tang, J., Emmanouilidis, C., Salonitis, K.: Reconfigurable manufacturing system characteristics in digital twin context. IFAC PapersOnLine **53**(2), 10585–10590 (2020)
20. Yang, J., Son, Y.H., Lee, D., Noh, S.D.: Digital twin-based integrated assessment of flexible and reconfigurable automotive part production lines. Machines **10**(2), 1–29 (2022)
21. Mo, F., et al.: A framework for manufacturing system reconfiguration and optimisation utilising digital twins and modular artificial intelligence. Robot. Comput.-Integr. Manuf. **82**, 1–19 (2023)
22. Leng, B., et al.: Digital twin monitoring and simulation integrated platform for reconfigurable manufacturing systems. Adv. Eng. Inform. **58**, 1–14 (2023)
23. Caesar, B., Barton, K., Tilbury, D.M., Fay, A.: Digital twin framework for reconfiguration management: concept & evaluation. IEEE Access **4**, 1–25 (2023)
24. Flores-García, E., Barrera Díaz, C.A., Wiktorsson, M., Ng, A., Aslam, T.: Enabling CPS and simulation-based multi-objective optimisation for material handling of reconfigurable manufacturing systems (2023)
25. Darrow, W.P.: A Survey of Flexible Manufacturing Systems Implementations. U.S. Department of Commerce. July (1986)
26. MacCarthy, B.L., Liu, J.: A new classification scheme for flexible manufacturing systems. Int. J. Prod. Res. **31**(2), 299–309 (1993)
27. Chen, F.F., Adam, E.E., Jr.: The impact of flexible manufacturing systems on productivity and quality. IEEE Trans. Eng. Manage. **38**(1), 33–45 (1991)
28. Ivanov, D., Dolgui, A.: Stress testing supply chains and creating viable ecosystems. Oper. Manag. Res. **15**, 475–486 (2022)
29. Chan, F.T.S., Chan, H.K.: Dynamic scheduling for a flexible manufacturing system – the pre-emptive approach. Int. J. Adv. Manuf. Technol. **17**, 760–768 (2001)
30. Zakaria, Z., Deris, S., Othman, M.R., Kasim, R.: Non-reshuffle-based approach for rescheduling of flexible manufacturing system. Int. J. Adv. Sci. Eng. Inf. Technol. **7**, 1543–1552 (2017)
31. Min, Q., Lu, Y., Liu, Z., Su, C., Wang, B.: Machine learning based digital twin framework for production optimization in petrochemical industry. Int. J. Inf. Manage. **49**, 502–519 (2019)

A Dynamic Fit-Out Scheduling Framework for Digital Twin-Enabled Modular Integrated Construction

Qiqi Chen[1], Jiyuchen Ding[1,2], Mingyue Sun[1], Zhiheng Zhao[1](✉), and George Q. Huang[1]

[1] Research Institute of Advanced Manufacturing, Department of Industrial and Systems Engineering, The Hong Kong Polytechnic University, Hung Hom, Hong Kong, China
zhiheng.zhao@polyu.edu.hk

[2] School of Internet of Things, Nanjing University of Posts and Telecommunications, Nanjing, China

Abstract. Modular Integrated Construction (MiC) is a novel construction method for its approved space-saving and time-saving for the construction industry. Fit-out construction, as the last essential process of MiC module assembly, significantly affects the delivery of MiC modules for its complexity and flexibility in operations. Emerging technologies, such as the Internet-of-Things (IoT) and Digital Twin (DT), capture real-time fit-out operation data and worker information, which reshape the fit-out job scheduling problem. This study proposes a dynamic fit-out scheduling framework by integrating job scheduling and worker assignment for DT-enabled MiC. First, this study defines the workflow of fit-out operations and the fit-out scheduling problem in MiC. The new fit-out scheduling problem considers job scheduling and worker assignment simultaneously. Second, this work designs a DT-enabled framework for cyber-physical operational synchronisation in MiC regarding workers, operations and jobs. Third, this work implements a dynamic fit-out scheduling method based on Deep Reinforcement Learning. This work intends to conduct experimental analysis using real-life case data. This work aims to provide a real-time data-driven solution for fit-out scheduling problems in MiC.

Keywords: Digital Twin · Scheduling · Worker Assignment · Fit-out Construction · Modular Integrated Construction

1 Introduction

Modular Integrated Construction (MiC) is a novel construction method for its approved space-saving and time-saving for the construction industry. MiC produces the repeatable modules, following the factory assembly and on-site installation mode, in which the modules are produced and assembled in the factories and transported to construction on-site installation. MiC is commonly used in healthcare buildings, educational institutions, and industrial and residential construction.

Published by Springer Nature Switzerland AG 2024
M. Thürer et al. (Eds.): APMS 2024, IFIP AICT 733, pp. 168–179, 2024.
https://doi.org/10.1007/978-3-031-71645-4_12

Fit-out construction is the last essential process of MiC module assembly in the factories to enable the functions of the completed infrastructures for living and working. Fit-out construction significantly affects the delivery of MiC modules for its complexity and flexibility in operations. Many studies have introduced advanced technologies, such as the Internet-of-Things (IoT) and Digital Twin (DT), for MiC fit-out construction. The technical solutions apply plentiful smart devices in the logistics and assembly operations to collect the operational data and resource information in a real-time manner, which improves the efficiency and transparency of MiC. However, these studies reveal the research gap in how these real-time data can be extracted and organised for decision-making in fit-out construction.

Current studies regard fit-out scheduling problems as job shop scheduling problems of fixed-position assembly. However, fit-out workers significantly influence MiC fit-out efficiency, considering worker factors such as process proficiency, fatigue, and working hours. Fit-out workers require specific certified skills for different types of jobs, which causes complexity in assigning proper workers for fit-out operations. Fit-out construction follows a fixed-position assembly model with a semifinished module in a workstation, requiring workers and materials to simultaneously arrive at the fit-out workstation. Three challenges still exist in fit-out construction in practice. First, the fit-out operations are complex due to their multiple jobs. For example, fitting an operating room MiC module generally consists of four phases: plumbing, electrical and plumbing, interior walls, installation of fixed equipment and installation, and movable equipment. Each phase contains various types of jobs with specific sequences. Second, the fit-out operations are flexible in jobs. Third, fit-out operations require workers with specific certificated skills. However, integrating worker assignments with job scheduling is difficult for the fit-out scheduling problem.

This study proposes a novel dynamic framework to improve the efficiency of fit-out scheduling, integrating job scheduling and worker assignment for DT-enabled fit-out operations in the MiC. This study introduces the typical fit-out operations in MiC production using the case of fitting out operating room modules. Further, this work defines the fit-out scheduling problem in MiC, which simultaneously integrates job scheduling and worker assignment. This work proposes a digital twin-enabled framework for real-time data collection and execution decision distribution. Then, this work proposes a dynamic fit-out scheduling method based on Deep Reinforcement Learning (DRL) to solve the problem. Finally, this work intends to evaluate the proposed framework and method using real-life case data.

The research framework is shown in Fig. 1. The rest of this paper is organised as follows. Section Two conducts the literature review about MiC and its relevant emerging technologies and reinforcement learning for job shop scheduling to identify the research gaps. Section Three introduces the motivating case of the MiC fit-out scenario and designs a DT-enabled MiC fit-out framework. Section Four develops a dynamic fit-out scheduling framework for DT-enabled MiC fit-out. Section Five discusses this work's experiment design and the proposed framework's potential contribution.

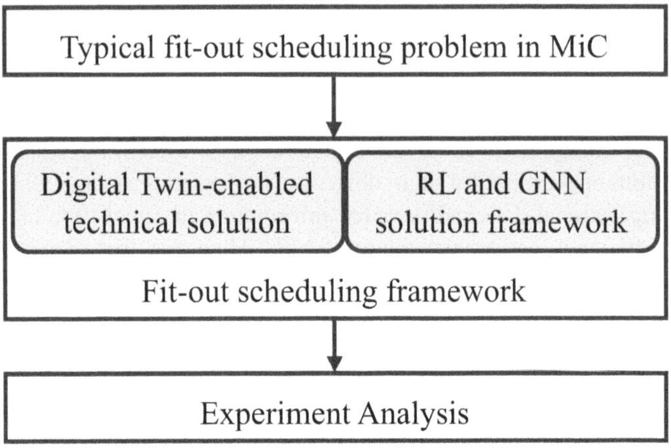

Fig. 1. Research framework.

2 Literature Review

2.1 Modular Integrated Construction and its Relevant Emerging Technologies

Modular Integrated Construction (MiC) refers to dividing a building into multiple modular units according to specific functions. The modular units are prefabricated in factories to assemble columns, panels, walls, doors, windows, internal equipment and pipework, and then they are transported to the construction site to complete the stacking and installation of the modules [1]. MiC has gained much attention in the megacity construction industry due to its improved quality, efficiency, site safety, and reliability, as seen in China, mainland China, Hong Kong, and Singapore [2]. The Housing Department of Hong Kong has proposed the introduction of advanced construction techniques and new procurement modes to increase the speed of housing construction, for example, the adoption of the Design for Manufacture and Assembly (DfMA) and Modularized Component (MiC) construction methods when building public housing [3].

Emerging technologies, such as Digital Twin (DT), Internet-of-Things (IoT) and generative artificial intelligence, catch much attention in MiC supply chain management [4]. IoT has been applied for MiC part tracking and tracing for many years, and many cases prove IoT's effectiveness and efficiency in improving MiC production operations [1, 5]. IoT is a critical technical solution for data collection in DT. Due to their advantages in virtual-reality fusion and real-time data interaction, DT has been widely researched in recent years, such as supply chain management, manufacturing, maintenance, construction, and agriculture [6–8]. DT first originated in the military aerospace and aviation field and was proposed by Professor Grieves in the product lifecycle management course. In the latest research about the digital twin model, Tao et al. propose a complete five-dimensional DT model, which contains five primary elements, namely physical entity, virtual entity, service, twin data, and connectivity, to improve the infrastructure of DT [9]. Tao et al. further investigated the application of DT in product use and maintenance and proposed nine principles to improve maintenance efficiency and reduce maintenance failures [10]. DT is an emerging technology that improves MiC manufacturing and logistics for process monitoring and data collection [11]. Pan and Zhang propose a DT-enabled framework integrating various technologies for process mining and time series analysis in MiC project management [12]. However, these studies discuss less about how the collected real-time data can help the decision-making of MiC logistics and manufacturing. This study fills the gap by considering real-time collected data in fit-out scheduling problems.

2.2 Reinforcement Learning for Job Shop Scheduling

Job shop scheduling problems confront various dynamic changes in the production environment, and Reinforcement Learning (RL) for job shop scheduling problems has been a popular research direction in recent years [13]. The earliest research on the application of reinforcement learning in manufacturing dates back to 1995, when Zhang and Dietterich introduced reinforcement learning methods to achieve optimal decision-making in task scheduling in manufacturing workshops [14]. In recent years, deep reinforcement learning (DRL), a fusion of reinforcement learning and deep learning, has been widely used in manufacturing workshop scheduling problems, especially in cloud manufacturing, due to its advantages in solving problems with high complexity and high uncertainty. Zhang et al. develop a deep reinforcement learning framework to model and solve a collaborative problem of batch production and logistics in a cloud manufacturing scenario. Luo et al. use Double DQN to optimise dynamic multi-objective scheduling to implement a flexible workshop scheduling problem [15, 16]. Liu et al. design DDPG to be applied to the service combination problem in cloud manufacturing [17]. Wang et al. develop multi-intelligent reinforcement learning and graph convolutional networks in a cloud manufacturing scheduling problem [18]. Park et al. apply graph neural networks (GNN) to learn the node characteristics of the spatial structure of the embedded job shop scheduling problem by designing the optimal scheduling features and strategy [19]. RL method for MiC prefab fit-out scheduling has yet to be studied. Ding et al. propose a multistage decision-making method for the prefab assembly problem [5]. Lee and Hyun develop a framework for MiC project management, including on-site construction, manufacturing, transport and installation plans, and optimising resource allocation decisions by genetic algorithms [20].

3 Digital Twin-Enabled Fit-Out Operations in MiC Factory Assembly

3.1 Motivating Case

The typical fit-out operations in MiC module manufacturing are shown in Fig. 1, motivated by the case of producing operating room modules for hospital construction. MiC Fit-out construction planning and control follows a make-to-order (MTO) mode. The builders order MiC modules for on-site construction and send the requirements to the MiC factories several months before the construction project starts. Each customer order contains MiC module types, the quantity of each MiC module, and the delivery date. Once the MiC factory receives the order, it starts to purchase the specific raw parts for constructing the ordered MiC modules. MiC factory plans the production orders by arranging the MiC module requirements to its manufacturing capability and order lead time. Then, the MiC factory splits the MiC production plans into monthly or weekly-based schedules for batch delivery to the construction sites. The production planning and scheduling criteria include the number of MiC modules, production due date, and the capability of Fit-out resources, such as workstations, materials, workers, and tools.

In the MiC fit-out scheduling stage, the fit-out of an operating room module follows fixed-position assembly, requiring material, worker, and tool to come to the workstation simultaneously. The process of fitting out operating room modules generally consists of various phases and jobs in each phase and the jobs are completed at different processing times. As shown in Fig. 1, the fit-out of an operating room module contains five stages, including installing pipelines, process surfaces, mounting fixtures, installing fittings, and drying. The first four phases include the following jobs:

- Install Pipelines: This phase includes the design, installation, and acceptance of plumbing, electrical wiring, purification systems, ventilation systems, and medical gas pipelines within the operating room.
- Process Surfaces: This phase includes the installation of suspended ceilings, painting, floors, walls, power sockets, and switches. The wall construction contains sub-steps of fixed equipment holes, lamps and vents, end equipment locations, and mobile equipment tracks.
- Mount Fixture: this phase includes installing fixed equipment, fixed operation equipment, lights, towers, doors, monitors, and windows.
- Install Fitting: This phase involves installing storage cabinets, moveable equipment, hanging mirrors, medical waste bins, tables, seats, and scalpels.

The MiC fit-out scheduling also assigns workers for the workstations. Each worker is certificated to complete various job types in fit-out operations. Based on the fit-out phase and fit-out job type, a worker is selected based on worker skill, tardiness, and the number of processed jobs.

3.2 Digital Twin-Enabled MiC Fit-Out Framework

This section proposes a DT-enabled MiC fit-out framework for cyber-physical operational synchronisation in MiC fit-out, as shown in Fig. 2. The proposed framework follows the typical DT framework, including the physical and twin layers.

The physical layer implements cutting-edge technologies, such as IoT devices, Bluetooth beacons, and mobile operating devices, in the MiC fit-out working environment. The fit-out resources, such as parts, workers, and workstations, are monitored and managed promptly. For example, a workstation's completion progress and status are easily tracked through an RFID reader, with workers signing in and out by tapping their RFID staff cards.

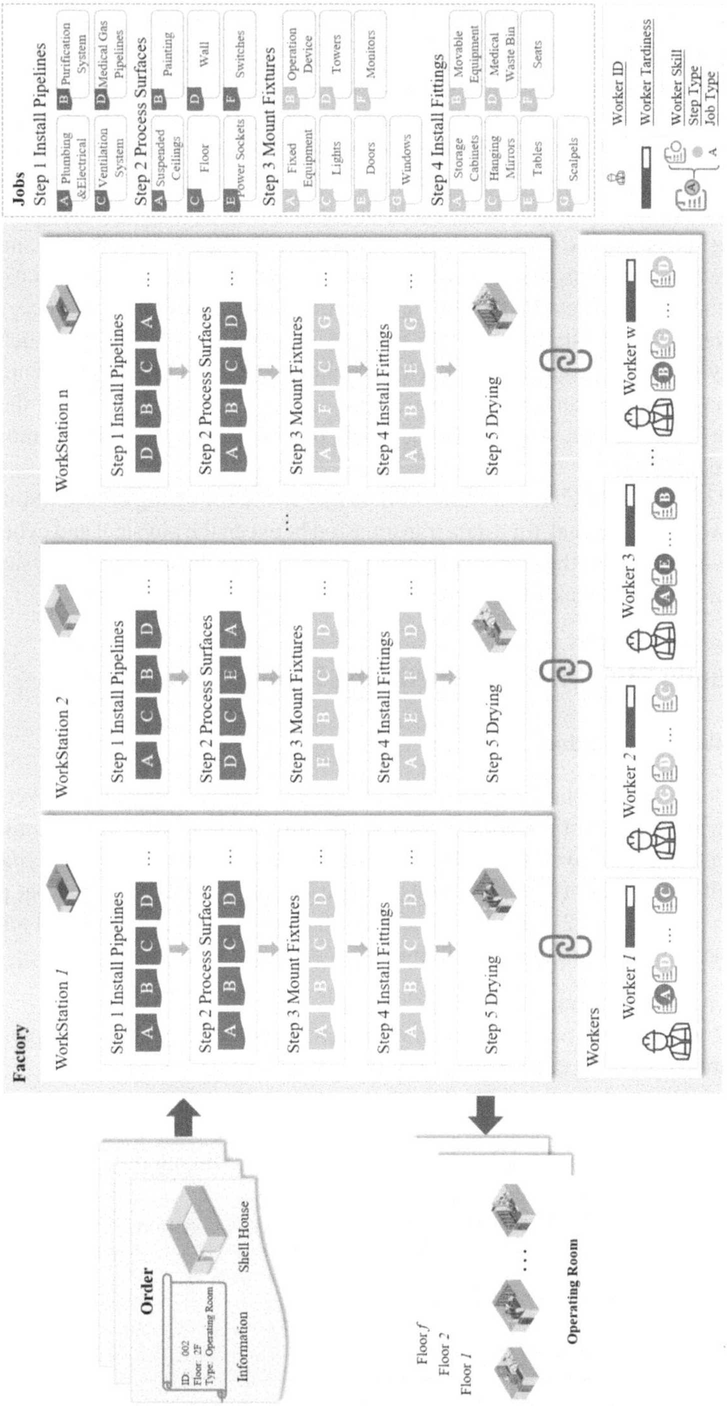

Fig. 2. Typical fit-out operations in MiC production.

The twin layer contains the fit-out execution and fit-out scheduling components. Fit-out scheduling completes the execution decisions and transfers them to twin entities in fit-out execution. In contrast, fit-out execution performs the executions and updates real-time results to fit-out scheduling. Fit-out scheduling processes the decision-making based on RL and GNN. Fit-out scheduling defines the primary and real-time information of job and worker as the scheduling states and constructs a disjunctive graph with job and worker nodes. Further, fit-out scheduling processes RL-GNN training using the scheduling states and the disjunctive graph. It generates execution decisions, including selecting which job to process and which worker to process this job. Fit-out execution manages the twin entities. Each twin entity corresponds to a fit-out resource, such as part, worker, and workstation, to keep track of its real-time information, such as the availability status, working status, current job, and processed jobs. The twin entities can execute the operations and communicate between the physical entities and the fit-out scheduling. The real-time data is uploaded and updated to fit-out scheduling module for decision-making.

The DT-enabled MiC fit-out framework achieves the following contributions. First, the framework realises real-time data transmission between the physical and cyber layers of MiC fit-out. Second, the framework integrates real-time data from the physical layer for dynamic decision-making in fit-out scheduling and execution (Fig. 4).

4 Dynamic MiC Fit-Out Scheduling Framework

4.1 Problem Description

The traditional fit-out scheduling problem considers job scheduling and worker assignment using multistage optimisation methods. This study proposes a novel dynamic fit-out scheduling framework to consider these two decisions simultaneously. The dynamic fit-out scheduling problem (DFSP) answers two questions: 1) which job in what phase is selected to process, and 2) which worker is assigned to complete the selected job. Some common assumptions are listed as follows.

1. Each workstation produces one MiC module.
2. ch MiC module is fit out in one fixed workstation, and the fit-out phase types and job types are the same in each workstation.
3. MiC modules vary from one production order to another.
4. MiC module fit-out phases are the same in each workstation.
5. Job types of the fit-out phases are the same for each MiC module, but the jobs are selectable for each module.
6. The processing time of one job type for each worker is constant.
7. The notations of DFSP are defined in Table 1.

Table 1. Notations for dynamic fit-out scheduling framework.

Notation	Description
P	Set of fit-out phases, $P = \{P_p \| P_1, P_2, \ldots, P_{\|P\|}\}$
J	Set of jobs, $J_p = \{J_{pj} \| J_{p1}, J_{p2}, \ldots, J_{p\|J\|}\}$
W	Set of workers, $W = \{W_w \| W_1, W_2, \ldots, W_{\|W\|}\}$
g_w	The grade of worker W_w
$x_{pjp'j'}$	$x_{pjp'j'} = 1$ if job J_{pj} is processed before $J_{p'j'}$, otherwise, $x_{pjp'j'} = 0$
y_{pjw}	$y_{pjw} = 1$ if fit-out job J_{pj}. is processed by worker W_w
t	Current decision time
W_{pj}	The worker will process the job J_{pj}
t_{pj}^{PT}	The processing time of fit-out job job J_{pj}
t_p^{ST}	The start time of fit-out phase P_p
t_{pj}^{ST}	The start time of fit-out job J_{pj}
t_{pj}^{DT}	The due time of fit-out job J_{pj}
t_p^{DT}	The due time of fit-out phase P_p
t_{pj}^{CT}	The completion time of fit-out job J_{pj}
t_p^{CT}	The completion time of fit-out phase P_p
t_p^{TD}	The tardiness of fit-out phase P_p
t_p^{ETA}	The estimated tardiness of fit-out phase P_p

4.2 Dynamic Fit-Out Scheduling Framework

This section proposes a dynamic fit-out scheduling framework (DFSF) based on RL and GNN to address the integrated job scheduling and worker assignment decisions, shown in Fig. 3.

DFSF defines several basic elements following the basic definition of the RL framework. DFSF defines the scheduling states $S(t)$ using information on all the fit-out jobs and workers. The action $A(t)$ represents the job scheduling rules and worker assignment rules. The state transition uses real-time data gained through the DT-enabled fit-out framework, such as the current job in each workstation and the available workers at the decision point. The reward function is defined using the tardiness of all the fit-out jobs in completing one MiC module. The policy $\pi(a_t|s_t)$ is the probability distribution of taking action a_t at each state s_t.

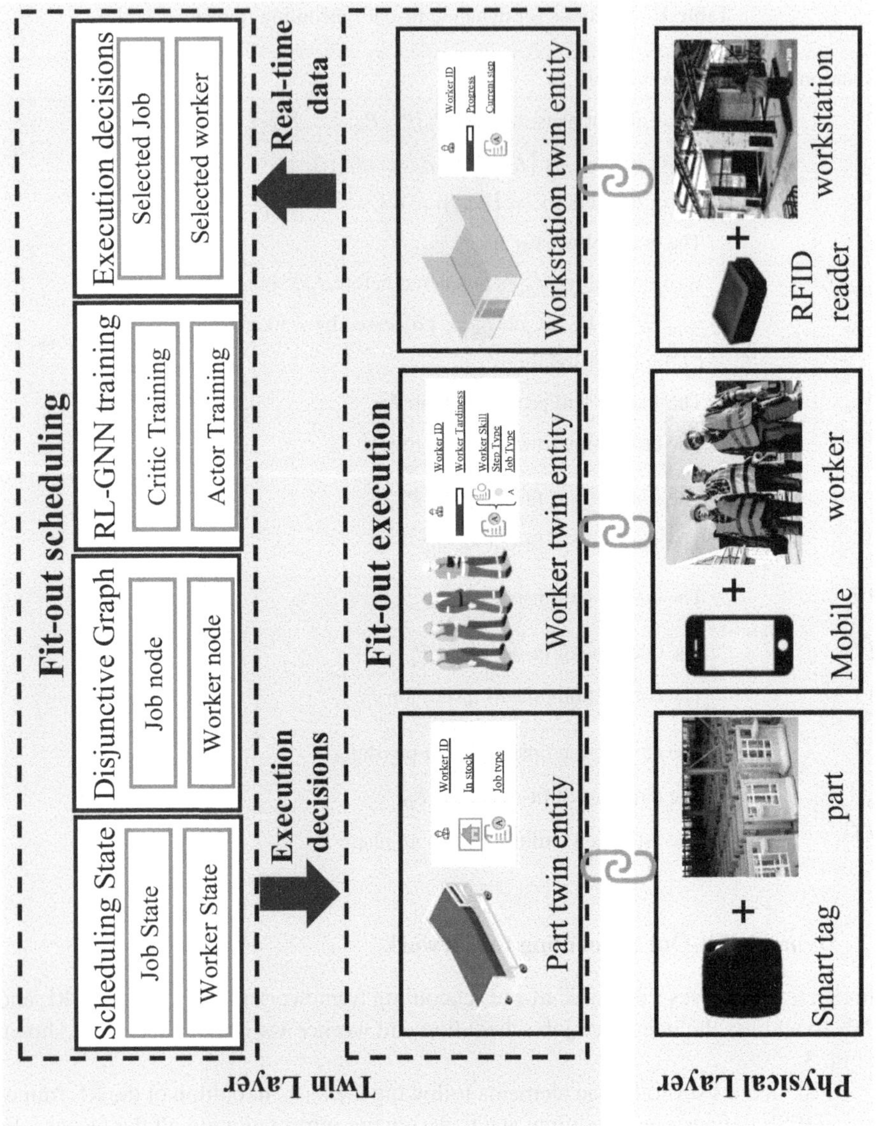

Fig. 3. Digital twin-enabled fit-out scheduling framework.

The detailed scheduling method is described as follows. First, DFSF is a sequenced decision-making process that iteratively takes scheduling actions and considers real-time information dynamically. At each scheduling action, DFSF picks the next job to be performed and assigns a worker to perform this job. The scheduling states contain job states and worker states. In each iteration, the scheduling states are converted into a heterogeneous graph, representing workers and jobs by nodes and workers' skills by the connectivity between them. Second, DFSF uses a disjunctive Graph (DG) to represent

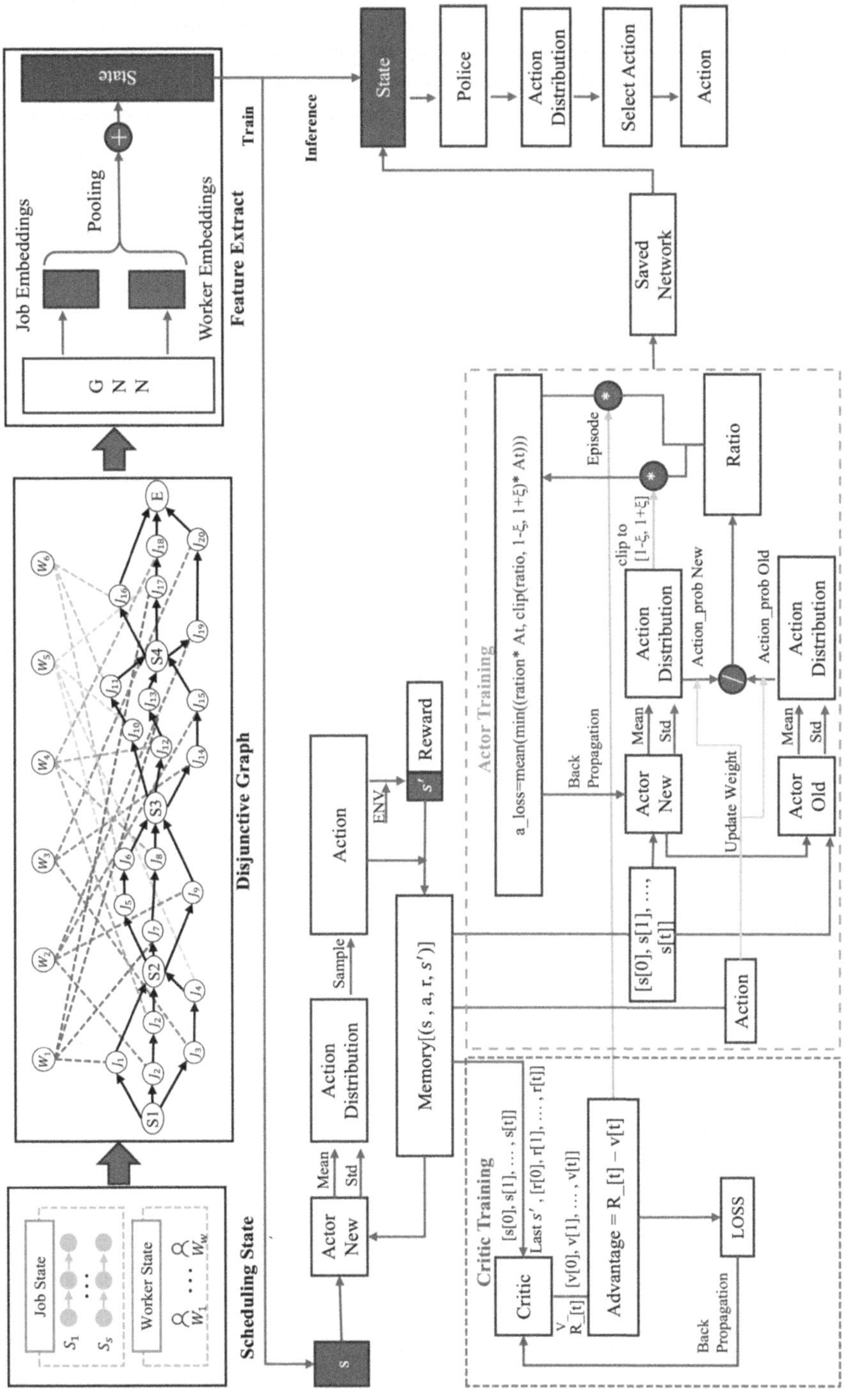

Fig. 4. Dynamic fit-out scheduling framework based on RL and GNN.

and store the fit-out scheduling problem, including job and worker nodes. The job nodes represent the MiC fit-out jobs, and the worker nodes indicate the available workers to complete the fit-out jobs. The arc between two job nodes indicates the job sequence, and the arc between a job node and a worker node represents the worker who can process the job. Third, DFSF uses Graph Neural Network (GNN) with a two-stage embedding process for feature extraction. The extracted job and worker features are fed to the states of the RL algorithm for training and inference.

DFSF applies Proximal Policy Optimization (PPO) to solve the scheduling problem. PPO is a typical RL algorithm for the discrete decision problem. The core concept of PPO is for policy optimisation to ensure stability in policy updates and avoid significant policy changes that could lead to instability or catastrophic performance degradation. The details of the PPO algorithm for DFSF are shown in Fig. 3.

5 Discussion and Conclusion

This study proposes a novel dynamic framework to improve the efficiency of fit-out scheduling, integrating job scheduling and worker assignment for DT-enabled fit-out operations in the MiC. This work contributes from two aspects.

First, this work proposes a digital twin-enabled framework for cyber-physical operational synchronisation in MiC. In the physical layer, all the fit-out resources, including parts, workers, and workstations (jobs), are equipped with sensors, gateways, beacons and smart devices. With innovative IoT technologies, these physical resources can be detected and perceived in the digital layer. In the digital layer, the physical resources are modelled into twin entities, and the twin entities are used for real-time monitoring. In addition, the twin entities transmit the fit-out execution decisions and real-time execution results between the cyber layer and the physical environment of MiC fit-out.

Second, this study defines the fit-out scheduling problem in MiC, which simultaneously integrates job scheduling and worker assignment. The dynamic fit-out scheduling framework based on DRL achieves integrated decision-making for fit-out operations. DFSF constructs the jobs and workers into one disjunctive graph and extracts their features for DRL training. DFSF achieves dynamic decision-making by integrating real-time job and worker information at each decision point.

The future work of this study is to evaluate and estimate the proposed framework using real-life case data, including constructing a DT-enabled fit-out laboratory environment to test the cyber-physical operational synchronisation and numerical analysis of the proposed DFSF.

Acknowledgements. The authors would like to express their sincere thanks to partial financial support from the HK RGC TRS project (No. T32-707/22-N), the National Natural Science Foundation of China (Grant No. 52305557), the HK RGC RIF project (No. R7036-22), and PolyU RIAM.

Disclosure of Interests. The authors have no competing interests to declare relevant to this article's content.

References

1. Zhai, Y., et al.: An Internet of Things-enabled BIM platform for modular integrated construction: a case study in Hong Kong. Adv. Eng. Inf. **42**, 100997 (2019)
2. Xu, Z., Zayed, T., Niu, Y.: Comparative analysis of modular construction practices in mainland China, Hong Kong and Singapore. J. Clean. Prod. **245**, 118861 (2020)
3. Lu, W., et al.: Design for manufacture and assembly (DfMA) in construction: the old and the new. Arch. Eng. Des. Manag. **17**(1–2), 77–91 (2021)
4. Jackson, I., et al.: Generative artificial intelligence in supply chain and operations management: a capability-based framework for analysis and implementation. Int. J. Prod. Res. 1–26 (2024)
5. Ding, H., et al.: Multistage self-adaptive decision-making mechanism for prefabricated building modules with IoT-enabled graduation manufacturing system. Autom. Constr. **148**, 104755 (2023)
6. Ivanov, D.: Conceptualisation of a 7-element digital twin framework in supply chain and operations management. Int. J. Prod. Res. **62**(6), 2220–2232 (2024)
7. Grieves, M., Vickers, J.: Digital twin: mitigating unpredictable, undesirable emergent behavior in complex systems. In: Transdisciplinary Perspectives on Complex Systems: New Findings and Approaches, pp. 85–113 (2017)
8. Ivanov, D.: Intelligent digital twin (iDT) for supply chain stress-testing, resilience, and viability. Int. J. Prod. Econ. **263**, 108938 (2023)
9. Tao, F., et al.: Five-dimension digital twin model and its ten applications. Comput. Integr. Manuf. Syst. **25**(1), 1–18 (2019)
10. Tao, F., et al.: Digital twin-driven product design, manufacturing and service with big data. Int. J. Adv. Manuf. Technol. **94**(9–12), 3563–3576 (2017)
11. Opoku, D.-G.J., et al.: Digital twin application in the construction industry: a literature review. J. Build. Eng. **40**, 102726 (2021)
12. Pan, Y., Zhang, L.: A BIM-data mining integrated digital twin framework for advanced project management. Autom. Constr. **124**, 103564 (2021)
13. Ivanov, D., et al.: A control approach to scheduling flexibly configurable jobs with dynamic structural-logical constraints. IISE Trans. **53**(1), 21–38 (2021)
14. Zhang, W., Dietterich, T.G.: A reinforcement learning approach to job-shop scheduling. In: IJCAI. Citeseer (1995)
15. Zhang, L., et al.: Distributed real-time scheduling in cloud manufacturing by deep reinforcement learning. IEEE Trans. Ind. Inf. **18**(12), 8999–9007 (2022)
16. Luo, S., Zhang, L., Fan, Y.: Dynamic multi-objective scheduling for flexible job shop by deep reinforcement learning. Comput. Ind. Eng. **159**, 107489 (2021)
17. Liu, Y., et al.: Logistics-involved service composition in a dynamic cloud manufacturing environment: a DDPG-based approach. Rob. Comput.-Integr. Manuf. **76**, 102323 (2022)
18. Wang, X., et al.: Dynamic scheduling of tasks in cloud manufacturing with multi-agent reinforcement learning. J. Manuf. Syst. **65**, 130–145 (2022)
19. Park, J., et al.: Learning to schedule job-shop problems: representation and policy learning using graph neural network and reinforcement learning. Int. J. Prod. Res. **59**(11), 3360–3377 (2021)
20. Lee, J., Hyun, H.: Multiple modular building construction project scheduling using genetic algorithms. J. Constr. Eng. Manag. **145**(1), 04018116 (2019)

Comparing Digital Twins and Virtual Engineering in Buyer Supplier Relationships for Complex Production Facilities

Luca Janecki[1(✉)], Oliver Antons[2], Daniel Reh[1], and Julia C. Arlinghaus[2]

[1] Assembly Planning, Volkswagen Commercial Vehicles, Mecklenheidestrasse 74, 30419 Hannover, Germany
luca.janecki1@volkswagen.de

[2] Chair of Production Systems and Automation, Otto-von-Guericke-University Magdeburg, Universitätsplatz 2, 39104 Magdeburg, Germany

Abstract. In today's highly competitive and interconnected global marketplace, the effective management of buyer-supplier relationships is essential for organizations seeking to stay ahead. With the rise of complex production facilities, ensuring product quality and optimizing communication between buyers and suppliers presents an escalating challenge. This complexity is further compounded by the integration of Digital Twins and Virtual Engineering, requiring innovative solutions to navigate the intricacies of modern supply chain dynamics. This article explores the role of Digital Twins and Virtual Engineering as a strategic approach to enhance buyer-supplier relationship management within the context of complex production facilities using insights from the Principal Agent Theory. Based on a systematic literature review, we explore current approaches to the use of virtual engineering and digital twins to overcome existing tensions from principal-agent theory.

Keywords: Virtual Engineering · Digital Twins · Buyer Supplier Relationship · Production facilities

1 Introduction

In the evolving context of contemporary industrial processes, the effective planning and execution of complex production facilities are of paramount importance for manufacturing enterprises aiming at sustainable growth and competitiveness. Concurrently, enterprises are prioritizing their fundamental capabilities [1], prompting their engagement with external suppliers specializing in the intricacies of facility planning projects. This collaboration leverages the specialized expertise of external entities dedicated to this specific domain [2]. Challenges in collaboration between buyers and suppliers during facility planning projects are pervasive, frequently resulting in operational inefficiencies and escalated costs. Virtual Engineering (VE) and especially Digital Twins (DT) create a basis for a collaboration, in which tests and verifications can be carried out using virtual

Published by Springer Nature Switzerland AG 2024
M. Thürer et al. (Eds.): APMS 2024, IFIP AICT 733, pp. 180–193, 2024.
https://doi.org/10.1007/978-3-031-71645-4_13

representations of the production facility before the physical facility has been built or even ordered [3].

Although specific functionalities of virtual engineering and Digital Twins have already been described in detail, there is a lack of consideration of the overall context in relation to the planning of complex production facilities. Particularly in view of the necessary cooperation between several contractual partners, there is a need for action to research optimal interaction between the parties [4].

In this article, we provide a systematic literature review to explore the potential of Virtual Engineering and Digital Twins in overcoming challenges within buyer-supplier relationships, particularly through the lens of principal-agent theory. Our goal is to categorize the insights in the context of complex production facilities. The remainder is structured as follows: Sect. 2 provides the theoretical background on the relevant research fields of virtual engineering, digital twin and principal agent theory. In Sect. 3, the selection of the research methodology is derived and the steps and framework conditions to be carried out are described. Section 4 evaluates and discusses the results of the systematic literature review. Section 5 classifies the results, discusses limitations and provides an outlook on the need for further research.

2 Theoretical Background

In the context of Industry 4.0 and increasing digitalization, manufacturing companies are increasingly using more complex, (partially) automated production facilities in order to achieve higher product quality, shorter lead times and efficient production processes. Efficient engineering of these production facilities is important in order to stay ahead of the global competition [5]. In the context of systems engineering, there are various definitions of the general engineering lifecycle of production facilities. Konstantinov et al. [6] divide it into the phases concept, design & engineering, layout & configuration, build, commissioning, and operation & maintenance, which seems to be generally applicable. The traditional engineering of complex production facilities comprises the phases of mechanical design, electrical design, program creation, commissioning, and the start of series production [7]. The first step is to define a rough production concept, which components are to be treated with which production process. Afterwards the specifications for robots, sensors, control systems and IT infrastructure, for example, are defined on this basis. The source code for the control system is then developed and the production system is physically constructed. Using the physical prototype, the control system can be validated, and the production facility put into operation [8, 9].

2.1 Virtual Engineering

The first attempts to improve the engineering process with virtual prototypes were already made in the early 2000s [10]. In the ongoing transition to Industry 4.0, virtual approaches have been continuously developed to achieve goals such as reducing planning and production costs, reducing planning lead times, improving product and factory quality and improving the distribution of knowledge within companies [11]. Nowadays, virtual engineering is a knowledge representation of the production process planning of an artifact that combines all shop floor information. It involves, for example, CAx tools, Gantt charts as well as the use of experience-based dynamic and kinematic simulations and modeling to design, test, and optimize products and processes before physical components are created. It allows organizations to visualize and analyze complex systems, enabling rapid iteration and cost-effective problem-solving [12]. This enables engineers to substitute physical prototypes with virtual counterparts, thereby expediting lead times in product development, streamlining production cycle times, and preempting potential technological and environmental challenges in the future [13].

2.2 Digital Twin

In the course of the digitalization, a theory called Digital Twin has developed over the last two decades. There are many definitions of the Digital Twin and it types in the literature [14, 15]. The concept was first outlined in 2002 in a presentation entitled "Conceptual Ideal for PLM" at the University of Michigan, which at the time was initially focused on product lifecycle management, but already highlighted characteristics of digital twins that have not changed significantly to this day, such as "real space" and "virtual space" linked by data and information exchange [16]. A more current definition in manufacturing defines the term as *"...a digital representation of an active unique product [...] or unique product-service system [...] that comprises its selected characteristics, properties, conditions, and behaviours by means of models, information, and data within a single or even across multiple life cycle phases."* [17].

Trauer et al. have conducted a highly cited literature review on the state of the art of Digital Twins, which was supplemented with impressions from the industry and developed the definition *"A Digital Twin is a virtual dynamic representation of a physical system, which is connected to it over the entire lifecycle for bidirectional data exchange"* which seems universally apt [18]. As a result of a Digital Twin, the properties and behaviors of its physical counterpart can be visualized, forecasted, and adjusted accordingly [19]. Digital Twin use increases connectivity and feedback between devices, in turn, improving reliability and performance. As part of virtual engineering, a physical system does not necessarily have to be available in advance. Instead, the digital twin is built up continuously over various phases and serves as the basis for the construction of the physical system [20]. For example, a production facility that is still in the planning phase can already have a Digital Twin which predicts its future behavior [21]. In the realm of planning complex production facilities, creating a digital twin becomes particularly challenging due to the necessity for integration and coordination with diverse suppliers [22].

2.3 Principal Agent Theory

The principal agent theory emerged in the 1970s, originating in the research fields of economics and institutional theory. It is based on Ross's [23] economic theory of agency and Mitnick's [24] institutional theory of agency. The exact origin of the theory is disputed, but the most cited paper is by Jensen and Meckling. The setting in a principal-agent situation is described as follows: One party (the principal) hires another party (the agent) to perform a task on its behalf. This can happen due to a lack of skill or time, among other things [25]. Originally, the focus was on the different attitudes of the parties to risk and the resulting different courses of action. It was later expanded to include the fact that there is a conflict of objectives between the two parties and that it is difficult or expensive for the principal (due to so-called agency costs) to monitor the agent's actions [26]. The application focus of agency theory is on determining optimal contractual situations (e.g. in construction projects) for the relationship. [27]. The principal-agent problem can be further differentiated into sub-areas. On the one hand, there is a risk of adverse selection for the principal (e.g. the selection of unsuitable agents) even before the parties formalize the contract. On the other hand, there is the risk of so-called moral hazard, which implies that the agent acts in self-interest and opportunism, as the principal does not have full insight into the actions (hidden action) after the relationship has been established by the contract. In both cases, diverging goals and information asymmetries are seen as the core source of the problem [28] (Fig. 1).

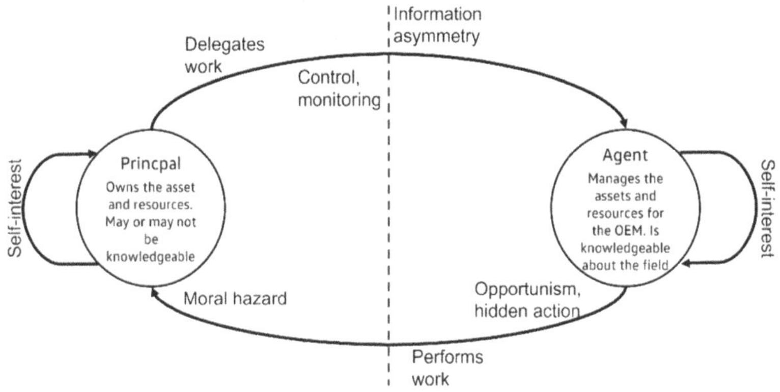

Fig. 1. Principal Agent Theory [29]

In order to reduce information asymmetries and reconcile the divergent goals, the principal agent theory offers solution mechanisms such as monitoring and signaling. Signaling emerges as one potential strategy, wherein agents utilize available resources to portray themselves as the most qualified candidates [30]. Through monitoring, the principal will regularly review or inspect the agents work to ensure that the work is being done properly and to the agreed standards, but this also incurs additional costs. The costs of the resolution mechanisms are called agency costs. It encompasses, for example, the expenses associated with selecting an appropriate agent, gathering information to establish performance benchmarks, monitoring to supervise the agent's actions, bonding costs, and the losses incurred as a result of the agents' inefficient decisions. [25]. In order to still be effective and economical, these mechanisms must therefore be selected and implemented in such a way that they justify the agency costs. Since Industry 4.0 the use of virtual engineering methods and tools are expected to open up potential for companies [31] that can act as possible headmasters or agents with regard to the efficiency of their process execution. It remains an important questions whether these can help to minimize agency costs and thus make their contribution to justifying the agency costs for signaling and monitoring.

3 Methodology

In order to study virtual engineering and digital twins for buyer supplier relationships using PAT, we identified the relevant research streams through a systematic literature review. For this purpose, the method of literature research according to WEBSTER AND WATSON [32] is applied. To conduct unbiased research, the steps recommended by XIAO AND WATSON [33] are consistently followed:

(1) Search the literature: We selected the Google Scholar, Sciencedirect and IEEE databases as a foundation These were examined for relevant studies in German and English using the search strings from the core areas of "Virtual Engineering", "Digital Twin" and "Principal Agent". We combined those key words with Boolean operators to the search string *("digital planning" OR "virtual engineering" OR "Digital Twin" OR "Cyber physical systems") AND "Principal Agent"* to identify literature from different research streams. This resulted in 536 studies, which were evaluated for relevance.
(2) Screen for inclusion: After removing duplicates, the articles were sorted based on predefined inclusion criteria such as publication year and the topics of manufacturing and production facility engineering. Exclusion criteria, such as the topics of blockchain and medicine, were applied because these research fields are far removed from the paper's focus, ensuring that our review concentrates on recent and relevant research streams. This resulted in 92 publications.

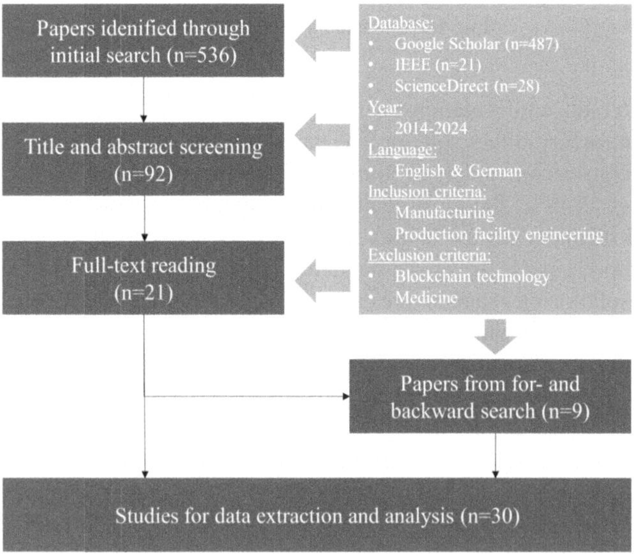

Fig. 2. Systematic literature review process

(3) Title and abstract screening: After minimizing the results due to the exclusion and inclusion criteria we checked the titles and abstracts which results in a list of 10 publications which were completed by forward and backward searches. In total, a list of 18 publications remains, which are included in the work.
(4) Assess quality: We obtained the full texts and validated the inclusion/exclusion by full text reading.
(5) Data extraction: After collecting the relevant publications, they were structured in terms of their content.
(6) Analyzing and synthesizing data: We organized the data according to the chosen review generating an overview of the thematic coverage of the research areas as proposed by Teiwes [34] (Fig. 2).

4 Results

The systematic literature search yielded a total of 30 articles that met the inclusion criteria outlined in the methodology. Of these, 21 are journal articles and 9 are conference papers. Figure 3 illustrates the chronological distribution of these articles across the years 2014 to 2023. The publication trends demonstrate the topic's longstanding relevance and a marked increase in interest, particularly evident from a peak in 2023, indicating its growing importance in the engineering field.

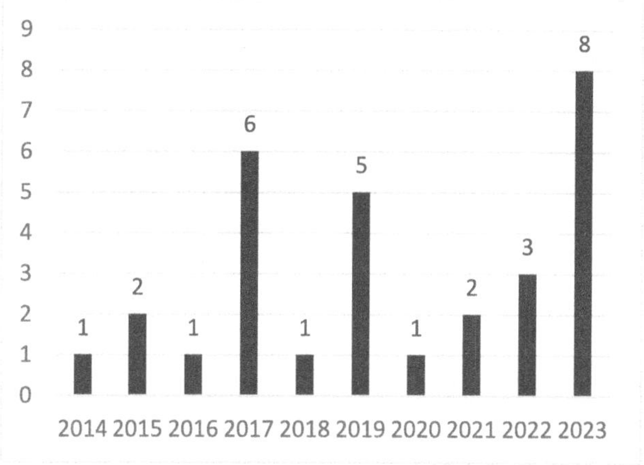

Fig. 3. Chronological distribution of the articles

The thematic analysis of the retrieved articles revealed a diverse range of research fields exploring the intersection of digital planning, virtual engineering, digital twin, cyber-physical systems, and principal-agent relationships. Figure 4 presents the distribution of articles across various research fields. The majority of the articles were focused on production systems, supply chain management, and production economics, indicating a significant interest in leveraging digital technologies for enhancing manufacturing processes and optimizing resource allocation. Additionally, articles were found in diverse fields such as systems engineering, supply chain management, and construction engineering, highlighting the interdisciplinary nature of research in this area.

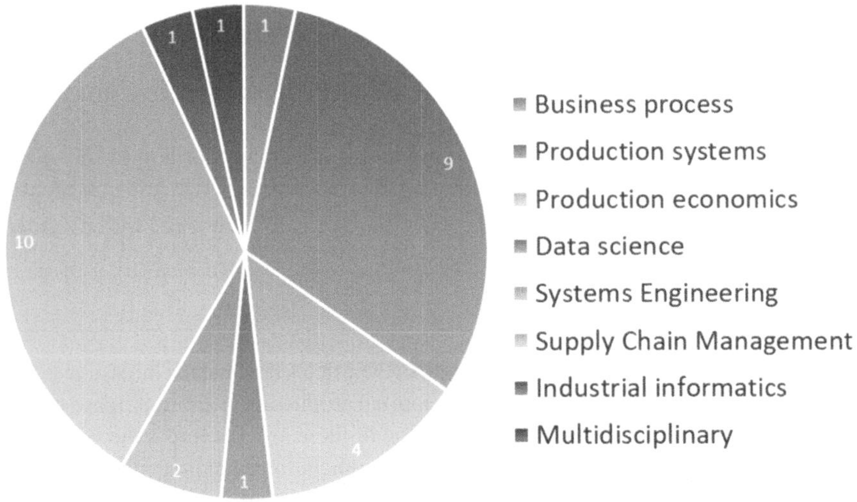

Fig. 4. Classification of publications generated from the SLR in research streams

The systematic literature review synthesized findings from various studies addressing Virtual Engineering, Digital Twin, and Principal-Agent Theory. Each study was evaluated based on the coverage and important aspects within these domains. As can be seen in Table 1, it can be said that the areas of virtual engineering and digital twin in particular are important to advances in the engineering process. The Principal Agent Theory is considered rather specifically in depth in the articles found. There are also approaches that describe the improvement of the buyer-supplier relationship through digital technologies such as these, without specifically addressing principal agent theory.

4.1 Virtual Engineering

Across the examined literature, VE emerged as a focal point in several studies providing comprehensive insights [35–37]. Notably, Auris et al. highlighted VE's significance in early design phases allowing precise cycle time estimations and effective adjustments in component dimensioning, emphasizing its crucial role in the product development process [35]. Similarly, Cadet et al. emphasized the importance of a consistent digital model across all phases of the product lifecycle, stressing its value for internal and external collaboration within buyer supplier settings [36]. Additionally, Eigner et al. highlighted the necessity of consistent digital models throughout various phases for effective collaboration within the supply chain [20]. Konstantinov et al. envisioned a VE approach consistent with the real system throughout various lifecycle stages, emphasizing its role in design, engineering, and operational phases [6, 37]. Shafiq et al. emphasized VE's role as an experiential knowledge representation in engineering, covering critical process planning information [12]. Asif et al. explore the impact and potential of Industry

Table 1. Evaluation of systematic literature research

Publications	Virtual Engineering	Digital Twin	Principal Agent Theory
Abideen et al., 2023	○	○	◑
Ahi et al., 2022	◑	○	◑
Asif et al., 2022	◑	○	●
Auris et al., 2017	●	◑	○
Becker at al., 2021	○	◑	○
Cadet et al., 2017	●	◑	○
Dalenogare et al., 2018	◑	○	○
Fuller et al., 2020	◑	●	○
Grieves and Vickers, 2017	◑	●	○
Halenar et al., 2019	◑	●	○
Ji et al., 2023	◑	◑	◑
Konstantinov et al., 2017	●	◑	○
Konstantinov et al., 2022	●	◑	○
Maier et al., 2015	◑	○	●
Mardenli and Sackmann, 2023	◑	○	●
Maier et al., 2014	◑	○	●
Yacoob et al., 2021	◑	○	●
Eigner et al., 2017	◑	◑	◑
Mulchandani et al., 2023	◑	○	◑
Nieto et al., 2023	◑	○	◑
Panda and Leepsa, 2017	○	○	●
Pfäff et al., 2023	◑	○	●
Radke et al., 2019	◑	◑	◑
Reim et al., 2023	◑	●	○
Schroeder et al., 2016	◑	●	○
Shafiq et al., 2015	●	◑	○
Spieske et al., 2023	○	○	◑
Srai and Lorentz, 2019	◑	○	◑
Stark and Damerau, 2019	◑	●	○
Tao et al., 2019	◑	●	○
● = fully covered	◑ = partly covered	○ = not covered	

4.0 on supplier evaluation and verification, also provide comprehensive insights into VE [38].

Other studies offered partial coverage of VE, which acknowledge VE but do not delve deeply into its performance impact or full lifecycle integration or mainly focus on digitalization and collaboration but not exclusively on VE.

Several publications, including Panda and Leepsa, Abideen et al., Ahi et al., and Spieske et al., do not cover VE at all, highlighting a gap in the discussion of VE in buyer supplier relationship contexts.

4.2 Digital Twin

Digital Twin emerged as a prominent area of investigation in the reviewed literature. Several studies provided comprehensive insights into DT emphasizing its application throughout the product lifecycle and its critical role in modern industrial processes [16, 17, 39]. Grieves and Vickers outlined the lifecycle of a system, emphasizing its virtual emergence, physical realization, operational phase, and eventual retirement [16]. Halenar et al. advocated for an inclusive virtualization process spanning all product lifecycle stages, albeit acknowledging the complexity of implementation [40]. Reim et al. focusses on digital platforms and DT collaboration [41].

There are also articles provided partial coverage of DT [20, 35–37, 42–44]. Becker et al. highlighted challenges in building valid digital twins of organizations but did not delve engineering aspect [42]. Ji et al. discussed the enhancement of industrial performance through digital technologies in buyer-supplier relationships, acknowledging DT's role but not extensively exploring its implementation [43].

Several publications do not cover DT, indicating significant gaps in integrating this concept into buyer supplier relationship contexts. This is in line with Tao et al. who identify collaborations based on DTs as important future research needs [19].

4.3 Principal-Agent Theory

Principal-Agent Theory is less frequently discussed compared to VE and DT. Studies by Maier et al. and Mardenli and Sackmann extensively discussed agency problems in the context of cyber-physical systems [45, 46]. Maier et al. addressed agency problems within supply chain management lifecycles [45], while Mardenli and Sackmann identified digitalization as a factor influencing information asymmetries between buyers and suppliers [46]. Panda and Leepsa focusing on agency costs and factors mitigating them [47]. Furthermore Asif et al. and Yaacob et al. address PAT in the context of reducing information asymmetry and solving agency problems through digitalization and cyber-physical systems [38].

Conversely, other studies only partially covered PAT but explored the impact of digital technologies on industrial performance without specifically addressing PAT [17, 20, 41, 43, 44, 48].

Several studies do not cover PAT at all so it should be noted that there is a gap in the literature when it comes to going into detail on the application of digitization technologies and, for example, examining the potential for reducing information asymmetries through virtual engineering or digital twin.

4.4 Classification of the Research Areas in the Engineering Lifecycle

Based on literature insights, we derived a scope comparison of the relevant research areas. In order to classify the results in a comparable way, we have taken the results of Konstaninov et al. as a basis, who deal with the engineering lifecycle in their work and consider various definitions and arrive at the universal classification into Concept, Design & Engineering, Layout & Configuration, Build, Commissioning and Operation & Maintenance [6]. It was noticeable that the fields of application were categorized similar in the engineering lifecycle. Although the terms and definitions are not always identical and clear-cut, an overview of the different classifications is provided in Fig. 5. Virtual engineering can already be used in concepts and particularly early phases [35], the topic of digital twin is mostly considered from the design & engineering or build phase phase at the earliest [19] but also publications discusses DTs in concept phase [19]. For example, a production facility that is still in the design and engineering process can already have a Digital Twin describing the future characteristics of the facility [21]. However, there are also different approaches within the research areas. In the area of virtual engineering, for example, either the entire lifecycle [12] can be considered or only the early phases [35]. The topic of digital twin is also increasingly being considered across all phases from design to operation [19], but there is also literature that only considers the topic from physical construction onwards [39]. Once the concept has been created, an external supplier is often brought in for complex production systems and works closely with the buyer until commissioning. Digital tools can therefore improve collaboration between these phases.

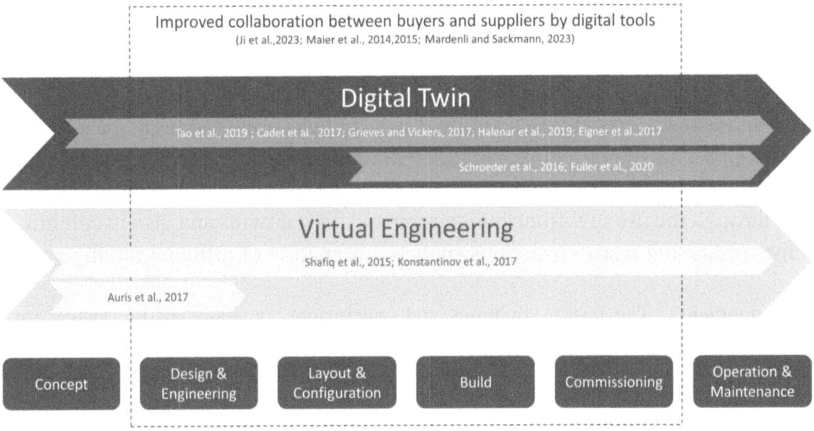

Fig. 5. Overview temporal definitions Horizon of DT, VE with regards to production facility lifecycle

5 Theoretical and Managerial Contributions, Limitations and Outlook

Virtual engineering and digital twins are powerful approaches to enhance buyer-supplier relationship management within complex production facilities. By integrating virtual engineering and Digital Twin methods and technologies, organizations can realize benefits such as enhanced collaboration, cost reduction, and improved product quality. However, there is a lack of a holistic view of virtual engineering and digital twins and their effects on known areas of tension with regard to information asymmetries within the principal agent theory. This article addresses the theoretical gap by examining the application from virtual engineering and digital twins as interaction between contractual partners in the planning of complex production facilities, thereby providing a comprehensive understanding of virtual engineering within the broader context of collaborative project planning. The literature review reveals a strong focus on Virtual Engineering and Digital

Twin technologies in the literature, with significant discussions around their application and challenges. However, Principal-Agent Theory is less frequently integrated, suggesting a gap in understanding the agency problems within digital transformations in supply chain management. Future research should aim to bridge this gap by exploring the intersections of VE, DT, and PAT more thoroughly, particularly in addressing how digital technologies such as VE and DT can mitigate agency problems and enhance collaboration in industrial processes.

This showcases potentials for researchers in the field of production planning and virtual engineering for future research and application in practice. The results can be useful in practice for responsible planners and their management on both buyer and supplier side to provide them with an understanding of how quality assurance of the planning process can be carried out time and cost-effectively in early project phases by means of virtual engineering and digital twins. While these preliminary results look promising, the results are limited by a small sample of publications. It would be important to validate these results with insights from practice and industry and, if necessary, expand them. Here, the results should be further validated on practical cases with a broader base of views on the research field. In addition, Case studies on the researched challenges in the planning process of production facilities should be carried out. The question remains as to how the processes between buyers and suppliers can be designed as efficiently as possible through the use of virtual engineering and digital twins and also in combination with other promising research streams such as automation or artificial intelligence.

Acknowledgements. The results, opinions and conclusions expressed in this paper are not necessarily those of Volkswagen Aktiengesellschaft.

References

1. Meyer, L., Seiz, M.: Industrie 4.0 im Mittelstand: Deloitte (2019)
2. Baraldi, E., Proença, J.F., Proença, T., et al.: The supplier's side of outsourcing: taking over activities and blurring organizational boundaries. Ind. Mark. Manag. **43**, 553–563 (2014). https://doi.org/10.1016/j.indmarman.2014.02.012
3. Konstantinov, S., Hansen, J., Assad, F., et al.: An analysis of the available virtual engineering tools for building manufacturing systems digital twin. Procedia CIRP **116**, 570–575 (2023). https://doi.org/10.1016/j.procir.2023.02.096
4. Albers, A., Weissenberger-Eibl, M.A., Duehr, K., et al.: Literature-based identification of success-relevant influencing factors of distributed product development. Procedia CIRP **91**, 415–420 (2020). https://doi.org/10.1016/j.procir.2019.11.007
5. Papulová, Z., Gažová, A., Šufliarský, Ľ.: Implementation of automation technologies of industry 4.0 in automotive manufacturing companies. Procedia Comput. Sci. **200**, 1488–1497 (2022). https://doi.org/10.1016/j.procs.2022.01.350
6. Konstantinov, S., Assad, F., Ahmad, B., et al.: Virtual engineering and commissioning to support the lifecycle of a manufacturing assembly system. Machines **10**, 939 (2022). https://doi.org/10.3390/machines10100939
7. Verein Deutscher Ingenieure. Digital Factory: Digital Factory Operations ICS 03.100.50 (4499) (2011). https://www.vdi.de/richtlinien/details/vdi-4499-blatt-2-digitale-fabrik-digitaler-fabrikbetrieb. Accessed 06 Dec 2022

8. Albo, A., Svedlund, L., Falkman, P.: Modular Virtual preparation method of production systems using a digital twin architecture. In: 2021 26th IEEE International Conference on Emerging Technologies and Factory Automation (ETFA): Proceedings, Västerås, Sweden - online, 07–10 September 2021, pp. 1–8. IEEE, Piscataway (2021)
9. Janecki, L., Reh, D., Arlinghaus, J.C.: Challenges of quality assurance in early planning and ramp up of production facilities - potentials of planning automation via virtual engineering. Procedia Comput. Sci. **232**, 2498–2507 (2024). https://doi.org/10.1016/j.procs.2024.02.068
10. Cecil, J., Kanchanapiboon, A.: Virtual engineering approaches in product and process design. Int. J. Adv. Manuf. Technol. **31**, 846–856 (2007). https://doi.org/10.1007/s00170-005-0267-7
11. Dombrowski, U., Reiswich, A., Alexander, K.: Designing digital tools for factory planning: Integrating requirements for usability on a meta-level. In: 2018 IEEE 23rd International Conference on Emerging Technologies and Factory Automation (ETFA): Politecnico di Torino, Torino, Italy, 04–07 September 2018, Proceedings, pp. 99–106. IEEE, Piscataway (2018)
12. Shafiq, S.I., Sanin, C., Szczerbicki, E., et al.: Virtual engineering object/virtual engineering process: a specialized form of cyber physical system for Industrie 4.0. Procedia Comput. Sci. **60**, 1146–1155 (2015). https://doi.org/10.1016/j.procs.2015.08.166
13. Bysko, S., Krystek, J., Bysko, S.:X Automotive paint shop 4.0. Comput. Ind. Eng. **139**, 105546, pp. 99–106. https://doi.org/10.1016/j.cie.2018.11.056
14. Boyes, H., Watson, T.: Digital twins: an analysis framework and open issues. Comput. Ind. **143**, 103763 (2022). https://doi.org/10.1016/j.compind.2022.103763
15. Ivanov, D.: Conceptualisation of a 7-element digital twin framework in supply chain and operations management. Int. J. Prod. Res. **62**, 2220–2232 (2024). https://doi.org/10.1080/00207543.2023.2217291
16. Grieves, M., Vickers, J.: Digital twin: mitigating unpredictable, undesirable emergent behavior in complex systems. In: Kahlen, F.-J., Flumerfelt, S., Alves, A.C. (eds.) Transdisciplinary perspectives on complex systems: New findings and approaches, Softcover reprint of the hardcover 1st edition 2016, pp. 85–113. Springer, Cham (2017)
17. Stark, R., Damerau, T.: Digital twin. In: Chatti, S., Tolio, T. (eds.) CIRP Encyclopedia of Production Engineering, pp. 1–8. Springer, Heidelberg (2019)
18. Trauer, J., Schweigert-Recksiek, S., Engel, C., et al.: What is a digital twin? – definitions and insights from an industrial case study in technical product development. Proc Des Soc.: Des Conf. **1**, 757–766 (2020). https://doi.org/10.1017/dsd.2020.15
19. Tao, F., Sui, F., Liu, A., et al.: Digital twin-driven product design framework. Int. J. Prod. Res. **57**, 3935–3953 (2019). https://doi.org/10.1080/00207543.2018.1443229
20. Eigner, M., Detzner, A., Schmidt, P.H., et al.: Definition des Digital Twin im Produktlebenszyklus. Zeitschrift für wirtschaftlichen Fabrikbetrieb **114**, 345–350 (2019). https://doi.org/10.3139/104.112107
21. Kuhn, T.: Digitaler Zwilling. Informatik Spektrum **40**, 440–444 (2017). https://doi.org/10.1007/s00287-017-1061-2
22. Layer, M., Leidich, J., Schwoch, S., et al.: Data management of process plants as complex systems: systematic literature review and identification of challenges and opportunities. Rev. Chem. Eng. 40, 329–349 (2023). https://doi.org/10.1515/revce-2022-0077
23. Ross, S.A.: The economic theory of agency: the principal's problem. Am. Econ. Rev. **63**, 134–139 (1973)
24. Mitnick, B.M.: The theory of agency. Public Choice **24**, 27–42 (1975). https://doi.org/10.1007/BF01718413
25. Jensen, M.C., Meckling, W.H.: Theory of the firm: managerial behavior, agency costs and ownership structure. J. Finan. Econ. **3**, 305–360 (1976). https://doi.org/10.1016/0304-405X(76)90026-X
26. Eisenhardt, K.M.: Agency theory: an assessment and review. Acad. Manag. Rev. **14**, 57 (1989). https://doi.org/10.2307/258191

27. Nwajei, U.O.K., Bølviken, T., Hellström, M.M.: Overcoming the principal-agent problem: the need for alignment of tools and methods in collaborative project delivery. Int. J. Project Manag. **40**, 750–762 (2022). https://doi.org/10.1016/j.ijproman.2022.08.003
28. Steinle, C., Schiele, H., Ernst, T.: Information asymmetries as antecedents of opportunism in buyer-supplier relationships: testing principal-agent theory. J. Bus. Bus. Mark. **21**, 123–140 (2014). https://doi.org/10.1080/1051712X.2014.903457
29. Lee, B.D., Paredis, C.J.: A Conceptual framework for value-driven design and systems engineering. Procedia CIRP **21**, 10–17 (2014). https://doi.org/10.1016/j.procir.2014.06.147
30. Mei, Q., Zhou, Q., Liu, S., et al.: A safety signaling game for small and medium-sized manufacturing enterprises in the supply chain. Int. J. Inf. Syst. Supply Chain Manag. **13**, 118–135 (2020). https://doi.org/10.4018/IJISSCM.2020010106
31. Dalenogare, L.S., Benitez, G.B., Ayala, N.F., et al.: The expected contribution of Industry 4.0 technologies for industrial performance. Int. J. Prod. Econ. **204**, 383–394 (2018). https://doi.org/10.1016/j.ijpe.2018.08.019
32. Webster, J., Watson, R.T.: Analyzing the past to prepare for the future: writing a literature review. MIS Q. **26**, xiii–xxiii (2002)
33. Xiao, Y., Watson, M.: Guidance on conducting a systematic literature review. J. Plan. Educ. Res. **39**, 93–112 (2019). https://doi.org/10.1177/0739456X17723971
34. Teiwes, H.: Stand der Forschung. In: Teiwes, H. (ed.) Systematische Steigerung der Energieeffizienz im Karosseriebau, vol. 150, pp. 35–62. Springer Fachmedien Wiesbaden, Wiesbaden (2020)
35. Auris, F., Suss, S., Schlag, A., et al.: Towards shorter validation cycles by considering mechatronic component behaviour in early design stages. In: 2017 22nd IEEE International Conference on Emerging Technologies and Factory Automation, Limassol, Cyprus, 12–15 September 2017, pp. 1–4. IEEE, Piscataway (2017)
36. Cadet, M., Sinnwell, C., Fischer, J., et al.: Referenzentwicklungsprozess für cybertronische Produkte und Produktionssysteme. In: Muggeo, C., Eigner, M., Koch, W. (eds.) Modellbasierter Entwicklungsprozess cybertronischer Systeme: Der PLM-unterstützte Referenzentwicklungsprozess für Produkte und Produktionssysteme, pp. 45–62. Springer Vieweg, Heidelberg (2017)
37. Konstantinov, S., Ahmad, M., Ananthanarayan, K., et al.: The cyber-physical E-machine manufacturing system: virtual engineering for complete lifecycle support. Procedia CIRP **63**, 119–124 (2017). https://doi.org/10.1016/j.procir.2017.02.035
38. Asif, M., Searcy, C., Castka, P.: Exploring the role of industry 4.0 in enhancing supplier audit authenticity, efficacy, and cost effectiveness. J. Clean. Prod. **331**, 129939 (2022). https://doi.org/10.1016/j.jclepro.2021.129939
39. Schroeder, G., Steinmetz, C., Pereira, C.E., et al.: Visualising the digital twin using web services and augmented reality. In: 2016 IEEE 14th International Conference on Industrial Informatics (INDIN), pp. 522–527 (2016)
40. Halenar, I., Juhas, M., Juhasova, B., et al.: Virtualization of production using digital twin technology. In: Kot, A., Nawrocka, A. (eds.) Proceedings of the 2019 20th International Carpathian Control Conference (ICCC), Kraków, Wieliczka, Hotel Turówka, Poland, 26–29 May 2019, pp. 1–5. IEEE, Piscataway (2019)
41. Reim, W., Andersson, E., Eckerwall, K.: Enabling collaboration on digital platforms: a study of digital twins. Int. J. Prod. Res. **61**, 3926–3942 (2023). https://doi.org/10.1080/00207543.2022.2116499
42. Becker, M.C., Pentland, B.T. (eds.): Digital Twin of an Organization: Are You Serious? Springer, Heidelberg (2021)
43. Ji, K., Liu, X., Xu, J.: Digital economy and the sustainable development of China's manufacturing industry: from the perspective of industry performance and green development. Sustainability **15**, 5121 (2023)

44. Radke, A.M., Isa, J.B., Tseng, M.M., et al.: A framework for production planning and control as a service. In: 2019 IEEE 7th Conference on Systems, Process and Control (ICSPC 2019): Proceeding, Malaysia: Hatten Hotel, Hatten Square, Jalan Merdeka, Bandar Hilir, 75000 Melaka, Malaysia, 13–14 December 2019, pp. 14–19. IEEE, Piscataway (2019)
45. Maier, M.A., Korbel, J.J., Brem, A.: Innovation in supply chains - solving the agency dilemma in supply networks by using industry 4.0 technologies. IJCNDS **15**, 235 (2015). https://doi.org/10.1504/IJCNDS.2015.070976
46. Mardenli, A., Sackmann, D.: Identification of asymmetric information in the supplier–buyer relationship: a structural review. Supply Manag. Res., 169–218 (2023). https://doi.org/10.1007/978-3-658-42635-4_9
47. Panda, B., Leepsa, N.M.: Agency theory: review of theory and evidence on problems and perspectives. Indian J. Corp. Gover. **10**, 74–95 (2017). https://doi.org/10.1177/0974686217701467
48. Nieto, M.J., Santamaria, L., Bammens, Y.: Digitalization as a facilitator of open innovation: are family firms different? Technovation **128**, 102854 (2023). https://doi.org/10.1016/j.technovation.2023.102854

Optimization

A Cost Effective Two-Step Exploration for Constrained Two-Dimensional Packing Problems

Hyeongoo Pyeon[1] and SookYoung Son[1,2(✉)]

[1] AI Center, HD Korea Shipbuilding and Ocean Engineering Co., 477, Bundangsuseo-ro, Seongnam-si, Gyeonggi-do, Republic of Korea
{hgpyeon,sson}@hd.com

[2] Department of Naval Architecture and Ocean Engineering, Seoul National University, 1, Gwanak-ro, Gwanak-gu, Seoul, Republic of Korea
sson527@snu.ac.kr

Abstract. A combinatorial optimization problem, particularly an NP-hard challenge, aims to find an optimal solution within a discrete search space and has been applied across diverse industries. However, practical use-cases often involve numerous constraints, making the optimization problem significantly more complex as the search space expands. To address this issue, this paper presents a new methodology that efficiently solves two-dimensional bin packing problems with multiple constraints. We propose a two-step exploration technique, consisting of combination and permutation, and apply it to the practical design of LNG vessels. The proposed methodology not only enhances efficiency in search time and space but also ensures the discovery of global optimal solutions without requiring additional exploration. Our results demonstrate remarkable improvements in both time and spatial efficiency compared to brute-force global searches. The methodology proposed in this paper is expected to contribute significantly to technological advancements in solving combinatorial optimization problems.

Keywords: combinatorial optimization · packing problem · NP-hard · constrained optimization problem · two-step exploration · time and spatial efficiency · optimal solution

1 Introduction

Combinatorial optimization problems entail identifying the optimal solution that satisfies the criteria of a given objective function within a discrete search space. These problems encompass complex challenges such as job-shop scheduling, the traveling salesman problem (TSP), and bin packing problems. Among these, the bin packing problem is particularly significant. It involves packing items in fixed bins (or spaces, containers) and can be defined in various ways depending on the shapes, dimensions, and positioning rules of the items and bins. A

Published by Springer Nature Switzerland AG 2024
M. Threr et al. (Eds.): APMS 2024, IFIP AICT 733, pp. 197–212, 2024.
https://doi.org/10.1007/978-3-031-71645-4_14

notable instance of the bin packing problem is the nesting problem (or cutting problem), which focuses on arranging non-standard or irregularly shaped items within a space without overlapping to minimize wasted space. Another example is the efficient loading of items into three-dimensional bins in storage facilities, illustrating the widespread occurrence of bin packing problems across diverse industries with varying constraints and objectives.

The bin packing problem is classified as NP-hard, prompting extensive research into its resolution using metaheuristic methods such as Genetic Algorithms (GA) and Simulated Annealing (SA). For instance, Jacobs [6], Liu, and Teng [11] designed a GA algorithm wherein the order of items is represented as permutations, and the item placement is determined using a bottom-left strategy. Kapmke [7] applied SA, while Blazewicz [1] was the first to employ Tabu search to address this problem.

In the context of the shipbuilding industry, the bin packing problem manifests during the design of LNG storage tanks for transportation. LNG tanks are constructed by welding insulation panels and membrane sheets onto the hull to maintain the cryogenic state of LNG. To reduce production costs, shipyards seek optimal designs that minimize the number of panels attached to the tank's ten walls and reduce the welding lines of the membrane sheets on top of the panels. This can be defined as a two-dimensional packing problem involving the arrangement of rectangular-shaped items, aiming to fully pack the bin without waste. The solution must adhere to predefined item positioning rules and output an optimal arrangement that meets the objective function. The constraints considered include production safety, equipment specifications, and classification society regulations, with different optimal solutions possible depending on the combination of constraints. The key challenge is to rapidly identify the optimal item arrangement under various constraints. In practice, the optimal arrangement is typically sought through brute-force methods, which exhaustively search all possible combinations to determine the best solution. However, this approach is time-consuming, inflexible to changes in constraints, and often does not guarantee an optimal solution (Fig. 1).

Therefore, this study introduces a two-step exploration methodology, designed to swiftly explore optimal solutions under varying constraints. This approach segments the two-dimensional space into one-dimensional backbone and rib sections, facilitating systematic exploration through combinations and permutations within each segment. The backbone represents a reference direction in which fewer combinations can be formed from the horizontal and vertical directions of the two-dimensional space, derived from the combination of existing items. By storing the exploration results sorted according to the objective function for each constraint, the optimal solution can be quickly identified without re-searching, even if constraints are added or changed. We validated our methodology through experiments involving the design of LNG vessel tanks in the shipbuilding industry. The results demonstrated that our proposed method significantly outperforms traditional brute-force search methods and GA in terms of time and memory efficiency.

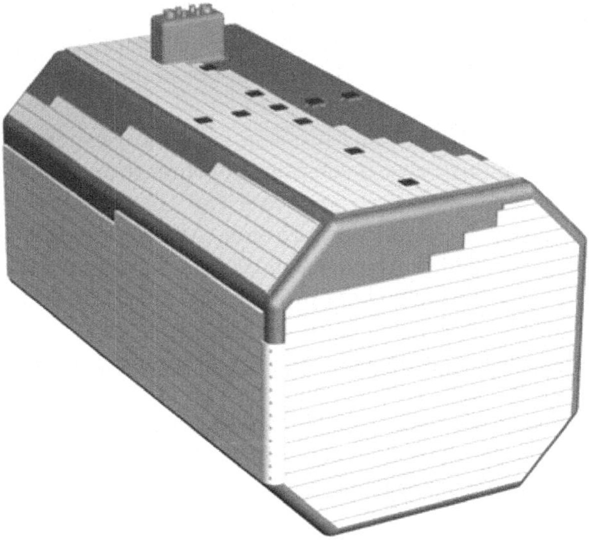

Fig. 1. Schematic arrangement for LNG CCS tank design [14]

This paper is structured as follows. Firstly, we review existing studies related to the bin packing problem. Next, we propose the specific problem definition and methodology employed in this study. Subsequently, we present the experimental results, including a case study involving the design of LNG vessel tanks in the shipbuilding industry. Finally, we summarize the contributions of this research and outline potential avenues for future work.

2 Background

Empty packaging issues involve loading items into a given space or cutting materials of specific sizes such as paper, steel, and glass. The bin packing problem can be classified into 1-dimensional, 2-dimensional, or 3-dimensional bin packing problems depending on the dimensions of the bin and items. In this paper, we will focus on the 2-dimensional bin packing problem (2D-BPP). Various approaches have been attempted in the literature to solve the 2D-BPP based on the shape, stacking, and cutting standards of the items defined in the problem. Given that these problems are NP-hard and easily converge to local solutions, numerous studies have employed metaheuristics to address them [9].

Ferreira evaluated four heuristic algorithms-first fit decreasing, local search, SA, and GA-on a rectangular 2D-BPP problem with various constraints. These algorithms showed some limitations in terms of time complexity and optimality, meaning that the optimal solution is not guaranteed [5]. Jean-François addressed the 2D-BPP using a divide-and-conquer strategy known as Benders decomposition [2]. This algorithm breaks down the problem and variable sets into two parts and iteratively solves each part based on the results of the previous step. However, it does not guarantee a global optimal solution as it terminates when the difference between the lower bound and upper bound solutions is acceptable, which their experimental results also confirmed [4]. S. Böhm compared a greedy search approach for the 2D packing problem of placing wooden pieces into trolleys in the furniture industry with constraint optimization. They adopted a best fit heuristic for the greedy search and used the Gurobi solver to test numerous metaheuristics for constraint optimization. Although both techniques found optimal solutions, constraint optimization required more time [3].

However, these methods generally find good solutions for some problems and only average ones for others. To address this, Lopez et al. proposed a unified hyper-heuristic framework that produces a deterministic algorithm capable of generating results quickly and deterministically. This approach yields results comparable to those obtained using the best problem-specific heuristics without needing to try every heuristic individually [12].

With the success of deep learning in various fields, recent research has begun to tackle combinatorial optimization problems using neural networks. Zhao et al. discretized the 2D space to create a height map and recorded the maximum height on each grid. They used two convolutional neural networks: one to predict the probability of each action and another to eliminate illegal actions [15]. The BPP problem generally consists of three stages: item selection, orientation selection, and position selection. If these stages are processed simultaneously, the BPP problem can be treated as a Markov Decision Process (MDP). Otherwise, the last two stages are not strict MDPs as they are influenced by the previous stage. To resolve this, Li et al. used the prediction of the last stage as a conditional query to formulate a strict MDP, although it remains vulnerable to the geometric representation of placed items [10]. Alexandre et al. introduced Ranked Reward (R2) algorithms inspired by adversarial self-play in two-player games like AlphaZero [13]. Since the BPP problem is akin to a single-player game, they ranked the rewards obtained by a single agent over multiple games to create a relative performance metric. Their experiments demonstrated that this algorithm outperforms generic Monte Carlo tree search [8].

Despite the diversity of research on combinatorial optimization problems, applying these solutions to real-life problems, particularly industrial ones, remains challenging. Industrial problems differ significantly from well-defined classical problems due to their detailed conditions and frequent variable constraints. Most importantly, the optimality of a solution is critical to profitability in industry. However, previous studies only guarantee acceptable solutions rather than optimal ones, which could result in substantial financial costs. To overcome these issues, we have developed a clear and robust exploration method that ensures the optimal solution in a short time and explains the reasons for failure when the optimal solution is not found. This method is particularly powerful when multiple constraints are present, as changing constraints does not require re-exploration. A detailed description of our method is provided in Sect. 3.

3 TETP: Two-Step Exploration for Two-Dimensional Packing Problem

This section defines the two-dimensional bin packing problem with the specific constraints this paper aims to address and introduces the methodology used to solve it. Detailed explanations of the objective function and constraints in LNG design will be provided in the case study Sect. 4.

3.1 Problem Definition

The two-dimensional bin packing problem addressed in this paper involves arranging n similarly rectangular-shaped items, each with width w and height h, into a rectangular bin with fixed dimensions W and H, aiming to pack them without leaving empty spaces. Unlike the general packing problem, which often focuses on minimizing empty space and reducing waste, our goal is to completely fill the space by combining items seamlessly.

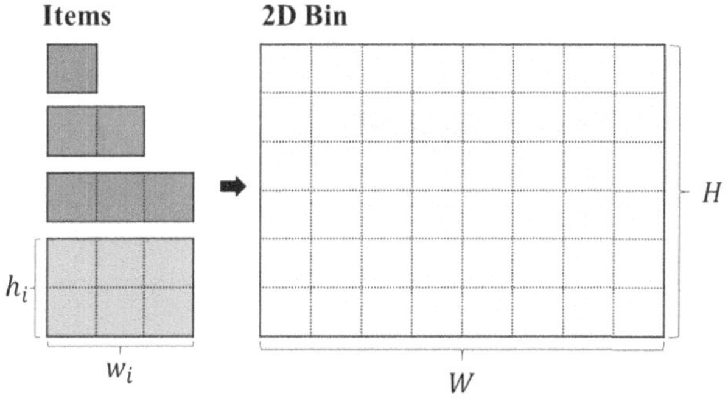

Fig. 2. Two-dimensional bin packing problem

Objective functions and constraints can vary depending on specific application cases. Generally, the bin packing problem seeks to minimize the number of items required to fill a bin. Within this framework, multiple sets of combinations sharing the same objective function can arise, depending on the sorting and placement order of the items. Therefore, the goal is to identify the optimal placement combination within these sets, considering additional placement conditions. Our focus extends beyond simple item count minimization to incorporate specific constraints and conditions that enhance the overall optimization of the placement arrangement (Table 1).

Define the set of unique widths and height of items as $\bar{W}$ and $\bar{H}$ subject to combination of items for backbone B^C and ribs R^C. Now we can formulate the objective function for width and height as follows:

$$W = a_0\bar{w}_0 + a_1\bar{w}_1 + \ldots + a_k\bar{w}_k$$

Table 1. Notation

Indices:	
M	Number of bins ($m \in \{1, \ldots, M\}$)
N	Number of items ($n \in \{1, \ldots, N\}$)
Parameters:	
W	Width of Bin
H	Hegith of Bin
w_i	Width of ith item
h_i	Height of ith item
$\bar{W}$	unique width $\{\bar{w}_1, \ldots, w_k\}$ where $k \leq i$
$\bar{H}$	unique height $\{\bar{h}_1, \ldots, \bar{w}_l\}$ where $l \leq i$
B	Backbone (vertical direction in which the width of the item is placed)
R	Rib (vertical direction in which the length of the item is placed)
B^C	combination of items for backbone
R^C	combination of items for rib
B^σ	permutation of items for backbone
R^σ	permutation of items for rib
C	set of constraints
$\bar{A}$	solution set of linear combinations
$\tilde{A}$	optimized solution set of linear combinations
Output:	
X	set of optimized item coordinates

$$H = b_0\bar{h}_0 + b_1\bar{h}_1 + \ldots + b_l\bar{h}_l$$

where a_j and b_j are non-negative integers such that $j = 0, 1, \ldots, k$ and $j = 0, 1, \ldots, l$ respectively. Thus, the solution sets of linear combinations can be defined as follows:

$$\bar{A} = \{A_0, A_1, \ldots, A_n\}$$

where

$$A_m = \{a_0^m, \ldots, a_k^m\} \quad \text{where} \quad m = 0, \ldots, n$$

$$\bar{B} = \{B_0, B_1, \ldots, B_p\}$$

where

$$B_q = \{b_0^q, \ldots, b_l^q\} \quad \text{where} \quad q = 0, \ldots, p$$

To find the optimal solution, we need to satisfy both width and height constraints:

$$\tilde{W} = a_0\bar{w}_0 + a_1\bar{w}_1 + \ldots + a_k\bar{w}_k$$

$$\tilde{H} = b_0\bar{h}_0 + b_1\bar{h}_1 + \ldots + b_l\bar{h}_l$$

Thus, we need to find the set $\tilde{A}$ that satisfies both constraints:

$$\tilde{A} = \{(A, B) \mid A \in \bar{A}, B \in \bar{B}\}$$

To solve this problem, item positioning rules adhere to several fundamental principles: items must not overlap or extend beyond the designated space, and altering the shape of the items, such as cutting or rotating them, is strictly prohibited. Items may be reused, and some items may remain unused. Additional placement options, acting as constraint variants, may depend on specific application considerations, which will be detailed in the Case Study section. For instance, in the insulation panel arrangement of the LNG tank discussed in this paper, a crucial condition is that the outer edge of the panel must not overlap with the membrane sheet line installed on the panel. An optional condition could be that larger panels are preferred on the outer side of the bin and that using more items of the same type is advantageous. These optional conditions can be selectively applied while ensuring the required conditions are met.

Consider an example where items can range in width from 2 to 8, and the goal is to find a combination of items that fill a width of 37. There could be two item combinations that achieve this: [8, 7, 7, 7, 8] and [8, 8, 5, 8, 8]. Both combinations fulfill the objective function, as the number of items and item types are identical. However, if the constraint favors smaller size differences between consecutive items, then [8, 7, 7, 7, 8] might be considered the optimal arrangement.

3.2 Methodology

Therefore, this paper requires an algorithm that can quickly find the optimal arrangement according to changing selection conditions. Achieving an optimal solution in the 2D-BPP problem necessitates exploring all possible arrangements. However, due to its NP-hard nature, solving this problem within polynomial time is impractical, and significant memory space is needed to store the multitude of search results. While backtracking can provide some improvements, determining appropriate backtracking logic becomes challenging when diverse constraints are involved.

To address these challenges, we propose a method that divides the 2D space into two 1D spaces, namely the backbone and rib, which are then combined to form the complete 2D space arrangement. To minimize time and memory wastage on incorrect arrangements, the search process is bifurcated into two steps. In the initial step, a candidate combination suitable for arrangement in the backbone and rib directions is identified. Subsequently, in the second step, a permutation is generated and checked for constraint satisfaction. Unlike the brute-force search method, which demands $O(mn)$ memory space per arrangement for all arrangements, our proposed method (TETP) requires only $O(m + n)$ memory space. This efficiency is attributed to the division of 2D space and the avoidance of storing all arrangement results, where m and n represent the number of items used in both directions.

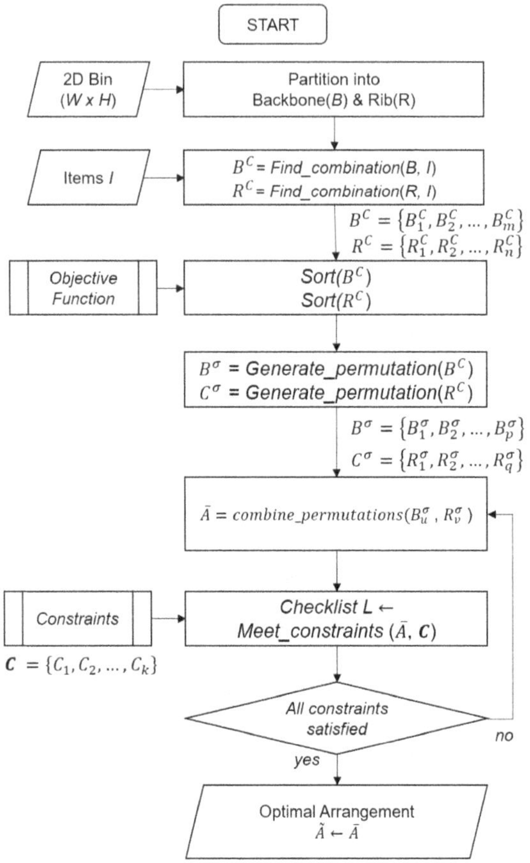

Fig. 3. Flow chart of TETP method

A detailed process is depicted in Fig. 3 and Algorithm 1.

1. Divide the 2D space into the backbone and rib directions. In our problem, the vertical direction in which the width of the item is placed is defined as the backbone, and the horizontal direction in which the length of the item is placed is defined as the rib. This may vary depending on the problem definition or constraints.
2. Calculate combinations that can fill each direction with the width or length of an item. For example, if the length of the rib is 5 and the lengths of the items are [1, 2, 3], the possible combinations are as follows:
   $\{[3, 2], [2, 2, 1], [2, 1, 1, 1], [1, 1, 1, 1, 1]\}$.
   The order is not considered at this step. That is, if [3, 2] is calculated, [2, 3] would not be counted.

Algorithm 1. TETP: Two-step Exploration for Two-dimensional Packing problem

Input: Backbone B, Rib R, Items $I = \{(w_i, h_i)\}_{i=1}^{N}$,
Set of constraints $\mathbf{C} = \{C_1, C_2, \cdots, C_k\}$

Output: Optimal arrangement $\tilde{A}$,
Checklist of constraints L

Determine Backbone B and Rib R
Find combinations of B and R with Items I
  $\mathbf{B^c} = \mathit{find\_combinations}(B, I)$
  $\mathbf{R^c} = \mathit{find\_combinations}(R, I)$
Sort the combinations $\mathbf{B^c}$ and $\mathbf{R^c}$ based on an objective function
while more combinations available **do**
  Generate permutations B_u^σ, R_v^σ for next combination
  Make an arrangement $\bar{A}$ by combining B_u^σ and R_v^σ
  for C_r in Constraints $\mathbf{C}$ **do**
    $L[r] \leftarrow \mathit{meet\_constraints}(\bar{A}, C_r)$
  end for
  if all constraints are satisfied **then**
    $\tilde{A} \leftarrow \bar{A}$
    break
  end if
end while

3. Sort the combinations based on an objective function. The previous example is the result of sorting when the number of items is defined as the objective function.
4. Generate permutations for the combination of each direction, and combine them to complete the 2D arrangement.
5. Check that the arrangement meets all constraints. Whether the arrangement satisfies each constraint or not is recorded on a checklist as seen in Fig. 4.
6. If the arrangement satisfies all constraints, return it as the optimal arrangement. If not, go back to step (4) and make the next arrangement with the next permutations. Repeat steps (4) to (6) until an optimal solution is achieved. Depending on whether the reason for not meeting the constraint is in the backbone or rib, the next permutation in the corresponding direction is used. If all permutations have been attempted within one combination, the permutation is calculated again by selecting the next combination.

Since the combinations are sorted with an objective function, and the search is terminated when the arrangement that satisfies all constraints is obtained, the optimal solution can be obtained considerably faster than conventional methods. However, there are cases where the optimal arrangement cannot be obtained depending on the problem and constraints. In such cases, by looking at the checklist generated in step (6), it is possible to understand why an optimal arrangement was not found, and the optimal arrangement can be obtained through appropriate modification of the constraint(s). For example, if the constraint C_3 is turned off in Fig. 4, A_1 will be obtained as the optimal arrangement.

A simple application of the TETP is as follows. We have a 2D space of 10 by 16, and 4 items $I = \{(1,2),(1,4),(2,3),(2,4)\}$ as seen in Fig. 2. Firstly, divide the 2D space into a backbone of length 10 and a rib of length 16. Next, create the combinations of the backbone $\{[2,2,2,2,2],[2,2,2,2,1,1],...[1,1,...,1]\}$ and the combinations of the rib $\{[4,4,4,4],[4,3,3,3,3],...[1,1,...,1]\}$. Then, sort them based on an objective function, i.e., the number of items. Now, generate permutations on the first combination of each direction, and combine them to make a 2D arrangement. The first arrangement A_1 will be produced as depicted on the left in Fig. 5. Check constraints and return this arrangement if all the constraints are satisfied. If not, proceed to the next permutations and repeat the procedures of combining permutations to make a 2D arrangement and inspecting constraints. If the optimal solution is not achieved by the last combinations and permutations, A_p will be returned with a failure (Fig. 8 and Table 2).

Sorted Arrangement	Constraints C_1	C_2	C_3	...
A_1	O	O	X	
A_2	O	O	X	
A_3	O	O	X	
A_4	X	O	X	
A_5	X	O	O	
A_6	X	O	O	
A_7	X	O	O	
A_8	O	O	O	
...				

Fig. 4. A checklist of whether arrangements satisfy each of the constraints

4 Experiment Results

To verify the performance of the TETP methodology, we conducted an experiment using an LNG vessel design problem from the shipbuilding industry.

Due to the recent surge in global natural gas demand, there has been a substantial increase in the volume of LNG transportation worldwide. Consequently, there is a rising demand for LNG carriers, responsible for transporting LNG, and LNG dual-fuel ships, which use LNG as an environmentally friendly fuel for ship propulsion. LNG vessels require specifically designed Cargo Containment System (CCS) tanks with insulation panels to facilitate the storage and transportation of LNG at cryogenic temperatures. The sealed CCS tank, which consists of 10 walls, is manufactured by attaching rectangular insulation panels to the hull and attaching membrane sheets on top of the panels. Shipbuilding

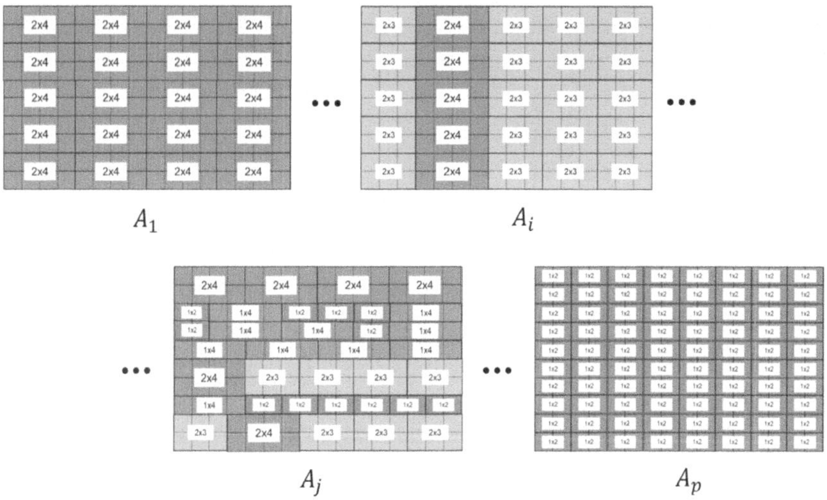

Fig. 5. Example of arrangement created in Fig. 2 ($1 < i < j < p$, p is the number of possible arrangements.)

companies face the challenge of optimizing the arrangement of panels to minimize the manufacturing costs of the CCS tanks and enhance productivity. The goal of optimal design is to reduce the type and number of insulation panels attached to the tank and to reduce the total welded length of the membrane sheet attached to the panels. During this process, ship designers must consider design constraints that take into account worker safety and production facility limitations. The design conditions used as constraints are diverse, including both mandatory conditions and selectively applicable conditions. Ship designers repeatedly modify the arrangement to find the optimal configuration according to various combinations of design conditions. Therefore, research is needed on how to quickly and easily find the optimal arrangement that can reduce the number of insulation panels and the total welding length within given design conditions.

In this context, numerous design considerations are treated as constraints, and the design process involves repetitive iterations to identify the optimal design by adjusting the placement and modifying the arrangement of panels within various tank dimensions. The problem at hand can be categorized as a 2-dimensional packing problem, the resolution of which is the primary focus of this paper. Particularly, the scenario where optimal placements must be rapidly confirmed while adapting to changes in constraints aligns well with the TETP problem proposed in this study.

We conducted experiments on four LNG vessels with different dimensions designed at a Korean shipyard. The membrane tank of an LNG vessel comprises 10 walls, as depicted in Fig. 6. Rectangular-shaped items of different sizes can be

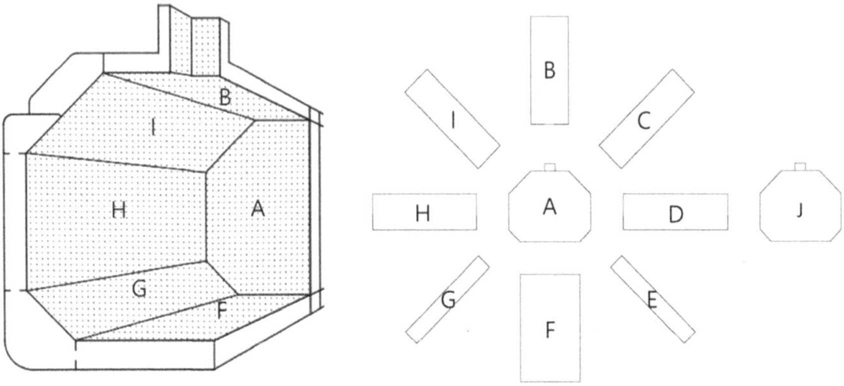

Fig. 6. Planar figure of CCS tank

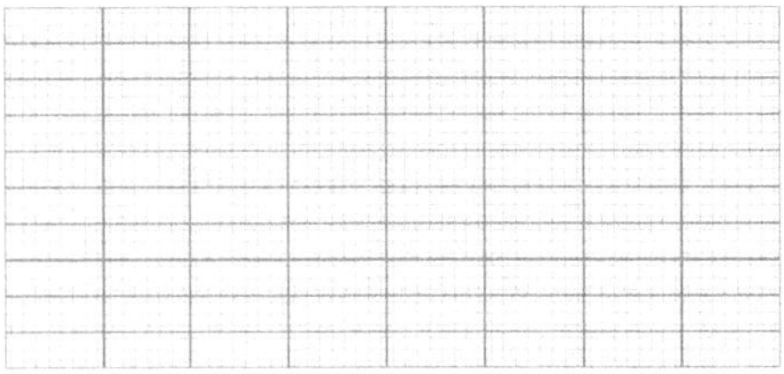

(a) Project:2, Wall: H, # of items: 80

(b) Project:3, Wall: B, # of items: 114

Fig. 7. Insulation panel placement results

placed on the 10 walls depending on the characteristics of the walls. To compare the arrangement performance of walls, we conducted an experiment on the B, F, H wall groups and I, G groups that share similarly shaped items. There is a difference in the types of items used in the two sets. Specifically, B, F, and H are filled by combining a total of 12 different item shapes, while I and G are completed by placing 28 different items. This differentiation in item shape reflects the unique requirements and characteristics of each set of walls within an LNG vessel's membrane tank. Through this experiment, we introduce the results of the tank design of the six ships using the existing brute-force global search method, the genetic algorithm, and the TETP methodology. We aim to prove

Table 2. Computational results of brute-force global search and TETP method

Case		# of items	Computation Time [s]			Computation Space		
Ship	Wall		Brute-force search	Genetic Algorithm	TETP	Brute-force search	Genetic Algorithm	TETP
1	B	114	2.77	1.3	0.18	19.1 MB	21.1 KB	8.1 KB
	F	95	1.18	1.51	0.11	8.0 MB	17.6 KB	4 KB
	I	38	0.70	0.98	0.09	0.6 MB	7 KB	1.2 KB
	H	91	10.59	1.67	0.27	50.1 MB	24.6 KB	20.4 KB
	G	57	21.71	1.21	0.36	4.5 MB	10.5 KB	5.3 KB
2	B	168	56.03	1.52	0.85	203.9 MB	28.1 KB	39.9 KB
	F	168	16.48	1.51	0.54	97.2 MB	28.1 KB	26.8 KB
	I	42	0.69	1.39	0.10	0.7 MB	7 KB	1.2 KB
	H	80	40.92	2.03	0.38	97.1 MB	28.1 KB	39.9 KB
	G	63	22.21	1.65	0.35	4.9 MB	10.5 KB	5.3 KB
3	B	114	3.13	0.49	0.36	19.1 MB	21.1 KB	5.7 KB
	F	152	56.53	1.23	0.78	184.5 MB	28.1 KB	39.9 KB
	I	76	8353.24	2.51	3.89	50.0 MB	14.1 KB	56.5 KB
	H	84	10.23	1.9	0.25	46.2 MB	24.6 KB	20.4 KB
	G	38	0.31	0.56	0.06	0.4 MB	7 KB	0.8 KB
4	B	140	5.81	1.66	0.31	36.3 MB	24.6 KB	14.2 KB
	F	180	119.53	2.14	1.33	306.0 MB	31.6 KB	61.2 KB
	I	80	3929.80	2.24	1.84	35.3 MB	14.1 KB	24.8 KB
	H	88	15.17	1.71	0.35	53.6 MB	28.1 KB	28.8 KB
	G	40	0.17	0.5	0.04	0.3 MB	7 KB	0.5 KB

the efficiency of the TETP methodology in searching for the optimal solution by comparing and analyzing the computation time and space requirements of the three methodologies targeting 30 walls.

The genetic algorithm includes the following components: generating an initial chromosome set, generating offspring, performing mutation operations, evaluating fitness, and determining termination conditions. The operation to generate the initial chromosome involves randomly selecting items one by one and listing them. Since our goal is to fill the 2D bin with as few items as possible, we used a random selection method that considers weights proportional to item size when selecting items. The crossover operation was used for generating offspring, and a point was randomly selected as the division point. Mutations were applied with a probability of 0.1, and new items were also selected at random with a weight proportional to their size. The goodness of fit was simply calculated as the area that filled the bin, but since the goal was to fill the bin without empty space, the fitness was set to 0 for unsuitable placements. Therefore, the fitness function operates like a binary function that determines whether the bin is filled exactly without empty space. Genetic algorithms do not guarantee an optimal solution, but for our problem, an optimal solution must exist, and the goal is

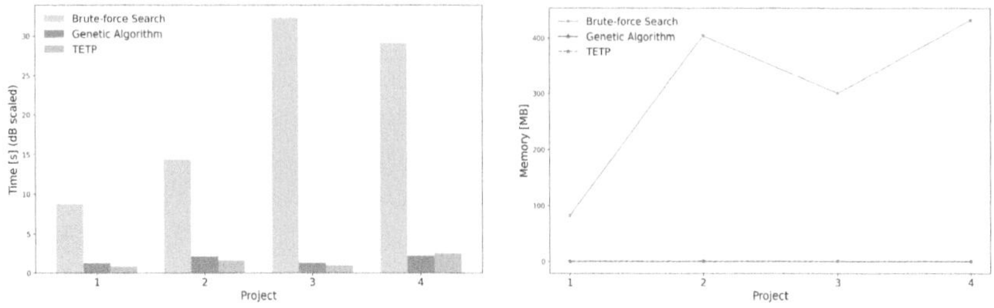

(a) Average computation time for ships (b) Total computation space for ships

Fig. 8. Experimental Results

to find it. Therefore, the termination condition was defined as when the genetic algorithm finds the optimal solution identified by the TETP. The population size and the number of generations were set to at least 30 to 100, depending on the bin size. If the genetic algorithm did not find the optimal solution, it was repeated until it did. The TETP algorithm was implemented in Python, and the experiment was carried out on a system with a 2.50 GHz Intel i9 CPU, 32 GB of RAM, and Windows 64-bit, under single core and single thread.

In this experiment, two evaluation indicators, time and space, were used to assess the performance of the proposed methodology. Time complexity refers to the time required to search for the optimal solution, and space complexity measures the memory needed to store the search results. The brute-force search method is calculated by multiplying the total number of possible batches (m), the number of items used per batch (n), and 36 bytes, which is the space required to store each item's information. Each item includes the item name, width, length, and location (x, y) information. The genetic algorithm is calculated by multiplying the number of population (m), the number of items used per population (n), the number of generations (r) and the integer data type byte. On the other hand, the TETP method is calculated by multiplying the total number of possible combinations (m), the length of the integer list representing the combination (n), and the integer data type byte. For example, if the combination is [1,3], it stores 2; if the combination is [1,2,3], it stores 3. The important assumption here is that the same combination is used at all backbone levels. If the combination varies for each backbone level, n must be multiplied by the number of backbone levels.

Figure 7 illustrates the data from the table in chart format, with (a) depicting the comparison of computation time and (b) representing computation space. To facilitate visualization, a logarithmic transformation is applied in (a). In cases where the execution time is less than 1 s, a negative number may result from the logarithmic transformation. To address this, we have adjusted the transformation to $10 \log(x + 1)$. As observed in the chart, the TETP method maintains a consistent level of memory requirement for the search even as the size of the bin space increases, as depicted in (b).

Through experiments, we proved that the TETP method has significantly better performance in both time and space compared to the existing brute-force method. Additionally, there is no need to re-search even if constraints are added or changed, so there is no loss in search time or memory due to changes in constraints. Robust results can be obtained with a single search. Moreover, this method has the advantage of identifying which constraints are causing problems when an optimal solution cannot be found.

5 Conclusion

In this paper, we introduced the Two-Step Exploration for Two-Dimensional Packing Problem (TETP) methodology to address the two-dimensional bin packing problem, specifically in the context of LNG vessel design. The primary objective was to optimize the arrangement of insulation panels within the Cargo Containment System (CCS) tanks of LNG vessels, aiming to reduce both the number of panels and the total welded length of membrane sheets. This optimization is crucial for minimizing manufacturing costs and enhancing productivity in the shipbuilding industry.

Our research demonstrated that the TETP methodology significantly outperforms traditional brute-force search methods and genetic algorithms in terms of both time and space efficiency. The experiments conducted on four LNG vessels with different dimensions showed that TETP could achieve optimal solutions rapidly, even when adapting to various design constraints. Unlike conventional methods, TETP efficiently manages memory space and consistently maintains performance regardless of the increasing size of the bin space. This is particularly important in industrial applications where the complexity and scale of the problem can lead to substantial computational burdens.

Through this research, we developed a novel two-step exploration method that effectively divides the 2D space into one-dimensional backbone and rib sections, allowing for systematic exploration of permutations and combinations. We proved substantial reductions in computational time and memory usage compared to brute-force and genetic algorithm methods, showcasing the practical applicability of TETP in real-world industrial scenarios. This paper demonstrated that TETP can quickly adapt to changes in design constraints without the need for re-searching, ensuring robust and flexible solutions. Additionally, we provided a mechanism to identify and address specific constraints that prevent finding an optimal solution, enhancing the overall robustness of the design process.

While this paper focused on optimizing the arrangement of insulation panels within the 2D space of individual walls, future research will expand this approach to consider the interconnections between walls in a 3D space, aligning with the characteristics of LNG tanks. Specifically, we plan to develop a decision-making model that considers the influence between walls using Graph Neural Networks (GNN). This advanced model will account for the connectivity of walls, optimizing the 3D placement of insulation panels across the entire tank. Such an approach aims to further enhance the efficiency and effectiveness of the design process, ensuring that the optimal solution is achieved not only within each wall but also across the entire structure of the LNG tank.

References

1. Błażewicz, J., Hawryluk, P., Walkowiak, R.: Using a tabu search approach for solving the two-dimensional irregular cutting problem. Ann. Oper. Res. **41**, 313–325 (1993)
2. BnnoBRs, J.: Partitioning procedures for solving mixed-variables programming problems. Numer. Math. **4**(1), 238–252 (1962)
3. Böhm, S., Neumayer, M., Kramer, O., Schiendorfer, A., Knoll, A.: Comparing heuristics, constraint optimization, and reinforcement learning for an industrial 2D packing problem. arXiv preprint arXiv:2110.14535 (2021)
4. Côté, J.F., Haouari, M., Iori, M.: Combinatorial benders decomposition for the two-dimensional bin packing problem. INFORMS J. Comput. **33**(3), 963–978 (2021)
5. Ferreira, D.N.G.: Rectangular bin-packing problem: a computational evaluation of 4 heuristics algorithms. U. Porto J. Eng. **1**(1), 35–49 (2015)
6. Jakobs, S.: On genetic algorithms for the packing of polygons. Eur. J. Oper. Res. **88**(1), 165–181 (1996)
7. Kämpke, T.: Simulated annealing: use of a new tool in bin packing. Ann. Oper. Res. **16**, 327–332 (1988)
8. Laterre, A., et al.: Ranked reward: enabling self-play reinforcement learning for combinatorial optimization. arXiv preprint arXiv:1807.01672 (2018)
9. Lee, S.H.: Application of tabu search to the two-dimensional bin packing problem. In: Proceedings of the Korean Operations and Management Science Society Conference, pp. 311–314. The Korean Operations Research and Management Science Society (2004)
10. Li, D., Ren, C., Gu, Z., Wang, Y., Lau, F.: Solving packing problems by conditional query learning (2019)
11. Liu, D., Teng, H.: An improved BL-algorithm for genetic algorithm of the orthogonal packing of rectangles. Eur. J. Oper. Res. **112**(2), 413–420 (1999)
12. López-Camacho, E., Terashima-Marin, H., Ross, P., Ochoa, G.: A unified hyper-heuristic framework for solving bin packing problems. Expert Syst. Appl. **41**(15), 6876–6889 (2014)
13. Silver, D., et al.: Mastering chess and shogi by self-play with a general reinforcement learning algorithm. arXiv preprint arXiv:1712.01815 (2017)
14. TIG: Panel type insulation for lng/lcg cargo tank. http://tig-groups.com/cryogenic/paneltype.htm
15. Zhao, H., She, Q., Zhu, C., Yang, Y., Xu, K.: Online 3D bin packing with constrained deep reinforcement learning. In: Proceedings of the AAAI Conference on Artificial Intelligence, vol. 35, pp. 741–749 (2021)

Energy Conscious Bi-objective Job Shop Scheduling: A New Formulation and Augmented ε-Constraint Method

S. Hocine Bouzegag and Saadettin Erhan Kesen(✉)

Department of Industrial Engineering, Faculty of Engineering and Natural Sciences, Konya Technical University, Selçuklu, Konya, Turkey
sekesen@ktun.edu.tr

Abstract. This paper addresses a job shop scheduling problem in which machines can operate at varying speeds and different energy efficient strategies known as speed scaling and switching on/off are incorporated as well. When a machine runs at high-speed, the amount of time to complete the job shortens but energy consumed by the machine increases. Selection of different speed modes of machines for different jobs (i.e. speed scaling) generates compromise solutions. To save energy further, one must decide whether to shut down the machine during idle periods of consecutive jobs. One option is to turn off the machine whenever the idle period occurs regardless of its duration, which may result in machine breakdown due to excessive opening and closing. Alternatively, a threshold or time limit can be determined below which the machine is kept in standby mode by consuming very little energy. We aim to minimize two conflicting objectives, energy consumption resulting from usage while the machine runs at a particular speed or in standby state and total tardiness emanating from late completions. To this end, we developed a MILP formulation for the problem and Augmented ε-Constraint (Augmecon) method is implemented to find pareto optimal solutions. The experimental result reveals that energy consumption and total tardiness objectives are conflicting. Based on payoff table, while the total energy consumed is 25000, total tardiness is 270. When energy consumption increases to 32186, total tardiness reduces to 36. Between the two, Augmented ε-Constraint (Augmecon) method provides non-dominated optimal solutions based on 11 grid points.

Keywords: Multi-Objective Optimization · Job Shop Scheduling · Sustainability · Switch on/off · Augmecon Method

1 Introduction

Today, energy is the most important resource in-demand all over the world. Global energy demand is in surge due to population growth and economic development. Global energy consumption substantially raised by 69.22% from 1990 to 2020 [1]. On the other hand, according to the International Energy Outlook 2023 by the U.S. Energy Information Administration, the industrial sector, comprises over 50% of end-use energy consumption [2].

Published by Springer Nature Switzerland AG 2024
M. Thürer et al. (Eds.): APMS 2024, IFIP AICT 733, pp. 213–228, 2024.
https://doi.org/10.1007/978-3-031-71645-4_15

As the American Industry Sector consumed 26.6 QBT (Quadrillion British Thermal) of energy in 2022, and it is expected to exceed 30.9 QBT in 2050. While the industrial sector in the Western European Countries recorded 26.3 QBT of energy consumption in 2022 and it is expected to reach 29.3 QBT in 2050. Meantime, the global energy consumption of industry sector will increase by more than 30% from 2022 to 2050 [2]. As depicted in Fig. 1, there is expected to be considerable rise in energy consumption over the next thirty years.

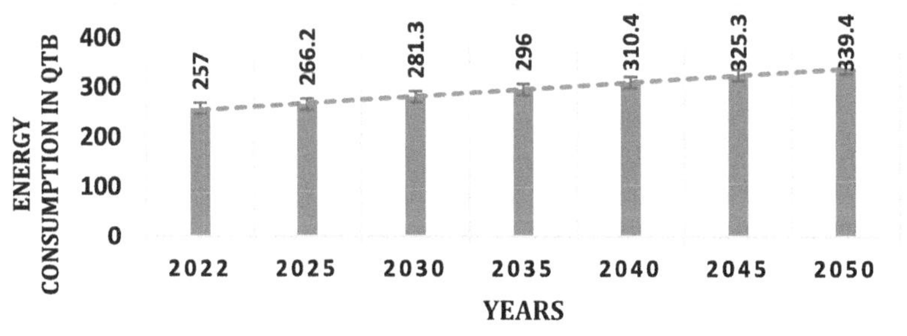

Fig. 1. Industrial Sector Energy Consumption Worldwide (from 2022 till 2050). (K_K1_r_230822.081459.pdf (eia.gov))

With the aim of saving energy and making good use of it, particularly within the manufacturing and production sectors, a number of research centered around the concept of sustainable production. Sustainable production involves the creation of products through economically viable processes that reduce adverse environmental impacts while conserving energy and natural resources.

Al-Yousfi [3], Tsoulfas and Pappis [4] and Lindsey [5] have primarily defined the principles of this concept from a broad and conceptual perspective, predominantly addressing environmental concerns. Numerous companies confront significant challenges associated with sustainable development policies, through which special measures are taken. Therefore, production companies have become more eager to develop soft methods that enable them to reduce energy consumption.

Recently, scheduling has emerged as the most suitable method for optimizing scarce production resources, incorporating green metrics alongside traditional performance indicators. This approach is commonly called energy-aware scheduling or energy-efficient scheduling. Gahm et al. [6], analyzed energy-efficient scheduling approaches that focus on enhancing energy efficiency, most widely encountered of which are switching on/off and speed-scaling.

Switching on/off approach: Mouzon et al. [7], introduced the method demonstrating that switching off machines during their idle periods enables them to save a significant amount of energy.

However, it may not be possible to completely switch off the machines during each idle period, either due to the significant additional energy required for repowering the machines or due to the potential damage caused by frequent switching on and off.

The switching on/off strategy may not be applicable in certain processes where the switching off the machines results in forbidden costs.

Speed-scaling approach: Originally proposed by Yao et al. [8]. This strategy enables machines to operate at varying speed levels during processing the operations of different jobs, meaning that the machine speed level is adjusted according to the status of the operation to be processed and the machine will not be switched off but instead will be in a standby state during the idle period. This does imply that the energy consumed is greater when idle time lasts longer, which does not fit the objective of minimizing the total energy consumption.

In this paper, we propose an alternative framework that joints switching on/off and Scaling-speed strategies. Here, the production environment considered is job shop in which, there are various jobs to be processed by multiple machines operating at different speed levels. Each job consists of multiple operations assigned to machines in a predetermined and strict order. In the processing phase, the machine speed level of processing is selected, and remains unchanged until the processing completes. In the idle phase, if the idle time lasts longer than the agreed threshold time, the machine will be switched off, and it will be switched on at the starting time of next operation (i.e., no energy consumption). Second, if the idle time is less than the agreed threshold time, the machine will remain idle during this period and consume relatively lower standby energy.

We, thus, determine two conflicting objectives, namely minimization of the total energy consumption and minimization of the total tardiness. Increasing machine speed level causes higher energy consumption and shorter processing times, resulting in earlier completion. Conversely, setting the machine to a low-speed level reduces energy consumption but prolongs processing times, resulting in tardiness. Therefore, we present a bi-objective mixed-integer linear programming formulation for job shop scheduling problem. Augmented ε-Constraint (Augmecon) method is applied to generate pareto optimal solutions.

Rest of this paper is constructed as follows: In Sect. 2, we outline recent studies concerning energy-efficient job scheduling problems and categorize them in a tabular form. In Sect. 3, the problem under study is formally defined and subsequently a bi-objective mathematical formulation is presented along with illustrative small-sized example. In Sect. 4, the exact solution procedure so-called Augmented ε-Constraint (Augmecon) method is presented. Pareto optimal solutions with their Gantt charts are demonstrated in Sect. 4.2. Finally, Sect. 5 concludes the paper.

2 Literature Review

The earliest works on energy-efficient scheduling problems date back to the 1990s [6, 9, 10]. However, energy-efficient scheduling has only recently raised as the subject of systematic research. The energy-efficient scheduling strategies in literature can be classified into three main groups: switching on/off, scaling-speed and off-peak production.

In switching on/off strategy, decision maker determines whether to switch off the underutilized machines or to remain it idle to reduce energy consumption.

In scaling-speed strategy, energy consumption affected by speed level of machine. That is adjusted according to the scheduling objectives by decision maker.

In the peak-off production strategy scheduling is conducted under time-dependent energy prices, where electricity costs are higher during peak time. Therefore, scheduling production during non-peak time can result in significant savings on electricity costs. As seen in Table 1, the recent studies addressing energy-efficient job shop scheduling problems with consideration of an energy saving strategy are classified.

As noticed below, the most studies regarding energy-efficient job scheduling problem addressed the makespan criterion together with energy consumption criterion whereas scant studies considering the tardiness (due date related) criterion as a common objective with the energy-related criterion due to its higher complexity. It is also observed that these previous works generally used single energy efficient strategy (i.e. either switching on/off strategy or scaling speed strategy or even pick off strategy).

Table 1. The recent studies on energy-efficient job shop scheduling with different strategies.

Study	Single/multi-obj	Objectives	EE-strategy	Solution app
[11]	Multi-objective	E, C_{max}	Scaling speed	GA + SA
[12]	Multi-objective	E, C_{max}	On/Off	moGA; SPEAII
[13]	Multi-objective	E, C_{max}	Scaling speed	GA
[14]	Multi-objective	E, C_{max}	Scaling speed	CP
[15]	Multi-objective	E, wT	Scaling speed	moGA + ILS
[16]	Multi-objective	EC, PC	Scaling speed	MILP
[17]	Multi-objective	E, C_{max}	Scaling speed	Simul
[18]	Multi-objective	E, N, C_{max}	Scaling speed	GA
[19]	Single Objective	EC	Scaling speed	
[20]	Single Objective	E	Peak off	GEP
[21]	Multi-objective	EC, PC	Scaling speed	WOA
[22]	Multi-objective	EC, T, RD	Scaling speed	GA
[23]	Multi-objective	E, wT	Scaling speed	MA
[24]	Multi-objective	E, C_{max}	Scaling speed	moEA
[25]	Multi-objective	EC, wT	Peak off	NSGAII
[26]	Multi-objective	E, C_{max}, T, LC	Scaling speed	moEA
[27]	Multi-objective	E, C_{max},wTE	On/Off	NSGAIII
[28]	Single objective	EC	Peak off, On/Off	

As opposed to these works, (1) we propose a new variant of integrating two different strategies (switching on/off and scaling-speed) which are simultaneously used. (2) we set two conflicting objectives to minimize: the total amount of energy consumption and the total tardiness. To the best of authors' knowledge, this paper is the first studying multiple energy efficient strategies on job shop scheduling problem with bi-objective.

3 Problem Definition

In this section, we present a formal definition, an illustrative example, and a mathematical formulation of the problem being addressed.

3.1 Formal Definition

The multi-objective job shop scheduling problem being investigated in this paper is formally introduced as follows:

There are n jobs to be processed by m machines operating at different speed levels $h_k(k = 1, \ldots, m)$. Each job consists of m operations assigned to machines in a predetermined order.

If any operation of job i is processed on machine k, the machine speed level of processing is selected from the set h_k, which remains unchanged until the processing completes. Accordingly, its actual processing time is simply the division of basic processing time to selected speed level h_k and is calculated as $AT_{ikh} = BT_{ik}/h_k$.

If $h' > \mathrm{h}$ then $P_{kh'} \times \mathrm{AT}_{ikh'} > P_{kh} \times AT_{ikh}$. This implies that if the machine is configured for higher speed level, the processing time will be shorter, but it will cause higher energy consumption. We also consider two potential states for the machine during idle periods. In the first scenario, if the idle time exceeds the threshold, the machine will be switched off, leading to no energy consumption. In the second scenario, if the idle time is less than the threshold, the machine will enter standby mode, incurring standby energy consumption. After the processing of all operations assigned to each machine completes, the machine will automatically shut down.

Different operations of any job processed on different machines cannot be performed at the same time. Machines are capable of processing at most single operation at the same time. Once the operation of any job starts, it cannot be preempted till the completion of the operation. We also assume that machine breakdown is not permitted.

3.2 Mixed Integer Linear Programming Formulation

In this section, a MILP formulation is introduced for the problem under study. The notations used to formulate this model are shown below:

Notation	Definition
Indices:	
i, l	: Indices of job $i, l = 1, \ldots, n$
k, q	: Indices of machine,$k, q = 1, \ldots, m$
h	: Index of machine speed level,$h = 1, \ldots, h_k$
Parameters:	
P_{kh}	: Instantaneous power of machine k for speed level h
SM_k	: Instantaneous power of machine k on standby mode

(continued)

(*continued*)

Notation	Definition
MB_k	: The maximum time that the machine k can remain in standby mode without switching off
OS_{ik}	: The visit order of job i on machine k
BT_{ik}	: The basic time required to process job i on machine k
AT_{ikh}	: The actual time required when job i is processed on machine k at speed level h
dd_i	: The due date of job i
Decision Variables:	
X_{ikh}	: 1 if job i is processed on machine k at speed level h, 0 otherwise
Z_{ilk}	: 1 if job i is processed immediately before job l on machine k, 0 otherwise
Y_{ilk}	: 1 if switching on/off is applied between consecutively processed jobs i and l on machine k, 0 otherwise
S_{ik}	: Starting time of job i on machine k
$IdleTime_{ilk}$	: The idle time between consecutively processed jobs i and l on machine k
$StandbyTime_{ilk}$	: The time spent as standby mode between consecutively processed jobs i and l on machine k .
T_i	: The tardiness of job i

Objective Function

$$inf_1 = \sum_{i=1}^{n}\sum_{k=1}^{m}\sum_{h=1}^{H_k} X_{ikh}P_{kh}AT_{ikh} + \sum_{m=1}^{k} SM_k \sum_{\substack{i=1 \\ l \neq i}}^{n} StandbyTime_{ilk} \tag{1}$$

$$Min\, f_2 = \sum_{i=1}^{n} T_i \tag{2}$$

Subject to

$$\sum_{h=1}^{H_k} X_{ikh} = 1 \quad i = 1, \ldots, n; k = 1, \ldots, m \tag{3}$$

$$S_{iq} - S_{ik} - \sum_{h=1}^{H_k} X_{ikh}AT_{ikh} \geq 0 \quad \begin{matrix} i = 1, \ldots, n; k = 1, \ldots, m; \\ (i, k) \rightarrow (i, q) \end{matrix} \tag{4}$$

$$S_{lk} - S_{ik} - \sum_{h=1}^{H_k} X_{ikh}AT_{ikh} + M(1 - Z_{ilk}) \geq 0 \quad \begin{matrix} i, l = 1, \ldots, n; i \neq l; \\ k = 1, \ldots, m \end{matrix} \tag{5}$$

$$\sum_{i=1}^{n}\sum_{\substack{l=1 \\ l \neq i}}^{n} Z_{ilk} = n - 1 \quad k = 1, \ldots, m \tag{6}$$

$$\sum_{\substack{l=1\\ l\neq i}}^{n} Z_{ilk} \leq 1 \; i = 1, ., n; k = 1, ., m \tag{7}$$

$$\sum_{\substack{l=1\\ l\neq i}}^{n} Z_{lik} \leq 1 \; i = 1, \ldots, n; k = 1, , m \tag{8}$$

$$IdleTime_{ilk} \leq S_{lk} - S_{ik} - \sum_{h=1}^{H_k} X_{ikh}AT_{ikh} + M(1 - Z_{ilk}) \quad \begin{matrix} i, l = 1, \ldots, n; i \neq l; \\ k = 1, \ldots, m \end{matrix} \tag{9}$$

$$IdleTime_{ilk} \geq S_{lk} - S_{ik} - \sum_{h=1}^{H_k} X_{ikh}AT_{ikh} - M(1 - Z_{ilk}) \quad \begin{matrix} i, l = 1, \ldots, n; i \neq l; \\ k = 1, \ldots, m \end{matrix} \tag{10}$$

$$IdleTime_{ilk} \leq M(Z_{ilk}) \quad \begin{matrix} i, l = 1, \ldots, n; i \neq l; \\ k = 1, \ldots, m \end{matrix} \tag{11}$$

$$IdleTime_{ilk} \leq MB_k - 1 + M(Y_{ilk}) \quad \begin{matrix} i, l = 1, \ldots, n; i \neq l; \\ k = 1, \ldots, m \end{matrix} \tag{12}$$

$$IdleTime_{ilk} \geq MB_k - M(1 - Y_{ilk}) \quad \begin{matrix} i, l = 1, \ldots, n; i \neq l; \\ k = 1, \ldots, m \end{matrix} \tag{13}$$

$$StandbyTime_{ilk} \leq IdleTime_{ilk} + M(Y_{ilk}) \quad \begin{matrix} i, l = 1, \ldots, n; i \neq l; \\ k = 1, \ldots, m \end{matrix} \tag{14}$$

$$StandbyTime_{ilk} \geq IdleTime_{ilk} - M(Y_{ilk}) \quad \begin{matrix} i, l = 1, \ldots, n; i \neq l; \\ k = 1, \ldots, m \end{matrix} \tag{15}$$

$$StandbyTime_{ilk} \leq M(1 - Y_{ilk}) \quad \begin{matrix} i, l = 1, \ldots, n; i \neq l; \\ k = 1, \ldots, m \end{matrix} \tag{16}$$

$$T_i \geq S_{ik'} + \sum_{h=1}^{H_{k'}} X_{ik'h}AT_{ik'h} - dd_i \quad i = 1, ..., n; k = 1, \ldots, m \tag{17}$$

$$S_{ik} \geq 0 \quad \text{Non negative constraint} \tag{18}$$

$$IdleTime_{ilk} \geq 0 \quad \text{Non negative constraint} \tag{19}$$

$$StandbyTime_{ilk} \geq 0 \quad \text{Non negative constraint} \tag{20}$$

$$T_i \geq 0 \quad \text{Non negative constraint} \tag{21}$$

The objective functions given in Eqs. (1) and (2) minimize the total energy consumption of the machines (energy spent times and idle in both processing time) and minimizes the total tardiness respectively. Equation (3) ensures that the selected speed level of machine k is not changed until the processing of job i completes. Equation (4) indicates that operations belonging to the same job must be processed sequentially. Equation (5) states that no machine can perform more than one operation simultaneously. Equations (6), (7) and (8) determine the Z_{ilk} value that indicates whether jobs i and l processed on machine k are consecutive or not. Equations (9), (10) and (11) calculate the idle times between jobs i and l consecutively processed on machine k. Equations (9) and (10) regulate the calculation of the idle time between jobs i and l. When Z_{ilk} equals 1, this value is found by subtracting the end time of job i from the start time of job l. Equation (11) ensures that when Z_{ilk} equals 0, the idle time between jobs i and l is zero. This means that jobs i and l are processed sequentially on machine k. Equations (12) and (13) express the basic idea of the switching on/off strategy. In the case where Y_{ilk} is 0, Eq. (12) guarantees that the idle time is less than the threshold value (i.e., implementation of standby mode). Otherwise, when Y_{ilk} is 1, Eq. (13) ensures that the idle time is not less than the threshold value (i.e., implementation of switching on/off). Equations (14), (15) and (16) calculate the standby time between jobs i and l, which are processed consecutively on machine k. Equations (14) and (15) show that when Y_{ilk} is 0, the $StandbyTime_{ilk}$ value is equal to the $IdleTime_{ilk}$ value. Equation (16) shows that when Y_{ilk} is 1, the $StandbyTime_{ilk}$ value equals zero. Equation (17) indicates the tardiness time of each job i. Finally, Eqs. (18), (19), (20) and (21) represent the non-negativity constraints.

3.3 Illustrative Numerical Example

To simplify the description of the problem, we present a small-sized numerical example in this section. In this job shop machining environment, we have 4 jobs processed by 3 machines. These machines can be adjusted to 3 different speed levels (Low, Medium, and High) and each job consists of 3 operations. The data presented in Table 2 are the parameters of the 3 machines used in the job shop.

Table 2. The parameters used for the machines.

	Low	Medium	High	Standby	*MB*
Machine 1	80	200	330	40	25
Machine 2	100	240	360	50	30
Machine 3	90	230	345	60	20

The data presented in Table 3 are the parameters related to the 4 jobs processed in this job shop.

4 Solution Procedure

In this section, we present the exact method for the multi-objective optimization problem being investigated, which is called AUGMECON used to produce Pareto optimal solutions. Here, the Augmecon method is explained, applied and the results obtained are shown.

Table 3. The parameters of jobs processed.

Job	Due date	Machine sequence	Processing Basic Time	Processing Actual Time
Job 1	20	M_3, M_2, M_1	$BT_{11} = 20$ $BT_{12} = 10$ $BT_{13} = 40$	$AT_{111} = 20$, $AT_{112} = 10$,$AT_{113} =$ 6.66, $AT_{121} = 10$,$AT_{122} =$ 5,$AT_{123} = 3.33$, $AT_{131} = 40$,$AT_{132} =$ 20,$AT_{133} = 13.33$
Job 2	20	M_2, M_3, M_1	$BT_{21} = 30$ $BT_{22} = 30$ $BT_{23} = 10$	$AT_{211} = 30$, $AT_{212} = 15$, $AT_{213} = 10$, $AT_{221} = 30$, $AT_{222} = 15$, $AT_{223} = 10$ $AT_{231} = 10$, $AT_{232} =$ 5,$AT_{233} = 3.33$
Job 3	10	M_1, M_2, M_3	$BT_{31} = 10$ $BT_{32} = 30$ $BT_{33} = 20$	$AT_{311} = 10$,$AT_{312} =$ 5,$AT_{313} = 3.33$ $AT_{321} = 30$,$AT_{322} =$ 15,$AT_{323} = 10$, $AT_{331} = 20$,$AT_{332} =$ 10,$AT_{333} = 6.66$
Job 4	40	M_1, M_3, M_2	$BT_{41} = 50$ $BT_{42} = 20$ $BT_{43} = 10$	$AT_{411} = 50$,$AT_{412} =$ 25,$AT_{413} = 16.66$ $AT_{421} = 20$,$AT_{422} =$ 10,$AT_{423} = 6.66$ $AT_{431} = 10$,$AT_{432} =$ 5,$AT_{433} = 3.33$

4.1 Augmented ε-Constraint (Augmecon) Method

There are various methods used to produce effective solutions to multi-objective optimization problems. Hwang and Masud [29] classified these methods into three categories called priori methods, interactive methods, and generation methods.

Here, both the ε-constraint method and the Augmecon method will be explained. The ε-constraint method is one of the most widely used generation methods for solving multi-objective optimization problems. In 1971, Haimes developed the traditional ε-constraint method. In general, while the researchers optimized one of the objective functions, they used the other objective functions as constraints in the model. Let us assume the multi-objective optimization problem in the following example:

$$max\big(f_1(x), f_2(x), \ldots f_p(x)\big) \tag{22}$$

st

$$x \in S$$

Here x is the vector of decision variables, $f_1(x), \ldots f_p(x)$, p are the objective functions and S is the feasible region.

$$max f_1(x) \tag{23}$$

st

$$f_2(x) \geq e_2 \tag{24}$$

$$\vdots$$

$$f_p(x) \geq e_p \tag{25}$$

The effective solutions to the problem obtained by parametric variation in the (e_i) RHS of restricted objective functions. Although the ε-constraint method has advantages over other methods, it also includes drawbacks. Using this traditional method enable researchers to produce weakly efficient solutions. To eliminate this drawback, Mavrotas [30] proposed the Augmecon method, which enables researchers avoid producing weak Pareto optimal solutions and minimize solution time by eliminating unnecessary iterations. Augmecon method operates as follows:

$$min(f_1(x) - \delta(s_2/r_2 + s_3/r_3 + \cdots + s_p/r_p)) \tag{26}$$

st

$$f_i(x) + s_i = \varepsilon_i \quad i = 1, 2, \ldots p \tag{27}$$

$f_i(x)$ represents the objective function i to be optimized. $\boldsymbol{\varepsilon_i}$ is the value of the right-hand side, s_i is the slack variable, and $\boldsymbol{r_i}$ is the range associated with the objective. δ is a small number.

Applying Augmecon method requires at first the creation of payoff table by lexicographic optimization ensuring the generation of only efficient solutions.

In the lexicographic optimization technique, the problem is first solved to minimize (or maximize) the first objective function. Then, the resulting objective value is considered as the upper (or lower) bound and added to the existing set of constraints as a new constraint. After the first objective function value is optimized, the second objective function value is also optimized. In similar way, the procedure is applied for the subsequent objective function and the payoff table is created.

Then, the ranges of each objective and the grid points of the problem are found, and pareto optimal solutions are produced.

4.2 The Obtained Results Using GAMS Application

Based on the data provided in the illustrative example, Augmecon method is applied, and the obtained results are shown step by step. As mentioned before, the first step of the AUGMECON method is to create the payoff table using lexicographic optimization. The obtained payoff table is given in following Table 4.

Table 4. The Obtained Payoff Table.

	f_1	f_2
$\min f_1$	25000	270
$\min f_2$	32186	36

After constructing the payoff table, it is necessary to determine the range of each objective in the constraints section of the table. In this instance, only the second objective (f_2) is incorporated as a constraint. Thus, the value of $r_2 = 233.37$ was derived from the return table. Then, r_2 is divided into 10 equal intervals, each referred to as a grid point.

Grid points are used to parametrically change the right-hand side value (e_i) of a binding constraint. So, the AUGMECON method process with $Step_2 = 23.33$ is as follows:

$$For\ i = 0\ \ to\ 10$$

$$e_2\ = 270 + i * 23.33$$

$$\text{Solve (1)}$$

$$Next\ (i)$$

The resulting grid points are listed in the following Table 5.

Table 5. Grid Points.

	Grid Points										
	0	1	2	3	4	5	6	7	8	9	10
f_2	270	246	223	199	176	153	129	106	83	59	36

Finally, we present the pareto optimal solutions of this illustration multi-objective problem on Table 6 accompanied with Fig. 2 clarifying Pareto front.

Table 6. Values of The Effective Solutions Obtained.

	S_1	S_2	S_3	S_4	S_5	S_6	S_7	S_8	S_9	S_{10}	S_{11}
E	25000	25696	26297	26796	27495	28096	28746	29393	30041	30936	32186
T	270	243	223	199	176	151	128	106	81	59	36

As can be shown in Fig. 2, the "Pareto front" is represented by the Pareto optimal solutions obtained for the instance in order to clarify the behavior of both the total energy consumed by the machines and the total tardiness of jobs. These results show

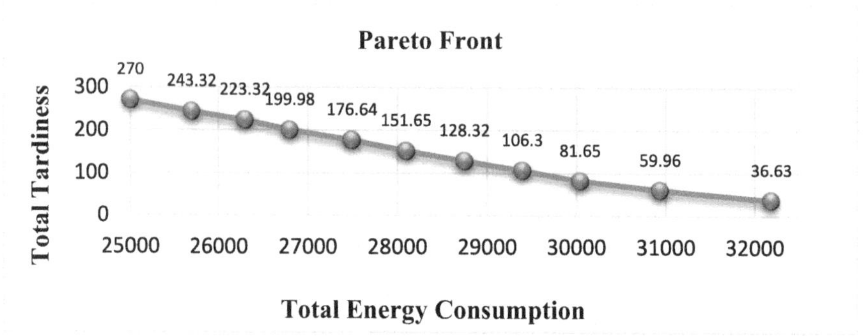

Fig. 2. Pareto Front.

an increasing in the total energy consumed by the machines, starting from 25,000 to 32,186. On the other hand, the total tardiness of jobs decreases from 270 min to 36 min. These behaviors are due to changing the speed modes of the machines, as whenever the machine is set to a faster speed mode, it consumes more energy while completing the work in a shorter period, and vice versa.

4.3 Model Status in Different Time Periods

In the process of producing effective solutions for the multi-objective model, the states of the basic variables constantly change over time. This causes different scenarios to occur in the model. Consequently, we identified three active solutions as distinct stages wherein different scenarios of the model can be demonstrated as follows:

The model variables status on the effective solution (25000; 270). According to the efficient solution (25000; 270), the model being investigated is presented in Gantt chart as follows (see Fig. 3):

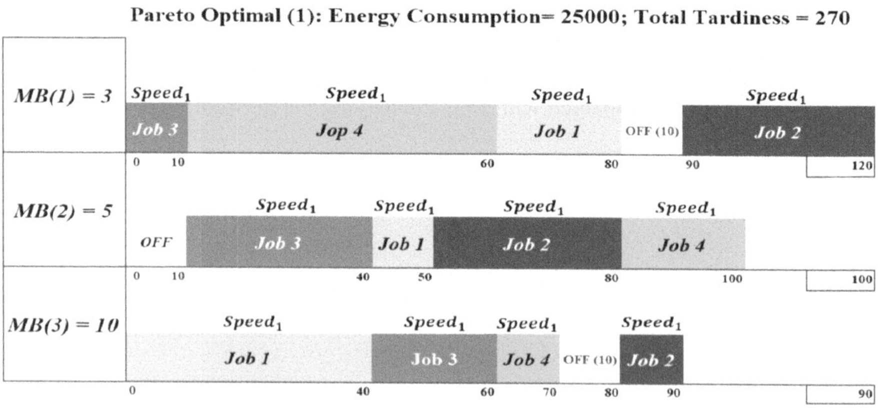

Fig. 3. GANTT chart given according to (25000; 270).

The model variables status on the effective solution (32186; 36). According to the efficient solution (32186;36), the model being investigated is presented in GANTT chart as follows (see Fig. 4):

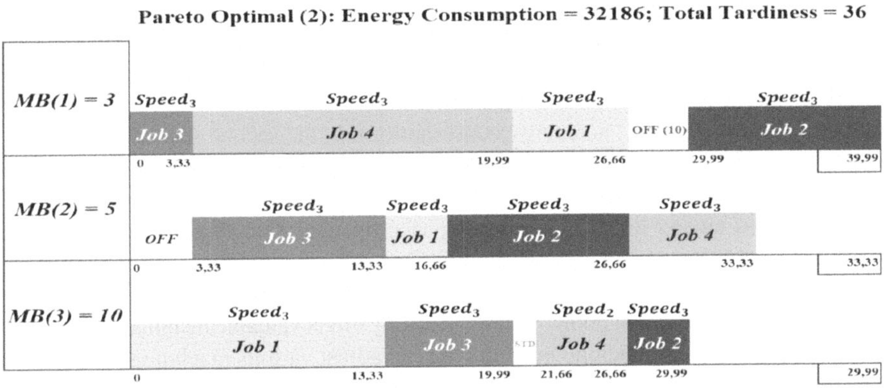

Fig. 4. GANTT chart given according to (32186; 36).

The model variables status on the effective solution (26796; 199). According to the efficient solution (26796;199), the model being investigated is presented in GANTT chart as follows (see Fig. 5).

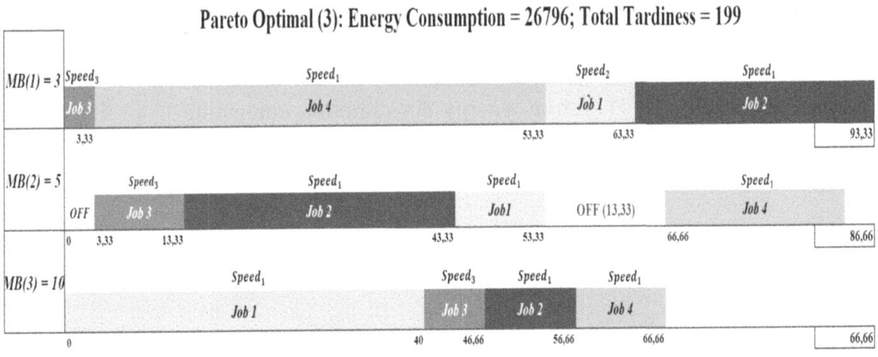

Fig. 5. GANTT chart given according to (26796; 199).

5 Conclusion

In this paper, we study job shop scheduling problem in which efficient-energy related new framework integrating two different strategies (switching on/off and scaling-speed) are simultaneously jointly used. Within the problem setting, there exist two opposing objectives that need to be minimized: the total energy consumption and the total tardiness.

We assume job shop with various jobs to be processed by multiple machines operating at different speed levels. In the processing phase, the machine speed level of processing is selected, and remains unchanged until the processing completes. Increasing machine speed level causes higher energy consumption and shorter processing times, resulting in earlier completion. Conversely, setting the machine to a low-speed level reduces energy consumption but prolongs processing times, resulting in tardiness. In the idle phase, there are two potential states for the machine. In the first scenario, if the idle time exceeds the threshold, the machine will be switched off, and it will be switched on at the starting time of next operation (i.e., no energy consumption). In the second scenario, if the idle time is less than the threshold, the machine will remain idle during this period and consume relatively lower standby energy. To this end, we formulate the problem as bi-objective mixed-integer linear programming accompanied with small-sized illustrative example and apply the so-called Augmented ε-Constraint method for generating efficient solutions. We present the obtained Pareto optimal solutions (11 solutions), Pareto front of this problem. From practical point of view, this study offers valuable insights. Companies suffering from frequent machine breakdowns can adjust various machine speeds while respecting pre-determined energy consumption cost and tardiness levels. In addition, they can utilize their machines by selecting the most appropriate threshold levels considering their working conditions and avoiding them from excessive failures. Finally, they can manage their operations by observing the non-dominated solutions and choosing the best one regarding the volatile conditions.

Though the problem is handled for the first time in the literature, the study can be extended in the future research: First, the higher intractability of the problem prohibits Augmecon to find pareto optimal solutions for big sized instances. Therefore, it is advisable to use a heuristic method the appropriate technique for such type of problems. Second, it is possible to rely on a dual-resource production environment can be realized and transportation equipment (e.g., automated guided vehicles) can be considered in addition to machines. Therefore, energy consumption emanating from transportation equipment usage can be integrated. Third, other energy efficient strategy known as off-peak can also be regarded as standalone or combination with the ones we utilized.

Acknowledgments. The authors are indebted many thanks to two anonymous referees for their constructive and valuable comments which lead to big improvement on the paper.

Disclosure of Interests. The authors have no competing interests to declare that are relevant to the content of this article.

References

1. Chanthakett, A., Arif, M., Khan, M.M.K., Subhani, M.: Hydrogen production from municipal solid waste using gasification method. In: Hydrogen Energy Conversion and Management, pp. 103–131. Elsevier (2024)
2. INTERNATIONAL ENERGY OUTLOOK 2023. https://www.eia.gov/outlooks/ieo/
3. Al-Yousfi, A.B.: Cleaner production for sustainable industrial development: concept and applications. Pract. Period. Hazardous Toxic Radioactive Waste Manag. **8**(4), 265–273 (2004)

4. Tsoulfas, G.T., Pappis, C.P.: Environmental principles applicable to supply chains design and operation. J. Clean. Prod. **14**(18), 1593–1602 (2006)
5. Lindsey, T.C.: Sustainable principles: common values for achieving sustainability. J. Clean. Prod. **19**(5), 561–565 (2011)
6. Gahm, C., Denz, F., Dirr, M., Tuma, A.: Energy-efficient scheduling in manufacturing companies: a review and research framework. Eur. J. Oper. Res. **248**(3), 744–757 (2016)
7. Mouzon, G., Yildirim, M.B., Twomey, J.: Operational methods for minimization of energy consumption of manufacturing equipment. Int. J. Prod. Res. **45**(18–19), 4247–4271 (2007)
8. Yao, F., Demers, A., Shenker, S.: A scheduling model for reduced CPU energy. In: Proceedings of IEEE 36th Annual Foundations of Computer Science. IEEE. (1995)
9. Gao, K., Huang, Y., Sadollah, A., Wang, L.: A review of energy-efficient scheduling in intelligent production systems. Complex Intell. Syst. **6**, 237–249 (2020)
10. Akbar, M., Irohara, T.: Scheduling for sustainable manufacturing: a review. J. Clean. Prod. **205**, 866–883 (2018)
11. Tang, D., Dai, M.: Energy-efficient approach to minimizing the energy consumption in an extended job-shop scheduling problem. Chin. J. Mech. Eng. **28**(5), 1048–1055 (2015)
12. May, G., Stahl, B., Taisch, M., Prabhu, V.: Multi-objective genetic algorithm for energy-efficient job shop scheduling. Int. J. Prod. Res. **53**(23), 7071–7089 (2015)
13. Escamilla, J., Salido, M.A., Giret, A., Barber, F.: A metaheuristic technique for energy-efficiency in job-shop scheduling. Knowl. Eng. Rev. **31**(5), 475–485 (2016)
14. Salido, M.A., Escamilla, J., Barber, F., Giret, A., Tang, D., Dai, M.: Energy efficiency, robustness, and makespan optimality in job-shop scheduling problems. AI EDAM **30**(3), 300–312 (2016)
15. Zhang, R., Chiong, R.: Solving the energy-efficient job shop scheduling problem: a multi-objective genetic algorithm with enhanced local search for minimizing the total weighted tardiness and total energy consumption. J. Clean. Prod. **112**, 3361–3375 (2016)
16. Giglio, D., Paolucci, M., Roshani, A.: Integrated lot sizing and energy-efficient job shop scheduling problem in manufacturing/remanufacturing systems. J. Clean. Prod. **148**, 624–641 (2017)
17. Raileanu, S., Anton, F., Iatan, A., Borangiu, T., Anton, S., Morariu, O.: Resource scheduling based on energy consumption for sustainable manufacturing. J. Intell. Manuf. **28**, 1519–1530 (2017)
18. Yin, L., Li, X., Gao, L., Lu, C., Zhang, Z.: Energy-efficient job shop scheduling problem with variable spindle speed using a novel multi-objective algorithm. Adv. Mech. Eng. **9**(4), 1687814017695959 (2017)
19. Jiang, T., Zhang, C., Zhu, H., Gu, J., Deng, G.: Energy-efficient scheduling for a job shop using an improved whale optimization algorithm. Mathematics **6**(11), 220 (2018)
20. Zhang, L., Li, Z., Królczyk, G., Wu, D., Tang, Q.: Mathematical modeling and multi-attribute rule mining for energy efficient job-shop scheduling. J. Clean. Prod. **241**, 118289 (2019)
21. Jiang, T., Zhang, C., Sun, Q.-M.: Green job shop scheduling problem with discrete whale optimization algorithm. IEEE Access **7**, 43153–43166 (2019)
22. Luo, J., El Baz, D., Xue, R., Hu, J.: Solving the dynamic energy aware job shop scheduling problem with the heterogeneous parallel genetic algorithm. Futur. Gener. Comput. Syst. **108**, 119–134 (2020)
23. Abedi, M., Chiong, R., Noman, N., Zhang, R.: A multi-population, multi-objective memetic algorithm for energy-efficient job-shop scheduling with deteriorating machines. Expert Syst. Appl. **157**, 113348 (2020)
24. Wang, L., Peng, Z.-P.: Solving energy-efficient distributed job shop scheduling via multi-objective evolutionary algorithm with decomposition. Swarm Evol. Comput. **58**, 100745 (2020)

25. Kurniawan, B., Song, W., Weng, W., Fujimura, S.: Distributed-elite local search based on a genetic algorithm for bi-objective job-shop scheduling under time-of-use tariffs. Evol. Intel. **14**, 1581–1595 (2021)
26. Li, W., He, L., Cao, Y.: Many-objective evolutionary algorithm with reference point-based fuzzy correlation entropy for energy-efficient job shop scheduling with limited workers. IEEE Trans. Cybern. **52**(10), 10721–10734 (2021)
27. Wei, H., et al.: Unified multi-objective genetic algorithm for energy efficient job shop scheduling. IEEE Access **9**, 54542–54557 (2021)
28. Xu, E., Li, Y., Liu, Y., Du, J., Gao, X.: Energy saving scheduling strategy for job shop under TOU and tiered electricity price. Alex. Eng. J. **61**(1), 459–467 (2022)
29. Hwang, C.-L., Masud, A.S.M.: Multiple Objective Decision Making—Methods and Applications: A State-of-the-Art Survey, vol. 164. Springer, Heidelberg (2012). https://doi.org/10.1007/978-3-642-45511-7
30. Mavrotas, G.: Effective implementation of the ε-constraint method in multi-objective mathematical programming problems. Appl. Math. Comput. **213**(2), 455–465 (2009)

Routing Heuristics for in-House Transportation in Assembly Systems

Ebenezer Olatunde Adenipekun[1,2](✉), Veronique Limère[1,2], and Nico André Schmid[1,3]

[1] Department of Business Informatics and Operations Management, Ghent University, Ghent, Belgium

[2] Flanders Make@UGent-core lab CVAMO, Ghent, Belgium
ebenezerolatunde.adenipekun@ugent.be

[3] IESEG School of Management, Univ. Lille, CNRS, UMR 9221 - LEM - Lille Economie Management, 59000 Lille, France

Abstract. This research addresses routing challenges in high-variety mass customization environments, focusing on in-house transportation problems involving traversal between supermarket cells and consecutive assembly stations. Supermarket cells are organized separately from assembly stations in parallel rows, each cell having a designated pickup point and each assembly station a drop-off point. The routing problem aims to establish an efficient path, starting from the depot and visiting all pickup and drop-off points exactly once before returning. This study introduces a range of heuristic methodologies tailored to address the problem. These heuristics, such as nearest neighbor, S-shape, modified S-shape, and combined heuristics, are characterized by practical implementability, offering practitioners defined guided rules to navigate the shop floor. Artificial problem instances of varying sizes, characterized by the number of cells and pickup/drop-off points, are analyzed, comparing heuristic performance against an optimal benchmark solution. Results indicate that heuristics perform better in larger problem sizes, particularly with 100 rows, showing average gaps below 11.48% and 1.02% for 7-row and 100-row instances, respectively.

Keywords: Transportation · Manufacturing · Heuristics · Optimization

1 Introduction

The continuous rise in mass customization, driven by consumers' growing desire for personalized end products, has made managing assembly systems increasingly challenging. Consequently, there is an escalating need for efficient logistics solutions within the manufacturing environment. Proper feeding policies are essential to present parts to assembly stations optimally [16]. In response, researchers have offered practitioners various descriptive and optimization-based models for the

Published by Springer Nature Switzerland AG 2024
M. Threr et al. (Eds.): APMS 2024, IFIP AICT 733, pp. 229–243, 2024.
https://doi.org/10.1007/978-3-031-71645-4_16

assembly line feeding problem [4,5,11,15]. The assembly line feeding problem is concerned with the distinctive assignment of every part to a feeding policy in order to minimize the total feeding costs. Feeding policies, e.g., line stocking, sequencing, and stationary kitting, refer to how parts are presented to the assembly line [2]. Supermarkets, acting as storage areas located separately from the assembly stations, are used for presorting and repackaging assembly items such that they can be delivered in smaller space-saving quantities to the assembly lines [7,8]. Efficiently coordinating the transportation between the supermarket(s) and assembly stations is of paramount importance as a means to minimize in-house transportation efforts.

In recent years, manufacturing companies have been utilizing centralized and decentralized supermarkets for part preprocessing and repackaging (see e.g., [6]). This study centers its attention on a specific configuration of centralized supermarkets, partitioned into various supermarket cells (see Fig. 1a). This configuration aims to expedite the repackaging process into smaller load carriers or kits. The utilization of multiple supermarket cells allows for the aggregation of parts, functionally equivalent components, or those required for a specific end-product within a supermarket cell. Subsequently, after the preprocessing stage, parts are conveyed to the assembly stations using a dedicated transportation vehicle. With the presence of multiple supermarket cells, the determination of optimal routes for the transportation operation becomes imperative.

This study focuses on developing heuristic techniques to optimize in-house transportation routing, considering a configuration with a vehicle depot, multiple supermarket cells, and a sequence of consecutive assembly stations within a single route. Each supermarket cell has a pickup point, and each assembly station is designated a drop-off point. The routing problem aims to establish efficient procedures starting from the depot, visiting all pickup and drop-off points exactly once before returning to the depot in a single route, ensuring the visitation of all pickup points before any drop-off point. These heuristic approaches emphasize practical implementability, offering practitioners clear guidelines for navigating the shop floor. The ultimate objective is to enhance the efficiency of transportation logistics, leading to improved productivity and cost-effectiveness for manufacturing operations.

The remainder of this paper is structured as follows: Sect. 2 presents the problem description. Section 3 provides a selected overview of the literature on warehousing, vehicle routing, and assembly systems discussing related problems. Section 4 introduces various heuristic solution procedures and a comparison to the optimal solution. Within Sect. 5, the creation of problem instances and computational results are presented. Section 6 concludes the paper with a summary and further provides recommendations for the future.

2 Problem Description

The assembly system's shop floor typically comprises a warehouse, a supermarket, and an assembly line. The supermarket is partitioned into supermarket cells in this study (see Fig. 1a). Initially, parts are stored in bulk at the warehouse, usually on pallets. The replenishment process involves transporting these parts from the warehouse to the supermarket cells. At the cells, the parts are repackaged into smaller load carriers or kits. Figure 1a illustrates the schematic representation of the shop floor, comprising essential components such as the warehouse, supermarket (housing the supermarket cells), and assembly stations. The arrangement shows that the supermarket cells are positioned between the warehouse and the assembly stations. The supermarket is first partitioned into rows, and each row is partitioned into supermarket cells. To illustrate, the supermarket in Fig. 1a is partitioned into five rows. The In-House Transportation Problem (ITP) pertains to the delivery of smaller load carriers or kits collected from various supermarket cells to a sequence of consecutive assembly stations, all within a single route. Table 1 introduces the notations used in this study.

Table 1. Definition of notations

Sets		Parameters	
$\mathcal{C}$	Set of supermarket cells (indexed c)	d_{ab}	Distance between any point a and b
$\mathcal{C}_w$	Set of supermarket cells in row w (indexed c)	e_c	Starting point of supermarket cell c
$\mathcal{D}$	Set of drop-off points	l_c	Length of supermarket cell c
$\mathcal{D}^*$	Set of drop-off points in a given route	l_s	Length of assembly station s
$\mathcal{M}$	Set of dummy points	L	Cumulative length of all supermarket cells in any row
$\mathcal{P}$	Set of pickup points	w_c	Width of supermarket cell c
$\mathcal{P}^*$	Set of pickup points in a given route	w_w	Width of row w
$\mathcal{S}$	Set of assembly stations (indexed s)	x_a	x-coordinate value for point a
$\mathcal{W}$	Set of rows (indexed w)	y_a	y-coordinate value for point a

Let $\mathcal{S}$ denote the set of assembly stations, and $\mathcal{W}$ represent the set of rows on the shop floor. For an example of a row, one may consider the green area in Fig. 1a, highlighting row $w = 4$. Additionally, let $\mathcal{C}$ be the set of supermarket (or pre-assembly) cells. It is assumed that on each row $w \in \mathcal{W}$, a specific collection of supermarket cells, denoted by $\mathcal{C}_w$, has been installed, thereby making $\mathcal{C}$ the aggregate set of all supermarket cells across all rows. Each supermarket cell is equipped with a designated pickup point, and correspondingly, each assembly station possesses a designated drop-off point.

The transportation process involves a dedicated vehicle that commences its journey from an initial position, namely the depot d. The aim is to devise optimal routes for the vehicle, such that each route originates from the depot, and subsequently visits all the predetermined pickup and drop-off points assigned to it. For any given route, all pickup points must be visited before the drop-off points can be visited and may be described as a precedence requirement. This requirement ensures that the vehicle adheres to an organized sequence during its transportation operations. This is rooted in the widespread practice of using one-way paths next to the assembly stations. The primary objective of this optimization effort is to achieve the minimization of the total distance covered by the designated transportation vehicle along each route. Through the development of efficient route planning, the logistics of the shop floor can be effectively streamlined, thereby resulting in heightened productivity and notable reductions in operational costs. Each supermarket cell possesses distinct dimensions, with a length denoted by l_c and a width represented by w_c, as well as the starting point e_c (see Fig. 1a). The pickup point is centrally located on the width of each supermarket cell, while the drop-off point is situated at the midpoint of every assembly station. The collection of pickup points is denoted as $\mathcal{P}$, while the set of drop-off points is represented as $\mathcal{D}$. To facilitate the optimization process, additional dummy points are introduced to serve dual purposes: (*i*) to function as the meeting points for any combination of lengths l_c and widths w_c of the supermarket cells, and (*ii*) to act as the terminal points (both starting and ending) of the assembly stations. These dummy points are visually depicted as black vertices in Fig. 1b, and their collective set is denoted as $\mathcal{M}$. The dummy points are necessary for the computation of shortest path distances between any pair of points belonging to the set $d \cup \mathcal{P} \cup \mathcal{D}$ (see Sect. 4).

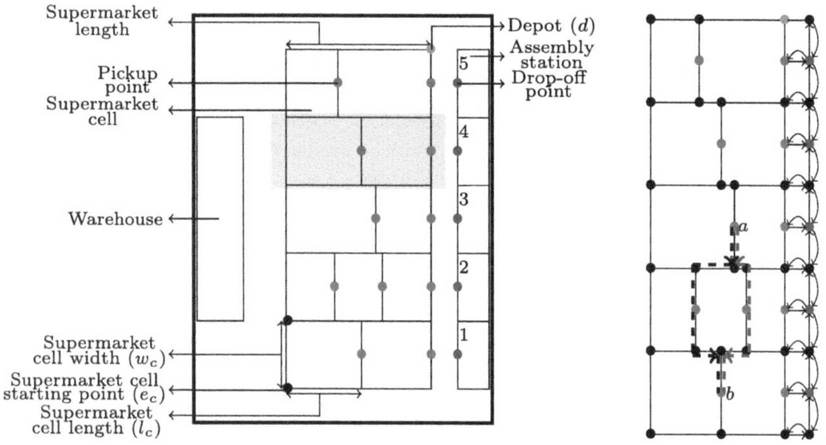

(a) Schematic representation of the shop floor

(b) Graph representing the routing problem

Fig. 1. Shop floor and graph representation

Although we describe the ITP in the context of assembly line feeding, the same transportation problem may also arise in other manufacturing scenarios involving part coupling, such as in small pre-assembly cells, where sub-assemblies are required to be picked up from multiple pre-assembly cells and delivered to a sequence of assembly stations along a single route. In this case, a complete route starts at the depot, and visits all pre-assigned pre-assembly cells prior to navigating the sequence of assembly stations before returning to the depot.

The ITP is formulated on a connected mixed graph $\mathcal{G} = (\mathcal{V}, \mathcal{E})$. A mixed graph is characterized by the composition of directed and undirected edges on a single graph. The set of vertices $\mathcal{V}$ comprises distinct elements, including the depot denoted as d, the set $\mathcal{P}$ encompassing all pickup points, the set $\mathcal{D}$ containing all drop-off points, and the set $\mathcal{M}$ consisting of dummy points, i.e., $\mathcal{V} = d \cup \mathcal{P} \cup \mathcal{D} \cup \mathcal{M}$. Edges in $\mathcal{E}$ establish connections between pairs of neighboring vertices in $\mathcal{V}$. The structure of the graph $\mathcal{G} = (\mathcal{V}, \mathcal{E})$ is visualized in Fig. 1b, with distinctive vertex colors: the depot depicted in brown, pickup and drop-off points represented by red and blue vertices, respectively, and dummy points illustrated as black vertices. The edges between any blue vertex and its adjacent red vertex characterize the distances between the assembly stations and the respective supermarket cells. By defining the problem within this mixed graph, the ITP can be systematically addressed, and suitable techniques can be applied to obtain efficient and optimal routes for transportation operations within the considered system.

3 Literature Review

In warehousing systems, researchers have explored various routing methods to optimize order-picking processes. [14] investigated heuristics such as S-shape, largest gap, aisle-by-aisle, and a combined heuristic, comparing their performance with a branch-and-bound algorithm designed to generate the shortest feasible order picking routes. Their results demonstrated that the combined heuristic outperforms other heuristics in diverse warehouse layouts with two or more cross aisles and varying order sizes. Similarly, [17] addressed the routing problem of order pickers in multi-parallel-aisle warehouse systems by proposing hybrid heuristics and algorithms based on the Traveling Salesman Problem (TSP). Their evaluation showed that the TSP-based algorithms significantly reduced routing distance, achieving up to 47% improvement compared to other heuristics.

Furthermore, [18] advanced the field by introducing a mixed integer programming formulation to address the joint problem of order batching and picker routing in warehouses, incorporating valid inequalities to enhance computational efficiency across multiple instances. The authors achieved optimal solutions for scenarios involving up to 20 orders. [19] provided a comprehensive review of order-picking systems, focusing on tactical and operational planning in picker-to-parts systems. They explored the interconnections between various planning problems and identified optimal policy combinations, offering practical guidelines for warehouse managers to design efficient order-picking systems and improve customer service standards.

In the area of general pickup and delivery problems, [13]'s extensive survey on pickup and delivery problems includes discussing the Traveling Salesman Problem with Clustered Backhauls (TSPCB), where a route commences at the depot, servicing delivery customer clusters before traversing pickup customer clusters, and returns to the depot. Each customer is visited once. [10] have notably adapted GENIUS, a set of construction-improvement heuristics originally devised for the standard TSP by [9], to address the intricacies of TSPCB, highlighting the versatility of these heuristics.

In assembly systems, researchers have proposed optimal methods for routing and scheduling challenges, such as determining the number of consecutive assembly stations to be served by each tow train and establishing start times for just-in-time part delivery. [6] addressed these issues, while [12] combined the supermarket location problem with vehicle type selection for mixed-model assembly lines. They defined routes based on vehicle deployment through a supermarket to service consecutive assembly stations. Similarly, [2] developed a mixed integer linear programming model integrating vehicle type selection and assembly line feeding decisions to minimize total feeding costs, defining routes as paths between the warehouse or supermarket and consecutive assembly stations, following the approach in [1].

Despite the significant focus on (*i*) routing problems in warehousing systems using various heuristics and optimal methods, particularly where the depot coincides with the drop-off point; (*ii*) TSPCB with Euclidean distances and complex heuristics, the ITP remains unexplored. This study addresses this gap by examining the routing problem in an in-house environment involving multiple supermarket cells and consecutive assembly stations within a single route. Unlike existing literature focusing on routing problems with a single supermarket, this research employs practical heuristics to navigate a centralized supermarket partitioned into multiple cells and a sequence of assembly stations. The distance between any two points in ITP results from the sum of horizontal and vertical distances between the two points (see Sect. 4). The proposed routing approach aims to minimize travel distances and transportation costs by addressing this specific scenario.

4 Description of Various Solution Procedures

The ITP can be regarded as a variant of the NP-hard TSP, as the routes originate from the depot and visit each pickup and drop-off point exactly once before returning to the depot. Furthermore, it is required that all pickup points must be visited before the drop-off points for each route. Within this section, we will provide different approaches for addressing the ITP.

Since the routing involved in the ITP is a special case of TSP, a similar approach to the classical TSP can be used when solving the ITP. First, through Dijkstra's algorithm, the length of the shortest path distance between any two points $a, b \in \{d\} \cup \mathcal{P} \cup \mathcal{D}$ must be calculated using the distance resulting from the sum of horizontal and vertical distances between the two points. Since there could be many paths between points a and b, the resulting path with the shortest length is kept. For instance, in Fig. 1b, movement between pickup points a and b could be achieved through different paths. Two of such paths (black and blue dotted lines) are depicted. The shortest path distance (blue) is kept as the path between pickup points a and b. Once these distances have been computed, the set $\mathcal{M}$ of dummy vertices are not considered any further. Suppose $\mathcal{P}^*$ and $\mathcal{D}^*$ respectively represent the set of pickup and drop-off points for a given route, so that $\{d\} \cup \mathcal{P}^* \cup \mathcal{D}^*$ represents the set of vertices that will be traversed in the route. The resulting adjacency matrix for the ITP exhibits partial asymmetry due to the following conditions: (*i*) the distance $d_{ab} = d_{ba}$ when a and b represent any two pickup points; (*ii*) the distance $d_{da} = d_{ad}$ when d denotes the depot and a corresponds to any pickup or drop-off point; (*iii*) the distance $d_{ab} \neq d_{ba}$ when a is a pickup point and b is a drop-off point, with d_{ba} assigned a large value to discourage visiting drop-off points before pickup points; and (*iv*) the distance $d_{a,a+1} \neq d_{a+1,a}$ when a and $a+1$ are any two consecutive drop-off points, with $d_{a+1,a}$ assigned a large value to discourage visiting a drop-off point, closer to the depot, before its preceding one in any route (see Fig. 1b). This asymmetric adjacency matrix serves as input to the employed heuristics for solving the ITP, which will be further elucidated in subsequent sections. Additionally, we provide a detailed description on how to transform the problem for solving it optimally. An example for each heuristic technique and the optimal solution is given in Fig. 2a to 2h In the figures, the brown vertex represents the depot, the red vertices represent the pickup points in the route, the blue vertices represent the drop-off points in the route, and the green vertices represent pickup or drop-off points not present in the route. Algorithm 1 presents a comprehensive summary of the solution approach for the ITP.

Algorithm 1. Algorithm for ITP

procedure ITP PROCEDURE
  Generate the graph
  Using shortest path, compute the distance: between any 2 pickup points, between the depot and any point in $\mathcal{P} \cup \mathcal{D}$, from any pickup to any drop-off point, from any drop-off point a to another drop-off point $a+1$, where a and $a+1$ are farther and closer to the depot, respectively
  Input a large value for the distance: from any drop-off point to any pickup point, from any drop-off point $a+1$ to another drop-off point a, a and $a+1$ are farther and closer to the depot, respectively
  Generate sets $\mathcal{P}^*$ and $\mathcal{D}^*$ for a route
  for each combination generated **do**
    Starting from the depot
    Using the optimal solution procedure and the heuristics described in Section 4.1 to 4.6, find the shortest possible tour that traverses the pickup and drop-off points once through the distances computed in 3 and 4
    End at the depot
    Compute the tour cost
  end for
end procedure

4.1 Nearest Neighbor (NN)

The route commences at the depot and iteratively visits the nearest pickup point until all points in $\mathcal{P}^*$ have been covered. The nearest pickup point is found by considering the shortest path from the current pickup point to any unvisited pickup point. Subsequently, the route proceeds to visit the drop-off points in sequence, beginning with the drop-off point farthest from the depot, and continues until reaching the drop-off point closest to the depot. Finally, the route concludes by returning to and ending at the depot. An example route for this heuristic is depicted in Fig. 2a.

4.2 S-Shape

The S-shape algorithm is defined by the following sequential steps: (*i*) initiation of the route from the depot; (*ii*) determination of the closest row including pickup points in $\mathcal{P}^*$; (*iii*) selection of the two terminal pickup points situated at the outer left and right of the identified row; (*iv*) visiting the nearest terminal pickup point prior to systematically traversing the rest of the points along the row in order of proximity until reaching the other terminal point.

This aforementioned route configuration is reiterated by revisiting steps (*ii*) to (*iv*) iteratively until all pickup points have been visited. Subsequently, the route progresses to encompass the drop-off points, commencing from the initial point and culminating at the terminal drop-off point. Ultimately, the route concludes its itinerary by returning to the depot. Figure 2b delineates an example route for the S-shape heuristic.

4.3 Modified S-Shape (MS)

The modified S-shape algorithm integrates two distinct sub-routes, namely: (*i*) Let $i, j \in \mathcal{D}^*$ represent the drop-off points farthest and closest from the depot, respectively. Suppose $e \in \mathcal{P}^*$ is the nearest pickup point to i such that $y_e \leq y_i$. The initial sub-route originates from the depot and follows an S-shape trajectory (see Sect. 4.2) while encompassing all pickup points except e; (*ii*) a complementary sub-route commences at the unvisited pickup point e described above and progresses from the first drop-off point i to the last drop-off point j. Subsequently, the route returns to the depot. The algorithm provides a final solution by establishing a concatenation of the first route described in (*i*) and the complementary route described in (*ii*). In Fig. 2c, an example route for the heuristic is presented.

4.4 SRNN

This algorithm describes the fusion of three discrete sub-routes, created by an S-shape (S) approach and a reversed nearest neighbor approach (RNN). Let $i, j \in \mathcal{D}^*$ signify the drop-off points farthest and closest from the depot, respectively. (*i*) The S-shape sub-route commences its journey from the depot and undertakes a systematic traversal of all pickup points whose y-coordinate values are strictly greater than y_i. This traversal is characterized by an S-shape trajectory as described above in Sect. 4.2. (*ii*) In the second sub-route, the sub-route proceeds to survey all pickup points characterized by y-coordinate values smaller or equal than y_i. Within this subset of points, the algorithm starts by visiting the nearest pickup point to the drop-off point i and subsequently engages in visiting pickup points with the minimum x-coordinate value difference. This procedure is sustained until the pickup point situated within the first row is encompassed. Thereafter, the route iteratively targets the nearest pickup point until the entirety of the remaining pickup points have been visited. (*iii*) a third sub-route commences from the first drop-off point i and progresses through the

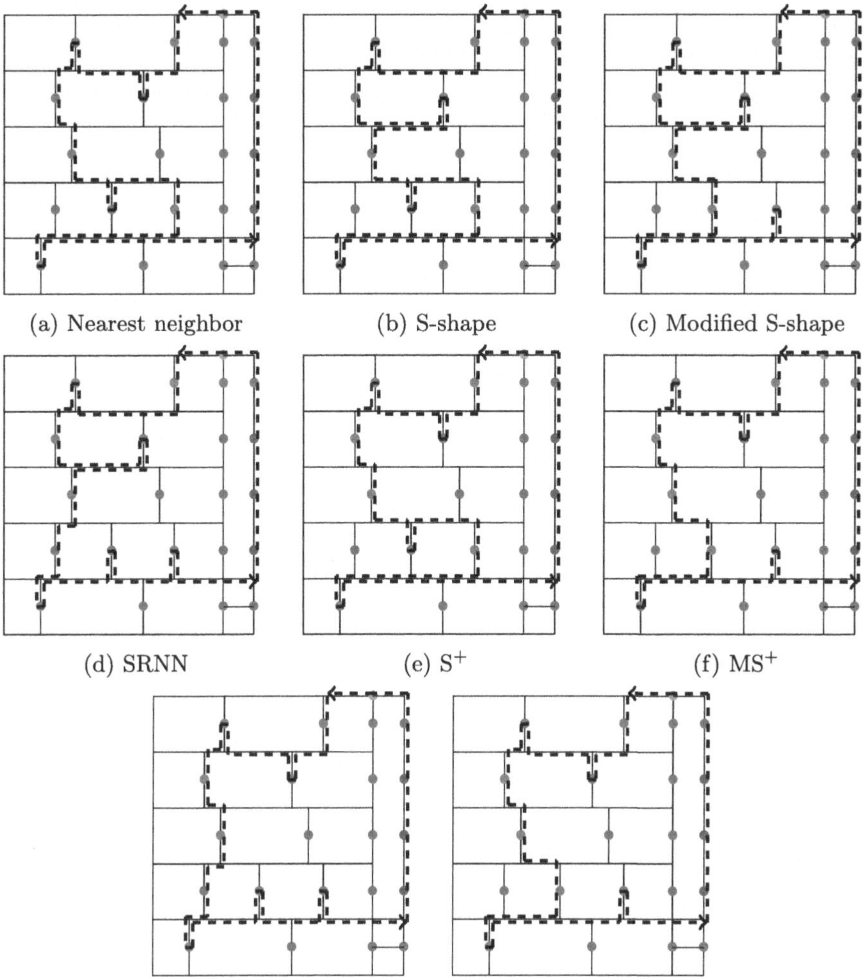

(a) Nearest neighbor (b) S-shape (c) Modified S-shape

(d) SRNN (e) S$^+$ (f) MS$^+$

(g) SRNN$^+$ (h) Optimal

Fig. 2. Illustrative example for the eight routing techniques

last drop-off point j. Subsequently, the route returns to the depot. The final step of the algorithm concatenates the first route described in (i), the inverted route described in (ii), and the complementary route described in (iii). An example route for this heuristic is shown in Fig. 2d.

4.5 Combined Heuristics

Three additional algorithms are created by combining some of the above algorithms with an additional procedure, the so-called $^+$ procedure, to further improve performance. The $^+$ procedure, which can be applied during the creation of a route or sub-route, is described as follows. Let $a, b \in \mathcal{P}^*$ represent the most recently visited pickup points, with b being the most recently visited pickup point. After the inclusion of b via the routing algorithm, a series of conditions are evaluated among the unvisited pickup points, encompassing: (i) the presence of another pickup point $c \in \mathcal{P}^*$ situated in the closest row w to b; (ii) adherence to the inequality relationships $x_a \leq x_c \leq x_b$ or $x_b \leq x_c \leq x_a$; (iii) fulfillment of the distance criterion $d_{ac} + d_{cb} \leq d_{ab} + d_{bc}$; ($iv$) the absence of any other unvisited pickup point $d \in \mathcal{P}^*$ within the nearest row w to c, wherein $x_d \geq x_c$ if $x_c \geq x_b$ and $x_d \leq x_c$ if $x_c \leq x_b$. When all of these conditions are met, c is inserted between a and b in the route. Conversely, if the conditions are not met, the routing procedure continues without the insertion of c.

The S^+ algorithm combines the algorithm described in Sect. 4.2 with the $^+$ procedure. In Fig. 2e, an example route of the S^+ heuristic is presented. The MS^+ algorithm describes the formation and concatenation of two routes similar to the ones described in Sect. 4.3. However, during the creation of the first sub-route the $^+$ procedure is included. An example route for the MS^+ heuristic is delineated in Fig. 2f. The $SRNN^+$ algorithm describes the formation and concatenation of three routes similar to the ones described in Sect. 4.4. However, during the creation of the first sub-route the $^+$ procedure is included. Figure 2g depicts an example route for this heuristic.

4.6 Optimal Solution

The problem at hand is addressed using the exact Concorde (TSP) solver proposed by [3] to attain the optimal solution. Given that the ITP involves asymmetric distance matrices, while the solver is configured for symmetric distance matrices, a transformation is executed to convert the asymmetrical distance matrix into a symmetric counterpart. The transformation methodology is delineated in the subsequent exposition. To facilitate this procedure, each point encompassed within the set $d \cup \mathcal{P}^* \cup \mathcal{D}^*$ is partitioned into a pair of auxiliary vertices (referred to as "dummy" vertices). To illustrate, let $a \in d \cup \mathcal{P}^* \cup \mathcal{D}^*$; it is subsequently denoted as a' and a^*, respectively signifying the in-vertex and out-vertex for point a. By virtue of this redefinition, a' encompasses any edge entering into point a, whereas a^* corresponds to any arc emanating from point a. The resultant symmetric distance matrix bears the ensuing attributes: (i) for any point $a \in d \cup \mathcal{P}^* \cup \mathcal{D}^*$, the distance between a' and a^* is rendered as zero, symbolized by $d_{a'a^*} = d_{a^*a'} = 0$; ($ii$) for any pair of points a and b encompassed within $d \cup \mathcal{P}^* \cup \mathcal{D}^*$, the values of $d_{a'b'}$ and $d_{a^*b^*}$ are assigned to be large; (iii) for any two points a and b within the set $d \cup \mathcal{P}^* \cup \mathcal{D}^*$, the distance from a to b is stipulated as $d_{a^*b'}$, whereas the distance from vertex b to a is denoted as $d_{b^*a'}$. When $a, b \in d \cup \mathcal{P}^*$, then $d_{a^*b'} = d_{b^*a'}$. However, when $a \in d \cup \mathcal{P}^*$ and $b \in \mathcal{D}^*$, or $a, b \in \mathcal{D}^*$, then $d_{a^*b'} \neq d_{b^*a'}$ (see the second paragraph of this Section for more explanation). This symmetric distance matrix is subsequently given to the Concorde TSP solver. In Fig. 2h an example route for the optimal route is given.

5 Computational Study

This section provides a comprehensive overview of the computational experiment conducted in this study. All computational tasks, including the generation of problem instances and the implementation of algorithms, were performed using the Python programming language. The computer was equipped with 8 GB of RAM and an Intel Core i5-8250 processor running at 1.8 GHz speed. Sections 5.1 and 5.2 will present detailed information regarding the problem instances creation and the presentation of computational results, respectively.

Algorithm 2. Generation of Problem Instances

procedure GENERATEPROBLEMINSTANCES
    Determine the set of rows (indexed w)
    Determine width of each row (w)
    Determine the maximal length of the supermarket
    Determine the set of assembly stations (indexed s)
    Determine the length of each assembly station (s)
    Generate the drop-off point associated with each assembly station
    for each row $w \in \mathcal{W}$ **do**
        Generate the number of supermarket cells
        Assign a non-negative length ($l_c \geq 0$) to each supermarket cell
        Assign a non-negative starting point ($e_c \geq 0$) to each supermarket cell
        : $e_c + l_c = e_{c+1}$
        Compute the total row length by summing up all l_c in the row
        Normalize the row length : row length = supermarket length
        Generate the pickup point associated with each supermarket cell
    end for
    Generate the edge d_{ab} between any vertex a at the supermarket border and its adjacent vertex b at the assembly station
end procedure

5.1 Problem Instances Creation

The problem instance generation process is outlined in Algorithm 2.

The test bed is established according to the following specifications:

- The number of rows $|\mathcal{W}|$ is set to 7, 15, 20, 30, 50, or 100 rows.
- The number of stations $|\mathcal{S}|$ is set equal to $|\mathcal{W}|$.
- The cumulative length L of all cells in any row is specified as 60 units.
- The width w_w of any row is defined as 8 units.
- The width w_w of any row, which is defined as 8 units, is equivalent to the width w_c of any supermarket cell c and the length l_s of any station s.
- The number of supermarket cells $|\mathcal{C}_w|$ in row w is randomly selected from the interval $\left[\lceil \frac{L}{L} \rceil, \lceil \frac{2L}{w_w} \rceil\right]$. The interval is further partitioned into intervals $\left[\lceil \frac{L}{L} \rceil, \lceil \frac{2L}{3w_w} \rceil\right]$, $\left[\lceil \frac{2L+3w_w}{3w_w} \rceil, \lceil \frac{4L}{3w_w} \rceil\right]$, and $\left[\lceil \frac{4L+3w_w}{3w_w} \rceil, \lceil \frac{2L}{w_w} \rceil\right]$, so that integers within interval $\left[\lceil \frac{2L+3w_w}{3w_w} \rceil, \lceil \frac{4L}{3w_w} \rceil\right]$ have a higher likelihood of selection compared to the integers within the other two intervals.
- The resulting edge d_{ab} between any vertex a at the supermarket border and its adjacent vertex b at the assembly station is specified to be 10 units.

- The number of pickup points in a route $|\mathcal{P}^*|$ is randomly selected from the intervals [1, 5], [6, 10], [11, 15], [16, 20], and [21, 25].
- The number of consecutive assembly stations corresponding to the drop-off points in a route $|\mathcal{D}^*|$ is set to $min\left(\lceil\frac{1}{2}|\mathcal{P}^*|\rceil, \lceil\frac{1}{2}|\mathcal{S}|\rceil\right)$

The number of assembly stations or rows in the problem instance generation process is a crucial determinant of the problem's size and complexity. These values are sourced from existing literature and firsthand observations from a machinery manufacturing company visit. For example, we include a configuration with seven assembly stations, reflecting the setup of the pre-assembly line observed at the company. Additionally, [2] utilized a dataset with 15 assembly stations. To represent real-world scenarios comprehensively, we analyze instances with 20, 30, 50, and 100 assembly stations, generating 30 distinct instances for each size category, with findings detailed in Sect. 5.2.

5.2 Computational Results

In this section, our primary objective is to conduct a comparative analysis between the outcomes derived from the optimal solution and those obtained through utilizing heuristic methods for the ITP. Table 2 provides a comprehensive summary of various key parameters, including (*i*) the employed heuristics and the optimal algorithm; (*ii*) the problem size, represented by the number of rows on the shop floor ($|\mathcal{W}|$); (*iii*) the number of pickup points (# pkup); (*iv*) the mean and median of the percentage change in objective value between the heuristic and optimal solutions (denoted as avg. gap (%) and med. gap (%), respectively); (*v*) The average solving time across all instances (avg. time (s)), and; (*vi*) the median solving time, required to solve all instances (med. time (s)). This analytical framework is designed to provide a comprehensive assessment of the heuristic and optimal solutions, enabling a nuanced evaluation of their respective performance in addressing the ITP problem.

The comparative analysis systematically evaluates the performance disparity between heuristics and the optimal solution across various problem sizes, focusing on shop floor configurations characterized by different numbers of rows and pickup points. For a seven-row configuration, the average gap between the least effective heuristic and the optimal solution remains negligible (0.00%) for 1–5 pickup points, but increases to 1.59% for 6–10 pickup points, 1.05% for 11–15 pickup points, 8.31% for 16–20 pickup points, and 11.48% for 21–25 pickup points. We observe a similar pattern in a 15-row configuration, with gaps not exceeding 9.04% across all intervals and specific gaps of 0.07%, 0.97%, 3.82%, 6.49%, and 9.04% for the corresponding intervals. For larger problem sizes of 20 and 30 rows, the average gaps also increase with the number of pickup points, but less pronouncedly compared to the smaller configurations. In contrast, configurations with 50 and 100 rows exhibit relative stability in the average gaps, with respective gaps of 0.02%, 0.11%, 0.42%, 1.36%, and 1.67% for 50 rows, and 0.00%, 0.20%, 0.21%, 0.44%, and 1.02% for 100 rows, indicating more consistent heuristic performance in these larger setups.

Table 2. Average times and gaps

$\|\mathcal{W}\|$	#pkup	gap (%)		time (s)		gap (%)		time (s)		gap (%)		time (s)		gap (%)		time (s)	
		avg.	med.	avg.	med.	avg.	med.	avg.	med.	avg.	med.	avg.	med.	avg.	med.	avg.	med.
		NN				S				MS				SRNN			
	1-5	0.00	0.00	0.0001	0.0001	0.00	0.00	0.0001	0.0001	0.00	0.00	0.0001	0.0001	0.00	0.00	0.0001	0.0001
	6-10	0.64	0.00	0.0003	0.0003	1.59	0.00	0.0003	0.0003	1.09	0.00	0.0004	0.0003	0.95	0.00	0.0004	0.0004
7	11-15	0.92	0.00	0.0005	0.0005	1.05	0.00	0.0004	0.0004	0.52	0.00	0.0004	0.0003	0.48	0.00	0.0006	0.0005
	16-20	8.06	8.25	0.0013	0.0011	7.96	7.41	0.0013	0.0011	8.26	8.10	0.0012	0.0009	8.31	8.63	0.0013	0.0011
	21-25	7.75	5.43	0.0017	0.0016	11.48	11.32	0.0012	0.0010	10.38	9.56	0.0012	0.0013	9.98	8.99	0.0023	0.0026
	1-5	0.07	0.00	0.0001	0.0001	0.07	0.00	0.0001	0.0001	0.00	0.00	0.0001	0.0001	0.00	0.00	0.0001	0.0001
	6-10	0.97	0.00	0.0003	0.0003	0.96	0.00	0.0003	0.0003	0.35	0.00	0.0003	0.0003	0.29	0.00	0.0003	0.0003
15	11-15	2.92	2.78	0.0010	0.0010	3.82	3.86	0.0009	0.0008	3.22	2.79	0.0011	0.0010	2.80	2.02	0.0012	0.0010
	16-20	4.61	4.24	0.0020	0.0021	6.49	7.44	0.0017	0.0018	5.55	5.99	0.0017	0.0016	5.07	5.75	0.0022	0.0017
	21-25	8.59	8.53	0.0031	0.0029	9.04	9.06	0.0025	0.0024	8.92	9.10	0.0027	0.0024	8.88	9.15	0.0031	0.0031
	1-5	0.03	0.00	0.0001	0.0000	0.03	0.00	0.0001	0.0000	0.00	0.00	0.0001	0.0000	0.00	0.00	0.0001	0.0001
	6-10	0.48	0.00	0.0002	0.0002	0.32	0.00	0.0002	0.0002	0.17	0.00	0.0002	0.0002	0.17	0.00	0.0002	0.0002
20	11-15	1.72	0.45	0.0004	0.0004	2.21	1.58	0.0003	0.0003	2.22	1.66	0.0003	0.0003	2.05	1.66	0.0004	0.0004
	16-20	5.03	5.26	0.0008	0.0007	4.93	4.81	0.0006	0.0005	4.31	3.71	0.0005	0.0005	4.06	3.68	0.0007	0.0007
	21-25	6.16	6.70	0.0011	0.0009	7.46	7.23	0.0008	0.0008	6.58	7.28	0.0009	0.0008	6.27	6.82	0.0011	0.0010
	1-5	0.02	0.00	0.0001	0.0001	0.02	0.00	0.0001	0.0001	0.00	0.00	0.0001	0.0001	0.00	0.00	0.0001	0.0001
	6-10	0.15	0.00	0.0002	0.0002	0.21	0.00	0.0002	0.0001	0.18	0.00	0.0002	0.0001	0.18	0.00	0.0002	0.0002
30	11-15	1.46	0.43	0.0004	0.0004	1.12	0.20	0.0003	0.0003	0.75	0.00	0.0003	0.0003	0.60	0.00	0.0004	0.0004
	16-20	2.31	2.47	0.0007	0.0006	1.97	1.85	0.0005	0.0005	1.72	1.61	0.0005	0.0005	1.55	1.45	0.0007	0.0007
	21-25	4.30	4.15	0.0014	0.0013	4.08	4.52	0.0011	0.0009	4.19	4.62	0.0012	0.0009	3.99	3.89	0.0015	0.0013
	1-5	0.02	0.00	0.0001	0.0001	0.02	0.00	0.0001	0.0001	0.00	0.00	0.0001	0.0001	0.00	0.00	0.0001	0.0001
	6-10	0.08	0.00	0.0003	0.0003	0.11	0.00	0.0002	0.0002	0.11	0.00	0.0003	0.0002	0.11	0.00	0.0004	0.0003
50	11-15	0.37	0.00	0.0006	0.0006	0.42	0.00	0.0005	0.0004	0.36	0.00	0.0005	0.0004	0.36	0.00	0.0006	0.0006
	16-20	1.36	0.99	0.0010	0.0009	1.32	0.86	0.0007	0.0006	0.99	0.55	0.0006	0.0006	0.75	0.45	0.0008	0.0008
	21-25	1.66	1.49	0.0013	0.0011	1.67	1.69	0.0010	0.0010	1.44	1.19	0.0009	0.0008	1.29	1.02	0.0013	0.0011
	1-5	0.00	0.00	0.0002	0.0001	0.00	0.00	0.0001	0.0000	0.00	0.00	0.0001	0.0001	0.00	0.00	0.0001	0.0001
	6-10	0.20	0.00	0.0002	0.0002	0.18	0.00	0.0001	0.0001	0.11	0.00	0.0002	0.0001	0.11	0.00	0.0002	0.0002
100	11-15	0.21	0.00	0.0004	0.0004	0.20	0.00	0.0003	0.0003	0.13	0.00	0.0003	0.0003	0.14	0.00	0.0004	0.0004
	16-20	0.44	0.00	0.0007	0.0006	0.29	0.00	0.0006	0.0006	0.27	0.00	0.0006	0.0005	0.28	0.00	0.0008	0.0008
	21-25	1.02	0.78	0.0015	0.0012	0.66	0.35	0.0009	0.0009	0.64	0.23	0.0010	0.0010	0.63	0.23	0.0015	0.0014
		S^+				MS^+				$SRNN^+$				Optimal			
	1-5	0.00	0.00	0.0001	0.0001	0.00	0.00	0.0001	0.0001	0.00	0.00	0.0001	0.0001			0.1054	0.0933
	6-10	0.63	0.00	0.0005	0.0005	0.15	0.00	0.0006	0.0005	0.41	0.00	0.0006	0.0005			0.1376	0.1424
7	11-15	0.82	0.00	0.0011	0.0011	0.56	0.00	0.0011	0.0012	0.42	0.00	0.0007	0.0006			0.1262	0.1148
	16-20	6.18	7.08	0.0017	0.0015	6.49	7.08	0.0017	0.0015	6.45	6.81	0.0025	0.0021			0.2534	0.1922
	21-25	7.91	6.36	0.0027	0.0026	6.54	6.94	0.0037	0.0049	7.69	7.73	0.0027	0.0032			0.8518	0.2148
	1-5	0.07	0.00	0.0002	0.0001	0.00	0.00	0.0002	0.0001	0.00	0.00	0.0001	0.0001			0.1189	0.1190
	6-10	0.53	0.00	0.0004	0.0004	0.09	0.00	0.0004	0.0004	0.09	0.00	0.0005	0.0004			0.1472	0.1464
15	11-15	2.03	1.48	0.0011	0.0011	1.10	0.89	0.0012	0.0011	1.10	0.72	0.0019	0.0017			0.1753	0.1647
	16-20	5.30	5.99	0.0030	0.0031	3.65	2.68	0.0032	0.0029	4.04	3.97	0.0033	0.0036			0.2197	0.1801
	21-25	6.64	6.96	0.0036	0.0030	6.70	7.01	0.0042	0.0037	6.79	7.22	0.0046	0.0044			0.3656	0.2420
	1-5	0.03	0.00	0.0001	0.0001	0.00	0.00	0.0001	0.0001	0.00	0.00	0.0001	0.0001			0.0827	0.0808
20	6-10	0.25	0.00	0.0003	0.0002	0.12	0.00	0.0003	0.0002	0.10	0.00	0.0003	0.0003			0.0947	0.0925
	11-15	0.77	0.00	0.0004	0.0004	0.54	0.00	0.0005	0.0004	0.56	0.00	0.0006	0.0005			0.1042	0.1066
	16-20	3.26	3.22	0.0008	0.0008	2.57	2.43	0.0010	0.0009	2.45	2.54	0.0010	0.0010			0.1490	0.1260
	21-25	4.83	4.15	0.0013	0.0012	4.15	4.04	0.0014	0.0013	4.01	4.39	0.0015	0.0014			0.2246	0.1627
	1-5	0.00	0.00	0.0001	0.0001	0.02	0.00	0.0001	0.0001	0.00	0.00	0.0001	0.0001			0.0822	0.0818
	6-10	0.03	0.00	0.0002	0.0002	0.00	0.00	0.0002	0.0002	0.00	0.00	0.0003	0.0002			0.0960	0.0942
30	11-15	0.61	0.00	0.0005	0.0005	0.34	0.00	0.0006	0.0006	0.26	0.00	0.0006	0.0006			0.1040	0.1040
	16-20	1.35	1.29	0.0008	0.0008	0.79	0.65	0.0010	0.0009	0.89	0.92	0.0010	0.0010			0.1176	0.1194
	21-25	3.13	3.37	0.0016	0.0015	3.11	3.18	0.0020	0.0017	2.96	3.18	0.0020	0.0019			0.1785	0.1525
	1-5	0.02	0.00	0.0002	0.0002	0.00	0.00	0.0002	0.0002	0.00	0.00	0.0002	0.0002			0.6261	0.1195
	6-10	0.03	0.00	0.0004	0.0003	0.03	0.00	0.0005	0.0004	0.03	0.00	0.0005	0.0004			0.1282	0.1236
50	11-15	0.04	0.00	0.0006	0.0006	0.06	0.00	0.0007	0.0007	0.06	0.00	0.0009	0.0008			0.6420	0.1290
	16-20	0.76	0.00	0.0010	0.0009	0.47	0.00	0.0010	0.0010	0.27	0.00	0.0012	0.0011			0.1530	0.1453
	21-25	1.09	0.72	0.0014	0.0013	0.90	0.57	0.0016	0.0015	0.76	0.63	0.0018	0.0017			0.1477	0.1346
	1-5	0.00	0.00	0.0001	0.0001	0.00	0.00	0.0001	0.0001	0.00	0.00	0.0001	0.0001			0.5857	0.0798
	6-10	0.09	0.00	0.0002	0.0002	0.03	0.00	0.0002	0.0002	0.03	0.00	0.0003	0.0003			0.0890	0.0876
100	11-15	0.07	0.00	0.0005	0.0005	0.00	0.00	0.0006	0.0005	0.01	0.00	0.0007	0.0006			0.1030	0.1032
	16-20	0.07	0.00	0.0010	0.0009	0.04	0.00	0.0012	0.0011	0.08	0.00	0.0012	0.0011			0.1179	0.1167
	21-25	0.24	0.00	0.0016	0.0015	0.17	0.00	0.0017	0.0017	0.16	0.00	0.0022	0.0021			0.6343	0.1369

An overarching analysis of the results indicates that the combined heuristics, particularly MS^+ and $SRNN^+$, consistently outperform other heuristic methodologies by achieving lower average gaps in objective values. The computational time for obtaining the optimal solution is significantly higher than for any heuristic technique (see Table 2), enhancing the appeal of heuristics. Furthermore, the primary aim of this study is to provide practitioners with heuristics that are practically implementable on the shop floor. Although the optimal solution yields a superior objective value, it may not be implementable practically due to its lack of adherence to predefined rules, potentially leading to inefficient resource utilization on the shop floor.

6 Conclusion and Recommendation

In the context of this research, we have delineated the ITP and introduced a suite of heuristic methodologies aimed at addressing this intricate logistical challenge. This study makes significant contributions by introducing and formally describing the ITP for navigating the shop floor and providing heuristic techniques that are practically applicable and easy to implement. A rigorous comparison of these heuristics against the optimal solution across various problem sizes reveals that smaller shop floor instances (seven or 15 rows) are effectively solved with up to 10 pickup points, yielding small average gaps. In contrast, larger shop floors (50 or 100 rows) exhibit smaller average gaps between the least effective heuristic and the optimal solution, not exceeding 1.67% and 1.02%, respectively. Notably, the heuristic methodologies demonstrate superior computational efficiency compared to the optimal procedure, which lacks defined guided rules and may confuse practitioners. This study provides valuable insights and actionable recommendations for enhancing operational efficiency and decision-making in in-house transportation logistics.

The validation of the ITP in this research is based on artificial problem instances, providing a foundational benchmark. Future research should incorporate real-life problem instances to enhance the realism and practical applicability of the ITP. The heuristic approaches presented are an initial exploration into methods addressing the ITP, and future investigations should develop more heuristics tailored to practitioners' needs. These new heuristics should aim to achieve proximity to the optimal solution while remaining easy to implement, thus improving their utility in real-world operational contexts.

References

1. Adenipekun, E.O., Limère, V., Schmid, N.: A conceptual framework for the integration of assembly line feeding and route building. In: The 34th Annual European Simulation and Modelling Conference, pp. 163–167. Eurosis (2020)
2. Adenipekun, E.O., Limère, V., Schmid, N.A.: The impact of transportation optimisation on assembly line feeding. Omega 102544 (2022)
3. Applegate, D., Bixby, R., Chvatal, V., Cook, W.: Concorde tsp solver (2006)

4. Bozer, Y.A., McGinnis, L.F.: Kitting versus line stocking: a conceptual framework and a descriptive model. Int. J. Prod. Econ. **28**(1), 1–19 (1992)
5. Caputo, A.C., Pelagagge, P.M.: A methodology for selecting assembly systems feeding policy. Ind. Manag. Data Syst. **111**(1), 84–112 (2011)
6. Emde, S., Boysen, N.: Optimally routing and scheduling tow trains for JIT-supply of mixed-model assembly lines. Eur. J. Oper. Res. **217**, 287–299 (2012)
7. Faccio, M.: The impact of production mix variations and models varieties on the parts-feeding policy selection in a JIT assembly system. Int. J. Adv. Manuf. Technol. **72**(1), 543–560 (2014)
8. Faccio, M., Gamberi, M., Bortolini, M., Pilati, F.: Macro and micro-logistic aspects in defining the parts-feeding policy in mixed-model assembly systems. Int. J. Serv. Oper. Manag. **31**(4), 433–462 (2018)
9. Gendreau, M., Hertz, A., Laporte, G.: New insertion and postoptimization procedures for the traveling salesman problem. Oper. Res. **40**(6), 1086–1094 (1992)
10. Gendreau, M., Hertz, A., Laporte, G.: The traveling salesman problem with backhauls. Comput. Oper. Res. **23**(5), 501–508 (1996)
11. Limère, V., Van Landeghem, H., Goetschalckx, M., Aghezzaf, E.H., McGinnis, L.F.: Optimising part feeding in the automotive assembly industry: deciding between kitting and line stocking. Int. J. Prod. Res. **50**(15), 4046–4060 (2012)
12. Nourmohammadi, A., Eskandari, H., Fathi, M., Ng, A.H.: Integrated locating in-house logistics areas and transport vehicles selection problem in assembly lines. Int. J. Prod. Res. **59**(2), 598–616 (2021). https://doi.org/10.1080/00207543.2019.1701207
13. Parragh, S.N., Doerner, K.F., Hartl, R.F.: A survey on pickup and delivery problems: Part i: transportation between customers and depot. J. für Betriebswirtschaft **58**, 21–51 (2008)
14. Roodbergen, K.J., Koster, R.: Routing methods for warehouses with multiple cross aisles. Int. J. Prod. Res. **39**(9), 1865–1883 (2001)
15. Sali, M., Sahin, E.: Line feeding optimization for just in time assembly lines: an application to the automotive industry. Int. J. Prod. Econ. **174**, 54–67 (2016)
16. Schmid, N.A., Limère, V.: A classification of tactical assembly line feeding problems. Int. J. Prod. Res. **57**(24), 7586–7609 (2019)
17. Theys, C., Bräysy, O., Dullaert, W., Raa, B.: Using a tsp heuristic for routing order pickers in warehouses. Eur. J. Oper. Res. **200**(3), 755–763 (2010)
18. Valle, C.A., Beasley, J.E., Da Cunha, A.S.: Optimally solving the joint order batching and picker routing problem. Eur. J. Oper. Res. **262**(3), 817–834 (2017)
19. Van Gils, T., Ramaekers, K., Caris, A., De Koster, R.B.: Designing efficient order picking systems by combining planning problems: state-of-the-art classification and review. Eur. J. Oper. Res. **267**(1), 1–15 (2018)

A Flexible Job Shop Scheduling Problem Involving Reconfigurable Machine Tools Under Industry 5.0

Hessam Bakhshi-Khaniki[1], Reza Tavakkoli-Moghaddam[1](✉), Zdenek Hanzalek[2], and Behdin Vahedi-Nouri[1]

[1] School of Industrial Engineering, College of Engineering, University of Tehran, Tehran, Iran
tavakoli@ut.ac.ir

[2] Industrial Informatics Department, Czech Institute of Informatics Robotics and Cybernetics, Czech Technical University in Prague, Prague, Czech Republic

Abstract. The rise of Industry 5.0 has introduced new demands for manufacturing companies, requiring a shift in how production schedules are managed to address human-centered, environmental, and economic goals comprehensively. The flexible job shop scheduling problem (FJSSP), which involves processing operations on various capable machines, accurately reflects the complexities of modern manufacturing settings. This paper investigates the FJSSP involving reconfigurable machine tools with configuration-dependent setup times, while integrating human aspects like worker assignments, moving time, and rest periods, as well as minimizing total energy consumption. A mixed-integer programming (MIP) model is developed to simultaneously optimize these objectives. The model determines the assignment of operations to machines, workers, and configurations while sequencing operations, scheduling worker movements, and respecting rest periods, and minimizing overall energy consumption. Given the NP-hard nature of the FJSSP with worker assignments and reconfigurable tools, a memetic algorithm (MA) is proposed. This meta-heuristic evolutionary algorithm features a three-layer chromosome encoding method, specialized crossover and mutation strategies, and neighborhood search mechanisms to enhance solution quality and diversity. Comparisons of MA with MIP and genetic algorithms (GA) on benchmark instances demonstrate the MA's efficiency and effectiveness, particularly for larger problem instances where MIP becomes impractical. This research paves the way for sustainable and resilient production schedules tailored for the factory of the future under the Industry 5.0 paradigm. The work bridges a crucial gap in current literature by integrating worker and environmental impact into the FJSSP with reconfigurable machine models.

Keywords: Flexible Job Shop Scheduling · Reconfigurable Manufacturing Systems · Human Centric · Sustainable Manufacturing · Industry 5.0

Published by Springer Nature Switzerland AG 2024
M. Thürer et al. (Eds.): APMS 2024, IFIP AICT 733, pp. 244–257, 2024.
https://doi.org/10.1007/978-3-031-71645-4_17

1 Introduction

As manufacturing enterprises transition from the Industry 4.0 era towards the emerging Industry 5.0 paradigm, they are faced with new imperatives around human-centricity, environmental sustainability, and operational resilience. Scheduling, which lies at the heart of production operations, can no longer purely focus on economic objectives like minimizing makespan and costs. There is a critical need to holistically account for human factors related to worker well-being and social sustainability, as well as environmental impact factors like energy consumption and emissions.

The flexible job shop scheduling problem (FJSSP), in which operations can be processed on multiple capable machines, accurately represents the modern production environment [1]. Integrating human aspects (e.g., worker skills, ergonomics, fatigue management, and the human dimension) into the FJSSP is essential for enabling human-centric manufacturing. Simultaneously, it is vital to incorporate environmental criteria like energy efficiency, emissions reduction, and waste minimization to drive sustainable production schedules. The shift towards Industry 5.0 necessitates manufacturing companies to redefine their priorities, placing equal emphasis on economic, human, and environmental criteria. By holistically incorporating these multi-faceted objectives into scheduling decisions, companies can drive operational excellence while fostering workforce prosperity and environmental stewardship. The survey paper [2] underscores vital research opportunities to develop scheduling techniques that harmoniously balance these key dimensions for the factory of the future.

The well-being of workers is optimized through ergonomic scheduling, rest periods, and safe conditions. Skills development includes versatile scheduling and on-the-job learning. Human-machine collaboration enhances tasks, while decentralized decision-making empowers workers. Social sustainability fosters teamwork and work-life balance. Together, these practices prioritize workers' needs and skills, promoting adaptability and resilience [3].

The job shop scheduling problem (JSSP) has extensively been studied over recent decades, with researchers proposing various approaches to support scheduling decisions in evolving production environments. The FJSSP, which allows operations to be processed on multiple capable machines, better represents real-world manufacturing systems and has gained significant attention. Traditional JSSP and FJSSP research has predominantly focused on optimizing economic objectives like minimizing makespan, production costs, and tardiness. However, with the advent of Industry 4.0 and now Industry 5.0, new priorities around human-centricity, sustainability, and resilience have emerged [4].

Many researchers are exploring FJSSP models that incorporate human factors, such as worker skills [4], ergonomic risks [5], fatigue management [6], and labor costs [7]. He et al. [8] developed an improved African vulture optimization algorithm to solve the dual-resource constrained FJSSP, which minimizes makespan and total delay. Tan et al. [6] proposed an enhanced NSGA-II for solving a fatigue-conscious dual resource-constrained FJSSP, which minimizes worker fatigue and makespan by jointly scheduling machines and workers, which has practical relevance in flexible manufacturing systems. Müller et al. [4] addressed an FJSSP with workforce constraints to minimize makespan.

It considers a heterogeneous workforce, in which workers may possess different qualifications for operating machines, affecting processing times. They proposed filter-and-fan heuristic approaches that combine local search with tree search, decomposing the problem into machine allocation/sequencing and worker assignment decisions. Hongyu et al. [5] proposed an enhanced survival duration-guided NSGA-III for a sustainable FJSSP with dual resources, aiming to optimize makespan, energy consumption, and ergonomic risk while scheduling machines and workers.

In parallel, prompted by global environmental concerns, researchers have investigated integrating objectives and constraints related to energy consumption [9], carbon emissions [10], and noise pollution [11] into green FJSSP formulations. Jia et al. [9] developed an enhanced GA utilizing a four-layer chromosome encoding scheme and improved initialization, crossover, and mutation operations for solving the green FJSSP that considers cost, carbon emissions, and customer satisfaction under the time of use electricity pricing. Wang et al. [10] addressed concerns regarding the low-carbon, many-objective FJSSP in the context of global warming and energy crises. It aims to minimize completion time, total delay time, processing load rate of the bottleneck machine, and total carbon emissions of the system. Hajibabaei et al. [11] introduced a mathematical model and a Lagrangian relaxation algorithm for an FJSSP incorporating an assembly stage, machine breakdowns, and batch transportation to customers. The model addresses tardiness/earliness costs, fuzzy transportation/makespan costs, fuzzy CO_2 emissions, and noise pollution resulting from increased machine speeds. However, simultaneous optimization of economic, environmental, and human factors remains largely unexplored.

In recent years, reconfigurable manufacturing systems and reconfigurable machine tools (RMTs) have emerged as a new manufacturing paradigm to meet rapidly changing production requirements. The modular structure of RMTs enables easy reconfiguration for performing different operations or changing their functional capabilities. This reconfigurability allows RMTs to process various operation types by changing tool attachments, adding/removing functional modules, etc. [12].

Scheduling problems involving RMTs pose new challenges compared to the classical FJSSP. In addition to the normal decisions of operation-to-machine assignment and operation sequencing, scheduling with RMTs requires determining the configuration of each machine for processing its assigned operations. Furthermore, setup times for switching a machine's configuration are often sequence-dependent. Fan et al. [13] proposed an enhanced GA for the FJSSP with machine reconfigurations, in which machines adapt with auxiliary modules for various tasks. Fan et al. [14] investigated the FJSSP that considers lot-streaming and machine reconfigurations to minimize total weighted tardiness. To address this problem involving both discrete and continuous optimization attributes, they proposed a matheuristic algorithm that combines a GA framework with two mixed integer linear programming-based local search strategies for optimizing lot-sizing plans. Guo et al. [15] proposed a novel dynamic FJSSP with reconfigurable manufacturing cells, wherein each cell is capable of executing various operations, requiring reconfiguration time during transitions between operations. The authors devised an enhanced genetic programming hyper-heuristic method, integrating an individual simplification policy to address the FJSSP, accounting for objectives like completion time, delay time, and reconfiguration time.

The existing body of research concerning integrated sustainable FJSSP models is notably scarce when it comes to considering both worker well-being and environmental impact concurrently. This gap in the literature is particularly pronounced in the realm of reconfigurable machine tools. Reconfigurable machine tools, with their ability to adapt to various manufacturing tasks and environments, offer a promising avenue for enhancing sustainability in manufacturing processes. However, the dearth of studies addressing their role within integrated sustainable FJSSP models represents a significant lacuna in current research efforts. Addressing this gap could yield valuable insights into optimizing manufacturing processes not only in terms of efficiency but also in promoting worker welfare and minimizing environmental harm.

2 Problem Description

The FJSSP, in which machines are reconfigurable with configuration-dependent setup times, considers both human and environmental factors, specifically minimizing total energy consumption while limiting worker assignment. This problem involves sets of jobs J, machines K, workers L, and configurations C_K. Each job requires a sequence of operations that need to be processed with a predefined sequence and each operation can be completed in configurations by a subset of workers and machines. The assumptions and limitations of our model are as follows:

- Jobs are independent.
- There is a set of reconfigurable machines on the shop floor, each with multiple configurations.
- Each operation can be processed on at least one configuration of one of the reconfigurable machines.
- Setup time between changing configurations is considered.
- Each machine can be in only one configuration and process one operation at a time.
- Preemption of operations is not allowed.
- Both human (i.e., workers) and machine resources are considered.
- Worker moving time is considered when they need to move to another machine.
- Workers rest after each operation is completed.
- Workers rest for up to 10% of their work time.

The notations used in the proposed mathematical model are summarized as follows:

Indices:

i, i'	Jobs index
j, j'	Operations index
k, k'	Machines index
l	Worker index
c, c1, c2	Configuration index

Sets:

J Set of jobs
K Set of machines
L Set of workers
C_k Set of configurations of machine k

Parameters:

N Total number of jobs
M Total number of machines
L Total number of workers
n_i Total number of operations of job i
U A big positive number
N Total number of jobs
O_{ij} Operation j of job i
M_{ij} Subsets of machines for O_{ij}
W_k Subsets of workers for machine k
Pt_{ij} Processing time of O_{ij}
$Se_{c1,c2,k}$ Setup time of changing configuration from C_1 to C_2 on machine k
$Mt_{kk'}$ Worker moving time between machine k and k'
Rt_{ijl} Rest time of worker l after performing operation O_{ij}
E_{ijkl} Energy consumption of operation O_{ij} on machine k performed by worker l
AE Auxiliary energy

Decision variables:

TE Total energy consumption
C_{max} Makespan
C_{ij} Completion time of O_{ij}
St_{ij} Starting time of processing of O_{ij}
U_{ijkc} 1 if machine k on configuration c can process operation O_{ij}; 0, otherwise
X_{ijklc} 1 if machine k and worker l on configuration c are selected for operation O_{ij}; 0, otherwise
$Y_{iji'j'}$ 1 if operation O_{ij} is performed before operation $O_{i'j'}$; 0, otherwise
$Z_{i'j'kk'l}$ 1 if worker l moves from machine k to perform operation $O_{i'j'}$ on machine k'; 0, otherwise

The mathematical model is defined as follows:

$$MinTE = \mathrm{AE}C_{max} + \sum_{i \in I} \sum_{j \in J} \sum_{k \in K} \sum_{c \in C_k} \sum_{l \in W_k} E_{ijkl} Pt_{ijklc} X_{ijklc} \tag{1}$$

s.t.

$$C_{ij} = St_{ij} + Pt_{ijklc} X_{ijklc} \; \forall i \in J, j \in n_i, k \in K, l \in L, c \in C_K \tag{2}$$

$$C_{max} \geq C_{in_i} \; \forall i \in J \tag{3}$$

$$\sum\nolimits_{k \in M_{ij}} \sum_{c \in C_k} \sum_{l \in W_k} X_{ijklc} = 1 \, \forall i \in J, j \in O_{ij} \tag{4}$$

$$X_{ijklc} \leq r_{ijklc} \; \forall i \in J, j \in O_{ij}, k \in M_{ij}, l \in W_K, c \in C_K \tag{5}$$

$$St_{ij} - St_{i,j-1} \geq \sum\nolimits_{k \in M_{ij}} \sum_{c \in C_k} \sum_{l \in W_k} Pt_{ijklc} X_{ijklc} \forall i \in J, j \in O_{ij} \tag{6}$$

$$\begin{aligned} & St_{i'j'} - St_{ij} \geq Pt_{ijklc} X_{ijklc} + Mt_{kk'} Z_{i'j'kk'l} + Rt_{ijl}\left(1 - Z_{i'j'kk'l}\right) \\ & \forall i, i' \in J, j \in O_{ij}, j' \in O_{i'j'}, O_{ij} \neq O_{i'j'}, k, k' \in M_{ij}, l \in W_K, c \in C_K \end{aligned} \tag{7}$$

$$\begin{aligned} & St_{ij} - St_{i'j'} \geq Pt_{ijklc_2} + Se_{c_1,c_2,k} - \left(2 - X_{ijklc_2} - X_{i'j'klc_1} + Y_{iji'j'}\right)U \\ & \forall i, i' \in J, j \in O_{ij}, j' \in O_{i'j'}, O_{ij} \neq O_{i'j'}, k \in M_{ij}, l \in W_K, c_1, c_2 \in C_K \end{aligned} \tag{8}$$

$$\begin{aligned} & St_{i'j'} - St_{ij} \geq Pt_{i'j'klc_2} + Se_{c_1,c_2,k} - \left(3 - X_{ijklc_1} - X_{i'j'klc_2} - Y_{iji'j'}\right)U \\ & \forall i, i' \in J, j \in O_{ij}, j' \in O_{i'j'}, O_{ij}, \neq O_{i'j'}, k \in M_{ij}, l \in W_K, c_1, c_2 \in C_K \end{aligned} \tag{9}$$

$$St_{ij}, C_{ij}, C_{max} \geq 0 \forall i \in J, j \in O_{ij} \tag{10}$$

$$\begin{aligned} X_{ijklc}, Y_{iji'j'}, Z_{i'j'kk'l} \in \{0, 1\} \quad & \forall i \in J, j \in J, j \in O_{ij}, k \in M_{ij}, \\ & l \in W_K, c \in C_K \end{aligned} \tag{11}$$

Equation (1) represents the objective function, aiming to minimize the total energy consumption. Equation (2) calculates the completion time of each operation based on the starting time and processing time. Constraint (3) is used to determine makespan, while constraint (4) ensures that each operation is processed by one machine and one worker in the one-machine configuration. Constraint (5) guarantees that operations can be processed only on machine configurations, for which they are eligible. Constraint (6) ensures that each operation of the same job should not be processed before the previous operation is completed. Constraint (7) determines the starting time of an operation considering worker rest time and moving time, while Constraints (8) and (9) prevent operations from overlapping. Lastly, Constraints (10) and (11) define the decision variables.

3 Methodology

The FJSP with worker assignment and machine reconfigurable tools is an NP-hard combinatorial optimization problem, making it challenging to find optimal solutions for medium and large instances in a short time. As a result, obtaining optimal solutions using exact methods within a reasonable timeframe is extremely difficult. To address this issue, memetic algorithm (MA), which is a meta-heuristic evolutionary algorithm, is well-suited. The key steps of the developed MA are as follows:

Step 1: Set parameters.

Step 2: Chromosome encoding and decoding.

Step 3: Initialize population. Set the initial generation count $i = 0$ and initialize population P(i) with n individuals.

Step 4: Generate offspring population $F_1(i)$ using the crossover method.

Step 5: Generate offspring population $F_2(i)$ using the mutation method.

Step 6: Generate population $Q(i)$ by combining F_1(i), $F_2(i)$, and $P(i)$.

Step 7: Perform neighborhood strategies N_1, N_2, and N_3 to improve the quality of individuals in $Q(i)$ and set $P(i + 1) = Q(i)$.

Step 8: If the termination criteria are not satisfied, return to Step 4; otherwise, output $P(i + 1)$.

The FJSSP with machine reconfigurable tools employs a three-layer chromosome encoding method, as illustrated in Fig. 1. The three layers are as follows:

Operation Sequencing (OS) Layer: Represented as [1–3], where each element corresponds to operations O_{11}, O_{31}, O_{32}, O_{12}, O_{33}, O_{21} and O_{22}, respectively.

Configuration Selection (CS) Layer: This layer includes both machine assignment and machine configuration. It is represented as [(1,1), (2,2), (2,2), (1,1), (2,1), (1,1), (1,1)]. Here, the first element denotes the machine, and the second element denotes its configuration, corresponding to operations O_{11}, O_{12}, O_{21}, O_{22}, O_{31}, O_{32}, and O_{33}, respectively.

Worker Selection (WS) Layer: Represented as [1, 1], where each element indicates the worker assigned to the respective operations O_{11}, O_{12}, O_{21}, O_{22}, O_{31}, O_{32}, and O_{33}.

For the initial population, the smallest energy consumption (SEC) method is used to design the OS, CS, and WS layers. In this method, workers, machines, and configurations are assigned based on the smallest energy consumption for each operation.

The crossover generator is employed to improve search efficiency. In this paper, we utilize a random selection process, where parents (i.e., P_1 and P_2) produce offspring (i.e., F_1 and F_2) as shown in Fig. 2(a). We develop a job-based crossover (JBX) mechanism specifically for the OS layer. This method involves creating two distinct sets of jobs. For offspring F_1, the job set from parent P_1 is directly assigned to F_1, preserving the exact positions of the jobs. The remaining positions in F_1 are filled with jobs from parent P_2 that are not present in the job set of P_1, maintaining the order in which these jobs appear in P_2. Similarly, for offspring F_2, the job set from parent P_2 is directly assigned to F_2, maintaining the exact positions of these jobs. The remaining positions in F_2 are filled with jobs from parent P_1 that are not included in the job set of P_2, preserving their original order from P_1. This method ensures that the offspring inherit a mix of job assignments from both parents while maintaining the structure and integrity of job positions from each parent's respective job set. By doing so, the JBX method enhances the diversity and quality of the solutions generated in the search process, potentially leading to better optimization outcomes.

Furthermore, a multi-configuration crossover (MCX) and a multi-worker crossover (MWX) are developed for the CS and WS layers as shown in Figs. 2(b) and 2(c), respectively. This process begins with the generation of a random chromosome composed of binary values (i.e., 0 and 1). For offspring F_1, genes located at positions marked by a value of 0 are directly inherited from parent P_1. Conversely, genes at positions marked by a value of 1 are swapped with the corresponding genes in parent P_2. In parallel, for Offspring F_2, positions marked by a value of 0 inherit genes from parent P_2, while positions marked by a value of 1 are exchanged with the corresponding genes in parent P_1.

To perform mutations in our algorithm, we employ different strategies for each layer. In the OS layer, we utilize the swapping method, which involves selecting genes at random and exchanging their positions. For instance, in Fig. 3(a), genes at positions [2, 5] are swapped with each other. In the CS layer, we randomly select several genes and replaced them with eligible machines and configurations. As depicted in Fig. 3(b), genes at positions [2, 5, 7] are substituted with new, eligible machine and configuration pairs. Similarly, in the WS layer, multiple genes are randomly selected and replaced with eligible workers. As demonstrated in Fig. 3(c), genes at positions [2, 4, 5] are replaced with eligible workers.

To improve our algorithm with neighborhood search, a method is integrated that systematically explores the neighborhood of a solution to uncover enhancements. Three strategies for neighborhood search are devised as follows:

N_1: In this strategy, the operations of two different jobs are randomly exchanged. Then, the corresponding genes in the CS layer at the same positions are selected. Subsequently, the machines and configurations for these genes are altered accordingly, and an eligible worker is selected.

N_2: This strategy entails two genes being selected in the OS layer, and the order of the genes between them being reversed. Next, the corresponding genes in the CS layer at the same positions are selected, their machines and configurations are adjusted, and then an eligible worker is selected.

N_3: In this approach, two genes are randomly selected in the OS layer. Then, the position of the second gene is moved to immediately follow the position of the first gene, shifting all intervening genes to the right. Following this adjustment, the corresponding genes in the CS layer at the same positions are selected, their machines and configurations are modified, and then an eligible worker is selected.

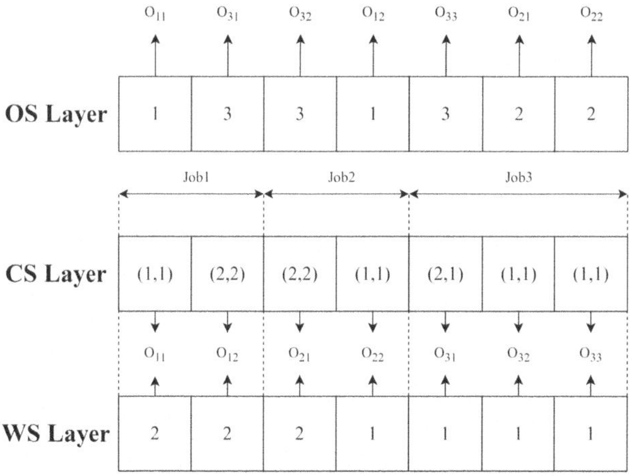

Fig. 1. Encoding instance.

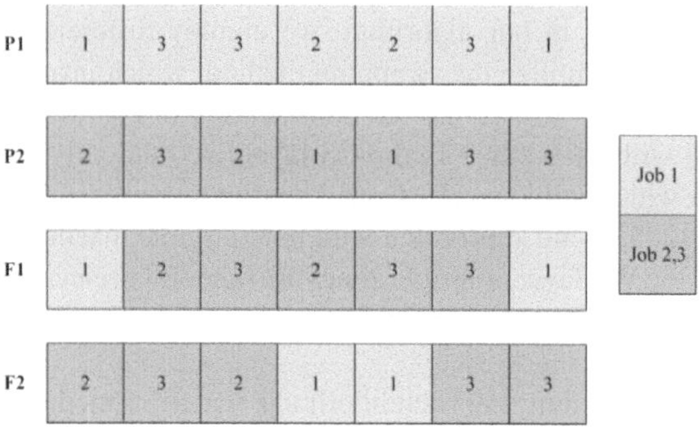

(a) JBX crossover for OS layer.

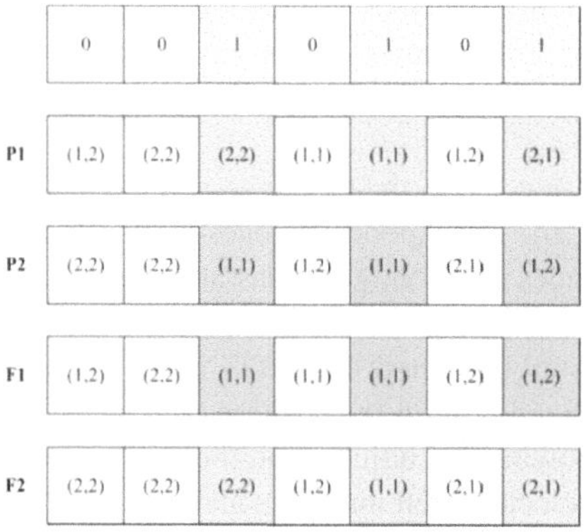

(b) MCX crossover for CS layer.

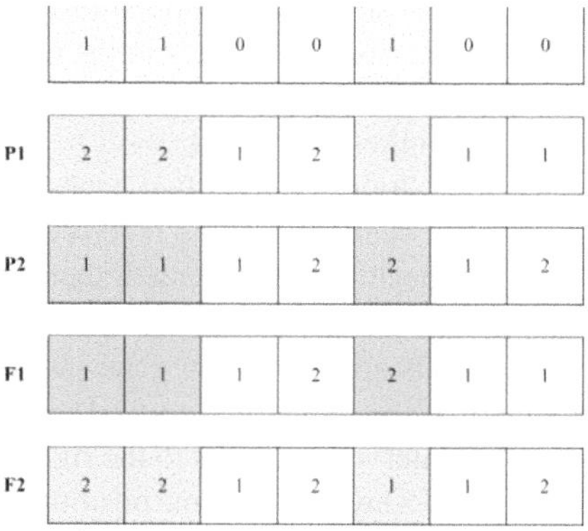

(c) MWX crossover for WS layer.

Fig. 2. Crossover generator.

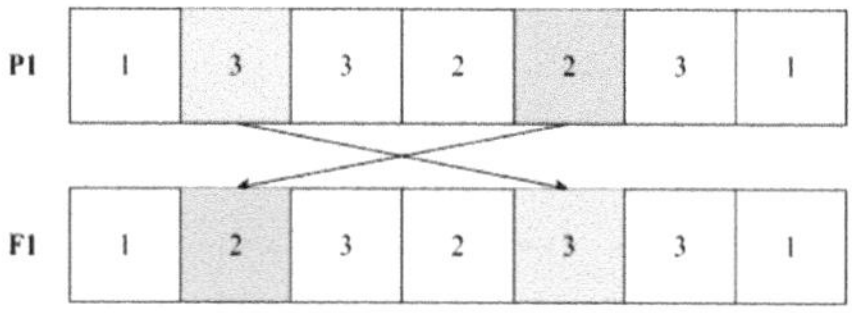

(a) Mutation for OS layer.

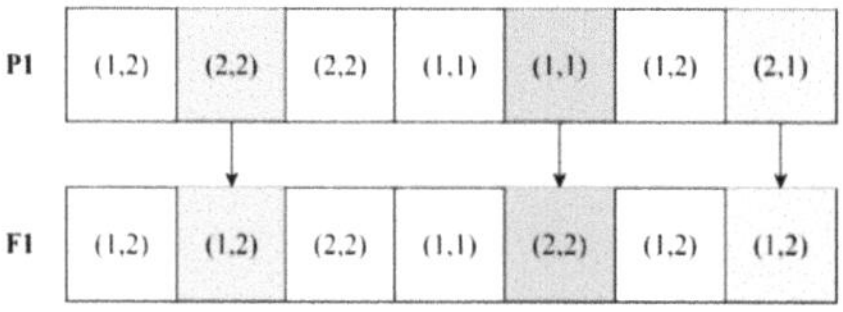

(b) Mutation for CS layer.

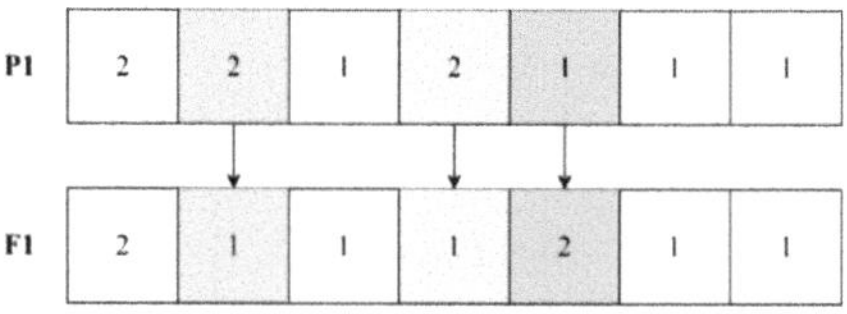

(c) Mutation for WS layer.

Fig. 3. Mutation generator.

4 Discussions

To validate the presented mathematical model, we adopt 20 instances [16–18] and extend them for our problem as shown in Table 1. Here, n denotes the number of jobs, m denotes the number of machines, l represents the number of workers, and c represents the number of configurations. The energy consumption values are randomly generated within the range of 3 to 30 units. The number of configurations, c, is randomly generated as an integer within the range [m/4, m/2].

Table 1. Computational results.

Instance	Size $n \times m \times l \times c$	RPD			CPU time (s)		
		MA	GA	MIP	MA	GA	MIP
E01	$5 \times 6 \times 3 \times 2$	4.77	10.34	**0.00**	**63.12**	**63.12**	91.35
E02	$5 \times 7 \times 3 \times 2$	4.94	12.46	**0.00**	**76.55**	79.12	113.97
E03	$6 \times 7 \times 3 \times 2$	5.56	14.06	**0.00**	**93.01**	103.16	204.36
E04	$7 \times 7 \times 3 \times 2$	7.95	9.77	**0.00**	**121.06**	154.33	801.03
E05	$7 \times 7 \times 3 \times 3$	8.32	10.31	**0.00**	**145.87**	177.89	975.74
E06	$8 \times 7 \times 3 \times 2$	8.64	11.03	**0.00**	**204.38**	241.55	1964.66
E07	$8 \times 7 \times 3 \times 3$	9.55	11.79	**0.00**	**243.47**	278.00	2133.73
E08	$9 \times 8 \times 4 \times 3$	**0.00**	20.44	-	**289.65**	300.54	-
E09	$10 \times 6 \times 4 \times 2$	**0.00**	17.45	-	**322.02**	394.04	-
E10	$10 \times 6 \times 4 \times 3$	**0.00**	16.30	-	**360.04**	421.23	-
E11	$11 \times 8 \times 4 \times 4$	**0.00**	17.66	-	**408.58**	488.02	-
E12	$12 \times 8 \times 4 \times 4$	**0.00**	25.01	-	**453.97**	533.33	-
E13	$15 \times 4 \times 5 \times 2$	**0.00**	25.48	-	**675.63**	702.22	-

(continued)

Table 1. *(continued)*

Instance	Size $n \times m \times l \times c$	RPD			CPU time (s)		
		MA	GA	MIP	MA	GA	MIP
E14	$15 \times 8 \times 6 \times 2$	**0.00**	25.37	-	**689.33**	797.50	-
E15	$15 \times 8 \times 6 \times 3$	**0.00**	14.41	-	**766.09**	866.69	-
E16	$10 \times 15 \times 8 \times 4$	**0.00**	29.39	-	**705.98**	846.07	-
E17	$20 \times 5 \times 4 \times 2$	**0.00**	19.23	-	**1358.75**	1541.55	-
E18	$20 \times 10 \times 6 \times 3$	**0.00**	20.07	-	**1397.43**	1596.90	-
E19	$20 \times 10 \times 6 \times 4$	**0.00**	21.31	-	**1495.08**	1788.02	-
E20	$20 \times 15 \times 8 \times 5$	**0.00**	20.54	-	**1683.33**	1901.25	-

The MIP model is implemented using CPLEX 12.6.3 solver, while the MA and GA are executed in Python 3.12.3 and run 20 times. All computations are performed on a personal computer equipped with 12 GB RAM and a Core-i9 CPU.

To compare MA with MIP and GA, the Relative Percentage Difference (RPD) measure is utilized [19]. The RPD is calculated as follows:

$$RPD = \frac{F_{2-}F_1}{F_1} \times 100\% \tag{12}$$

where F_1 represents the best-known solution obtained by the MA and F_2 denotes the best-known solution obtained by other algorithms, such as MIP and GA. The parameters are as follows: population size of 100, crossover rate of 0.8, mutation rate of 0.3, and neighborhood search rate of 0.1.

Figure 4 illustrates a comparative box plot for the MA and GA over twenty instances. The results presented in Table 1 indicate that the CPLEX yields the best solutions for the first seven instances. However, it becomes impractical for larger instances due to excessive computation times. Although the MIP approach outperforms the MA in terms of solution quality, the solutions provided by the MA are still quite close to those obtained by the MIP. Moreover, the MA has the advantage of significantly shorter running times compared to the MIP. Therefore, the MA proves to be an efficient method for solving smaller instances of the problem.

For the remaining instances (E08 – E20), the MA outperforms other methods in achieving the best solutions. Furthermore, its CPU times are shorter than those of the GA. While the GA is effective in solving the FJSSP, our problem's inclusion of worker and machine configurations expands the solution space, making it challenging to identify the best solution within a limited runtime. The proposed MA is particularly effective due to its three neighborhood search mechanisms, which enhance its capability to escape local optima and find superior solutions.

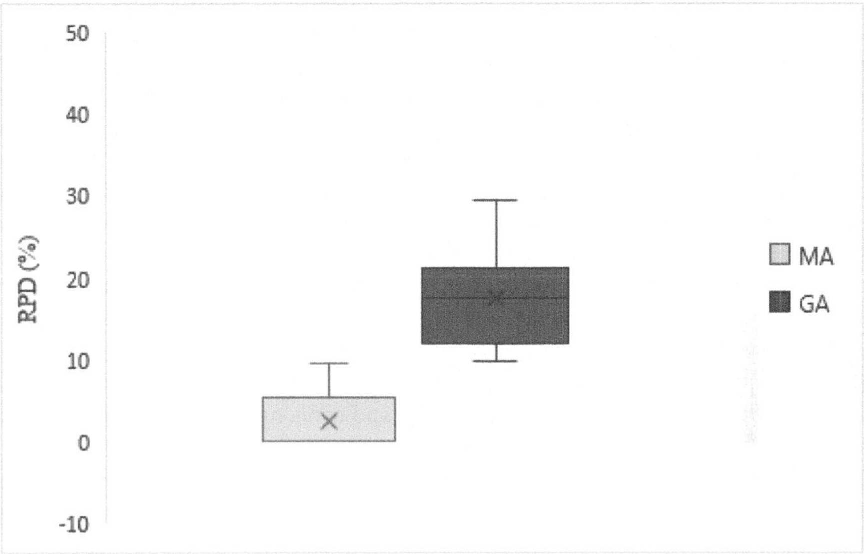

Fig. 4. Box plot for the RPD.

5 Conclusion

The transition towards Industry 5.0 has necessitated a paradigm shift in manufacturing scheduling approaches. Traditional methods that solely focus on economic objectives are no longer sufficient, as human-centric and environmental considerations have become paramount. The flexible job shop scheduling problem (FJSSP) with reconfigurable machine tools (RMTs) offers a compelling framework to address these multi-faceted priorities. This paper has reviewed integrated FJSSP models that account for human factors like worker skills, ergonomics, and fatigue management, as well as environmental criteria such as energy consumption and emissions. The review has highlighted the scarcity of research efforts that simultaneously optimize these economic, human, and environmental dimensions, especially in the context of reconfigurable machine tools.

The proposed mathematical model contributes by integrating worker assignment constraints, rest periods, and energy consumption objectives into the FJSSP with reconfigurable machines and sequence-dependent setup times. The primary objective of the model is to minimize total energy consumption. Solving large-sized, real-world instances will require the development of efficient meta-heuristic algorithms.

To address this challenge, we developed a memetic algorithm (MA) tailored for this NP-hard combinatorial optimization problem. The key steps of the developed MA include initializing a population, generating offspring using crossover and mutation methods, and applying neighborhood strategies to improve the quality of individuals. This algorithm employs a three-layer chromosome encoding method, including operation sequencing (OS), configuration selection (CS), and worker selection (WS) layers.

The methodology and results of our study demonstrate that the MA provides a competitive alternative to the MIP model, particularly for larger instances, in which exact methods become impractical due to excessive computation times. Although the MIP approach outperforms the MA in terms of solution quality for smaller instances, the MA achieves solutions that are close in quality while significantly reducing computational times. For larger instances, the MA outperforms both the GA and MIP, achieving the best solutions with shorter CPU times. This efficiency is attributed to the three neighborhood search mechanisms integrated into the MA, which enhance its capability to escape local optima and find superior solutions.

Future research can explore multi-objective formulations that capture the trade-offs between economic, human, and environmental criteria, enabling the decision-makers to derive schedules that harmonize these competing objectives. As manufacturing firms navigate the transition to Industry 5.0, holistic scheduling approaches that prioritize human well-being, environmental stewardship, and economic viability will be crucial drivers of operational resilience and sustainable competitiveness. The research directions outlined in this paper pave the way for developing robust scheduling techniques that can unlock the full potential of human-centric, eco-efficient manufacturing in the factory of the future.

Acknowledgment. The research has been supported by the European Union under the project ROBOPROX (reg. no. CZ.02.01.01/00/22 008/0004590).

Disclosure of Interests. The authors have no competing interests to declare that are relevant to the content of this article.

References

1. Nessari, S., Tavakkoli-Moghaddam, R., Bakhshi-Khaniki, H., Bozorgi-Amiri, A.: A hybrid simheuristic algorithm for solving bi-objective stochastic flexible job shop scheduling problems. Decis. Anal. J. **11**, 100485 (2024)
2. Destouet, C., Tlahig, H., Bettayeb, B., Mazari, B.: Flexible job shop scheduling problem under Industry 5.0: a survey on human reintegration, environmental consideration and resilience improvement. J. Manuf. Syst. **67**, 155–173 (2023)
3. Aheleroff, S., Huang, H., Xu, X., Zhong, R.Y.: Toward sustainability and resilience with Industry 4.0 and Industry 5.0. Front. Manuf. Technol. **2**, 951643 (2022)
4. Müller, D., Kress, D.: Filter-and-fan approaches for scheduling flexible job shops under workforce constraints. Int. J. Prod. Res. **60**, 4743–4765 (2022)
5. Hongyu, L., Xiuli, W.: A survival duration-guided NSGA-III for sustainable flexible job shop scheduling problem considering dual resources. IET Collab. Intell. Manuf. **3**, 119–130 (2021)
6. Tan, W., Yuan, X., Wang, J., Zhang, X.: A fatigue-conscious dual resource constrained flexible job shop scheduling problem by enhanced NSGA-II: An application from casting workshop. Comput. Ind. Eng. **160**, 107557 (2021)

7. Liu, Y., Zhang, L., Sun, T.: An improved nondominated sorting genetic algorithm-II for multi-objective flexible job-shop scheduling problem considering worker assignments. In: IEEE 3rd International Conference on Communications, Information System and Computer Engineering (CISCE), Beijing, China, 14–16 May 2021, pp. 837–843. IEEE (2021)
8. He, Z., Tang, B., Luan, F.: An improved African vulture optimization algorithm for dual-resource constrained multi-objective flexible job shop scheduling problems. Sensors **2023**(23), 90 (2023)
9. Jia, S., et al.: The green flexible job-shop scheduling problem considering cost, carbon emissions, and customer satisfaction under time-of-use electricity pricing. Sustainability **2024**(16), 2443 (2024)
10. Wang, Z., He, M., Wu, J., Chen, H., Cao, Y.: An improved MOEA/D for low-carbon many-objective flexible job shop scheduling problem. Comput. Ind. Eng. **188**, 109926 (2024)
11. Hajibabaei, M., Behnamian, J.: Fuzzy cleaner production in assembly flexible job-shop scheduling with machine breakdown and batch transportation: Lagrangian relaxation. J. Comb. Optim. **45**, 1–26 (2023)
12. Mahmoodjanloo, M., Tavakkoli-Moghaddam, R., Baboli, A., Bozorgi-Amiri, A.: Flexible job shop scheduling problem with reconfigurable machine tools: An improved differential evolution algorithm. Appl. Soft Comput. **94**, 106416 (2020)
13. Fan, J., Zhang, C., Liu, Q., Shen, W., Gao, L.: An improved genetic algorithm for flexible job shop scheduling problem considering reconfigurable machine tools with limited auxiliary modules. J. Manuf. Syst. **62**, 650–667 (2022)
14. Fan, J., Zhang, C., Shen, W., Gao, L.: A matheuristic for flexible job shop scheduling problem with lot-streaming and machine reconfigurations. Int. J. Prod. Res. **61**, 6565–6588 (2023)
15. Guo, H., Liu, J., Wang, Y., Zhuang, C.: An improved genetic programming hyper-heuristic for the dynamic flexible job shop scheduling problem with reconfigurable manufacturing cells. J. Manuf. Syst. **74**, 252–263 (2024)
16. Fattahi, P., Bagheri Rad, N., Daneshamooz, F., Ahmadi, S.: A new hybrid particle swarm optimization and parallel variable neighborhood search algorithm for flexible job shop scheduling with assembly process. Assem. Autom. **40**, 419–432 (2020)
17. Brandimarte, P.: Routing and scheduling in a flexible job shop by Tabu search. Ann. Oper. Res. **41**, 157–183 (1993)
18. Lei, D., Guo, X.: Variable neighbourhood search for dual-resource constrained flexible job shop scheduling. Int. J. Prod. Res. **52**, 2519–2529 (2014)
19. Hatami, S., Ruiz, R., Andrés-Romano, C.: Heuristics and metaheuristics for the distributed assembly permutation flowshop scheduling problem with sequence dependent setup times. Int. J. Prod. Econ. **169**, 76–88 (2015)

Makespan Minimisation in Hybrid Flexible Flowshops with Buffers and Machine-Dependent Transportation Times

Stavros Vatikiotis(✉), Ilias Mpourdakos, Dimitrios Papathanasiou, and Ioannis Mourtos

ELTRUN Research Lab, Department of Management Science and Technology, Athens University of Economics and Business, Athens, Greece
stvatikiotis@aueb.gr
https://eltrun.org/

Abstract. Hybrid Flexible Flowshop Scheduling (HFFS) is the problem where a set of jobs must be processed in a given sequence of stages and each stage has a set of (typically identical) parallel machines. The flexibility of HFFS allows a job to skip some stages. Modern production environments, e.g., assembly lines, exhibit additional structure, namely limited-capacity buffers and transportation times between subsequent stages, while the layout also imposes that such times are machine-to-machine dependent. We propose two formal models, namely a Mixed-Integer Linear Program (MILP) that incorporates transportation times but not buffers and a Constraint Program (CP) that handles both, given a sequence of all jobs per machine. This sequence is provided by the MILP or constructive heuristics or a Genetic Algorithm (GA). The scalability and performance of all methods is evaluated computationally on large-scale real-life instances of about 500 jobs on 15 stages with up to 5 machines per stage and 30 machines in total.

Keywords: Hybrid Flexible Flowshop · Transportation Times · Buffers · Mixed Integer Linear Programming · Constraint Programming

1 Introduction and Background

Within the blooming domain of manufacturing operations, plant managers continuously seek novel tools and methods to achieve greater degrees of productivity and, thus, maximise the efficiency of their shop floor. Regarding these goals, scheduling algorithms have long been a broad research area [1], ranging from the simple case of a single machine [2] to parallel machine environments [3]. However, several products require multiple operations to be performed in subsequent stages. To investigate such realistic conditions, research has focused on shop scheduling [4] and, within them, Flow Shop scheduling (FS), which originally assumes a single

Published by Springer Nature Switzerland AG 2024
M. Threr et al. (Eds.): APMS 2024, IFIP AICT 733, pp. 258–273, 2024.
https://doi.org/10.1007/978-3-031-71645-4_18

machine per stage and the same stage sequence for all jobs. Additional features include the time needed to transfer jobs between stages (transportation times), resources typically renewable (e.g., human workers) and limited capacity buffers, where jobs are temporarily stored between subsequent stages.

A prevailing variation of the problem is the Hybrid Flow Shop (HFS), where each stage has multiple parallel (identical) machines. HFS exhibits a greater proximity to practical cases, thus adapting to various environments and objectives [5]. Notable industries where the aforementioned variant is employed, include steel-making manufacturing [6], transformer industry [7] and chemical production [8]. Research-wise, [9] propose a Simulated Annealing (SA) algorithm to deal with transportation time dependencies, while a similar approach is examined in [10]. Sequence-dependent setup times are addressed in [11], where metaheuristic algorithms i.e., Neighbourhood Search (NS) and Genetic Algorithm (GA) are employed. [12] focus on tardiness related objectives and propose an MILP formulation and two constructive heuristics, while a similar approach is followed in [13] aiming at makespan minimisation. The inclusion of transportation times is studied in [14], while similar frameworks of metaheuristic algorithms (PSO and SA) are examined in [15] under renewable resource constraints. Finally, [16] tackle the HFS with zero and infinite capacity buffers, transportation times and restricted resources, using constructive heuristics and GA.

A job may not be obliged to undergo the operation provided by a stage, i.e., it can skip it, in which case we have the Hybrid Flexible Flow Shop problem (HFFS). A variety of MILP models are compared in [17] under makespan minimisation, to conclude on their limited scalability and, thus, propose a Particle Swarm Optimisation (PSO) metaheuristic, while (meta-)heuristics (GA, PSO, hybrid PSO-GA) are, also, employed in [18]. [19] examine the problem under a robust probabilistic programming approach with limited resources. Finally, [20] propose a CP formulation, an MILP, and local search to deal with the transportation times property. The extension of this work includes a two-phased heuristic algorithm, consisting of a metaheuristic based on a greedy method and a CP model [21].

Motivated by this interplay among MILP, CP and (meta-)heuristics, we examine an HFFS problem applicable on a motor manufacturing plant. In our context, transportation times are machine-to-machine dependent and not stage dependent, as opposed to the current literature (to our knowledge). In addition, on the entry and on the exit of each machine, there are limited capacity buffers, where jobs are obliged to wait, if the next machine to process each job is not available. The final property considered in our problem is the obligation of scheduling jobs of the same order subsequently (commonly referred to as batch scheduling). In our case, this constraint is derived by the necessity of instantly packaging all jobs belonging to the same order. Even though such restrictions have been studied in the literature, e.g., [22,23], an approach tackling all the properties presented in our problem has not yet appeared, again to our knowledge.

To sum up, Table 1 presents related literature. The standard notation by [24] is utilised to denote the machine environment (α) and problem objectives (γ). The table also compares the inclusion of batches (*Batch*), Transportation Times

(*Tr. Times*) and Limited Capacity Buffers or Work-In-Progress Inventories (*WIP*), while also mentioning the method employed.

Table 1. Literature Review (2014 -)

References	α	γ	Batch	Tr. Times	WIP	Method
[10]	HFS	C_{max}	-	✓	–	MILP, SA
[17]	HFFS	C_{max}	–	–	–	MILP, PSO
[19]	HFFS	C_{max}	–	–	–	MILP
[13]	HFS	C_{max}, Other	–	–	–	MILP, Heur.
[16]	HFS	C_{max}	-	✓	✓	MILP, Heur., GA
[15]	HFS	C_{max}	–	–	–	MILP, PSO, SA
[20]	HFFS	C_{max}	-	✓	–	CP, MILP
[14]	HFS	C_{max}, Other	–	✓	–	MILP, Heur.
[18]	HFFS	C_{max}	–	✓	–	MILP,PSO,GA
[21]	HFFS	C_{max}	–	✓	-	CP, Heur.
[12]	HFS	Other	–	–	–	MILP, Heur.
[22]	HFS	C_{max}, Other	✓	✓	–	MILP, Heur
Current	HFFS	C_{max}	✓	✓	✓	Heur., GA

HFS: Hybrid Flow Shop, HFFS: Hybrid Flexible Flow Shop, PSO: Particle Swarm Optimisation, GA: Genetic Algorithm, CP: Constraint Programming, MILP: Mixed-Integer Linear Programming, SA: Simulated Annealing, Heur.: Heuristics.

Our Contribution. Our contribution is multi-faceted. We present a novel variant of HFFS that originates from a real case. In technical terms, we introduce: i) an MILP formulation that modifies one among the formulations in [17] to incorporate the machine-to-machine dependent transportation times, ii) a CP model taking as input a sequence of jobs per machine to then deal with both transportation times and limited capacity buffers, iii) two index-based heuristics that, coupled with the CP model, provide feasible solutions in negligible time, and iv) a GA coupled again with the CP model to trade off time with improved solution quality. We provide results from our extensive experimentation process on synthetic datasets of large scale, as provided by a real case.

Outline. In Sect. 2, we formally describe our HFFS variant and present the MILP and CP formulations. Section 3 presents our (meta-)heuristic algorithms. Section 4 describes our case study and demonstrates the results derived from our experiments. Section 5 concludes our work and lists ideas for further research.

2 Problem Definition and Mathematical Modeling

A set of orders O must be processed on a set of stages I. Each order $o \in O$ consists of a set of jobs $j \in J_o$, while each stage i consists of a set of identical machines M_i. It is possible for jobs to skip some stages depending on their necessary operations. The jobs of each order need to be scheduled consecutively, i.e., without any other job in between. Transportation times to transfer jobs between stages are determined by the floor layout hence they depend on the machine of stage i from which the job is taken from and the machine of stage $i+1$ the job is taken to; i.e., such times are machine-to-machine dependent. Finally, each machine is associated with a set of limited capacity buffers both at its entry and at its exit. Table 2 summarises the above.

Table 2. Model Parameters

Parameters	
O	The set of orders
o	Index for orders where $\{1, 2, ..., \|O\|\}$
J_o	The set of jobs connected to order $o \in O$
j, k	Index for jobs of order o, i.e., $j, k \in \{1, 2, ..., \|J_o\|\}$
I	The set of stages
i	Index for stages, i.e., $i \in \{1, 2, ..., \|I\|\}$
I_j	The set of required stages for job j
M_i	The set of machines of stage i
m, l	Index for machines on stage i, i.e., $m, l \in \{1, 2, ..., \|M_i\|\}$
W_m^-	The buffer capacity on the entry of machine m
W_m^+	The buffer capacity on the exit of machine m
p_{ji}	The processing time of job j on stage i
$t_{mm'}$	The transportation time from machine m of stage i to m' of stage $i+1$
V	A large positive number
C_{max}	The makespan of the schedule

Let us illustrate our notation through a small example. The instance depicted in Fig. 1 includes three stages, hence $I = \{s_1, s_2, s_3\}$, with the number of machines varying across stages. The set of orders $O = \{o_1, o_2\}$ must be scheduled, where $J_{o_1} = \{j_1, j_2\}$ and $J_{o_2} = \{j_3\}$. Each machine has an input and an output buffer of varying capacities. Each edge between a pair of machines denotes the transportation time needed to transfer a job between these two machines. Note that it is possible to transfer a job from e.g. m_1 to m_4, but such connections have been omitted in the diagram for visualisation purposes (Fig. 1).

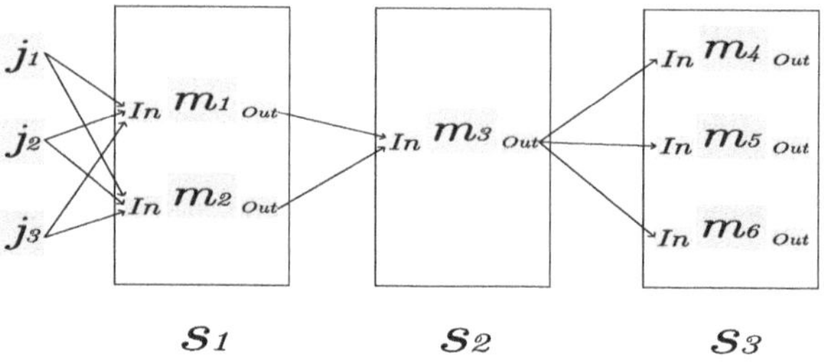

Fig. 1. Problem Representation

As mentioned in Sect. 1, we propose an MILP formulation, which modifies the fourth model proposed in [17] to deal with the machine-to-machine dependent transportation times and, also, the consecutive scheduling of each job enclosed within an order. Note, also, that in [25], it is stated that, when transportation times are considered, a sequence-based (against a slot-based) mathematical formulation is to be preferred. Table 3 presents the decision variables of the model.

Table 3. MILP Variables

Variables	
x_{jik}	Binary variable: 1 if job j is assigned after job k on stage i, 0 otherwise
y_{jil}	Binary variable: 1 if job j is assigned on machine l of stage i, 0 otherwise
C_{ji}	The completion time of job j on stage i

The MILP formulation $\mathcal{R}$ follows:

$\mathcal{R}$:

$\min C_{max}$

$$\sum_{l \in M_i} y_{jil} = 1 \qquad j \in J, i \in I_j \tag{1}$$

$$x_{kij} + x_{j'ik} \leq 1 \qquad o, o' \in O, j, j' \in J_o, k \in J_{o'}, i \in I_j \cap I_k \cap I_{j'} \tag{2}$$

$$C_{ji} \geq C_{ji-1} + p_{ji} + t_{mm'} - V \cdot (2 - y_{ji-1m} - y_{jim'}) \quad j \in J, i \in I_j, m \in M_{i-1}, m' \in M_i \tag{3}$$

$$C_{ji} \geq C_{ki} + p_{ji} - V \cdot (3 - x_{jik} - y_{jil} - y_{kil}) \qquad j, k \in J, i \in I_j \cap I_k, l \in M_i \tag{4}$$

$$C_{ki} \geq C_{ji} + p_{ki} - V \cdot x_{jik} - V \cdot (2 - y_{jil} - y_{kil}) \qquad j, k \in J, i \in I_j \cap I_k, l \in M_i \tag{5}$$

$$C_{ji} \geq C_{j0} \qquad j \in J, i \in I_j \tag{6}$$

$$C_{max} \geq C_{ji} \qquad j \in J, i \in I_j \tag{7}$$

$$y_{jil}, x_{jik} \in \{0, 1\} \qquad j, k \in J, i \in I, l \in M_i$$

$$C_{ji} \geq 0 \qquad i \in I, j \in J$$

Constraints (1) ensure that each job is assigned to exactly one machine per stage it should pass through. By (2), if j and j' are jobs of the same order $o \in O$, then a job k of order $o\prime$ cannot be assigned between j and $j\prime$. This ensures that jobs of the same order will be processed sequentially. By (3), the completion time C_{ji} of job j on stage i is associated with its completion time on stage $i-1$, the processing time p_{ji} and the transportation time $t_{mm\prime}$ to transfer a job from machine m to $m\prime$. Constraints (4) and (5) associate the completion time of each job to the completion time of the previous job on the same machine. Constraints (6) enforce a lower bound on the completion time of job j, equal to the completion time of j on the first stage. Constraints (7) calculate the makespan objective (C_{max}) with respect to the completion times of all jobs.

The aforementioned MILP is able to deal with all the features of our problem, except limited capacity buffers, which are essentially resources hence would require a time-indexed formulation. As a result, it is obvious to deduce that the optimal solution of the MILP is a lower bound to the problem we examine. To accommodate such constraints, we opt for a CP formulation that receives as input an already defined assignment and sequencing of jobs to machines, without the possibility of altering any of these already made decisions. Then, the CP component appropriately adjusts the start and completion times of each job on each machine in order to minimise the makespan under the actual buffer capacities. We present our CP model following the nomenclature of CP Optimizer [28].

Let $(\bar{x}, \bar{y})$ be an optimal solution to $\mathcal{R}$ that is taken as input to the CP model listed below (Table 4).

$\mathcal{CP}$:

min C_{max}

subject to:

$$\texttt{startAtEnd}(\tau_{jim}, \beta_{jim}) \qquad j \in J, i \in I_j, m \in M_i : \bar{y}_{jim} = 1 \tag{8}$$

$$\texttt{startAtEnd}(\alpha_{jim}, \tau_{jim}) \qquad j \in J, i \in I_j, m \in M_i : \bar{y}_{jim} = 1 \tag{9}$$

$$\texttt{startAtEnd}(\beta_{j(i+1)m\prime}, \alpha_{jim}, -t_{mm\prime}) \qquad j \in J, i \in I_j, m \in M_i : \bar{y}_{jim} = 1, m\prime \in M_{(i+1)} : \bar{y}_{j(i+1)m\prime} = 1 \tag{10}$$

$$\sum_{j \in J : \bar{y}_{jim}=1} \texttt{Pulse}(\beta_{jim}) \leq W_m^- \qquad i \in I, m \in M_i \tag{11}$$

$$\sum_{j \in J : \bar{y}_{jim}=1} \texttt{Pulse}(\alpha_{jim}) \leq W_m^+ \qquad i \in I, m \in M_i \tag{12}$$

$$\sum_{j \in J : \bar{y}_{jim}=1} \texttt{Pulse}(\tau_{jim}) \leq 1 \qquad i \in I, m \in M_i \tag{13}$$

$$\texttt{endBeforeStart}(\tau_{jim}, \tau_{j\prime im}) \qquad i \in I, m \in M_i, j, j\prime \in J : \bar{x}_{j\prime ij} = \bar{y}_{jim} = \bar{y}_{j\prime im} = 1 \tag{14}$$

$$C_{max} \geq \texttt{endOf}(\alpha_{jim}) \qquad j \in J, i \in I_j, m \in M_i : \bar{y}_{jim} = 1 \tag{15}$$

τ_{jim} : interval at least p_{ji} $\qquad j \in J, i \in I_j, m \in M_i : \bar{y}_{jim} = 1$

α_{jim} : interval $\qquad j \in J, i \in I_j, m \in M_i : \bar{y}_{jim} = 1$

β_{jim} : interval $\qquad j \in J, i \in I_j, m \in M_i : \bar{y}_{jim} = 1$

C_{max} : integer

Table 4. CP Variables

Variables	
τ_{jim}	Interval variable, indicating the start and end time of the process of job j assigned to machine $m \in M_i$ of stage $i \in I_j$
α_{jim}	Interval variable, indicating the start and end time of the wait of job j on the exit buffer of machine $m \in M_i$ of stage $i \in I_j$
β_{jim}	Interval variable, indicating the start and end time of the wait of job j on the entry buffer of machine $m \in M_i$ of stage $i \in I_j$

In the above formulation, expressions like '$j \in J, i \in I_j, m \in M_i : \bar{y}_{jim} = 1$' mean that we select the unique element in a subset indicated by the relevant variable being 1 in the optimal solution of $\mathcal{R}$, e.g., given a job j and a stage i among the stages that job passes through, m is the single machine of stage i for which $\bar{y}_{jim} = 1$ (i.e., the single machine in which j is processed in stage i). Also, as we assume flexibility in terms of a job possibly skipping some stage, the above formulation slightly abuses notation by denoting as $i+1$ the stage following stage i for any job. We opt for simplicity and avoid using notation like $next_j(i)$ for the stage in which job j goes after i or $last_j$ for the last stage encountered by job j. Constraints (8) ensure that when a job stops waiting at the entry buffer of a stage, then it has to be processed on the corresponding machine. Similarly, Constraints (9) ensure that when the processing of a job has finished, then this job waits at the exit buffer of the machine. Constraints (10) imply that a job cannot be transferred to its next scheduled stage, without imposing the appropriate transportation time. By (11) and (12) the capacities on the entry and exit buffers are respected, while by (13), each machine is restricted to process only a single job at each time. Constraints (14) enforce that the provided sequence of jobs on each machine are respected, while the makespan is calculated by (15).

3 (Meta-)heuristics

A common restriction on the employment of exact methods on instances of realistic scale is their limited scalability. This is why we propose three different (meta-)heuristic algorithms aiming at a successful trade-off between solution quality and time. Our approach includes two index-based constructive heuristics and an evolutionary algorithm. Since Genetic Algorithms (GA) are commonly used in related literature, we opt for employing this among other evolutionary (meta-)heuristics, e.g. PSO or Ant Colony Optimisation (ACO). That is, the proposed GA could be regarded as a sufficient representative of the aforementioned evolutionary algorithms.

3.1 Assignment Procedure

Before moving on with the detailed explanation of the designed algorithms, it is required to describe an *Assignment Procedure.* This procedure receives as input a sequence of orders and aims at assigning and sequencing all corresponding jobs $j \in J_o$ on a machine $m \in M_i$ of every eligible stage $i \in I_j$.

Contrary to the literature, the indices in our work are calculated per order due to the need of consecutive sequencing the underlying jobs of each order. As a result, when order o is selected to be scheduled, all jobs $j \in J_o$ composing order o are scheduled. Then, job j is selected and the stages $i \in I_j$ are identified. The machine $m' \in M_i$ of stage i with the minimum value of: $l_{m'} + p_{ji} + t_{mm'}$ is chosen, where $l_{m'}$ is the current load of machine m' and m is the machine where job j was assigned on stage $i - 1$. This concludes the assignment of a job. Machine loads are appropriately updated to efficiently track current assignments. This procedure stops when all jobs are scheduled. For clarity, this 'Assignment Procedure' is presented as Algorithm 1.

Algorithm 1. ASSIGNMENT PROCEDURE

```
Let S̄ be a sequence of orders;
for o ∈ S̄ do
    for j ∈ J_o do
        for i ∈ I_j do
            Calculate C_ji = l_m' + p_ji + t_mm' ∀m' ∈ M_i;
            Find m* with min C_ji;
            Assign job j to m*, where C_ji is minimum;
            Update l_m*;
            if l_m* > C_max then
                Update C_max;
return C_max;
```

3.2 Index-Based Constructive Heuristics

The execution of Algorithm 1 requires an initial sequence of orders. To obtain such a sequence, we calculate two different indices. The first is the *Average Job Eligibility Index - AJEI*, which is associated with the average number of stages, where each job of an order is eligible. By prioritising the orders with a greater value of $AJEI$, we aim at initially scheduling orders that require processing in more stages. As a result, the most 'demanding' orders are scheduled on an earlier part of the schedule, where a greater number of machine choices are available. The second index, namely, *Average Job Processing Time Index - AJPTI* is associated to the average time needed for every job on each order. Similarly, by scheduling jobs of orders with a greater value of $AJPTI$, we prioritise jobs of orders which, on average, require a greater portion of machine availability (due to their corresponding processing times). As a result, these jobs are scheduled first, hoping that machine loads remain balanced. Formally, we define $AJEI_o = \frac{\sum |I_j|}{|J_o|}$ and $AJPTI_o = \frac{\sum p_{ji}}{|J_o|}$. Hence, using these two indices and the *Assignment Procedure* of Algorithm 1, we obtain the sequencing and assignment of jobs on machines on each stage in two possible manners. However, to provide complete solutions, it is required to couple the above method with the CP component to include limited-capacity buffers. Algorithm 2 formalises this process.

Algorithm 2. CONSTRUCTIVE HEURISTIC

Let order sequence $\bar{S}$ derived by *AJEI* or *AJPTI*;
Employ Algorithm 1 of subsection 3.1;
Run $\mathcal{CP}$ of Section 2 and get C_{max};
return C_{max};

3.3 Genetic Algorithm (GA)

The GA originally showed up in [26], and since then, has been widely implemented on several applications, including scheduling [27]. Solutions are represented in the form of chromosomes, which are decoded into solutions by a so-called 'Decoding Algorithm' - this being Algorithm 1 in our approach - while each gene corresponds to an order. After the creation of an initial population P of chromosomes of size P_s, and their solution quality calculation, chromosomes are mixed (crossover procedure) and mutated (mutation procedure), based on the parameter values. To ensure chromosomes consist of unique orders, we employ the 'repairing' operation of [27]. Then, the algorithm merges the new set of solutions with the population and passes to the next generation the best P_s solutions. The detailed procedure is presented as Algorithm 3.

Algorithm 3. GENETIC ALGORITHM (GA)

Initialise $G_c = 0$, where G_c is the generation counter;
Generate initial population P of size P_s;
Calculate C_{max} for each individual in P using Algorithm 2;
while $G_c \leq G_{max}$ **do**
  Create population $P' = P$ and an empty set of new offsprings, O;
  while $P' \neq \emptyset$ **do**
    Select uniformly at random two individuals x and y from P';
    Select uniformly at random r_c from $[0, 1]$;
    if $r_c < P_c$ **then**:
      Use the crossover operation on x and y and generate offsprings a and b;
      Select uniformly at random r_m from $[0, 1]$;
      if $r_m < P_m$ **then**:
        Use the mutation operation on a and b;
      Calculate C_{max} for a and b using Algorithm 1 and $\mathcal{CP}$ of Section 2;
      Insert a and b to O;
    Remove x and y from P';
  Merge P and O and recreate P with the smallest P_s chromosomes in terms of C_{max} values;
  Increase G_c by 1;
return the minimum C_{max} from population P;

4 Computational Evidence

4.1 An Actual Flowshop

The problem at hand has been inspired by a case study, which operates on the motor industry. This pilot aims at minimising the makespan of the schedules through the minimisation of machine idle times, with the goal of maximising their efficiency. The production environment includes 16 discrete stages of diverse operations with limited capacity buffers on the entry and exit. Each stage includes a different number of machines, while customer orders are provided as sets of different jobs. The latter batch creation requirement originates from the need of packaging and delivery to the customers. Indicative groups of processing stages consist of: i) Assembly operations, ii) Testing Operations iii) Special customer specification, all being operated by either manual (human resources) or automated (robotic resources) manner. The transfer of jobs between the machines of different stages is performed by a set of AGVs - which may be considered infinite due to their great number. Finally, note that due to the different scheduling needs required on the plant, we are obliged to provide a set of diverse algorithms in terms of the computational time needed. This difference originates from the urgency of implementation of the solutions provided for each application.

4.2 Experimental Setting

Our experiments are performed on 5 daily instances containing orders and the corresponding jobs, as provided by the use case. The average number of orders $|O|$ over all instances is 329, with 283 being the minimum and 350 the maximum. The average number of jobs per order is 1.91, with 1 being the minimum and 10 the maximum. For each job, we have obtained the processing times p_{ji} and we assume that, if for a job j, p_{ji} is equal to 0, then j will skip stage i. If a job j needs to be processed on a stage i, then we assume that it is eligible for all machines $m \in M_i$. Furthermore, we have been provided with the machine-to-machine transportation times and the limited capacity buffers employed on the plant. Note, however, that by pilot's request, data have been slightly differentiated.

As we aim at dealing with as many daily orders as possible, our experimentation is based on gradually incrementing the number of orders $|O|$ - instead of the more common, number of jobs $|J|$. Thus, for each different $|O|$, we randomly select such a number of orders from the provided daily instance. Note, however, that since this is a randomised procedure, we execute 3 different experiments on each daily instance, aiming at obtaining different samples of orders each time. After preliminary experimentation, the GA crossover (P_c) and mutation (P_m) parameters have been set to 0.4 and 0.1 respectively, while the Population Size (P_s) is set to 30 and the number of generations G_c is equal to 100. Finally, in order to remain within the bounds of reasonable computational times, we have imposed a timelimit of 3600 s. both for the $MILP$ and the GA - if exceeded, the

GA stops even without having completed the total number of generations. The computational time allocated to the *CP* model has been restricted to 30 s.

All experiments have been performed on a server with 4 Intel(R) Xeon(R) E-2126G @ 3.30GHz processors and 11 GB RAM, running CentOS/Linux 7.0. We used Python 3.8.5 for scripting and Pyomo 5.3. The solver of the MILP relaxation $\mathcal{R}$ was Gurobi Optimizer 9.1.5 [30], via the Pyomo library [29], compatible for Python 3.8.5. The $\mathcal{CP}$ model was solved by the CP Optimizer [28] of the DOCplex module, integrated into the provided Python API.

4.3 Findings

Table 5 summarises the results of our experimentation per number of orders $|O|$, while detailed results are provided in the Appendix. Column $|J|$ presents the number of jobs per instance with the corresponding minimum, average and maximum values. Columns *AJEI* and *AJPTI* correspond to the results obtained by the employment of each heuristic, while similar results are shown for the *GA*. Note that, since the *MILP* does not consider WIPs, we feed the solution obtained to the *CP* component to deal with the limited capacity buffers. Furthermore, to speed up the *MILP* solution procedure, we provide the best solution found by any of the (meta-)heuristic algorithms as its 'warm up'. *Time* is measured in seconds and, since it is not possible to calculate strong lower bounds for larger instances, we compare the efficiency of the proposed methods by calculating *Err* as $\frac{\text{ALG\_SOL - ALG\_BEST\_SOL}}{\text{ALG\_SOL}}$, where *Alg_sol* is the solution of an algorithm and *Alg_best_sol* is the best solution found by any of the (meta-)heuristics. For instances where it is possible to obtain lower bounds by the *MILP*, we, also, present the corresponding *Gap*% values as: $Gap = \frac{\text{ALG\_SOL - LOWER BOUND}}{\text{ALG\_SOL}}$.

Table 5. Experimental Results per $|O|$

$\|O\|$	$\|J\|$			*AJPTI*			*AJEI*			*GA*			*MILP*		
	min	Avg.	max	Err	Gap	Time	Err	Gap	Time	Err	Gap	Time	Err	Gap	Time
5	5	9	16	14.13	24.09	< 1	13.86	23.78	< 1	7.95	18.55	340	0.08	11.04	2063
10	10	17	30	17.25	39.77	< 1	15.22	38.08	< 1	8.77	33.51	530	1.57	26.5	2695
20	28	37	46	19.02	54.16	2	13.85	51.43	2	6.78	47.52	1341	4.85	46.16	3615
35	54	67	76	13.64	81.79	9	8.68	80.46	11	1.42	78.73	2810	6.03	79.62	3644
50	78	93	101	14.36	–	8	6.16	–	16	0.14	–	3747	–	–	–
100	156	185	204	16.43	–	9	5.26	–	29	0	–	3804	–	–	–
165	263	316	359	9.33	–	22	4.83	–	28	0.1	–	3792	–	–	–
329	475	531	587	8.09	-	33	5.93	–	32	0	–	3876	–	–	–

It is evident that in smaller size instances ($|O| = \{5, 10, 20\}$), lower values of %*Gap* and %*Err* are obtained by the *MILP* formulation, yet at the cost of longer computational times (approximately 2800 s.). However, as the instance size grows beyond 20 orders, GA outperforms the rest of the methods. Indicatively, when $|O| = 35$ the $\%Err$ value of GA stands at 1.42 %, with the second best being the *MILP* (6.03 %). On the other hand, note that the difference on solution quality between the (meta-)heuristics is reduced on large instances ($|O| > 100$), as GA needs more time to explore a sufficient chunk of the solution space.

When also noticing the significant difference on computational times (approximately 3700 s. vs less than 30 s.), the trade-off between solution quality and computational times should be more closely assessed. Note that this increase on the GA computational times is attributed to the large number of *CP* calls needed to be performed in order to evaluate each new solution obtained. Considering that the *CP* model may require up to 30 s. to provide such solutions, this sharp rise on the computational times becomes explainable.

Comparing the proposed indices, we notice that *Gap%* and *Err%* values of *AJEI* remain lower than the corresponding one of AJIPT on the majority of instances (90 %). In terms of computational times, both indices are proven to be very efficient on both small and large instances, with solutions returned in less than 15 s. on average. Finally, the great values of % Gaps - the maximum being on 84.37 % on large size instances of 35 orders and 74 jobs - do not necessarily reflect the solution quality. This behavior may also be attributed to the weak lower bounds obtained by the *MILP*, possibly due to the 'big-M' constraints.

As contemporary manufacturing considers various indicators, also in accordance with the ISO22400 standard [31], let us also compare the proposed algorithms in terms of machine utilization ($MU = \frac{\sum_{j \in J} p_{jm}}{C_{\max}}$), i.e., the average percentage of operating time over all machines and the throughput rate (TR) measuring the average amount of jobs completed within consecutive time intervals (for our purposes, one interval is 10% of the minimum C_{max}). Table 6 presents the average and standard deviation values per method for these two additional indicators.

Table 6. KPI Comparison Results per $|O|$

$\|O\|$	$\|J\|$			*AJPTI*				*AJEI*				*GA*				*MILP*			
				TR		MU		TR		MU		TR		MU		TR		MU	
	min	Avg.	max	Avg.	sd	Avg.	sd	Avg.	sd	Avg.	sd	Avg.	sd	Avg.	sd	Avg.	sd	Avg.	sd
5	5	9	16	0.67	0.96	0.25	0.13	0.44	0.77	0.26	0.13	0.66	0.99	0.27	0.14	0.91	1.42	0.27	0.15
10	10	17	30	1.56	1.69	0.32	0.17	0.94	1.2	0.34	0.18	1.33	1.68	0.36	0.19	1.69	2.03	0.37	0.18
20	28	37	46	2.99	2.4	0.42	0.21	2.27	2.52	0.47	0.23	2.96	2.85	0.49	0.25	3.12	2.65	0.47	0.22
35	54	67	76	5.89	3.73	0.51	0.28	5.27	4.48	0.55	0.3	6.33	4.37	0.58	0.31	6.16	3.87	0.54	0.3
50	78	93	101	8.48	4.58	0.47	0.3	8.09	6.56	0.56	0.33	9.24	6.27	0.57	0.34	–	–	–	–
100	156	185	204	17.27	7.98	0.52	0.33	16.63	11.19	0.65	0.37	18.46	9.45	0.63	0.36	–	–	–	–
165	263	316	359	30.16	11.14	0.55	0.36	28.63	17.6	0.68	0.4	31.71	12.21	0.65	0.4	–	–	–	–
329	475	531	587	50.51	16.82	0.6	0.39	47.85	27.78	0.7	0.42	53.08	16.57	0.67	0.41	–	–	–	–

In terms of TR, the most efficient methods seem to be the MILP (when $|O| \leq 20$) and the GA (when $|O| \geq 35$). However, it should be underlined that the $AJPTI$ heuristic provides results close to those of GA. Notably, $AJPTI$ exhibits the lowest standard deviation (sd) of TR on all instance sizes and this 'smooth' flow in the production line is commonly targeted by production planners. This difference on the performance of $AJPTI$ between the makespan and TR could be attributed to jobs being more evenly assigned due to the improved load balancing. Similar results show up regarding MU: the MILP achieves the greatest percentages of machine utilization when $|O| \leq 10$ (32 % on average) yet, as the instance grows to 50 orders, the GA exhibits the most promising results ($\sim 55\%$). However, on instances where $|O| \geq 100$, $AJEI$ is the prevailing method, as it appears more resource-efficient (regarding machines) than the GA, which, nevertheless, also provides satisfying results. Finally, note that MU values for all methods are relatively high, despite the flexibility of skipping some stages.

The above findings aid us in determining the way that the proposed methods may be applied on our case study. The combination of the methods within our setting is dependent to the time horizon of employing the solutions proposed. More precisely, in cases where the production schedule needs to be instantly (re-)optimised, the constructive heuristics should be preferred, due to the negligible computational time consumed even on large size instances. On the other hand, if a plant manager aims at generating a production schedule under longer planning horizons e.g. for the next couple of days, then the combination of the GA and CP may be employed. Finally, for small-size instances and on occasions where there is the luxury of investing more computational time to obtain a solution, the MILP model could be employed, warmed up by a heuristic solution and combined with the CP model. As a matter of fact, this dependence on the purpose of use by the stakeholders highlights the necessity of our versatile hybrid approach.

5 Concluding Remarks

We have focused on an elaborate scheduling problem motivated by an actual production environment and introduced a variety of models and methods. We have proposed two indeed simple, rule-based, heuristics which could well be replaced by alternative or more elaborate constructive procedures; their virtue is their simplicity and speed. Regardless of the constructive procedure though, it becomes difficult to overcome the handling of buffers through CP, which could therefore remain a necessary step of any such procedure.

Our extensive experimentation on real data has showed that the MILP formulation provides solutions with low error percentage when examined in small instances. When extending the instance size, it is obvious that our simple constructive heuristics lead to a significant gain. The implemented GA balances low % Err with a stabilised time frame of completion in large instances.

Possible extensions of this work may be along different directions. However, a common pillar should be the experimentation on benchmark datasets (e.g. [32]), to avoid overfitting the algorithmic design to the data of our case study. In technical terms, we aim at employing exact methods to deal with the entire problem at hand e.g. a complete CP model. Being equipped with both an MILP and a CP model allows for further experimentation on the hybrisation of the algorithms. Such experimentation may include the activation of diverse 'parts' of these components and compare the different combinations. As an example, we may remove constraints from the MILP formulation (e.g. the enforcement of sequencing jobs within an order) and add them to the CP.

Apart from the interplay of exact methods, it would be interesting to enhance our (meta-)heuristic framework by including search-based algorithms (e.g. local search, tabu search, variable-neighbourhood descent) or MILP-based constructive heuristics. Candidate solution neighbourhoods for the search methods may consist of swapping orders on the initial order sequence $\bar{S}$ or swapping already assigned jobs on different machines. However, ensuring the feasibility of produced solutions may lead to consecutive calls of the CP model and thus, the consumption of substantial computational time.

In terms of effective decision support, it would be interesting to examine different scenarios in order to estimate possible improvements occurring from different production parameters. Such parameters may refer to the change of the layout of the plant, the investment on machines on specific stages or changes on the capacity of the buffers. Additional enhancements include the maintenance issues that could potentially come up in the related environment. Such issues in scheduling problems have been thoroughly addressed in the scientific literature [33].

Acknowledgments. This research has been supported by the EU through the MODAPTO Horizon 2020 project, grant number 101091996.

Disclosure of Interests. The authors have no competing interests to declare that are relevant to the content of this article.

References

1. Lenstra, J.K., Kan, A.R., Brucker, P.: Complexity of machine scheduling problems. Ann. Disc. Math. **1**, 343–362 (1977)
2. Gupta, S.K., Kyparisis, J.: Single machine scheduling research. Omega **15**(3), 207–227 (1987)
3. Cheng, T.C.E., Sin, C.C.S.: A state-of-the-art review of parallel-machine scheduling research. Eur. J. Oper. Res. **47**(3), 271–292 (1990)
4. Márquez, C.R., Ribeiro, C.C.: Shop scheduling in manufacturing environments: a review. Int. Trans. Oper. Res. **29**(6), 3237–3293 (2022)
5. Neufeld, J.S., Schulz, S., Buscher, U.: A systematic review of multi-objective hybrid flow shop scheduling. Eur. J. Oper. Res. **309**(1), 1–23 (2023)
6. Lian, X., Zheng, Z., Wang, C., Gao, X.: An energy-efficient hybrid flow shop scheduling problem in steelmaking plants. Comput. Ind. Eng. **162**, 107683 (2021)
7. Jun, S., Park, J.: A hybrid genetic algorithm for the hybrid flow shop scheduling problem with nighttime work and simultaneous work constraints: A case study from the transformer industry. Expert Syst. Appl. **42**(15–16), 6196–6204 (2015)

8. Li, Z., Wan, G.: Batching and scheduling in a continuous-discrete hybrid flowshop: Lagrangian relaxation-based heuristic algorithms. Int. J. Prod. Res. **61**(17), 5934–5955 (2023)
9. Naderi, B., Zandieh, M., Balagh, A.K.G., Roshanaei, V.: An improved simulated annealing for hybrid flowshops with sequence-dependent setup and transportation times to minimize total completion time and total tardiness. Expert Syst. Appl. **36**(6), 9625–9633 (2009)
10. Elmi, A., Topaloglu, S.: Scheduling multiple parts in hybrid flow shop robotic cells served by a single robot. Int. J. Comput. Integr. Manuf. **27**(12), 1144–1159 (2014)
11. Mousavi, S.M., Zandieh, M., Amiri, M.: Comparisons of bi-objective genetic algorithms for hybrid flowshop scheduling with sequence-dependent setup times. Int. J. Prod. Res. **50**(10), 2570–2591 (2012)
12. Dabiri, M., Yazdani, M., Naderi, B., Haleh, H.: Modeling and solution methods for hybrid flow shop scheduling problem with job rejection. Oper. Res. Int. J. **22**(3), 2721–2765 (2022)
13. Öztop, H., Tasgetiren, M.F., Eliiyi, D.T., Pan, Q.K.: Metaheuristic algorithms for the hybrid flowshop scheduling problem. Comput. Oper. Res. **111**, 177–196 (2019)
14. Gheisariha, E., Tavana, M., Jolai, F., Rabiee, M.: A simulation-optimization model for solving flexible flow shop scheduling problems with rework and transportation. Math. Comput. Simul. **180**, 152–178 (2021)
15. Abbaszadeh, N., Asadi-Gangraj, E., Emami, S.: Flexible flow shop scheduling problem to minimize makespan with renewable resources. Scientia Iranica **28**(3), 1853–1870 (2021)
16. Lei, C., Zhao, N., Ye, S., Wu, X.: Memetic algorithm for solving flexible flow-shop scheduling problems with dynamic transport waiting times. Comput. Ind. Eng. **139**, 105984 (2020)
17. Naderi, B., Gohari, S., Yazdani, M.: Hybrid flexible flowshop problems: models and solution methods. Appl. Math. Model. **38**(24), 5767–5780 (2014)
18. Amirteimoori, A., Mahdavi, I., Solimanpur, M., Ali, S.S., Tirkolaee, E.B.: A parallel hybrid PSO-GA algorithm for the flexible flow-shop scheduling with transportation. Comput. Ind. Eng. **173**, 108672 (2022)
19. Mollaei, A., Mohammadi, M., Naderi, B.: A bi-objective MILP model for blocking hybrid flexible flow shop scheduling problem: robust possibilistic programming approach. Int. J. Manag. Sci. Eng. Manag. **14**(2), 137–146 (2019)
20. Armstrong, E., Garraffa, M., O'Sullivan, B., Simonis, H.: The hybrid flexible flowshop with transportation times. In: 27th International Conference on Principles and Practice of Constraint Programming (CP 2021). Schloss Dagstuhl-Leibniz-Zentrum für Informatik (2021)
21. Armstrong, E., Garraffa, M., O'Sullivan, B., Simonis, H.: A two-phase hybrid approach for the hybrid flexible flowshop with transportation times. In: Schaus, P. (ed.) International Conference on Integration of Constraint Programming, Artificial Intelligence, and Operations Research, vol. 13292, pp. 1–13. Springer, Heidelberg (2022)
22. Jiang, S.L., Xu, C., Zhang, L., Ma, Y.: A decomposition-based two-stage online scheduling approach and its integrated system in the hybrid flow shop of steel industry. Expert Syst. Appl. **213**, 119200 (2023)
23. Zhang, H.Y., Xi, S.H., Chen, Q.X., Smith, J.M., Mao, N., Li, X.: Performance analysis of a flexible flow shop with random and state-dependent batch transport. Int. J. Prod. Res. **59**(4), 982–1002 (2021)

24. Graham, R.L., Lawler, E.L., Lenstra, J.K., Kan, A.R.: Optimization and approximation in deterministic sequencing and scheduling: a survey. Ann. Disc. Math. **5**, 287–326 (1979)
25. Naderi, B., Ruiz, R., Roshanaei, V.: Mixed-integer programming vs. constraint programming for shop scheduling problems: new results and outlook. INFORMS J. Comput. **35**(4), 817–843 (2023)
26. Holland, J.H.: Adaptation in Natural and Artificial Systems. University of Michigan Press, Ann Arbor (1975)
27. Avgerinos, I., Mourtos, I., Vatikiotis, S., Zois, G.: Weighted tardiness minimisation for unrelated machines with sequence-dependent and resource-constrained setups. Int. J. Prod. Res. **62**(1–2), 359–379 (2024)
28. IBM: IBM ILOG CPLEX 12.7 User's Manual. IBM ILOG CPLEX Division, Incline Village (2017)
29. Bynum, M.L., et al.: Pyomo-Optimization Modeling in Python, 3rd edn. Springer, Heidelberg (2021). https://doi.org/10.1007/978-3-319-58821-6
30. Gurobi Optimization, LLC: Gurobi Optimizer Reference Manual (2024)
31. Bauer, M., Lucke, M., Johnsson, C., Harjunkoski, I., Schlake, J.C.: KPIs as the interface between scheduling and control. IFAC-PapersOnLine **49**(7), 687–692 (2016)
32. Vallada, E., Ruiz, R., Framinan, J.M.: New hard benchmark for flowshop scheduling problems minimising makespan. Eur. J. Oper. Res. **240**(3), 666–677 (2015)
33. Geurtsen, M., Didden, J.B., Adan, J., Atan, Z., Adan, I.J.B.F.: Production, maintenance and resource scheduling: a review. Eur. J. Oper. Res. **305**(2), 501–529 (2023)

Multi-depot Electric Vehicle Routing Problem with Half-Open Routes and Rotations: A Mathematical Formulation

Hakan Erdeş(✉) and Saadettin Erhan Kesen

Department of Industrial Engineering, Faculty of Engineering and Natural Sciences, Konya Technical University, Selçuklu/Konya, Turkey
{herdes,sekesen}@ktun.edu.tr

Abstract. The electrification spreads all around the world with the goal of global net zero emission, especially in transportation sector. As a natural consequence of this phenomenon, many countries and organizations announce their targets relating to the Electric Vehicles (EVs). Although shifting from fossil-fuel towards EVs is a major step forward, there are other precious actions to be taken, one of which is to integrate EVs into scientific studies along with the routing improvements. To this end, we introduce the Multi-Depot Electric Vehicle Routing Problem with Half-Open Routes and Rotations (MDEVRP-HORR) as an extension of the MDEVRP where the rotation refers to set of all routes assigned to an EV. EVs can be recharged at the depots or at the recharging stations along the route with full recharging strategy. The problem is formulated as a 0–1 mixed integer linear program (MILP) and existing procedures are applied to improve the model. The proposed model relaxes the restriction that the departure depot of the first route and arrival depot of the last route of an EV must be the same. It also avoids from creating dummy nodes for recharging stations and replenishment depots. We conducted experimental studies by using small-sized instances from the recent literature due to problem intractability. Out of 36 instances, results reveal that inclusion of half-open routes leads to a cost reduction for 20 instances. Besides, inclusion of both half-open rotations and half-open routes results in cost reduction for 28 instances.

Keywords: Electric Vehicle · Multi-Depot Electric Vehicle Routing Problem · Half-Open Route and Rotation

1 Introduction

Transportation is one of the foremost sectors that contribute to carbon emissions all around the world. As internal combustion engines dominated the transportation sector, concerns about their environmental impact raised. At this point, whilst the community converts to alternative fuel vehicles, Electric Vehicles (EVs) emerge as an effective way for fighting against air pollution and the climate change [1]. These recent developments have led to integration of environmental aspects into academic studies. When it comes

Published by Springer Nature Switzerland AG 2024
M. Thürer et al. (Eds.): APMS 2024, IFIP AICT 733, pp. 274–290, 2024.
https://doi.org/10.1007/978-3-031-71645-4_19

to environmental aspects about the transportation, the Vehicle Routing Problem (VRP) and its variants offer worthwhile solutions, therefore there is a growing interest in this phenomenon in the literature [2]. In this regard, Electric VRP (EVRP) is one of the main variations since EVs are promising not only for reducing the pollution but also for reducing the transportation cost. Many countries like Norway, Netherlands, Germany, United Kingdom, Türkiye, France and Italy contemplate enforcing to brand new sales to be EV and ban the sales of internal combustion engine vehicle in the following years, even the purchase subsidies or tax benefits have already been applied recently and so on [3, 4]. However, the limited driving range, long recharging times and sparse available recharging options entail the EVRP to find optimal solutions by considering battery capacity, recharging times, and recharging options along with the locations. Regarding EVRP, there are growing bodies of literature studying the minimization of traveled distance [5–7], recharging cost [8, 9] or total time (travel, recharging etc.) [10, 11] or finding optimal locations for depots and recharging stations [12, 13].

Since the electricity is generated using thermal power in many countries, reduction in carbon emission resulting from penetration of EVs is limited [14]. For this reason, effective operations of EVs like resource sharing play an important role on curbing these indirect and adverse effects. The approach of half-open routes and rotations consists of sharing the vehicles among the depots and allows a vehicle to return to any depot contrary to the standard Multi-Depot VRP (MDVRP) [15]. Owing to this approach, the number of unnecessary and underloaded vehicles and a great deal of distribution cost could be significantly reduced [16]. On the other hand, integrating EVs to this approach leads to sharing not only the EVs but also the recharging stations among the depots. In this way, the opportunities of reducing the recharging costs and the number of recharging stations arise. While some studies include the sharing of EVs, others consider sharing the recharging stations or depots [2, 12, 17–20]. For this reason, sharing resources among multiple depots could be evaluated as a prominent strategy for logistics companies to gain competitive advantages [16], to reduce the traveled distance and energy consumption and to improve the distribution efficiency in the long run [14].

Yet, another usual extension of sharing vehicles is using the vehicles multiple times, which means a vehicle docks a depot at the end of a route and then can start a new route. This type of situation is often encountered when the duration of a single route is shorter as compared to working hours due to limited vehicle capacity or limited fleet size [21]. In practice, scheduling vehicles for multiple trips can provide a significant cost reduction, especially if the number of vehicles and hence drivers could be reduced or utilized with a higher rate. This situation may lead to a trade-off between scheduling/ routing related costs and number of vehicle/driver related costs [22].

Although there exist papers considering MDEVRP and resource sharing [7, 18], half-open approach in EVRP [14, 16] and even including good replenishment [23, 24], to the best of author's knowledge, the Multi-Depot Electric Vehicle Routing Problem with Half-Open Routes and Rotations (MDEVRP-HORR) is studied for the first time where the rotation refers to the set of all routes assigned to single vehicle [21]. We relax the restriction that the departure depot of the first route and arrival depot of the last route of an EV must be the same. To this end, a novel 0–1 Mixed-Integer Linear Programming (MILP) model is introduced, avoiding from generating dummy nodes of recharging stations and replenishment depots. This approach requires to add equations into mathematical model. The model also includes a decision variable controlling whether to receive recharging service at the depots.

The remainder of this paper is organized as follows. First, the related literature review outlining the extensions of our problem is presented in Sect. 2. Section 3 includes the problem description, illustrative example, and 0–1 MILP formulation. Computational results are given in Sect. 4. Finally, Sect. 5 concludes the paper.

2 Literature Review

The MDEVRP-HORR is a variation of a diverse well known VRPs in the literature. One of them is EVRP where EVs need to visit recharging stations to continue servicing the customers along the route. In this regard, Schneider et al. [5] expanded the Green VRP [25] by including EVs, introduced EVRP with time windows (EVRPTW) and proposed MILP model as well as hybrid heuristic including variable neighborhood search algorithm and tabu search. However they assume that the EVs can leave recharging stations only with full battery which is later relaxed by Keskin and Çatay [6] and partial recharging is allowed, a MILP model as well as Adaptive Large Neighborhood Search algorithm is introduced. Yet, they both employed dummy nodes to allow the recharging stations to be used multiple times.

Other extension is MDEVRP where each EV is deployed at one of the depots, and each route starts and ends at the same depot [26]. Muñoz-Villamizar et al. [27] proposed a two-stage solution for collaboration among depots of logistics companies, analyzed the effect of EVs usage and introduced MILP model. Paz et al. [7] studied the MDEV location routing problem (LRP) with time windows and introduced three mathematical models where only the small sized-instances are solved. Koç et al. [2] addressed the EVRP where multiple companies collaboratively invest in the installation of recharging stations containing different technologies. Accordingly, each company has its own depot and EV fleet. Yet, the customers are pre-assigned, implying that customers have a single option for receiving service. Almouhanna et al. [12] addressed the LRP with a constrained distance by employing EVs and allowed them to be shared among the depots. They proposed both multi-start heuristic and a metaheuristic. Karakatič [10] introduced MDVRPTW and EVs with partial nonlinear recharging times and proposed a genetic algorithm. Londoño et al. [28] proposed a hybrid methodology of exact solution method and meta-heuristic for the MDEVRP and evaluated three different costs with trade-offs among each other. Wang et al. [17] studied the EV Charging Station LRP and resource sharing problem which consists of sharing the resources among depots within multiple service periods. Wang et al. [18] introduced collaborative MDEVRPTW and shared recharging stations. They reported that sharing both customers and recharging stations among the depots outperforms other scenarios in terms of cost, number of EVs and number of recharging stations used. Vahedi-Nouri et al. [29] addressed the collaborative EVRP and developed two MILP models for collaborative and non-collaborative strategies for companies and reported that collaborative strategy provides cost reduction.

MDEVRP with half-open routes is an extension of MDEVRP where starting and ending depot of an EV are not necessarily the same. Lijun et al. [14] addressed the Half-Open Time-Dependent MDEVRP, considered both battery recharging and swapping technologies, developed MILP model and presented simulated annealing algorithm. Their experimental studies show that the half-open route strategy reduces the total cost and carbon emissions notably. Fan [16] proposed a MILP model for Multi-Depot Half-Open time dependent EVRP, aiming to minimize the total distribution cost as well as a two-stage hybrid ant colony algorithm. Nevertheless, in the last two references, EVs are dispatched only once.

There are also other VRP variants studied extensively in which resource replenishment is considered without EVs. Crevier et al. [21] introduced the MDVRP with inter-depot routes (MDVRPI) where vehicles can be replenished with goods at replenishment facilities and can be assigned multiple routes. Tarantilis et al. [30] addressed the VRP with intermediate replenishment facilities and proposed three-step algorithmic framework including cost-saving construction heuristic, tabu search with variable neighborhood search and guided local search. Muter et al. [31] later studied the MDVRPI by relaxing the closed route and developed a branch-and-price algorithm to solve the problem. Che et al. [32] studied the Multi-Depot Petrol Station Replenishment with Open Inter-Depot Routes where each depot can act as an intermediate facility and multiple trips for the trucks are allowed.

There also exist papers considering both resource replenishment and EVs. Schneider et al. [23] are the first to introduce the VRP with intermediate stops which includes stops for replenishment/disposal, refueling/recharging or combined ones and developed an adaptive variable neighborhood search. Schiffer et al. [24] studied on location-routing problem considering intra-route facilities that offers different replenishment services and presented an adaptive large neighborhood search. They reported that combined facilities are useful to decrease the fleet size and reduce the total cost. They also classified the replenished resources as arc-based like energy for EVs and as node-based like load of the vehicle. Nevertheless, the last two references utilize dummy nodes in order to allow multiple visits to replenishment facilities.

Our paper differs from the existing literature in several aspects. First, it is the first study combining the EVs with half-open rotations. Second, our paper contains multi-depot, arc-based and node-based resource replenishment by adopting a pre-computing procedure which eliminates some of the recharging stations and replenishment depots. Eventually, this paper introduces a MILP model for a specified variant of EVRP problem without dummy nodes for recharging stations and replenishment depots.

3 Problem Definition

MDEVRP-HORR deals with (i) a set of customers with known demands and service times, (ii) homogeneous EVs with limited battery and load capacity, and (iii) multiple depots and creation of both half-open routes and rotations. The battery charge level of each EV depletes proportional to the distance traveled. The battery could be fully recharged both at depots [5] and recharging stations en-route [2]. For the sake of simplicity, the linear recharging is adopted [5, 33]. Each EV can perform multiple trips and be replenished with goods at any depot [23]. The duration of good replenishment is ignored. Each customer must be visited exactly once. The total demand of the customers in each route cannot exceed the EV load capacity. Recharging stations are assumed to be available at all times and could be visited more than one by any EV and simultaneous recharging is permitted as well [33]. The objective of the problem is to minimize the sum of the total traveled distance and the fixed EV cost per use.

The MDEVRP-HORR is formulated as 0–1 MILP model. Let $V = \{1, \ldots, N\}$ be the set of customers and F be the set of recharging stations. Vertices DD, AD, RD denote the departure, arrival, and replenishment depots respectively and are located in the same place. To indicate that a set contains departure or arrival depot, it is subscripted with DD or AD, i.e., $V_{DD} = V \cup \text{DD}$ and $V_{AD} = V \cup \text{AD}$. Thus, MDEVRP-HORR can be defined on a complete directed graph $G = \left(V_{DD,AD}, A\right)$ with the set of arcs $A = \{\ (i, j), \, |i \in V_{DD}, j \in V_{AD}, i \neq j\}$ where $V_{DD,AD} = V \cup \text{DD} \cup \text{AD}$.

Between nodes i and j, there are associated distance of d_{ij} and a travel time of t_{ij}. Each customer $i \in V$ has a positive demand q_i and service time s_i. EVs can carry a maximum load of C and have battery capacity Q. The battery charge is consumed by $h \cdot d_{ij}$ for each arc traversed, where h represents constant charge consumption rate. So, one unit of travelling distance, therefore, equals one unit of energy consumption and takes one unit of time [9]. Battery is recharged with a rate of g, meaning that recharging time is the product of g and the amount of recharge. The duration of a rotation cannot exceed the maximum duration T_{max}. Variables τ_i, u_i *and* y_i specify the arrival time, remaining load, and remaining charge level at node $i \in V_{DD,AD}$ respectively.

The binary decision variable x_{ij} is equal to 1 if arc (i, j) is traveled, and 0 otherwise. We inspire from the study of Schneider et al. [5] in developing our mathematical model and thus stick to the notation they used with an exception that we do not generate dummy vertices to allow multiple visits to recharging stations, requiring significant augmentation of the network. So, we follow similar procedure to the study conducted by Koç and Karaoglan [34] and Bruglieri et al. [33] and introduce the decision variable z_{ijk}^F which equals 1 if a EV travels from node i to j through the recharging station k. They reported the superiority of this method over generation of dummy nodes. Meanwhile, we apply the same logic for replenishment depots and introduce the decision variable z_{ijl}^D which equals 1 if an EV travels from node i to j through the replenishment depot l. The binary decision variable z_i^C is also introduced being equal to 1 if an EV visits a replenishment depot and is recharged after node i, and 0 otherwise.

Furthermore, we also follow a pre-computing procedure which eliminates some of the recharging stations between each pair of arcs. Similar procedures are applied by Bruglieri et al. [33] and Keskin and Çatay [9]. They reported that this procedure decreases the number of variables employed, which in turn leads to better performance of the model. The procedure results in having sets $\hat{F}_{ij}$ and initially, each $\hat{F}_{ij} = F$ which means each recharging station is available between nodes i and j. Now consider two recharging stations k and k^*. If conditions $d_{ik^*} > d_{ik}$ and $d_{k^*j} > d_{kj}$ hold, then it is certain that, recharging station k^* is not visited between nodes i and j in the optimal solution. Therefore, k^* is removed from $\hat{F}_{ij}$. These eliminations are conducted for each pair of departure depot-to-customer, customer-to-customer, and customer-to-arrival depot nodes. The same procedure is applied for replenishment depots as well. This time, initially each $\widehat{RD}_{ij} = RD$ and if conditions $d_{il^*} > d_{il}$ and $d_{l^*j} > d_{lj}$ hold, replenishment depot l dominates the replenishment depot l^* and therefore l^* is removed from $\widehat{RD}_{ij}$. These eliminations are conducted only for each pair of customer-to-customer nodes, mainly because visiting replenishment depot between departure depot and customer or between customer and arrival depot is not permitted.

3.1 Illustrative Example

In this section, an illustrative example depicting the relation between closed route and half-open route as well as closed rotation and half-open rotation is visualized in Fig. 1. D, C and RS denote the depot, customer, and recharging station, respectively. Part a includes two EVs i.e., two rotations which are closed as follow: (i) **D1**-*RS2*-C2-C5-**D1**-C1-C10-C6-**D1**-C9-C8-*RS3*-C4-C7-**D1** (ii) **D2**-C3-**D2**. The first and second rotations consist of three and one closed routes respectively. Part b includes one closed rotation where the first and third routes are half-open, and the second one is closed: (i) **D2**-C3-C7-*RS3*-C4-*RS3*-C8-C9-**D1**-C6-C10-C1-**D1**-C5-C2-*RS2*-**D2**. Finally, part c includes a half-open rotation where the first two routes are closed and the last one is half-open: (i) **D1**-C9-C8-*RS3*-C4-C7-**D1**-C1-C10-C6-**D1**-C5-C2-*RS2*-C3-**D2**. Consequently, part a is limited with closed route and closed rotation, the half-open route is allowed in part b, and both half-open route and rotation are allowed in part c.

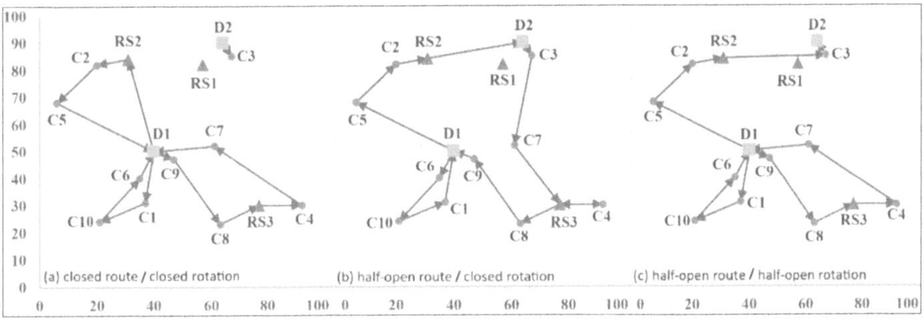

Fig. 1. Illustrations of closed and half-open routes and rotations

3.2 Mathematical Formulation

In what follows, first all sets, parameters, and decision variables of our model are given in Table 1, second the 0–1 MILP model is presented.

$$Min \sum_{i\in V_{DD}} \sum_{\substack{j\in V_{AD}\\ i\neq j}} \left(d_{ij}x_{ij} + \sum_{l\in \widehat{RD}_{ij}} (d_{il}+d_{lj}-d_{ij})z^{D}_{ijl} + \sum_{k\in \hat{F}_{ij}} (d_{ik}+d_{kj}-d_{ij})z^{F}_{ijk} \right) + \sum_{i\in DD}\sum_{j\in V_{AD}} c_{fix}x_{ij} \quad (1)$$

$$\sum_{\substack{j \in V_{AD} \\ i \neq j}} x_{ij} = 1 \qquad \forall i \in V \tag{2}$$

$$\sum_{\substack{i \in V_{AD} \\ i \neq j}} x_{ji} - \sum_{\substack{i \in V_{DD} \\ i \neq j}} x_{ij} = 0 \qquad \forall j \in V \tag{3}$$

$$\sum_{l \in \widehat{RD}_{ij}} z_{ijl}^{D} + \sum_{k \in \hat{F}_{ij}} z_{ijk}^{F} \leq x_{ij} \qquad \begin{matrix} \forall i \in V_{DD} \\ \forall j \in V_{AD} \\ i \neq j \end{matrix} \tag{4}$$

$$z_i^C \leq \sum_{\substack{j \in V_{AD} \\ i \neq j}} \sum_{l \in \widehat{RD}_{ij}} z_{ijl}^{D} \qquad \forall i \in V_{DD} \tag{5}$$

$$\begin{aligned} 0 \leq \tau_i + (t_{ij} + s_i)x_{ij} + \sum_{l \in \widehat{RD}_{ij}} (t_{il} + t_{lj} - t_{ij}) z_{ijl}^{D} \\ -T_{max}\left(1 - x_{ij} + z_i^C + \sum_{k \in \hat{F}_{ij}} z_{ijk}^{F}\right) \leq \tau_j \end{aligned} \qquad \begin{matrix} \forall i \in V_{DD} \\ \forall j \in V_{AD} \\ i \neq j \end{matrix} \tag{6}$$

$$\begin{aligned} 0 \leq \tau_i + \sum_{l \in \widehat{RD}_{ij}} (t_{il} + t_{lj} + s_i) z_{ijl}^{D} + \sum_{k \in \hat{F}_{ij}} (t_{ik} + t_{kj} + s_i) z_{ijk}^{F} \\ +g\left[Q - \left(y_i - \sum_{l \in \widehat{RD}_{ij}} d_{il} z_{ijl}^{D} - \sum_{k \in \hat{F}_{ij}} d_{ik} z_{ijk}^{F}\right)\right] \\ -(T_{max} + gQ)\left(2 - x_{ij} - z_i^C - \sum_{k \in \hat{F}_{ij}} z_{ijk}^{F}\right) \leq \tau_j \end{aligned} \qquad \begin{matrix} \forall i \in V_{DD} \\ \forall j \in V_{AD} \\ i \neq j \end{matrix} \tag{7}$$

$$\tau_i + t_{ij} + s_i - T_{max}\left(1 - x_{ij} + \sum_{k \in \hat{F}_{ij}} z_{ijk}^{F}\right) \leq T_{max} \qquad \begin{matrix} \forall i \in V \\ \forall j \in AD \end{matrix} \tag{8}$$

$$\begin{aligned} \tau_i + s_i + \left(\sum_{k \in \hat{F}_{ij}} (t_{ik} + t_{kj}) z_{ijk}^{F}\right) + g\left[Q - \left(y_i - \sum_{k \in \hat{F}_{ij}} d_{ik} z_{ijk}^{F}\right)\right] \\ -T_{max}\left(1 - \sum_{k \in \hat{F}_{ij}} z_{ijk}^{F}\right) \leq T_{max} \end{aligned} \qquad \begin{matrix} \forall i \in V \\ \forall j \in AD \end{matrix} \tag{9}$$

$$0 \leq u_j \leq u_i - q_j x_{ij} + C\left(1 - x_{ij} + \sum_{l \in \widehat{RD}_{ij}} z_{ijl}^{D}\right) \qquad \begin{matrix} \forall i \in V_{DD} \\ \forall j \in V_{AD} \\ i \neq j \end{matrix} \tag{10}$$

$$0 \leq u_j \leq C - q_j + C\left(1 - \sum_{l \in \widehat{RD}_{ij}} z_{ijl}^D\right) \quad \begin{array}{c} \forall i \in V_{DD} \\ \forall j \in V_{AD} \\ i \neq j \end{array} \tag{11}$$

$$u_i = C \qquad \forall i \in DD \tag{12}$$

$$\begin{aligned} 0 \leq y_j \leq y_i - h d_{ij} x_{ij} - h \sum_{l \in \widehat{RD}_{ij}} (d_{il} + d_{lj} - d_{ij}) z_{ijl}^D \\ + Q\left(1 - x_{ij} + z_i^C + \sum_{k \in \hat{F}_{ij}} z_{ijk}^F\right) \end{aligned} \quad \begin{array}{c} \forall i \in V_{DD} \\ \forall j \in V_{AD} \\ i \neq j \end{array} \tag{13}$$

$$\begin{aligned} 0 \leq y_j \leq Q - h\left(\sum_{l \in \widehat{RD}_{ij}} d_{lj} z_{ijl}^D + \sum_{k \in \hat{F}_{ij}} d_{kj} z_{ijk}^F\right) \\ + Q\left(1 - z_i^C - \sum_{k \in \hat{F}_{ij}} z_{ijk}^F\right) \end{aligned} \quad \begin{array}{c} \forall i \in V_{DD} \\ \forall j \in V_{AD} \\ i \neq j \end{array} \tag{14}$$

$$y_i \geq h\left(\sum_{\substack{j \in V_{AD} \\ i \neq j}} \sum_{l \in \widehat{RD}_{ij}} \left(d_{il} z_{ijl}^D\right) + \sum_{\substack{j \in V_{AD} \\ i \neq j}} \sum_{k \in \hat{F}_{ij}} \left(d_{ik} z_{ijk}^F\right)\right) \qquad \forall i \in V_{DD} \tag{15}$$

$$y_i = Q \qquad \forall i \in DD \tag{16}$$

$$\sum_{i \in DD} \sum_{j \in AD} x_{ij} + \sum_{i \in DD} \sum_{j \in V} \sum_{l \in \widehat{RD}_{ij}} z_{ijl}^D + \sum_{i \in V} \sum_{j \in AD} \sum_{l \in \widehat{RD}_{ij}} z_{ijl}^D = 0 \tag{17}$$

$$\sum_{c \in V} x_{ic} - \sum_{c \in V} x_{jc} \leq 1 \qquad \begin{array}{c} \forall i, j \in DD \\ i \neq j \end{array} \tag{18}$$

$$\sum_{c \in V} x_{ic} - \sum_{c \in V} x_{jc} \leq 1 \qquad \begin{array}{c} \forall i, j \in AD \\ i \neq j \end{array} \tag{19}$$

$$\sum_{i \in D} \sum_{j \in V} x_{ij} + \sum_{i \in V} \sum_{\substack{j \in V \\ i \neq j}} \sum_{l \in \widehat{RD}_{ij}} z_{ijl}^D \geq \sum_{i \in V} q_i / C \tag{20}$$

$$x_{ij} \in \{0, 1\} \qquad \begin{array}{c} \forall i \in V_{DD}, \forall j \in V_{AD} \\ i \neq j \end{array} \tag{21}$$

$$z_{ijl}^D \in \{0, 1\} \qquad \begin{array}{c} \forall i \in V_{DD}, \forall j \in V_{AD} \\ \forall l \in \widehat{RD}_{ij}, i \neq j \end{array} \tag{22}$$

$$z_i^C \in \{0, 1\} \qquad \forall i \in V_{DD} \tag{23}$$

$$z_{ijk}^F \in \{0, 1\} \qquad \begin{array}{c} \forall i \in V_{DD}, \forall j \in V_{AD} \\ \forall k \in \hat{F}_{ij}, i \neq j \end{array} \tag{24}$$

The objective function in Eq. (1) minimizes the sum of total traveled distance and the fixed EV cost per use. Equation (2) ensures that each customer is visited exactly once. Equation (3) guarantees that the number of incoming and outgoing arcs to each node is equal. Equation (4) states that either replenishment depot or recharging station could be visited between nodes i and j providing that arc (i, j) is traversed. Equation (5) determines if the EV is recharged or not at replenishment depot during the good replenishment. Equation (6) ensures the time feasibility for arcs leaving customers in case of recharging is not provided, i.e., recharging station and replenishment depot is not visited, or recharging is not preferred while visiting the replenishment depot. The same argument is provided with Eq. (7) in the case that recharging is performed. Equations (6) and (7) prevent the formation of subtours as well. Equation (8) limits the arrival time at arrival depots to the maximum duration T_{max} providing that no recharging station is visited between the last customer and arrival depot. The same argument is ensured by Eq. (9) when a recharging station is visited between the last customer and arrival depot. Equation (10) guarantees that demands are met in the event that no replenishment depot is visited by assuring a nonnegative load level and same is ensured by Eq. (11) when a replenishment depot is visited. Equation (12) signifies that EVs depart from departure depots with full load capacity. Equation (13) guarantees that battery charge level is nonnegative and limited by maximum battery capacity in the case that EV is not recharged and the same is assured by Eq. (14) in the case that EV is recharged at replenishment depot or recharging station. Equation (15) ensures that the battery level is sufficient to reach the replenishment depot or recharging station if they are visited. Equation (16) guarantees that EVs depart from departure depots with full battery capacity. Equation (17) prohibits the travels between departure depots and arrival depots, departure depots and replenishment depots, replenishment depots and arrival depots. Equation (18) establishes a balance among departure depots and ensures that the difference between the number of EVs departing from each departure depot is at most 1. Equation (19) executes the same operation for arrival depots. Equation (20) provides a lower limit for the minimum number of EVs departing from departure depots and replenishment depots. Equations (21)–(24) define non-negativity constraints.

Table 1. Mathematical Notations and Definitions

Notation	Definition
Indices	
i, j, c	Customers
k	Recharging station
l	Replenishment depot
Sets	
DD	Set of departure depots
AD	Set of arrival depots
V	Set of customers
V_{DD}	Set of customers and departure depots ($V \cup DD$)
V_{AD}	Set of customers and arrival depots ($V \cup AD$)
F	Set of recharging stations
$\hat{F}_{ij}$	Set of available recharging stations between nodes i and j
RD	Set of replenishment depots
$\widehat{RD}_{ij}$	Set of available replenishment depots between nodes i and j
Parameters	
d_{ij}	Distance between nodes i and j
t_{ij}	Travel time between nodes i and j
q_i	Demand of customer i
s_i	Service time of customer i
C	Electric vehicle load capacity
Q	Electric vehicle battery capacity
h	Charge consumption rate
g	Recharging rate
T_{max}	Maximum duration
c_{fix}	Fixed cost per used electric vehicle
Decision Variables	
x_{ij}	1 if arc (i, j) is traversed and 0 otherwise
z_{ijl}^{D}	1 if an EV travels from node i to j through the replenishment depot l, and 0 otherwise
z_i^C	1 if an EV is recharged at depot after node i, and 0 otherwise
z_{ijk}^{F}	1 if an EV travels from node i to j through the recharging station k, and 0 otherwise
y_i	Remaining charge level at node i

(continued)

Table 1. (*continued*)

Notation	Definition
u_i	Remaining cargo level at node i
τ_i	Time of arrival at node i

4 Experimental Results

In this section, we present the results of numerical experiments. We conducted the experiments using the instances generated by Paz et al. [7]. In essence, their datasets are the modification of the instances generated by Schneider et al. [5]. Schneider et al. [5] modified the well-known VRPTW instances of Solomon [35] by adding recharging stations, battery capacity, energy consumption rate etc., and also derived the small sized instances from them. Later, Paz et al. [7] randomly added one depot for each of the small instances as well as one recharging station which is in the same location as the new depot to allow recharging at the depot.

To adapt those datasets to our problem, we first remove the recharging stations which are located at the same location of depots, since we already possess a decision variable controlling recharging at the depot. Second, since our problem does not involve time windows, we do not need them but the maximum duration T_{max}. So, we set the T_{max} to 600 unit of time for all the experiments. Third, since we investigate the multiple use of EVs, we set the EV load capacities as the two times of the highest customer demand for each instance. Besides, we set both routes and rotations to closed in Configuration-1 (C1), let routes be half-open in Configuration-2 (C2) and let both route and rotation be half-open in Configuration-3 (C3). These configurations correspond to parts a, b, and c respectively in Fig. 1 in Sect. 3.1. To set the configurations, we utilize some extra equations and notations. One is id_i, $i \in \text{DD} \cup \text{AD} \cup \text{RD}$ which denotes the depot identification number [7], and the id_i of each depot representing departure, arrival and replenishment are the same. Second, decision variable w_i, $i \in V_{AD}$ is used to follow the depot identification number of nodes. Last, the parameter NOD is utilized referring to the number of depots. To set C1, we add Eqs. (25)–(31) to our model which means EVs can return only to the same depot (closed rotation) and can only replenish their load at the same depot (closed route). For setting C2, we add Eqs. (25)–(29) to our model, making the rotations closed.

$$w_j \geq x_{ij}id_i - NOD(1 - x_{ij}) \quad \forall i \in DD, \forall j \in V \tag{25}$$

$$w_j \leq x_{ij}id_i + NOD(1 - x_{ij}) \quad \forall i \in DD, \forall j \in V \tag{26}$$

$$w_j \geq w_i - NOD(1 - x_{ij}) \quad \forall i \in V, \forall j \in V_{AD},\ i \neq j \tag{27}$$

$$w_j \leq w_i + NOD(1 - x_{ij}) \quad \forall i \in V, \forall j \in V_{AD},\ i \neq j \tag{28}$$

$$w_i = id_i \qquad \forall i \in AD \tag{29}$$

$$w_i \geq z_{ijl}^D id_l - NOD\left(1 - z_{ijl}^D\right) \qquad \begin{matrix} \forall i, j \in V, l \in \widehat{RD}_{ij} \\ i \neq j \end{matrix} \tag{30}$$

$$w_i \leq z_{ijl}^D id_l + NOD\left(1 - z_{ijl}^D\right) \qquad \begin{matrix} \forall i, j \in V, l \in \widehat{RD}_{ij} \\ i \neq j \end{matrix} \tag{31}$$

The experiments are carried out on a computer equipped with Intel Core i9-13900H Processor at 2.60 GHz and 16 GB of RAM. CPLEX solver embedded to GAMS version of 45.4.0 is used as the MILP-solver. The results of the experiments are presented in Table 2. While f denotes the objective function value, *td* denotes the total distance traveled, *t(s)* denotes the total run time in seconds, and *gap(%)* denotes the percentage gap between the lower and upper bound when the solver is complete in given maximum run time of 7200 s. The fixed cost per EV used c_{fix} is set to 100, 20 and 10 for the instances with 5, 10, and 15 customers, respectively. As the number of customers increases, we reduce the c_{fix} since we wish to reach optimal or near optimal solutions in a reasonable time to make a fair comparison between each pair of configurations. Such that, results of the pilot runs on 15 customers instances with c_{fix} equal to 50 reveal that no feasible solution is obtained in 3600 s, and it turns out that the average *gap(%)* is 31.42. Table 2 shows that the optimal solutions of all the instances with 5 and 10 customers are found. When customer number increases to 15, out of 12 instances, optimal solutions of 3, 4, and 4 instances are found for C1, C2, and C3 respectively. Out of 36 instances, results reveal that inclusion of half-open routes (i.e., from C1 to C2) leads to a cost reduction for 20 instances. Besides, inclusion of both half-open routes and rotations (i.e., from C1 to C3) results in cost reduction for 28 instances. On the other hand, in some instances, an improvement is observed although the distance is increased. For example, the total distance of C2 is higher than the one in C1 for RC204–15. However, the solution of C2 is better, which is because the number EV used in C2 is lower than the one in C1.

Table 2. Experimental Results for Small Sized Instances

Instance	Configuration-1			*gap(%)*	Configuration-2			*gap(%)*	Configuration-3			
	f	*td*	*t(s)*		*f*	*td*	*t(s)*		*f*	*td*	*t(s)*	*gap(%)*
C101-5	410.93	210.93	0.34	0	410.93	210.93	0.48	0	410.93	210.93	0.47	0
C103-5	358.96	158.96	0.36	0	358.96	158.96	0.36	0	358.96	158.96	0.33	0
C206-5	426.08	226.08	0.36	0	426.08	226.08	0.45	0	406.48	206.48	0.49	0
C208-5	382.45	182.45	0.8	0	382.45	182.45	0.83	0	382.45	182.45	1	0
R104-5	261.25	161.25	0.38	0	261.25	161.25	0.28	0	241.58	141.58	0.2	0
R105-5	256.08	156.08	0.38	0	240.70	140.70	0.34	0	222.95	122.95	0.47	0
R202-5	242.65	142.65	0.34	0	237.69	137.69	0.36	0	227.96	127.96	0.2	0
R203-5	295.63	195.63	0.23	0	279.68	179.68	0.22	0	273.05	173.05	0.2	0

(continued)

Table 2. (*continued*)

Instance	Configuration-1			*gap(%)*	Configuration-2			*gap(%)*	Configuration-3			
	f	*td*	*t(s)*		*f*	*td*	*t(s)*		*f*	*td*	*t(s)*	*gap(%)*
RC105-5	332.00	232.00	0.36	0	326.79	226.79	0.49	0	303.20	203.20	0.33	0
RC108-5	353.93	253.93	0.23	0	350.13	250.13	0.2	0	339.80	239.80	0.23	0
RC204-5	287.69	187.69	0.26	0	287.69	187.69	0.12	0	287.69	187.69	0.25	0
RC208-5	300.18	200.18	0.34	0	300.18	200.18	0.33	0	290.42	190.42	0.34	0
C101-10	417.19	337.19	2179	0	417.19	337.19	4389	0	406.60	326.60	1466	0
C104-10	376.42	316.42	785	0	376.42	316.42	1128	0	343.56	283.56	93	0
C202-10	299.46	239.46	175	0	299.46	239.46	213	0	297.14	237.14	178	0
C205-10	305.83	245.83	59	0	305.83	245.83	144	0	305.83	245.83	78	0
R102-10	301.84	281.84	134	0	290.54	270.54	36	0	289.45	269.45	34	0
R103-10	222.85	202.85	199	0	216.22	196.22	74	0	214.29	194.29	56	0
R201-10	261.36	241.36	145	0	244.91	224.91	20	0	244.91	224.91	15	0
R203-10	287.90	247.90	34	0	257.87	237.87	2	0	257.87	237.87	3	0
RC102-10	386.62	366.62	31	0	374.11	354.11	6	0	372.07	352.07	5	0
RC108-10	351.64	311.64	7	0	340.82	320.82	4	0	326.66	306.66	0.9	0
RC201-10	308.83	268.83	27	0	297.67	277.67	13	0	286.66	266.66	4	0
RC205-10	431.93	391.93	155	0	431.93	391.93	339	0	431.93	391.93	151	0
C103-15	356.53	306.53	7200	18	356.53	306.53	7200	15	356.53	306.53	7200	14
C106-15	344.97	294.97	7200	10	344.97	294.97	7200	10	337.21	287.21	7200	6
C202-15	429.15	389.15	7200	14	429.15	389.15	7200	17	429.15	389.15	7200	14
C208-15	549.13	499.13	7200	32	549.13	499.13	7200	31	510.68	460.68	7200	25
R102-15	313.31	293.31	7200	8	299.93	289.93	6624	0	299.93	289.93	5652	0
R105-15	297.76	287.76	2063	0	274.37	264.37	21	0	274.37	264.37	20	0
R202-15	347.98	327.98	687	0	342.09	322.09	391	0	342.09	322.09	290	0
R209-15	369.89	349.89	7200	8	369.44	349.44	7200	9	369.44	349.44	7200	9
RC103-15	411.29	391.29	7200	6	409.57	389.57	7200	5	409.57	389.57	7200	5
RC108-15	494.24	474.24	7200	13	463.47	443.47	7200	8	463.47	443.47	7200	8
RC202-15	380.36	360.36	3208	0	379.34	359.34	4567	0	379.34	359.34	3253	0
RC204-15	538.22	508.22	7200	25	529.80	509.80	7200	24	525.41	505.41	7200	23

Meanwhile, the comparisons between each pair of configurations based on the results can be seen in Table 3. In panel C1/C2, while *Δf(%)* denotes the percentage improvement of C2 over the C1 for the objective function value, *Δtd(%)* denotes the percentage improvement for the total distance traveled. The rest of the columns follow the same denotations. When C1 and C2 is comparatively investigated, one can observe that in 20 over 36 instances, C2 improved the solutions of C1. The maximum improvement with an 9.85 in traveled distance is a considerable one while the average improvements are 2.00 and 1.63 for the objective function and traveled distance, respectively. On the other hand, trade-off between routing cost and the fixed cost could also be observed. For instance, in RC108-10 while the *Δf(%)* equals 3.08, *Δtd(%)* equals -2.95. This is because while two EVs are employed in C1, only one EV is employed in C2. So, although

f improves, td increases. When the C2 and C3 is comparatively investigated, no trade-off is observed. However, there is an improvement in 19 over 36 instances. On the other hand, the panel C1/C3 shows the overall superiority of applying half-open routes and rotations simultaneously. While the solutions of 28 over 36 instances are improved up to 12.94 for $\Delta f(\%)$ and 21.23 for $\Delta td(\%)$, the average improvements are 3.94 for $\Delta f(\%)$ and 4.35 for $\Delta td(\%)$. As one can observe from the results that the trade-offs occur in instances with 10 and 15 customers due to c_{fix}. When the c_{fix} increases, the number of trade-offs decreases and eventually reaches 0.

As is stated before, if the duration of a single route is shorter comparing to the whole working hours, using vehicles multiple times can yield cost saving. Besides, if the related organization has several depots or distribution centers, allowing half-open routes and rotation can lead to extra cost savings. On the other hand, the magnitude of the trade-off may differ according to the ratio between routing and vehicle cost. As is seen in Table 3, no trade-off occurs in the instances with 5 customers, but in the ones with 10 and 15 customers. The level of trade-off may be at different levels depending on the related logistics company, the sector, the type of vehicle used, and both the load and battery capacity of the electric vehicle. It may also vary depending on different strategies used such as half-open routes & rotations, and the decision to receive recharging service at the depot. In other respects, this paper considers only the full recharging and one type of recharging option. However, allowing partial recharging as well as multiple recharging options with different speeds may further improve the solution in terms of distance traveled and the number of electric vehicles employed.

Table 3. Comparisons of Configurations for Small Sized Instances

Instance	C1/C2		C2/C3		C1/C3		Instance	C1/C2		C2/C3		C1/C3	
	Δf (%)	Δtd (%)	Δf (%)	Δtd (%)	Δf (%)	Δtd (%)		Δf (%)	Δtd (%)	Δf (%)	Δtd (%)	Δf (%)	Δtd (%)
C101-5	0.00	0.00	0.00	0.00	0.00	0.00	R201–10	6.29	6.82	0.00	0.00	6.29	6.82
C103-5	0.00	0.00	0.00	0.00	0.00	0.00	R203–10	10.43	4.05	0.00	0.00	10.43	4.05
C206-5	0.00	0.00	4.60	8.67	4.60	8.67	RC102–10	3.24	3.41	0.55	0.58	3.76	3.97
C208-5	0.00	0.00	0.00	0.00	0.00	0.00	RC108–10	3.08	-2.95	4.15	4.41	7.10	1.60
R104-5	0.00	0.00	7.53	12.20	7.53	12.20	RC201–10	3.61	-3.29	3.70	3.97	7.18	0.81
R105-5	6.01	9.85	7.37	12.62	12.94	21.23	RC205–10	0.00	0.00	0.00	0.00	0.00	0.00
R202-5	2.04	3.48	4.09	7.07	6.05	10.30	C103–15	0.00	0.00	0.00	0.00	0.00	0.00
R203-5	5.40	8.15	2.37	3.69	7.64	11.54	C106–15	0.00	0.00	2.25	2.63	2.25	2.63
RC105-5	1.57	2.25	7.22	10.40	8.67	12.41	C202–15	0.00	0.00	0.00	0.00	0.00	0.00
RC108-5	1.07	1.50	2.95	4.13	3.99	5.56	C208–15	0.00	0.00	7.00	7.70	7.00	7.70
RC204-5	0.00	0.00	0.00	0.00	0.00	0.00	R102–15	4.27	1.15	0.00	0.00	4.27	1.15

(*continued*)

Table 3. (*continued*)

Instance	C1/C2		C2/C3		C1/C3		Instance	C1/C2		C2/C3		C1/C3	
	Δf (%)	Δtd (%)	Δf (%)	Δtd (%)	Δf (%)	Δtd (%)		Δf (%)	Δtd (%)	Δf (%)	Δtd (%)	Δf (%)	Δtd (%)
RC208-5	0.00	0.00	3.25	4.88	3.25	4.88	R105–15	7.86	8.13	0.00	0.00	7.86	8.13
C101-10	0.00	0.00	2.54	3.14	2.54	3.14	R202–15	1.69	1.80	0.00	0.00	1.69	1.80
C104-10	0.00	0.00	8.73	10.38	8.73	10.38	R209–15	0.12	0.13	0.00	0.00	0.12	0.13
C202-10	0.00	0.00	0.77	0.97	0.77	0.97	RC103–15	0.42	0.44	0.00	0.00	0.42	0.44
C205-10	0.00	0.00	0.00	0.00	0.00	0.00	RC108–15	6.23	6.49	0.00	0.00	6.23	6.49
R102-10	3.74	4.01	0.38	0.40	4.10	4.40	RC202–15	0.27	0.28	0.00	0.00	0.27	0.28
R103–10	2.98	3.27	0.89	0.98	3.84	4.22	RC204–15	1.56	-0.31	0.83	0.86	2.38	0.55
							Average	2.00	1.63	1.98	2.77	3.94	4.35

5 Results and Discussion

In this paper, we investigated the effect of half-open routes and half-open rotations for the MDEVRP-HORR and introduced a novel 0–1 MILP model. For experimental studies, we used the small-sized instances up to 15 customers provided by Paz et al. [7]. These instances are the modification of the ones by Schneider et al. [5] based on well-known Solomon [35] instances. We first solved the instances with closed routes and rotations, second let routes be half-open and finally let them both be half-open. The results show that half-open routes and rotations lead to an improvement in terms of the total distance traveled, number of EV utilized and on 28 out of 36 instances total cost reduces. Further, in some instances, trade-off between the routing cost and fixed cost of using an EV is observed especially when the fixed cost is lower. This result may lead to interesting investigations for half-open routes and rotations in different sectors where the number of EVs utilized and the fixed cost vary. In this context, critical contributions of this paper can be listed as follows;

- The half-open rotation concept which allows EVs to be assigned multiple routes and flexibility to complete the tour at a different depot than the departure depot is adopted,
- A recharging station and replenishment depot elimination procedure is implemented in the MDEVRP which significantly reduces the number of variables employed,
- A procedure which helps avoid from using dummy recharging stations and depots and eventually obstruct network augmentation is applied.

Since the VRP belongs to class of hard combinatorial optimization problem and the MDEVRP-HORR is an extension of it, we carried out experimental studies on small sized instances. However, number of customers in small sized instances are lower as compared to real-life settings. Therefore, to solve our problem with medium and large sized instances, heuristic and metaheuristic algorithms including swarm intelligence (Particle swarm or ant colony optimization), evolutionary algorithms (genetic or memetic algorithm) or neighborhood structured algorithms (variable or adaptive large neighborhood search algorithm) must be developed. Besides, the trade-off between the routing cost and fixed EV cost per use could be examined according to the different logistics service areas. Besides, this paper adopted full recharging strategy. However, there are many other strategies in real life which are also often utilized in the literature such as partial recharging, battery swapping, or hybrid ones etc. So, integration of these different strategies to tackled problem is worth studying.

References

1. Global EV Outlook 2022, International Energy Agency (IEA). https://iea.blob.core.windows.net/assets/ad8fb04c-4f75-42fc-973a-6e54c8a4449a/GlobalElectricVehicleOutlook2022.pdf. Accessed 05 Oct 2023
2. Koç, Ç., Jabali, O., Mendoza, J.E., Laporte, G.: The electric vehicle routing problem with shared charging stations. Int. Trans. Oper. Res. **26**(4), 1211–1243 (2019)
3. Lopez, N.S., Allana, A., Biona, J.B.M.: Modeling electric vehicle charging demand with the effect of increasing EVSEs: a discrete event simulation-based model. Energies **14**(13), 3734 (2021)
4. Gönül, Ö., Duman, A.C.,Güler, Ö.: Electric vehicles and charging infrastructure in Turkey: an overview. Renew. Sustain. Energy Rev. **143** (2021). https://doi.org/10.1016/j.rser.2021.110913
5. Schneider, M., Stenger, A., Goeke, D.: The electric vehicle-routing problem with time windows and recharging stations. Transp. Sci. **48**(4), 500–520 (2014). https://doi.org/10.1287/trsc.2013.0490
6. Keskin, M., Çatay, B.: Partial recharge strategies for the electric vehicle routing problem with time windows. Transp. Res. Part C: Emerg. Technol. **65**, 111–127 (2016)
7. Paz, J., Granada-Echeverri, M., Escobar, J.: The multi-depot electric vehicle location routing problem with time windows. Int. J. Ind. Eng. Comput. **9**(1), 123–136 (2018)
8. Felipe, Á., Ortuño, M.T., Righini, G., Tirado, G.: A heuristic approach for the green vehicle routing problem with multiple technologies and partial recharges. Transp. Res. Part E: Logist. Transp. Rev. **71**, 111–128 (2014)
9. Keskin, M., Çatay, B.: A matheuristic method for the electric vehicle routing problem with time windows and fast chargers. Comput. Oper. Res. **100**, 172–188 (2018)
10. Karakatič, S.: Optimizing nonlinear charging times of electric vehicle routing with genetic algorithm. Expert Syst. Appl. **164**, 114039 (2021)
11. Montoya, A., Guéret, C., Mendoza, J.E., Villegas, J.G.: The electric vehicle routing problem with nonlinear charging function. Transp. Res. Part B: Methodol. **103**, 87–110 (2017)
12. Almouhanna, A., Quintero-Araujo, C.L., Panadero, J., Juan, A.A., Khosravi, B., Ouelhadj, D.: The location routing problem using electric vehicles with constrained distance. Comput. Oper. Res. **115**, 104864 (2020)
13. Schiffer, M., Walther, G.: The electric location routing problem with time windows and partial recharging. Eur. J. Oper. Res. **260**(3), 995–1013 (2017)
14. Lijun, F., Changshi, L., Zhang, W.: Half-open time-dependent multi-depot electric vehicle routing problem considering battery recharging and swapping. Int. J. Ind. Eng. Comput. **14**(1), 129–146 (2023)
15. Zhou, Z., Ha, M., Hu, H., Ma, H.: Half open multi-depot heterogeneous vehicle routing problem for hazardous materials transportation. Sustainability **13**(3), 1262 (2021)
16. Fan, L.: A two-stage hybrid ant colony algorithm for multi-depot half-open time-dependent electric vehicle routing problem. Complex Intell. Syst. 1–22 (2023)
17. Wang, Y., Zhou, J., Sun, Y., Wang, X., Zhe, J., Wang, H.: Electric vehicle charging Station location-routing problem with time windows and resource sharing. Sustainability **14**(18), 11681 (2022)

18. Wang, Y., Zhou, J., Sun, Y., Fan, J., Wang, Z., Wang, H.: Collaborative multidepot electric vehicle routing problem with time windows and shared charging stations. Expert Syst. Appl. **219**, 119654 (2023)
19. Ma, B.-S., Hu, D.-W., Sun, Q.: Research on route optimization of pure electric vehicle under joint distribution mode in city. In: 18th COTA International Conference of Transportation Professionals. 2018. American Society of Civil Engineers Reston, VA (2018)
20. Li, Y., Lim, M.K., Tan, Y., Lee, Y., Tseng, M.-L.: Sharing economy to improve routing for urban logistics distribution using electric vehicles. Resour. Conserv. Recycl. **153**, 104585 (2020)
21. Crevier, B., Cordeau, J.-F., Laporte, G.: The multi-depot vehicle routing problem with inter-depot routes. Eur. J. Oper. Res. **176**(2), 756–773 (2007)
22. Salhi, S., Petch, R.: A GA based heuristic for the vehicle routing problem with multiple trips. J. Math. Model. Algorithms **6**, 591–613 (2007)
23. Schneider, M., Stenger, A., Hof, J.: An adaptive VNS algorithm for vehicle routing problems with intermediate stops. OR Spectrum **37**, 353–387 (2015)
24. Schiffer, M., Schneider, M., Laporte, G.: Designing sustainable mid-haul logistics networks with intra-route multi-resource facilities. Eur. J. Oper. Res. **265**(2), 517–532 (2018)
25. Erdoğan, S., Miller-Hooks, E.: A green vehicle routing problem. Transp. Res. Part E: Logist. Transp. Rev. **48**(1), 100–114 (2012)
26. Li, J., Li, Y., Pardalos, P.M.: Multi-depot vehicle routing problem with time windows under shared depot resources. J. Comb. Optim. **31**, 515–532 (2016)
27. Muñoz-Villamizar, A., Montoya-Torres, J.R., Faulin, J.: Impact of the use of electric vehicles in collaborative urban transport networks: a case study. Transp. Res. Part D: Transp. Environ. **50**, 40–54 (2017)
28. Londoño, A., Gonzalez, W., Giraldo, O., Escobar, J.: A new matheheuristic approach based on Chu-Beasley genetic approach for the multi-depot electric vehicle routing problem. Int. J. Ind. Eng. Comput. **14**(3), 555–570 (2023)
29. Vahedi-Nouri, B., Arbabi, H., Jolai, F., Tavakkoli-Moghaddam, R., Bozorgi-Amiri, A.: Bi-objective collaborative electric vehicle routing problem: mathematical modeling and matheuristic approach. J. Ambient. Intell. Humaniz. Comput. **14**(8), 10277–10297 (2023)
30. Tarantilis, C.D., Zachariadis, E.E., Kiranoudis, C.T.: A hybrid guided local search for the vehicle-routing problem with intermediate replenishment facilities. Informs J. Comput. **20**(1), 154–168 (2008)
31. Muter, I., Cordeau, J.-F., Laporte, G.: A branch-and-price algorithm for the multidepot vehicle routing problem with interdepot routes. Transp. Sci. **48**(3), 425–441 (2014)
32. Che, A., Wang, W., Mu, X., Zhang, Y., Feng, J.: Tabu-based adaptive large neighborhood search for multi-depot petrol station replenishment with open inter-depot routes. IEEE Trans. Intell. Transp. Syst. **24**(1), 316–330 (2022)
33. Bruglieri, M., Mancini, S., Pezzella, F., Pisacane, O.: A new mathematical programming model for the green vehicle routing problem. Electron. Notes Discret. Math. **55**, 89–92 (2016)
34. Koç, Ç., Karaoglan, I.: The green vehicle routing problem: a heuristic based exact solution approach. Appl. Soft Comput. **39**, 154–164 (2016)
35. Solomon, M.M.: Algorithms for the vehicle routing and scheduling problems with time window constraints. Oper. Res. **35**(2), 254–265 (1987)

The Generalized One-to-One Pickup and Delivery Vehicle Routing Problem

Nurşah Yilmaz(✉) and İsmail Karaoğlan

Faculty of Engineering and Natural Sciences, Industrial Engineering Department, Konya Technical University, Konya, Turkey
{nyilmaz,ikaraoglan}@ktun.edu.tr

Abstract. Vehicle Routing Problem (VRP), one of the most important problems in the logistics industry, aims to find the most suitable route for vehicles that meet the demands of customers. VRP is examined under various variants, and one of these variants, the Pickup and Delivery Problem, deals with the pickup demands as well as the delivery demands of the customers. This problem is divided into three main categories according to demand type and route structure: one-to-many-to-one problems, many-to-many problems, and one-to-one problems. In one-to-many-to-one problems, there are goods at the depot to be delivered to customers as well as goods at customers to be carried back to the depot. In many-to-many problems, one node can be the origin or destination point for a good, and multiple nodes can be the origin or destination point for each good, while in one-to-one problems, there is a single origin and a single destination point for each unique demand. Within the scope of this study, a general case of one-to-one problems is introduced, in which multiple vehicles are allowed to leave the depot with goods and return to the depot with undeliverable goods, and a customer can be both a pickup and delivery node at the same time. A mathematical model is developed for the problem to obtain the optimal solution for the test instances. The model is tested on 45 randomly generated test instances. The results show that 40 out of 45 instances were solved using a commercial solver within one-hour computation time limit.

Keywords: Vehicle Routing Problem · One-to-one Pickup and Delivery · Logistic and Supply Chain Management

1 Introduction

Classic Vehicle Routing Problem (VRP) can be defined as obtaining the least-cost routes for vehicles with a particular capacity that leave a depot to meet the demands of customers and return to the depot at the end of the route. In its most basic form, VRP can be formally defined as follows: Let $G = (N, A)$ be a complete network where $N = \{0,1, \ldots, n\}$ denotes the set of all nodes. "0" represents the depot node, $N_c = \{1, \ldots, n\}$ represents the set of customers, and $A = \{(i, j)|i, j \in N, i \neq j\}$ represents the arcs between nodes. For each pair of nodes $(\forall i, j \in N)$, the distance between nodes is denoted by c_{ij}. The

Published by Springer Nature Switzerland AG 2024
M. Thürer et al. (Eds.): APMS 2024, IFIP AICT 733, pp. 291–306, 2024.
https://doi.org/10.1007/978-3-031-71645-4_20

demands for each customer ($\forall i \in N_c$) are expressed by d_i. There is an unlimited number of homogeneous vehicles located at the depot with a particular capacity (Q). In VRP, the goal is to find the least-cost routes that satisfy the following conditions:

- Each customer should be visited only once.
- Each route must start and end at the depot.
- The amount of load on the vehicle along the route should not exceed the capacity of the vehicle.

In this study, an important variant of VRP, known as the Pickup and Delivery Vehicle Routing Problem (PDVRP), will be discussed. In PDVRP, unlike classical VRP, good collection operations can be carried out from customers as well as transportation of goods to customers.

In the literature on PDVRP studies that deal with good transportation from depot to customers or from customers to the depot, and between customers. One-to-many-to-one (1-M-1) problems are examined for good transportation including depot, while many-to-many (M-M) problems and one-to-one (1-1) problems are examined for good transportation from customers to customers [1].

In one-to-many-to-one problems, some goods are dispatched from the depot to the customer, some goods are collected from the customer to transport the depot, and good transportation from customer to customer is not allowed. In many-to-many problems, customers' demands are common, and a point can be a delivery and/or pickup point for one or more goods. In one-to-one problems, there are many unique demands, and a demand is defined as the collection of a specific good from a customer and its delivery to another customer. Therefore, in this problem, there is a certain origin and destination point for each good.

Within the scope of this study, a real-world issue that will be evaluated under the topic of one-to-one problems will be discussed, in which multiple vehicles are allowed to leave the depot with goods and return to the depot with undeliverable goods, and a customer can be both a pickup and delivery node at the same time. The main goal of the study is to obtain a solution by identifying the problem that has an equivalent in practical life and has not been addressed in the literature so far.

The subsequent sections of the study are organized as follows: In the second section, literature research for one-to-one PDVRP (O-PDVRP) is given. In the third section, the problem addressed is explained in detail, and an example of the problem is presented. The fourth section includes experimental studies. The last section gives the conclusion and future directions.

2 Literature Review

The first research on the VRP in the literature was introduced by Dantzig and Ramser [2]. In this study, the problem of minimizing the distance traveled by a fleet of gasoline distribution trucks and the demands of gas stations is addressed. On the other hand, Clarke and Wright [3] have developed a heuristic method called the "savings algorithm" for the problem examined by Dantzig and Ramser [2] and emphasized that better results were obtained with this method. For interested readers, the studies in [4–6] are recommended as a comprehensive source for the VRP. Regarding PDVRP, which is the main topic of the problem addressed in this study, literature reviews can be examined in [7–10].

The literature on one-to-one problems is divided into two categories: single-vehicle and multiple-vehicle, depending on the number of vehicles in the system. For the Stacker Crane Problem, which is a special case of one-to-one problems with single-vehicle, Frederickson et al. [11] have developed a heuristic method. In another study published in 1988, an algorithm was discussed to reposition objects between stations via a robot arm and to minimize the total distance traveled by the robot arm [12]. The Pickup and Delivery Traveling Salesman Problem, which is a more general form of this problem, was first examined in the context of passenger transportation in 1978, and the heuristic method was used for the solution [13, 14].

Many exact and heuristic methods have been put forward to solve the studies that take into account the multiple-vehicle situation of one-to-one problems and are generally called the Pickup and Delivery Vehicle Routing Problem [1]. Lu and Dessouky [36] have developed a branch-cut algorithm that reached the optimal solution for test problems with up to 5 vehicles and 25 customers. In a study conducted by Korsvik et al. [37], a heuristic method was used to minimize the cost for a company that carries out maritime transportation by ships. More detailed information can be found about one-to-one pickup and delivery problems with multiple-vehicles as given in Table 1 [35].

In all the one-to-one problems in the literature, the customers are either collection or distribution customers, and the multiple vehicles leave and return the depot empty. Since these extensions are considered for the first time in this study, none of the existing solution methods can solve the problem. Therefore, a mathematical model is first developed to solve the problem.

3 Problem Description and Formulation

Within the scope of this study, the problem of transporting pickup and/or delivery demands between facilities located in different buildings within the production campus of a company operating in the defense industry sector will be introduced. In the problem under consideration, there is a depot that is used for locating vehicles and short-term storage of goods, and a number of facilities in which some semi-finished products are produced and transported to another facility for another operation or assembly. Vehicles departing from the depot are responsible for collecting the semi-finished products from facilities and delivering them to another facility for the next operation, and returning to the depot at the end of the pre-defined tour duration (*i.e.* three hours). In addition, goods that cannot be delivered during this tour duration are brought to the depot and temporarily stocked to be distributed in the next round.

Table 1. Studies related to one-to-one problems for the last 15 years.

Manuscript	Model	Objective Function	Vehicle Type
Cordeau et al. [15]	ILP, BCA, Heuristics	Cost Minimization	Homogeneous
Dumitrescu et al. [16]	ILP, BCA	Cost Minimization	Homogeneous
Öncan et al. [17]	MILP, BCA	Distance Minimization	Homogeneous
Demir and Van Woensel [18]	MILP	Fuel and Driver Cost Minimization	Heterogeneous
Mesa-Arango and Ukkusuri [19]	MILP, BBA	Cost Minimization	Homogeneous
Şahin et al. [20]	Heuristics, MILP	Distance Minimization	Homogeneous
Treleaven et al. [21]	Polynomial Time Algorithms	Waiting Time Minimization	Homogeneous
Ayadi et al. [22]	BBA	Travel Time Minimization	Homogeneous
Guerriero et al. [23]	Heuristics	Waiting Time Minimization	Homogeneous
Iori and Riera-Ledesma [24]	ILP, Heuristics	Routing Cost Minimization	Heterogeneous
Montero et al. [25]	ILP, Heuristics	Cost Minimization	Homogeneous
Haddad et al. [26]	Heuristics	Distance Minimization	Homogeneous
Soysal et al. [27]	MILP	Delivery Cost Minimization	Heterogeneous
Factorovich et al. [28]	MILP, BCA	Cost Minimization	Homogeneous
Castro et al. [29]	ILP, Lagrangian Approach	Distance Minimization	Homogeneous
Qi et al. [30]	MILP, Heuristics	Profit Maximization	Homogeneous
Xiong et al. [31]	ILP, Meta-heuristic	Utility Maximization	Homogeneous
Drexl [32]	Heuristics	Cost Minimization	Heterogeneous
Graf [33]	Heuristics	Distance Minimization	Homogeneous
Liu et al. [34]	GA	Cost Minimization	Heterogeneous
Taşdan et al. [35]	MILP	Cost Minimization	Heterogeneous

MILP: Mixed Integer Linear Programming, ILP: Integer Linear Programming, BCA: Branch and Cut Algorithms, BBA: Branch and Bound Algorithms, GA: Genetic Algorithms

In most of the studies published in the literature on one-to-one problems, a vehicle leaves and returns to the depot empty, a customer can only be a pickup or delivery node for only one good at the same time, and the single-vehicle case of the problem is addressed. However, in our study, vehicles are allowed to leave the depot with goods and return to the depot with undeliverable goods. In addition, the multiple-vehicle case and a customer can be both a pickup and delivery node at the same time is examined. These assumptions make the problem more relevant for real life scenarios, such as in the transportation of parts produced in different facilities within the same production campus or in the inter-machine semi-product transportation processes within a production facility. We call this problem as the Generalized One-to-One Pickup and Delivery Vehicle Routing Problem (GO-PDVRP).

Figure 1 provides an example for the problem, including 5 customers (in order to use the common terminology in the literature of VRP, from now on, the nodes where the products will be picked up and delivered will be called customers instead of facilities) and a depot. Customer requests are given on nodes, where requests $+(-)$ indicate the collection (destination) point for the goods and the capacity of the vehicles is accepted as 10 units. Distances and travel times between nodes are considered directly proportional to each other and are defined on the lines in kilometers and minutes, and given on edges, respectively. Total route periods are restricted to not exceeding the specified tour duration of 180 min, and service times at customers are ignored.

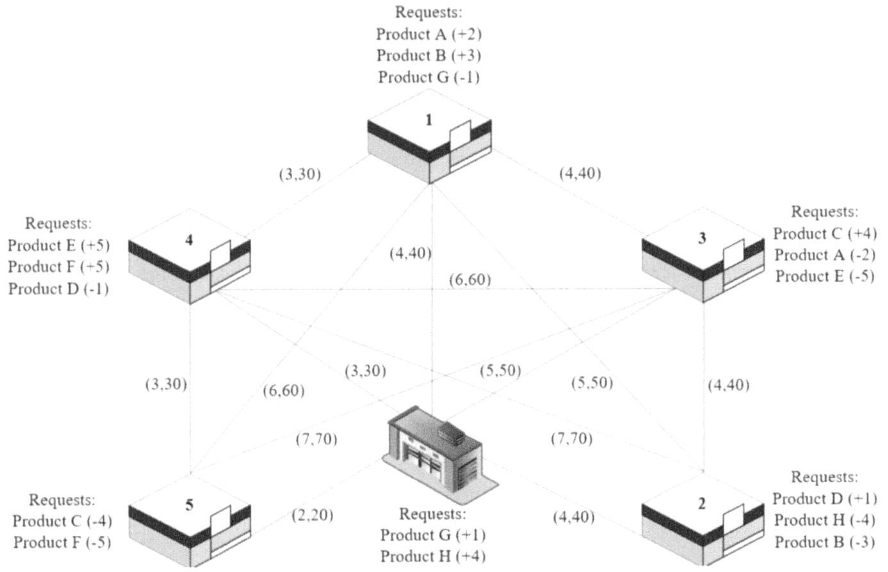

Fig. 1. An example for GO-PDVRP.

Figure 2 presents a feasible solution for this example. As can be seen in this figure, two vehicles leave the depot and follow two different routes: $< spanclass =' convertEndash' > 0 - 1 < /span > - < spanclass =' convertEndash' > 2 - 0 < /span >$ and $< spanclass =' convertEndash' > 0 - 4 < /span > - < spanclass =' convertEndash' > 3 - 5 < /span > -0$.

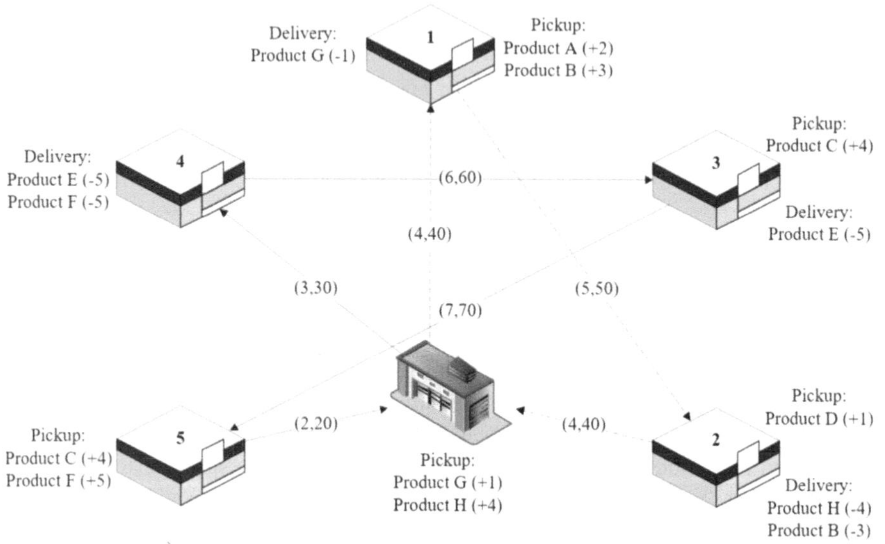

Fig. 2. Optimal solution for the example.

Table 2 gives the goods carried and the total amount of load on the vehicle while traveling from one node to another along the routes. In this table, (+) indicates that the good is collected from the node, (−) indicates that the good is delivered to the node, and other values indicate that the good continues to be carried on the vehicle at that node.

Table 2. The products carried in the vehicle and the total amount of load on the vehicle while traveling from one node to another node along the routes.

Product	ROUTE-1				ROUTE-2				
	0	1	2	0	0	4	3	5	0
A	-	+2	2	2	-	-	-	-	-
B	-	+3	−3	-	-	-	-	-	-
C	-	-	-	-	-	-	+4	−4	-
D	-	-	+1	1	-	-	-	-	-
E	-	-	-	-	-	+5	−5	-	-
F	-	-	-	-	-	+5	5	−5	-
G	+1	−1	-	-	-	-	-	-	-
H	+4	4	−4	-	-	-	-	-	-
Total Load on Vehicle	**5**	**9**	**3**	**3**	**0**	**10**	**9**	**0**	**0**

As can be seen from the table, goods A, B, D, G and H are carried on the vehicle along Route-1, while goods C, E and F are carried in the vehicle along Route-2. In addition, goods G and H (collected and cannot delivered in the previous tour and temporarily stocked at the depot) leave the depot on the first vehicle and are dropped off at the delivery points, while goods A and D return to the depot on the first vehicle before being delivered at the delivery points.

In the light of these explanations, GO-PDVRP can be defined formally as follows: Let $G = (N, A)$ be a complete network in which $N = \{0, 1, \ldots, n\}$ denotes the set of all nodes, "0" represents the depot node, $N_c = \{1, \ldots, n\}$ represents the set of customers, and $A = \{(i, j)|i, j \in N, i \neq j\}$ represents the arcs between nodes. For each pair of nodes ($\forall i, j \in N$), the distance and travel time between the nodes are denoted by c_{ij} and d_{ij}, respectively. A request $H = \{1, 2, \ldots, k + m\}$ is the collection of a specific good from a customer and its delivery to a specific customer, and each customer must appear at least one request. Among these requests, $H_0 = \{1, 2, \ldots, k\}$ is the requests that were collected from customers in the previous period and brought to the depot before being distributed (requests in progress), $H_c = \{k+1, k+2, \ldots, k+m\}$ represents requests that have never been processed. Each request $h \in H$ is defined as two different nodes $(i, j \in N, i \neq j)$ and the amount of goods to be transported between these nodes $\left(d_{ih}\ and\ d_{jh}\right)$. One of these nodes represents the origin node (i), the other represents the destination node (j), where the origin node has a positive demand ($d_{ih} > 0$), and the destination node has a negative demand $\left(d_{jh} < 0\right)$. A positive (negative) demand quantity indicates that pickup (delivery) will occur for a specific good at the node. The origin node of requests that were collected but could not be delivered in the previous period is the depot node. There is an unlimited number of homogeneous vehicles at the depot with a certain capacity (Q). Each route will be carried out by a vehicle, and the vehicle must return to the depot within a pre-defined tour duration (T).

There are two objectives in GO-PDVRP, the first of which is to minimize the amount of goods that are collected from customers but cannot be delivered (brought to the depot), and the second is to minimize the total route cost. GO-PDVRP aims to identify routes that meet the following conditions and optimize these objectives.

- Each customer should be visited only once.
- Each route must start and end at the depot.
- The amount of load on the vehicle along the route should not exceed the capacity.
- Each request should be picked up and delivered with the same vehicle.
- The origin node of each request must be visited before the destination node.
- The total time along the route should not exceed the allowed tour duration.

In the mixed integer linear programming (MIP) formulation to obtain the optimal solution of the GO-PDVRP, decision variables are defined as follows:

Decision Variables:

$$x_{ij}: \begin{cases} 1\ \textit{if the vehicle travels from node } i \textit{ to } j\ (\forall\, i, j \in N) \\ 0\ \textit{otherwise} \end{cases}$$

q_{ijh}: *The amount of product h on the vehicle while traveling from node i to j*

t_i: *service start time at node i*

By using these definitions and decision variables, GO-PDVRP can be formulated as follows:

MIP Formulation:

$$MinZ_1 = \sum_{i \in N_c} \sum_{h \in H} q_{i0h} \tag{1a}$$

$$MinZ_2 = \sum_{i \in N} \sum_{j \in N} c_{ij} x_{ij} \tag{1b}$$

$$\sum_{i \in N} x_{ij} = 1 \quad \forall j \in N_c \tag{2}$$

$$\sum_{j \in N} x_{ij} = 1 \quad \forall i \in N_c \tag{3}$$

$$\sum_{j \in N} q_{jih} + d_{ih} = \sum_{j \in N} q_{ijh} \quad \forall i \in N_c, \forall h \in H_0 \tag{4}$$

$$\sum_{j \in N} q_{jih} + d_{ih} \leq \sum_{j \in N} q_{ijh} \quad \forall i \in N_c, \forall h \in H_c \tag{5}$$

$$\sum_{h \in H} q_{ijh} \leq Q_{x_{ij}} \quad \forall i \in N, \forall j \in N, i \neq j \tag{6}$$

$$t_i + d_{i0} x_{i0} \leq T \quad \forall i \in N_c \tag{7}$$

$$t_i - t_j \leq (1 - x_{ij})M - d_{ij} \quad \forall i \in N, \forall j \in N_c, i \neq j \tag{8}$$

$$t_j - t_i \leq (1 - x_{ij})M + d_{ij} \quad \forall i \in N_c, \forall j \in N_c, i \neq j \tag{9}$$

$$t_i - d_{0i} \leq M(1 - x_{0i}) \quad \forall i \in N_c \tag{10}$$

$$x_{ij} \in \{0,1\} \quad \forall i \in N, \forall j \in N \tag{11}$$

$$q_{ijh} \geq 0 \quad \forall i \in N, \forall j \in N, i \neq j, h \in H \tag{12}$$

$$t_i \geq 0 \quad \forall i \in N \tag{13}$$

In this formulation, while the first objective function in (1a) minimizes the amount of goods returning to the depot, (1b) minimizes the total cost. Constraint sets (2) and (3) ensure that each customer must be visited exactly once. Constraint sets (4) and (5) determine the transportation quantities for the goods at the depot and customer, separately. Constraint set (6) prevents the transported goods from exceeding the vehicle capacity. It is guaranteed that the total route duration does not exceed the tour duration with (7), and it provides that there are no sub-tours with (8)–(10) where M is a sufficiently large number. Finally, constraint sets (11)–(13) are integrality constraints for decision variables. In this model, a priority is defined among the objective functions, and they are combined into a single objective function as given in (14).

$$MinZ = p_1 \sum_{i \in N_c} \sum_{h \in H} q_{i0h} + p_2 \sum_{i \in N} \sum_{j \in N} c_{ij} x_{ij} \tag{14}$$

In this constraint, p_1 and p_2 are the coefficients satisfying $p_1 \gg p_2$.

4 Computational Results

A number of test instances have been generated randomly to test the performance of the model. The instances consist of 10, 12, and 14 customers. The customer locations are randomly generated within the range of [−100,100], the depot is located in the center of the area, Euclidean distances are used as the distance (c_{ij}) and travel time (d_{ij}), and the tour duration is set to 300 seconds.

Three levels of pickup and delivery density are considered (*i.e.* 1, 2 and 3), where density is the number of occurrences of a customer in requests. While a density level of 1 means that each customer appears in exactly one request (pickup or delivery node), a density level greater than 1 means that each customer appears in exactly as many requests as the density. To ensure that all customers appear in a request, requests were generated by randomly sorting customers in a list and matching the first half of the list with the second half, and this process was repeated up to the density level. The quantity of each request is randomly generated between $[1, 5]$. In order to represent the requests that could not be delivered in the previous tour and were brought to the depot, a random number of requests equal to 20% of the number of customers was generated (in this computational environment, this equals two requests). For these requests, the depot node was set as the origin, and a randomly selected node among the customers was set as the destination node. Note that when the density level is 1 and there is no request originating from the depot, the problem converts to the classical one-to-one VRPDP. A total of 45 test instances were generated, including five random test instances for each customer and density level.

To evaluate the performance of the proposed model, randomly generated samples are solved with CPLEX 22.1 where the maximum solution time is set as 3600 s.

In this regard, instance "n10d3s1" is presented as an illustrative example below. In the example, the number of customers, density level, and the vehicle capacity are given as 10, 3, and 15 respectively.

While Table 3 contains the coordinates of the nodes, Table 4 includes pickup-delivery nodes of the requests and quantity of the demands.

Table 3. Coordinates of the nodes.

Node	X Coordinate	Y Coordinate
0	0	0
1	−74	−31
2	67	−83
3	−76	−52
4	1	−24
5	60	19

(continued)

Table 3. (*continued*)

Node	X Coordinate	Y Coordinate
6	35	33
7	−7	96
8	1	−73
9	−63	82
10	3	51

Table 4. Pickup-delivery nodes of the requests and quantity of the demands.

Request No	Pickup Node	Delivery Node	Demand
1	0	10	4
2	0	6	4
3	7	9	4
4	5	6	4
5	1	4	3
6	3	8	3
7	10	2	5
8	10	3	2
9	1	6	1
10	2	9	4
11	8	5	1
12	4	7	1
13	3	7	3
14	1	10	5
15	2	9	4
16	6	5	5
17	4	8	4

The optimal solution for the illustrative example is shown in Fig. 3. As can be seen, 4 vehicles leave the depot. Products that are indicated as underlined are the ones that cannot be delivered and are therefore returned to the depot.

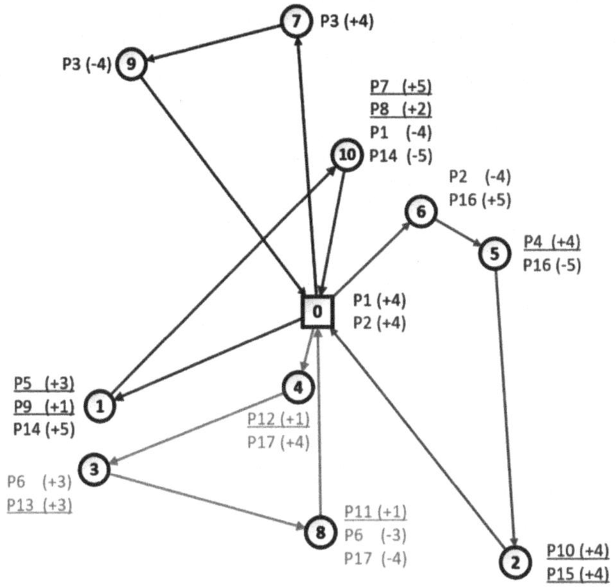

Fig. 3. Optimal solution for the illustrative example.

Considering the optimal solution, Table 5 gives the goods carried and the total amount of load on the vehicle while traveling from one node to another along the routes.

Table 5. The products carried in the vehicle and the total amount of load on the vehicle while traveling from one node to another node along the routes for illustrative example.

Product	ROUTE-1				ROUTE-2					ROUTE -3					ROUTE-4			
	0	1	10	0	0	4	3	8	0	0	6	5	2	0	0	7	9	0
1	+4	4	−4	-	-	-	-	-	-	-	-	-	-	-	-	-	-	-
2	-	-	-	-	-	-	-	-	-	+4	−4	-	-	-	-	-	-	-
3	-	-	-	-	-	-	-	-	-	-	-	-	-	-	-	+4	−4	-
4	-	-	-	-	-	-	-	-	-	-	-	+4	4	4	-	-	-	-
5	-	+3	3	3	-	-	-	-	-	-	-	-	-	-	-	-	-	-
6	-	-	-	-	-	-	+3	−3	-	-	-	-	-	-	-	-	-	-
7	-	-	+5	5	-	-	-	-	-	-	-	-	-	-	-	-	-	-
8	-	-	+2	2	-	-	-	-	-	-	-	-	-	-	-	-	-	-
9	-	+1	1	1	-	-	-	-	-	-	-	-	-	-	-	-	-	-
10	-	-	-	-	-	-	-	-	-	-	-	-	+4	4	-	-	-	-
11	-	-	-	-	-	-	-	+1	1	-	-	-	-	-	-	-	-	-
12	-	-	-	-	-	+1	1	1	1	-	-	-	-	-	-	-	-	-
13	-	-	-	-	-	-	+3	3	3	-	-	-	-	-	-	-	-	-

(continued)

The results of these 45 instances are carried out under the assumptions and constraints specified for GO-PDVRP are presented in Table 6. The first column in the table (Pname)

Table 5. *(continued)*

Product	ROUTE-1				ROUTE-2					ROUTE -3					ROUTE-4			
	0	1	10	0	0	4	3	8	0	0	6	5	2	0	0	7	9	0
14	-	+5	−5	-	-	-	-	-	-	-	-	-	-	-	-	-	-	-
15	-	-	-	-	-	-	-	-	-	-	-	-	+4	4	-	-	-	-
16	-	-	-	-	-	-	-	-	-	-	+5	−5	-	-	-	-	-	-
17	-	-	-	-	-	+4	4	-4	-	-	-	-	-	-	-	-	-	-
Total Load on Vehicle	**4**	**13**	**11**	**11**	**0**	**5**	**11**	**5**	**5**	**4**	**5**	**4**	**12**	**12**	**0**	**4**	**0**	**0**

includes the problem name, where *n* denotes the number of customers, *d* denotes the density level, and *s* denotes the sample number. The density level of requests (DL) is shown in the second column. The total number of customer requests (#Req) is shown in the third column, while the number of vehicles used (#Veh) is in the fourth column. The objective function values are shown in the fifth and sixth columns. In addition, the "CPU" column shows the total solution time for each instance. Finally, the Gap values calculated by CPLEX are provided in the last column (in percentage).

When the results are examined, it is seen that optimal solutions could be obtained for 40 out of 45 problems. Moreover, optimal solutions are reached in less than 3 min for all problems with 10 customers. In problems with 12 customers, optimal solutions are found in 14 out of 15 problems, and a feasible solution is obtained for "n12d2s2"

Table 6. Experimental results for GO-PDVRP.

Pname	DL	#Req	#Veh	$Z_1(1a)$	$Z_2(1b)$	CPU	Gap
n10d1s1	1	7	3	5	806	1.25	0.00
n10d1s2	1	7	3	7	754	0.80	0.00
n10d1s3	1	7	4	2	873	9.08	0.00
n10d1s4	1	7	3	6	698	3.00	0.00
n10d1s5	1	7	3	7	822	15.06	0.00
n10d2s1	2	12	3	14	806	4.62	0.00
n10d2s2	2	12	4	8	1048	154.39	0.00
n10d2s3	2	12	3	8	772	117.14	0.00
n10d2s4	2	12	4	10	1052	5.06	0.00
n10d2s5	2	12	3	11	716	2.22	0.00
n10d3s1	3	17	4	28	1054	6.38	0.00
n10d3s2	3	17	3	14	829	233.00	0.00
n10d3s3	3	17	4	17	1037	4.70	0.00
n10d3s4	3	17	4	18	909	8.42	0.00

(continued)

Table 6. *(continued)*

Pname	DL	#Req	#Veh	$Z_1(1a)$	$Z_2(1b)$	CPU	Gap
n10d3s5	3	17	5	34	1274	1.80	0.00
n12d1s1	1	8	4	6	1058	3.98	0.00
n12d1s2	1	8	5	0	1175	9.05	0.00
n12d1s3	1	8	6	11	1480	1.52	0.00
n12d1s4	1	8	5	1	1259	9.34	0.00
n12d1s5	1	8	4	5	1000	33.84	0.00
n12d2s1	2	14	4	25	1014	9.89	0.00
n12d2s2	2	14	4	15	927	3600.00	27.31
n12d2s3	2	14	5	22	1199	8.34	0.00
n12d2s4	2	14	4	11	1053	19.88	0.00
n12d2s5	2	14	6	16	1542	2.62	0.00
n12d3s1	3	20	5	39	1289	6.25	0.00
n12d3s2	3	20	4	28	1036	932.00	0.00
n12d3s3	3	20	5	39	1109	17.33	0.00
n12d3s4	3	20	5	32	1210	17.70	0.00
n12d3s5	3	20	6	32	1456	16.33	0.00
n14d1s1	1	9	5	9	1398	2973.00	0.00
n14d1s2	1	9	5	6	1245	18.22	0.00
n14d1s3	1	9	4	12	1012	10.80	0.00
n14d1s4	1	9	4	15	1072	3600.00	21.72
n14d1s5	1	9	4	9	1107	162.00	0.00
n14d2s1	2	16	4	20	1069	3600.00	51.68
n14d2s2	2	16	6	18	1552	8.25	0.00
n14d2s3	2	16	5	33	1279	10.03	0.00
n14d2s4	2	16	4	10	1045	3600.00	30.07
n14d2s5	2	16	5	26	1320	99.95	0.00
n14d3s1	3	23	5	36	1179	3600.00	20.15
n14d3s2	3	23	7	35	1602	51.17	0.00
n14d3s3	3	23	5	37	1371	175.69	0.00
n14d3s4	3	23	6	27	1441	321.69	0.00
n14d3s5	3	23	5	44	1269	85.58	0.00

within the given maximum solution time. For the problems with 14 customers, optimal solutions are observed in 11 out of 15 problems. Results indicate that as the problem size increases, it becomes more difficult to obtain an optimal solution.

5 Conclusion

This paper discusses a real-life problem that will be considered under the heading of one-to-one problems. In this direction, a mixed integer linear programming model was developed, and experimental studies were carried out on the randomly generated test instances. PDVRP, as the classical VRP, is an NP-Hard problem, optimal results have been achieved only in small-sized problems with the mathematical model.

In future studies, it is aimed to increase the performance of the model by strengthening constraint and adding valid inequalities to the mathematical model to be used for GO-PDVRP problems. It is aimed to develop an effective heuristic/meta-heuristic method that provides fast and optimal or near-optimal results for large-sized instances.

References

1. Toth, P., Vigo, D.: Vehicle routing: problems, methods, and applications. SIAM (2014)
2. Dantzig, G.B., Ramser, J.H.: The truck dispatching problem. Manage. Sci. **6**(1), 80–91 (1959)
3. Clarke, G., Wright, J.W.: Scheduling of vehicles from a central depot to a number of delivery points. Oper. Res. **12**(4), 568–581 (1964)
4. Toth, P., Vigo, D.: An overview of vehicle routing problems. The vehicle routing problem, pp. 1–26 (2002)
5. Pillac, V., Gendreau, M., Guéret, C., Medaglia, A.L.: A review of dynamic vehicle routing problems. Eur. J. Oper. Res. **225**(1), 1–11 (2013)
6. Braekers, K., Ramaekers, K., Van Nieuwenhuyse, I.: The vehicle routing problem: state of the art classification and review. Comput. Ind. Eng. **99**, 300–313 (2016)
7. Berbeglia, G., Cordeau, J.-F., Gribkovskaia, I., Laporte, G.: Static pickup and delivery problems: a classification scheme and survey. TOP **15**, 1–31 (2007)
8. Parragh, S.N., Doerner, K.F., Hartl, R.F.: A survey on pickup and delivery problems: Part I: transportation between customers and depot. Journal für Betriebswirtschaft **58**, 21–51 (2008)
9. Parragh, S.N., Doerner, K.F., Hartl, R.F.: A survey on pickup and delivery models Part II: transportation between pickup and delivery locations (2008)
10. Cai, J., Zhu, Q., Lin, Q., Ma, L., Li, J.,Ming, Z.: A survey of dynamic pickup and delivery problems. Neurocomputing 126631 (2023)
11. Frederickson, G.N., Hecht, M.S.,Kim, C.E.: Approximation algorithms for some routing problems. In: 17th Annual Symposium on Foundations of Computer Science (SFCS 1976). IEEE (1976)
12. Atallah, M.J., Kosaraju, S.R.: Efficient solutions to some transportation problems with applications to minimizing robot arm travel. SIAM J. Comput. **17**(5), 849–869 (1988)
13. Stein, D.M.: An asymptotic, probabilistic analysis of a routing problem. Math. Oper. Res. **3**(2), 89–101 (1978)
14. Stein, D.M.: Scheduling dial-a-ride transportation systems. Transp. Sci. **12**(3), 232–249 (1978)
15. Cordeau, J.-F., Laporte, G., Ropke, S.: Recent models and algorithms for one-to-one pickup and delivery problems. In: Golden, B., Raghavan, S., Wasil, E. (eds.) The Vehicle Routing Problem: Latest Advances and New Challenges, pp. 327–357. Springer, Cham (2008). https://doi.org/10.1007/978-0-387-77778-8_15
16. Dumitrescu, I., Ropke, S., Cordeau, J.-F., Laporte, G.: The traveling salesman problem with pickup and delivery: polyhedral results and a branch-and-cut algorithm. Math. Program. **121**, 269–305 (2010)

17. Öncan, T., Aksu, D.T., Şahin, G., Şahin, M.: A branch and cut algorithm for the multi-vehicle one-to-one pickup and delivery problem with split loads. In: 2011 IEEE International Conference on Industrial Engineering and Engineering Management. IEEE (2011)
18. Demir, E., Van Woensel, T.: Mathematical modeling of CO 2 e emissions in one-to-one pickup and delivery problems. In: 2013 IEEE International Conference on Industrial Engineering and Engineering Management. IEEE (2013)
19. Mesa-Arango, R., Ukkusuri, S.V.: Benefits of in-vehicle consolidation in less than truckload freight transportation operations. Procedia Soc. Behav. Sci. **80**, 576–590 (2013)
20. Şahin, M., Çavuşlar, G., Öncan, T., Şahin, G., Aksu, D.T.: An efficient heuristic for the multi-vehicle one-to-one pickup and delivery problem with split loads. Transp. Res. Part C Emerg. Technol. **27**, 169–188 (2013)
21. Treleaven, K., Pavone, M., Frazzoli, E.: Asymptotically optimal algorithms for one-to-one pickup and delivery problems with applications to transportation systems. IEEE Trans. Autom. Control **58**(9), 2261–2276 (2013)
22. Ayadi, M., Chabchoub, H., Yassine, A.: An exact method for the multi-vehicle static demand responsive transport problem based on service quality: the case of one-to-one. In: 2014 International Conference on Advanced Logistics and Transport (ICALT). IEEE (2014)
23. Guerriero, F., Pezzella, F., Pisacane, O., Trollini, L.: Multi-objective optimization in dial-a-ride public transportation. Transp. Res. Procedia **3**, 299–308 (2014)
24. Iori, M., Riera-Ledesma, J.: Exact algorithms for the double vehicle routing problem with multiple stacks. Comput. Oper. Res. **63**, 83–101 (2015)
25. Montero, A., José Miranda-Bront, J., Méndez-Díaz, I.: An ILP-based local search procedure for the VRP with pickups and deliveries. Ann. Oper. Res. **259**, 327–350 (2017)
26. Haddad, M.N., et al.: Large neighborhood-based metaheuristic and branch-and-price for the pickup and delivery problem with split loads. Eur. J. Oper. Res. **270**(3), 1014–1027 (2018)
27. Soysal, M., Çimen, M., Demir, E.: On the mathematical modeling of green one-to-one pickup and delivery problem with road segmentation. J. Clean. Prod. **174**, 1664–1678 (2018)
28. Factorovich, P., Méndez-Díaz, I., Zabala, P.: Pickup and delivery problem with incompatibility constraints. Comput. Oper. Res. **113**, 104805 (2020)
29. Castro, M.P., Cire, A.A., Beck, J.C.: An MDD-based Lagrangian approach to the multicommodity pickup-and-delivery TSP. INFORMS J. Comput. **32**(2), 263–278 (2020)
30. Qi, X., Fu, Z., Xiong, J., Zha, W.: Multi-start heuristic approaches for one-to-one pickup and delivery problems with shortest-path transport along real-life paths. PLoS ONE **15**(2), e0227702 (2020)
31. Xiong, J., Qi, X., Fu, Z., Zha, W.: Split demand one-to-one pickup and delivery problems with the shortest-path transport along real-life paths. IEEE Access **8**, 150539–150554 (2020)
32. Drexl, M.: On the one-to-one pickup-and-delivery problem with time windows and trailers. CEJOR **29**, 1115–1162 (2021)
33. Graf, B.: Preemptive stacker crane problem: extending tree-based properties and construction heuristics. Eur. J. Oper. Res. **292**(2), 532–547 (2021)
34. Liu, M., Zhang, C., Wu, Q., Meng, B.: Vehicle routing problem with soft time windows of cargo transport O2O platforms. Int. J. Simul. Model. **20**(2), 351–362 (2021)
35. Taşdan, M.E., Menten, C., Soysal, M., Çimen, M.: Heterojen Araçlar ile Sürdürülebilir Birebir Toplama ve Dağıtım Problemi Üzerine Bir Analiz. Verimlilik Dergisi **57**(1), 23–44 (2023)

36. Lu, Q., Dessouky, M.: An exact algorithm for the multiple vehicle pickup and delivery problem. Transp. Sci. **38**(4), 503–514 (2004)
37. Korsvik, J.E., Fagerholt, K., Laporte, G.: A large neighbourhood search heuristic for ship routing and scheduling with split loads. Comput. Oper. Res.ut. Oper. Res. **38**(2), 474–483 (2011)

An AUGMECON2VIKOR Algorithm for a Multi-objective Model in a Sustainable Manufacturing System Under Reliable Constraints

Reza Shahabi-Shahmiri[1], Reza Tavakkoli-Moghaddam[1], Alexandre Dolgui[2(✉)], Mohammad Ghasemi[3], and Seyed-Ali Mirnezami[4]

[1] School of Industrial Engineering, College of Engineering, University of Tehran, Tehran, Iran
[2] IMT Atlantique, LS2N - CNRS, La Chantrerie, Nantes, France
alexandre.dolgui@imt-atlantique.fr
[3] School of IT , Deakin University , 75 Pigdons Rd, Geelong, VIC 3216, Australia
[4] Department of Industrial Engineering, Faculty of Engineering, Shahed University, Tehran, Iran

Abstract. Recognizing suitable manufacturing methods for goods can be taken into account as a crucial manufacturing task resulting in the optimization of activities. It has gained significance in the food industry owing to the distinct limitations and assumptions posed by perishable as well as non-perishable goods in this issue. Therefore, this research presents a novel mathematical formulation based on knowledge discovery and an assignment model for the optimization of manufacturing systems for perishable as well as non-perishable tailored to satisfy their unique features. In the proposed formulation, three objectives of minimizing the makespan, the production costs, and the energy consumption are considered. Two numerical examples considering both small and medium sizes are applied to appraise the performance of the combination of an improved version of epsilon-constraint augmented (AUGMECON2) and VlseKriterijumska Optimizacija I Kompromisno Resenje in Serbia (VIKOR), namely AUGMECON2VIKOR. The sensitivity analysis represents the positive reliable constraint impacts, influencing objective functions.

Keywords: Sustainable manufacturing systems · Queuing system · Perishable and non-perishable products · Time-cost-green trade-offs · AUGMECON2VIKOR algorithm

1 Introduction

In the prevalent marketing context, there exists a growing tendency towards manufacturing adequate amounts for the satisfaction of customers with low expenditures, novel varieties, high-quality, and shorter completion times. These issues, which are chiefly associated with existing uncertainties, severe competition, and constant changes in the marketplace, make companies use systems that can present high flexibility to cope with various changes and to improve capability and effectiveness.

Published by Springer Nature Switzerland AG 2024
M. Thürer et al. (Eds.): APMS 2024, IFIP AICT 733, pp. 307–321, 2024.
https://doi.org/10.1007/978-3-031-71645-4_21

Same as other complicated systems, a flexible manufacturing system (FMS) might be subjected to different uncertainties, like random failures of components. Such uncertainties are able to result in the decline in FMS performances through making machines unavailable, alteration of due dates, abolishing particular orders, and accordingly giving rise to contractual challenges with customers. In addition, for justification of their considerable initial investment, an FMS needs to gain a high level of reliability for tackling unforeseen conditions [1].

Therefore, systems reliability is considered to be critical in intelligent manufacturing. Manufacturing expenditures, for instance, owing to defective products, are direct and crucial dimensions of the reliability benefits for corporations [2]. Real-time prognostic data and preventative maintenance treatments can become possible through applying system reliability investigations [3]. The significance of perishable products in a reliable manufacturing system lies in their potential to significantly impact on the performance of overall system and its efficiency, as any breakdown or inefficiency in the production processes can lead to substantial wastage of valuable resources and increase the likelihood of delivering unsafe or substandard products to consumers. In the study conducted by Chang et al. [4], a simulation approach was used to estimate system reliability of a multi-state manufacturing network (MSMN) with joint buffers. A novel integrated mission reliability modeling approach was introduced by Yao et al. [5] for manufacturing systems with multiple states considering heterogeneous feedstocks.

A review of the recent literature reveals critical gaps in addressing sustainable manufacturing systems. To fill these gaps, this article addresses the following contributions:

- Optimizing the assignment of perishable as well as non-perishable products to machines simultaneously.
- Considering the queue system under disruption in the reliable manufacturing system
- Analyzing the impacts of disruption rates as well as the rate of retrieval time on cost, energy consumption, and time.
- Investigating how queue system under disruption affects the assignment of products to machines
- Using the AUGMECON2VIKOR method to deal with a multi-objective model tailored for sustainable manufacturing systems.

2 Problem Definition and Formulation

2.1 Perishable and Non-Perishable Products Assignment

The main goal of the presented research is to introduce a novel mathematical model to improve production costs, quality, and environmental impacts by assigning the products to manufacturing systems. This assignment is carried out by discovering association relationships between manufacturing system capabilities and product features.

A food plant's packaging lines comprising three types of machines, namely perishable, non-perishable, and hybrid, are examined. Hybrid machines support both perishable and non-perishable products. In contrast with the third line, which is versatile, the first two lines are specialized and accurate. Employing the third line enhances the productivity, speed, and packaging for a specified volume of combined products, encompassing perishable and non-perishable products. It is worth noting that these combined products have a time-consuming process of separating. Additionally, there exists the possibility of exceeding the maximum capacity of the first and second lines due to a substantial amount of food products. As a result, this part of these food products can be transferred to the third line. Even though this line is applicable, given its potential capacity to accommodate a wide range of food products, it provides different quality attributes compared to the other two lines. One of the reasons necessitating the use of the third line packaging is that transferring food between the first and second lines to deal with a set of perishable and non-perishable food may result in a degradation of product quality due to temperature changes. The third line employs a Modified Atmosphere Packaging (MAP) method, reducing oxygen levels. This technique has crucial application in packaging various items such as pre-cooked foods, fish, fresh pasta, fruits, vegetables, coffee, tea, and ready-made foods. It is noteworthy that utilizing modern machinery is crucial for establishing an environmentally friendly manufacturing system because it enables advancements like better insulation and recuperation exhaust systems. These improvements help minimize energy waste and emissions, contributing to a more sustainable production process.

As illustrated in Fig. 1, raw materials enter the production systems. Bluish products present perishable goods while yellowish products present non-perishable goods. Products are assigned to each type of manufacturing system with the specified capabilities. The manufacture of each one of the products within a particular manufacturing system results in a specific level of cost and quality.

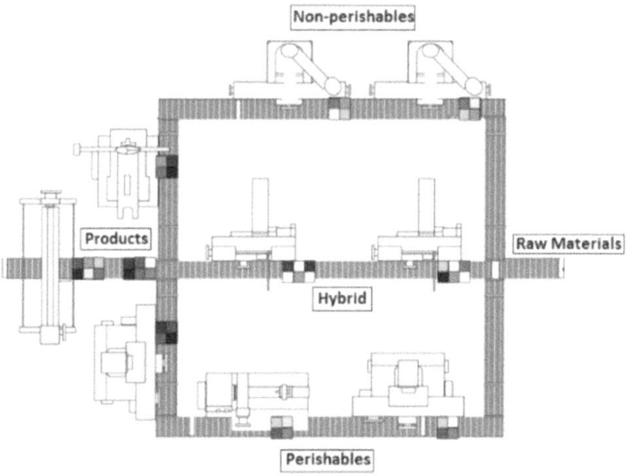

Fig. 1. Graphical representation of the proposed model.

2.2 Waiting Time of Manufacturing System Under Disruption

Due to the complications and numerous components of the manufacturing systems, a small amount of uncertainty might result in a tremendous disruption that leads to enormous expenditures as well as hard-to-recover impacts on the system [6]. These uncertainties (natural catastrophes, electricity disruptions, unforeseen work overloads, shortages, workforce strikes, etc.) are able to impact on costs, operation time, and energy consumption. Therefore, the occurrence of disruptions is probable and machines are likely to be unavailable for a period of time for provision of service. Retrieving of machines in such circumstances leads to additional expenditures, time, and energy in the manufacturing system.

Several investigations have addressed the use of queuing method for evaluating the operation in the manufacturing system (for instance, [1, 7–9], in which the obtained outcomes demonstrate that for calculating the waiting time of product flows, queuing methods are able to be appropriately utilized). According to these validations, a Poisson distribution was considered as an appropriate illustration of arrival rates in which there exists a variation around scheduled arrival times [10]. In general, this distribution is able to be taken into account in other service systems in which arrival rates and capacity levels vary considerably over time.

Concerning the previous research and to model the real-world conditions in the manufacturing operation, stochastic disruption of the queue system at each machine is taken into account and it is assumed that it is retrieved again with particular rates. Regarding the M/M/c queue system, independent service times are taken into account and distributed exponentially with rate ST. During disruptions, the number of operational servers reduces from c to c' and the service rates change from ST to $ST' \geq 0$. Immediately after retrieving the machine, the number of working servers and the respective service rates are restored to c and ST, respectively. It is considered that disruptions are based on a Poisson distribution with rate DR, and the retrieve times are i.i.d. exponentials with rate RT. The flow arrivals are based on a homogeneous Poisson distribution with intensity λ. According to the abovementioned explanations, the M/M/1 queue system is considered for all machines in this study.

2.3 Mathematical Model

In the following, a mixed-integer non-linear programming (MINLP) model is presented for the integration of assigning perishable/non-perishable products, and queue under disruption in the manufacturing system.

2.3.1 Notations

2.3.1. Notations

Indices

n	Non-perishable
p	Perishable
h	Perishable and non-perishable (hybrid)
i	Product
j, j'	Feature
m	Manufacturing system

Parameters

ϕ	Penalty cost for exceeding the average emission
ψ	Reward for each unit reduction below the average allowable emission
AG	Average allowable emission
G_{ijm}	Energy consumption amount of product *i* with feature *j* manufactured by system *m* with the ability to support feature *j*
Co_{ijm}	Cost of product *i* with feature *j* manufactured by system *m* with ability to support feature *j*
f_{ij}	1 if product *i* has the *j*-th feature; 0, otherwise
C_m	1 if system *m* has the ability to support feature *j*; 0, otherwise
Cap_m	Maximum capacity of product i by production line *m*
R_{ijm}	Amount of required resources for product *i* with feature *j* manufactured by system *m* with ability to support feature *j*
ST_m	Service time at manufacturing system *m*
DR_m	Disruption rate at manufacturing system *m*
RT_m	Retrieve time rate at manufacturing system *m*
γ	A big number

Binary Decision Variables

X_{ijm}	1 if product *i* with feature *j* is manufactured by system *m* with ability to support feature *j*; 0, otherwise
β	1 if energy consumption is less than average; 0, otherwise
α	1 if energy consumption exceeds average; 0, otherwise

Positive Decision Variables

$EGreen$	Amount of energy consumption less than the average allowable emission
$TGreen$	Excess energy consumption beyond the average
λ_m	Arrival rate of products to manufacturing system *m*
AW_m	Average waiting time at manufacturing system *m*
C_{max}	Makespan

2.4 Equations

$$MinZ_1 = C_{max} \tag{1}$$

$$\begin{aligned} MinZ_2 = & \sum_{i_p}^{i}\sum_{j_p}^{j}\sum_{m_p}^{m} CO_{i_pj_pm_p}X_{i_pj_pm_p} + \sum_{i_n}^{i}\sum_{j_n}^{j}\sum_{m_n}^{m} CO_{i_nj_nm_n}X_{i_nj_nm_n} \\ & + \sum_{i_n}^{i}\sum_{j_n}^{j}\sum_{m_h}^{m} CO_{i_nj_nm_h}X_{i_nj_nm_h} \\ & + \sum_{i_p}^{i}\sum_{j_p}^{j}\sum_{m_h}^{m} CO_{i_pj_pm_h}X_{i_pj_pm_h} + TGreen.\phi \\ & - EGreen.\psi \end{aligned} \tag{2}$$

$$\begin{aligned} MinZ_3 = & \sum_{i_p}^{i}\sum_{j_p}^{j}\sum_{m_p}^{m} G_{i_pj_pm_p}X_{i_pj_pm_p} + \sum_{i_n}^{i}\sum_{j_n}^{j}\sum_{m_n}^{m} G_{i_nj_nm_n}X_{i_nj_nm_n} \\ & + \sum_{i_n}^{i}\sum_{j_n}^{j}\sum_{m_h}^{m} G_{i_nj_nm_h}X_{i_nj_nm_h} \\ & + \sum_{i_p}^{i}\sum_{j_p}^{j}\sum_{m_h}^{m} G_{i_pj_pm_h}X_{i_pj_pm_h} \end{aligned} \tag{3}$$

Subject to:

$$\sum_{m_p}^{m} C_{m_p}X_{i_pj_pm_p} + \sum_{m_h}^{m} C_{m_h}X_{i_pj_pm_h} = f_{i_pj_p}; \forall i \in i_p, \forall j \in j_p \tag{4}$$

$$\sum_{m_n}^{m} C_{m_n}X_{i_nj_nm_n} + \sum_{m_h}^{m} C_{m_h}X_{i_nj_nm_h} = f_{i_nj_n}; \forall i \in i_n, \forall j \in j_n \tag{5}$$

$$\sum_{i_p}^{i}\sum_{j_p}^{j} X_{i_pj_pm_p}.R_{i_pj_pm_p} \le Cap_{m_p}; \forall m \in M_p \tag{6}$$

$$\sum_{i_n}^{i}\sum_{j_n}^{j} X_{i_nj_nm_h}.R_{i_nj_nm_h} + \sum_{i_p}^{i}\sum_{j_p}^{j} X_{i_pj_pm_h}.R_{i_pj_pm_h} \le Cap_{m_h}; \forall m \in M_h \tag{7}$$

$$\sum_{i_n}^{i}\sum_{j_n}^{j} X_{i_nj_nm_n} \le Cap_{m_n}; \forall m \in M_n \tag{8}$$

$$\sum_{i_p}^{i}\sum_{j_p}^{j} X_{i_pj_pm_p} \le \lambda_{m_p}; \forall m \in M_p \tag{9}$$

$$\sum_{i_n}^{i}\sum_{j_n}^{j} X_{i_nj_nm_h} + \sum_{i_p}^{i}\sum_{j_p}^{j} X_{i_pj_pm_h} \le \lambda_{m_h}; \forall m \in M_h \tag{10}$$

$$\sum_{i_n}^{i}\sum_{j_n}^{j} X_{i_n j_n m_n} \leq \lambda_{m_n}; \forall m \in M_n \tag{11}$$

$$AW_m = \frac{(RT_m + DR_m)^2 + ST_m DR_m}{(RT_m + DR_m)(RT_m(ST_m - \lambda_m) - \lambda_m DR_m)}; \forall m \in \mathrm{M} \tag{12}$$

$$t_{ijm}.X_{ijm} + AW_m.X_{ijm} \leq C_{max}; \forall i \in I, \forall j \in J, \forall m \in \mathrm{M} \tag{13}$$

$$\begin{aligned} &-\gamma.(1-\alpha) + \left(\sum_{i_p}^{i}\sum_{j_p}^{j}\sum_{m_p}^{m} G_{i_p j_p m_p} X_{i_p j_p m_p} + \sum_{i_n}^{i}\sum_{j_n}^{j}\sum_{m_n}^{m} G_{i_n j_n m_n} X_{i_n j_n m_n} \right. \\ &+ \sum_{i_n}^{i}\sum_{j_n}^{j}\sum_{m_h}^{m} G_{i_n j_n m_h} X_{i_n j_n m_h} \\ &\left. + \sum_{i_p}^{i}\sum_{j_p}^{j}\sum_{m_h}^{m} G_{i_p j_p m_h} X_{i_p j_p m_h} - AG \right) \leq TGreen \end{aligned} \tag{14}$$

$$\begin{aligned} &+\gamma.(1-\alpha) + \left(\sum_{i_p}^{i}\sum_{j_p}^{j}\sum_{m_p}^{m} G_{i_p j_p m_p} X_{i_p j_p m_p} + \sum_{i_n}^{i}\sum_{j_n}^{j}\sum_{m_n}^{m} G_{i_n j_n m_n} X_{i_n j_n m_n} \right. \\ &+ \sum_{i_n}^{i}\sum_{j_n}^{j}\sum_{m_h}^{m} G_{i_n j_n m_h} X_{i_n j_n m_h} \\ &\left. + \sum_{i_p}^{i}\sum_{j_p}^{j}\sum_{m_h}^{m} G_{i_p j_p m_h} X_{i_p j_p m_h} - AG \right) \geq TGreen \end{aligned} \tag{15}$$

$$\begin{aligned} &\gamma.(1-\beta) + \left(AG - \sum_{i_p}^{i}\sum_{j_p}^{j}\sum_{m_p}^{m} G_{i_p j_p m_p} X_{i_p j_p m_p} \right. \\ &+ \sum_{i_n}^{i}\sum_{j_n}^{j}\sum_{m_n}^{m} G_{i_n j_n m_n} X_{i_n j_n m_n} \\ &+ \sum_{i_n}^{i}\sum_{j_n}^{j}\sum_{m_h}^{m} G_{i_n j_n m_h} X_{i_n j_n m_h} \\ &\left. + \sum_{i_p}^{i}\sum_{j_p}^{j}\sum_{m_h}^{m} G_{i_p j_p m_h} X_{i_p j_p m_h} \right) \geq EGreen \end{aligned} \tag{16}$$

$$\begin{aligned} &\gamma.(1-\beta) + \left(AG - \sum_{i_p}^{i}\sum_{j_p}^{j}\sum_{m_p}^{m} G_{i_p j_p m_p} X_{i_p j_p m_p} \right. \\ &+ \sum_{i_n}^{i}\sum_{j_n}^{j}\sum_{m_n}^{m} G_{i_n j_n m_n} X_{i_n j_n m_n} \\ &+ \sum_{i_n}^{i}\sum_{j_n}^{j}\sum_{m_h}^{m} G_{i_n j_n m_h} X_{i_n j_n m_h} \\ &\left. + \sum_{i_p}^{i}\sum_{j_p}^{j}\sum_{m_h}^{m} G_{i_p j_p m_h} X_{i_p j_p m_h} \right) \geq EGreen \end{aligned} \tag{17}$$

$$\left(\sum_{i_p}^{i}\sum_{j_p}^{j}\sum_{m_p}^{m} G_{i_pj_pm_p}X_{i_pj_pm_p} + \sum_{i_n}^{i}\sum_{j_n}^{j}\sum_{m_n}^{m} G_{i_nj_nm_n}X_{i_nj_nm_n} + \sum_{i_n}^{i}\sum_{j_n}^{j}\sum_{m_h}^{m} G_{i_nj_nm_h}X_{i_nj_nm_h} + \sum_{i_p}^{i}\sum_{j_p}^{j}\sum_{m_h}^{m} G_{i_pj_pm_h}X_{i_pj_pm_h} - AG\right) \leq \gamma.\alpha \tag{18}$$

$$\left(AG - \sum_{i_p}^{i}\sum_{j_p}^{j}\sum_{m_p}^{m} G_{i_pj_pm_p}X_{i_pj_pm_p} + \sum_{i_n}^{i}\sum_{j_n}^{j}\sum_{m_n}^{m} G_{i_nj_nm_n}X_{i_nj_nm_n} + \sum_{i_n}^{i}\sum_{j_n}^{j}\sum_{m_h}^{m} G_{i_nj_nm_h}X_{i_nj_nm_h} + \sum_{i_p}^{i}\sum_{j_p}^{j}\sum_{m_h}^{m} G_{i_pj_pm_h}X_{i_pj_pm_h}\right) \leq \gamma.\beta \tag{19}$$

$$\alpha + \beta \leq 1 \tag{20}$$

$$\begin{aligned} &X_{ijm}, \beta, \alpha \in \{0, 1\}; \forall i, j, m \\ &EGreen, TGreen, \lambda_m, AW_m, C_{max} \geq 0; \forall m \end{aligned} \tag{21}$$

Three objectives are considered in the presented model. The makespan is minimized in Eq. (1). Assigning each product to each machine yields a distinct cost, which is a parameter that is required to be minimized by the second objective in Eq. (2). Equation (3) minimizes the total energy consumption generated by the machines for manufacturing the products. Equation (4) shows that perishable goods and features are able to be manufactured using a hybrid or a dedicated machine for perishable goods. Equation (5) states that non-perishable goods and features are able to be manufactured using a hybrid or a dedicated machine for non-perishable products. The capacity of each machine is represented in Eqs. (6)-(8) for perishable, hybrid, and non-perishable machines, respectively. Equations (9)-(11) calculate the arrival rate of products of each perishable, non-perishable, and hybrid machine. Equation (12) calculates the waiting time for each machine. Minimizing the maximum time is addressed in Eq. (13). Equations (14)-(17) compute the reward and penalty expenses associated with the green effects of the manufacturing system. It is shown in Eqs. (18)-(20) whether reward and penalty expenses are taken into account for the overall evaluation of the manufacturing system.

3 Solution Approach

The VIKOR (VlseKriterijumska Optimizacija I Kompromisno Resenje in Serbia) is a method for tackling multi-criteria decision-making (MCDM) problems taking into account diverse units in addition to conflicting criteria. The most suitable alternative among a set of choices is ranked and chosen by this method based on trade-offs, particularly for tackling problems taking into account conflicting criteria [11]. In the presented investigation, VIKOR in addition to AUGMECON2, namely AUGMECON2VIKOR, are applied for ranking and selecting, respectively. The choices are appraised separately based on a criterion, and ranked by making a comparison with the most suitable alternative. Various M choices are shown through $A_1, A_2, \ldots, A_m$. The rank of the j-th dimension can be indicated through f_{ij} for option A_i. In other words, j-th criterion for alternative A_i is represented using f_{ij} and N indicates the number of criteria. This approach was thoroughly explained in Shahabi-Shahmiri et al. [12]. Pseudo code of the introduced approach is illustrated in Fig. 2.

Start

Step 1. Initialize parameters, constraints, and data

1.1. Define the objective functions, constraints, and parameters
1.2. Set up the resource capacity constraints
1.3. Define the necessary fuzzy variables and membership functions

Step 2. AUGMECON2VIKOR Algorithm

2.1. Augmented epsilon-constraint (AUGMECON2) method
2.1.1. Define multiple objective functions ($f_1, f_2, \ldots, f_p$)
2.1.2. Set up the augmented ε-method model for multi-objective optimization
2.2. VIKOR method for ranking:
2.2.1. Calculate benefit values (S_i) and cost values (R_i) for each alternative
2.2.2. Determine the weights (W_j) for each criterion
2.2.3. Determine the best and worst values (f_j^* and f_j^-)
2.2.4. Calculate the S_i and R_i values for each alternative
2.3. Apply the AUGMECON2 method to obtain efficient solutions
2.4. Combine AUGMECON2 and VIKOR to rank and select the best solutions

Step 3. Output the final results

3.1. Display the Pareto solutions and ranking
3.2. Provide insights or recommendations based on the results

Step 4. End of the algorithm

Step 5. Additional notes and explanations

Provide any necessary comments and explanations for each step

End of the pseudo-code

Fig. 2. Pseudo code of the proposed solution approach

4 Experimental Results

4.1 Results

A numerical example is proposed to represent the capability of the introduced formulation for dealing with a manufacturing decision-making issue, which is vital for organizations endeavoring to conform to dynamic market as well as customer de-mands. Six products are considered to be allocated to the three machines. There exist six product features and abilities that three of them are related to perishable products and others are related to non-perishable products. For validation of the outcomes of the AUGMECON2VIKOR method, the introduced model has been coded using GAMS 24.1.2 taking into account five grid points. The needed CPU time in AUGMECON2VIKOR method to find Pareto optimal solutions was 175 s. The assignment of various products to the machines is represented in Fig. 3.

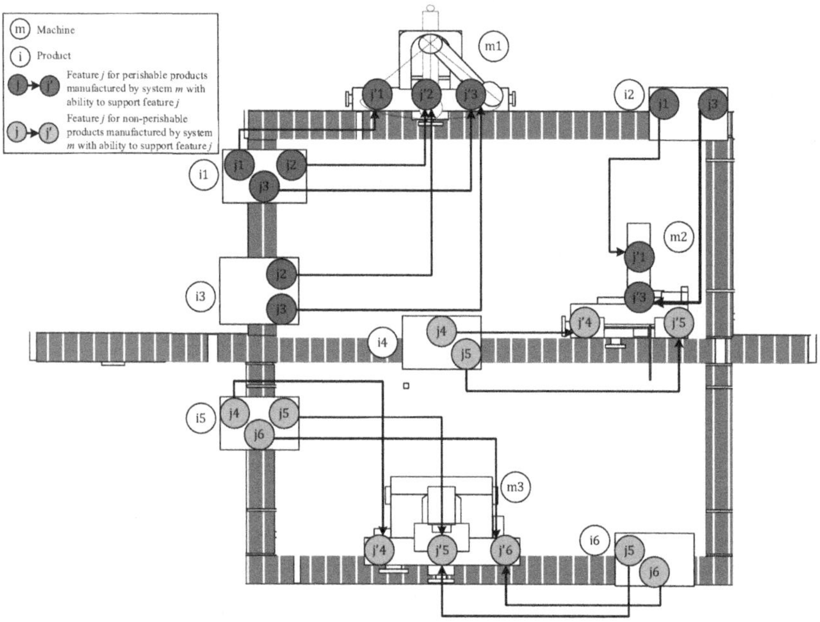

Fig. 3. Assignment of products to production systems according to the characteristics of the products and the systems.

4.2 Sensitivity Analysis

To validate the proposed mathematical model and the solution approach, the problem size increased. Considering 24 products, including 12 perishable, 12 non-perishable, 8 features, and six machines, the CPU time is reported to be 420 s. It should be noted that six machines, including two non-perishable, two perishable, and two hybrid and each machine has eight tasks. As it is represented in Fig. 4, the trade-off among three objective functions is reported in five grid points. Service time, disruption rate, and retrieve time in the manufacturing system are assumed [30, 40], 0.5, and 0.5, respectively. By increasing the time, most of the products are assigned to the hybrid machines. As a result, operational costs decreased. In these circumstances, more energy is consumed. As expected, Fig. 4 shows that the operation time increased from 4.010 in the first grid point to 4.275 in the fifth grid point. Second objective function (i.e., cost) decreased by 1.46% and energy consumption increased by 2.93%.

To demonstrate the impact of increasing rate of retrieve time and disruption rate in manufacturing system on objective functions, Tables 1 and 2 are represented. As the retrieve rate increases, the number of retrieve times increases. Hence, if the system is disrupted, it will be retrieved more quickly and the length of the queue will be reduced. As a result, the waiting time will also be significantly reduced. As the waiting time decreases, the system is more inclined to use the machines especially designed for each product and allocate them to the products. It leads to a reduction in cost and energy consumption. For instance, in grid point 3, cost and energy consumption were improved by 1.86% and 4.20%, respectively.

As the disruption rate increases, the number of disruptions increases and thus the length of the queue increases. As a result, the waiting time increases, and considering the first objective function, the products are also possible to be processed by hybrid machines. It leads to an increase in cost and energy consumption. For instance, in grid point 4, a 3.41% increase in cost and a 4.44% increase in energy consumption have been reported.

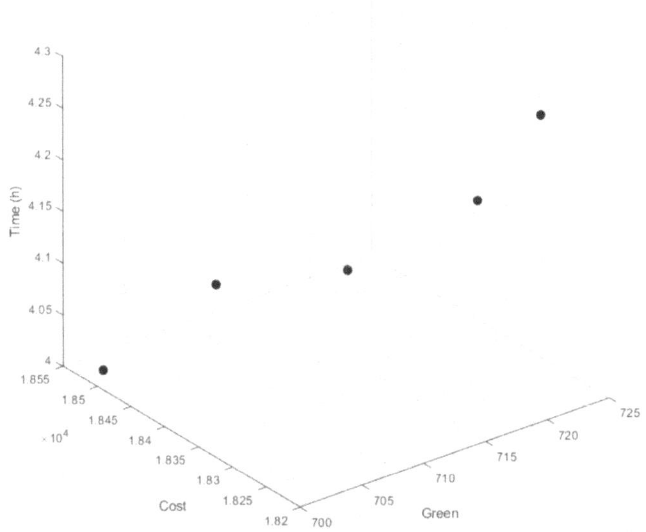

Fig. 4. Computational results of the trade-off between objective functions under

$ST_m = [30,40], DR_m = 0.5, RT_m = 0.5.$

Table 1. Comparative results of the objective functions under $ST_m = [30,40]$ by increasing RT_m

Grid points	$DR_m = 0.5, RT_m = 0.5$			$DR_m = 0.5, , RT_m = 0.7$		
	Z1 (Time)	Z2 (Cost)	Z3 (Green)	Z1 (Time)	Z2 (Cost)	Z3 (Green)
1	4.010	18505.699	700.825	3.146	18110.390	673.122
2	4.098	18434.443	706.113	3.161	18042.519	676.577
3	4.130	18332.639	711.154	3.491	17991.381	681.252
4	4.183	18283.745	718.970	3.731	17911.056	685.212
5	4.275	18234.721	721.384	3.939	17758.124	696.687

Table 2. Comparative results of the objective functions under $ST_m = [30,40]$ by increasing DR_m

Grid points	$DR_m = 0.5, RT_m = 0.5$			$DR_m = 0.7, RT_m = 0.5$		
	Z1 (Time)	Z2 (Cost)	Z3 (Green)	Z1 (Time)	Z2 (Cost)	Z3 (Green)
1	4.010	18505.699	700.825	4.347	19290.164	729.022
2	4.098	18434.443	706.113	4.362	19179.116	735.663
3	4.130	18332.639	711.154	4.434	19098.659	741.445
4	4.183	18283.745	718.970	4.511	18907.154	750.869
5	4.275	18234.721	721.384	4.616	18715.649	757.240

Table 3 shows the performance of the proposed algorithm in comparison to the previous method, namely epsilon-constrained, which was presented by Hwang and Masud [13]. Considering the AUGMECON2VIKO algorithm, operation time, cost, and energy consumption are approximately 5.09%, 6.59%, and 18.43% lower than the epsilon-constraint method. In addition to the abovementioned analysis, the proposed mathematical model with the presented algorithm is implemented considering two scenarios. In the first scenario, the presence or absence of reliable constraints is considered in the mathematical formulation. The consideration or lack of consideration of energy consumption as an objective function is investigated in the second scenario. As shown in Tables 4 and 5, considering reliable constraints as well as energy consumption, AUGMECON2VIKOR provides better Pareto solutions compared to the circumstances in which they are not considered.

Table 3. Comparative results of the objective functions under $ST_m = [30,40], DR_m = 0.5, RT_m = 0.5$ with AUGMECON2VIKOR and ε-constraint method

Grid points	AUGMECON2VIKOR			epsilon-constraint method		
	Z1 (Time)	Z2 (Cost)	Z3 (Green)	Z1 (Time)	Z2 (Cost)	Z3 (Green)
1	4.010	18505.699	700.825	4.171	19791.749	851.214
2	4.098	18434.443	706.113	4.226	19704.541	863.139
3	4.130	18332.639	711.154	4.342	19649.935	876.856
4	4.183	18283.745	718.970	4.404	19599.648	881.461
5	4.275	18234.721	721.384	4.833	19517.315	889.865
Average	4.1715	18358.25	711.6892	4.3952	19652.64	872.507

Table 4. Comparative results of the objective functions under $ST_m = [30,40], DR_m = 0.5, RT_m = 0.5$ with and without reliability

Grid points	With reliability			Without reliability		
	Z1 (Time)	Z2 (Cost)	Z3 (Green)	Z1 (Time)	Z2 (Cost)	Z3 (Green)
1	4.010	18505.699	700.825	6.224	23202.038	919.403
2	4.098	18434.443	706.113	6.583	23124.404	924.298
3	4.130	18332.639	711.154	6.650	23075.335	931.772
4	4.183	18283.745	718.970	6.914	22968.843	939.444
5	4.275	18234.721	721.384	7.592	22890.008	945.623

Table 5. Comparative results of the objective functions under $ST_m = [30, 40], DR_m = 0.5, RT_m = 0.5$ with and without energy consumption

Grid points	With energy consumption			Without energy consumption	
	Z1 (Time)	Z2 (Cost)	Z3 (Green)	Z1 (Time)	Z2 (Cost)
1	4.010	18505.699	700.825	4.110	24312.889
2	4.098	18434.443	706.113	4.180	24263.147
3	4.130	18332.639	711.154	4.236	24214.723
4	4.183	18283.745	718.970	4.423	24176.754
5	4.275	18234.721	721.384	4.642	24138.747

5 Conclusion

In the proposed research, a new MINLP model considering three objectives (i.e., minimizing makespan, cost, and energy consumption) was taken into account to deal with a complicated issue in the sustainable manufacturing system. The goods were classified in two classifications (namely perishable as well as non-perishable goods) and the machines were classified in three classifications (assigned machines to perishable as well as non-perishable goods and hybrid machines, which could carry out both tasks. Separate features were considered for each group of goods and each group of machines had their features. Due to the disruptions in complex manufacturing systems, the M/M/1 queue system was utilized to enhance the reliability of the proposed problem. To validate the model and demonstrate its applicability, a numerical example in a small size was implemented and the obtained results were presented. An AUGMECON2VIKOR method was applied for the first time in this issue to find Pareto solutions. The results obtained in five grid points show that as time increases, the cost decreases, and the third objective function (i.e., energy consumption) increases.

Moreover, the inclusion of reliable constraints in this study resulted in notable improvements in both time and energy consumption. By increasing the retrieve rate by 0.2, the time, cost, and energy consumption improved by 16.25%, 2.15%, and 4.09%, respectively. To maintain an unbiased perspective, it is important to highlight three limitations of this paper. Firstly, in the presented investigation, scheduling over just the current time period has been considered. Due to the impacts of disruption on manufacturing systems, scheduling over multiple periods of time plays a vital role in achieving the desired goals. Secondly, this article assumes that each product has been assigned to only one machine. However, in real-world systems, there is flexibility to assign each product to multiple machines, making scheduling highly dependent on the sequence of operations. Ultimately, while to address the complexity of real-life problems, it is essential to consider some uncertain parameters, this study presented a crisp formulation. Examining the issue considering multiple periods as well as assuming extra intrinsic complexities resulting from the uncertainty in real conditions are intriguing directions for future research. Developing heuristic/meta-heuristic methods for tackling these complex problems is another interesting research direction.

Acknowledgment. Alexandre Dolgui's work on this paper was supported by the European project ACCURATE https://accurateproject.eu, Grant Agreement number 101138269.

Disclosure of Interest. The authors have no competing interests to declare that are relevant to the content of this article.

References

1. Souier, M., Dahane, M., Maliki, F.: An NSGA-II-based multiobjective approach for real-time routing selection in a flexible manufacturing system under uncertainty and reliability constraints. Int. J. Adv. Manuf. Tech. **100**, 2813–2829 (2019)
2. Lazarova-Molnar, S., Mohamed, N.: Reliability assessment in the context of industry 4.0: data as a game changer. Procedia Comp. Sci. **151**, 691–698 (2019)
3. Bousdekis, A., Mentzas, G.: Condition-based predictive maintenance in the frame of industry 4.0. In: Lödding, H., Riedel, R., Thoben, KD., von Cieminski, G., Kiritsis, D. (eds.) Advances in Production Management Systems. The Path to Intelligent, Collaborative and Sustainable Manufacturing, APMS 2017, IFIP Advances in Information and Communication Technology, vol. 513, pp. 399–406. Springer, Cham (2017). https://doi.org/10.1007/978-3-319-66923-6_47
4. Chang, P.C., Lin, Y.K., Chiang, Y.M.: System reliability estimation and sensitivity analysis for multi-state manufacturing network with joint buffers — a simulation approach. Reliab. Eng. Syst. Saf. **188**, 103–109 (2019)
5. Yao, J., Gao, Z., He, Y., Peng, C.: Integrated mission reliability modeling for multistate manufacturing systems considering heterogeneous feedstocks based on extended stochastic flow manufacturing network. Reliab. Eng. Syst. Saf. **243**, 109840 (2024)
6. Song, H., Chang, R., Cheng, H., Liu, P., Yan, D.: The impact of manufacturing digital supply chain on supply chain disruption risks under uncertain environment - based on dynamic capability perspective. Adv. Eng. Infor. **60**, 102385 (2024)
7. Souza, M.L.H., da Costa, C.A., de Oliveira Ramos, G., da Rosa Righi, R.: A survey on decision-making based on system reliability in the context of Industry 4.0. J. Manuf. Syst. **56**, 133–156 (2020)
8. May, M.C., Albers, A., Fischer, M.D., Mayerhofer, F., Schäfer, L., Lanza, G.: Queue length forecasting in complex manufacturing job shops. Forecasting **3**(2), 322–338 (2021)
9. Yu, M., Tang, Y., Wei, Y.: Production fluid-type inventory model associated with a single vacation queue arising from additive manufacturing in the food sector. Appl. Math. Mod. **128**, 242–259 (2024)
10. Peterson, M., Bertsimas, D., Odoni, A.: Models and algorithms for transient queuing congestion at airports. Manag. Sci. **41**(8), 1279–1295 (1995)
11. Opricovic, S., Tzeng, G.H.: Extended VIKOR method in comparison with outranking methods. Eur. J. of Oper. Res. **178**(2), 514–529 (2007)
12. Shahabi-Shahmiri, R., Asian, S., Tavakkoli-Moghaddam, R., Mousavi, S.M., Rajabzadeh, M.: A routing and scheduling problem for cross-docking networks with perishable products, heterogeneous vehicles and split delivery. Comp. Ind. Eng. **157**, 107299 (2021)
13. Hwang, C.L., Masud, A.S.M.: Multiple objective decision making - methods and applications: a state-of-the-art survey, vol. 164, Springer, Cham (2012)

Additive Manufacturing

An Accessibility Assessment Algorithm for Support Structure Removal in Parts Produced by Powder Bed Fusion of Metal Using a Laser Beam

Leonie Pauline Pletzer-Zelgert(✉), Lukas Unruh, Martin Iza-Mendez, and Johannes Henrich Schleifenbaum

RWTH Aachen University, Campus-Boulevard 73, 52074 Aachen, Germany
pauline.pletzer.zelgert@dap.rwth-aachen.de

Abstract. In the additive manufacturing process of Powder Bed Fusion of Metals using a Laser Beam (PBF-LB/M), the removal of support structures represents a costly step, predominantly carried out manually in current industrial practices. Particularly challenging are areas of the parts where tools for support removal struggle to access. Therefore, in the data preparation of parts manufactured via PBF-LB/M, attempts are often made to orient the parts within the build chamber in such a way that no supports are needed in the difficult-to-access areas. However, identifying these areas is error-prone and typically relies on manual evaluation by engineers. To reduce susceptibility to errors, this paper introduces an algorithm called the Accessibility Assessment Algorithm (AAA), which automatically evaluates the accessibility of parts. The algorithm is implemented and tested on ten parts. Five experts experienced in removing support structures from PBF-LB/M-printed parts are interviewed and asked to evaluate the same parts for inaccessible areas, and then to assess the calculation results of the AAA. Subsequently, the results regarding inaccessible areas are compared between the algorithm's assessment and the experts' evaluations. The results show a 70.5% agreement, indicating its effectiveness in assessing the accessibility of parts.

Keywords: Additive Manufacturing · Powder Bed Fusion of Metal Using a Laser Beam · Support Structure Removability · Accessibility Assessment Algorithm

1 Introduction

The global impacts of the Covid-19 pandemic have significantly highlighted the vulnerability of supply chains [1]. Furthermore, the goal of enabling environmentally sustainable production of goods is gaining increasing importance [2, 3]. Especially in high-wage countries like Germany, a trend towards new business models is emerging, transitioning from mass production to individual manufacturing to utilize resources more efficiently [4, 5]. With this paradigm shift, there is a growing demand for technologies that enable

Published by Springer Nature Switzerland AG 2024
M. Thürer et al. (Eds.): APMS 2024, IFIP AICT 733, pp. 325–337, 2024.
https://doi.org/10.1007/978-3-031-71645-4_22

both higher resilience in logistics and the fulfillment of specific customer requirements that necessitate product customization. Additive Manufacturing (AM) is a promising technology in this regard. In AM, parts are produced from 3D model data by joining materials, typically layer by layer, as opposed to subtractive and forming manufacturing methods [6]. AM offers the flexibility to make design decisions more independent of the process chain and customize parts to meet customer requirements [7]. However, the increased implementation of AM in the private sector requires addressing various remaining technological challenges, such as reducing the production costs of AM parts, increasing applicability, and ensuring sufficient part quality for the application [8, 9]. Within the various AM processes for the production of metal parts, Powder Bed Fusion of Metals using a Laser Beam (PBF-LB/M) holds the largest market share [9]. In this process, metal powder is deposited layer by layer and then selectively melted using a laser beam [10].

The removal of support structures poses a particular challenge, especially for PBF-LB/M. Support structures are necessary for parts manufactured using PBF-LB/M whenever part surfaces have an overhang angle exceeding 45° relative to the build direction (as shown in Fig. 1). This is necessary because these areas of the parts could otherwise geometrically distort due to insufficient heat dissipation of the laser beam energy and gravitational forces [11].

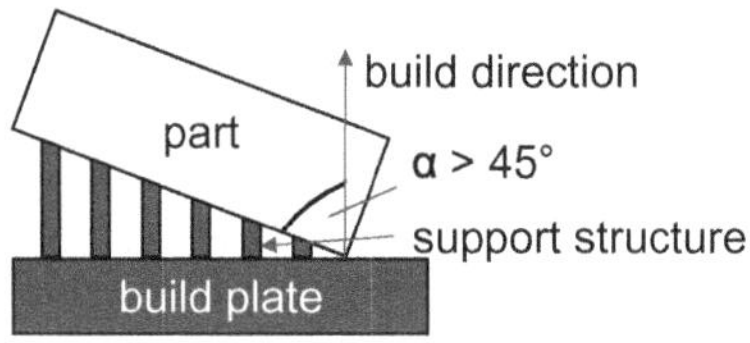

Fig. 1. PBF-LB/M part with support

Currently, support structure removal in the industry is mostly done manually and can be particularly time-consuming and costly, especially for hard materials like stainless steel [12]. Due to the geometric complexity of additively manufactured parts, inaccessible areas often arise, which present difficulties for conventional tools.

Data preparation offers the opportunity to optimize the effort required for support structure removal by avoiding the placement of support structures at locations that may later be difficult to access, through selecting an appropriate part orientation in the build chamber. Figure 2 illustrates how part orientation affects inaccessible part areas. However, to consider the inaccessible part areas during the decision-making process for part orientation, these areas must first be correctly identified by the engineer.

Manual assessment is particularly challenging and error-prone, especially for geometrically complex additive parts. This work aims to develop an Accessibility Assessment Algorithm (AAA) that automates the identification of part areas' accessibility. This is intended to reduce post-processing efforts in future work by incorporating the calculation results of the AAA algorithm into the orientation selection process. This way, orientations requiring support structures in hard-to-reach part areas can be avoided.

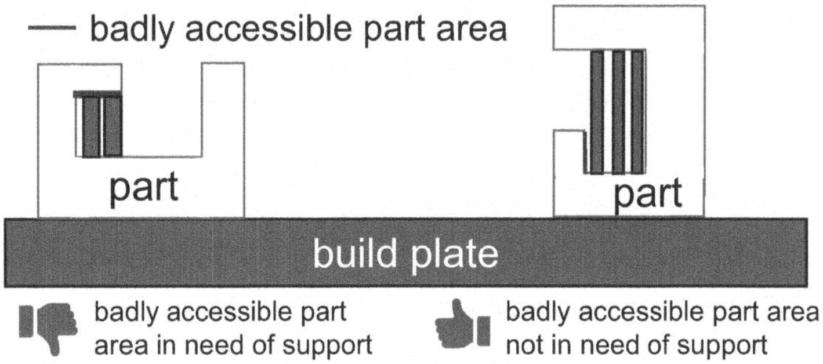

Fig. 2. Effect of part orientation on support structures and supported part areas.

2 State of the Art

The application of collision checks from CNC machines for part processing has already been explored through various algorithms, such as those by Bi et al. [13] and Ho et al. [14]. However, directly applying these techniques to support removal in the AM process presents a challenge because different tools are used for manual removal, and the movements for removing the support structures cannot be automatically read as in the case of CNC machines. Therefore, a direct transfer of methodologies is not possible. The literature review thus focuses on approaches directly developed for the removal of support structures in additively manufactured parts.

At the current state of the art, four main methods for evaluating the accessibility of tools for support structure removal can be identified:

1) **Voxelization and Ray Tracing**: This method utilizes voxelization and ray tracing techniques to evaluate support structure accessibility. The process involves creating a voxel grid that represents the support structure, and tracing six rays from each voxel along the coordinate axes. If all rays intersect with the printed object, the voxel is marked as inaccessible. [15, 16]
2) **Octree Data Structure**: An alternative approach employs an octree data structure to analyze accessibility. The method categorizes octree cubes into object parts, support structures, or empty spaces. For a support cube to be deemed accessible, a pathway devoid of part cubes to the external space must be identified. [17, 18]
3) **Surface Consideration**: Another method focuses on the accessibility of surfaces. Users mark specific surfaces and set a required distance parameter. These surfaces are checked for accessibility by computing the distance to the opposing surface. [19]
4) **Point-Based Accessibility Assessment**: This method involves defining a set of points surrounding the part and assessing accessibility based on the existence of a linear line from the examined point to predefined points around the part. [20]

These varied approaches highlight the diverse methods employed to evaluate the accessibility of support structure removal tools within AM processes.

A research gap persists in AAAs for support removal in additively manufactured parts. Firstly, the systematic validation of these algorithms, particularly regarding accessibility, remains largely unexplored. Additionally, existing methodologies often overlook the possibility of tools having forms other than linear shapes, such as T-nuts milling tools. Support removal via tools like T-slot milling machines is not considered, complicating the parameterization of the approach based on the tool used. As a result, the adaptation to different tool types lacks a systematic methodology. We therefore studied how the structure of an accessibility algorithm can be designed to effectively assess the removability of support structures in additively manufactured parts using a variety of tools.

3 Accessibility Assessment Algorithm

This work aims to develop an algorithm, referred to as the Accessibility Assessment Algorithm (AAA), which automatically detects the inaccessible areas of parts manufactured using PBF-LB/M. A ray tracing approach has been chosen for this purpose. There are two main reasons for selecting this method:

1) Ray tracing enables high flexibility. Firstly, the starting points of the rays can be freely chosen. Secondly, as mentioned in state of the art, some support structure removal tools can follow paths that are more than just linear. Successive rays can be utilized to represent various types/shapes of prop removal tools, thereby not automatically constraining the accessibility study to linear shape limitations so that tools which are not straight can also be examined.
2) Ray tracing is a well-explored topic in the field of 3D modelling. There are efficient methods for computing intersections of rays and meshes, leading to fast computations. By using existing frameworks (such as the library Geometry3Sharp [20]), which have already implemented these methods, an approach is ensured while also reducing development effort.

Initially, the starting points of the rays are selected. In contrast to the state of the art, the rays are shot from the triangular faces of the part, rather than from the voxelated supports. The advantage here is twofold: firstly, the assessment remains independent of orientation, and secondly, no supports need to be generated to make a general statement about the accessibility of the part. The ray tracing is performed in the Euclidean three-dimensional space $\mathbb{R}^3$. As a surface in this space has an infinite number of points, a representative set must be selected. Additionally, selecting too many starting points can lead to performance issues. Since parts are described as 3D models composed of a mesh of triangles, the centroid of each triangle is chosen as the starting point. The centroid is the point where the three medians of a triangle intersect as shown in Fig. 3.

From the centroid multiple rays are cast. As covering all directions is not feasible, a method has been devised to evenly distribute coverage across an adjustable number of directions. This method creates a hemisphere of rays starting at the centroid.

Then, a vector v_1 that is orthogonal to the normal is computed such that the vector and the triangle are parallel. Now, an angle $\alpha \in (0°, 360°)$ is chosen. With this angle, a circle of vectors is created as depicted in the left part of Fig. 3. When choosing α, the following must be ensured to achieve an even distribution: $360 \equiv 0 \pmod{\alpha}$. As an example, when $\alpha = 90°$, there are only four vectors created (v_1, v_2, $-v_1$, $-v_2$). This circle of vectors is called a layer. By creating multiple layers that move upwards – in the direction of the normal - a hemisphere is created. To do this, an additional angle $\beta \in$

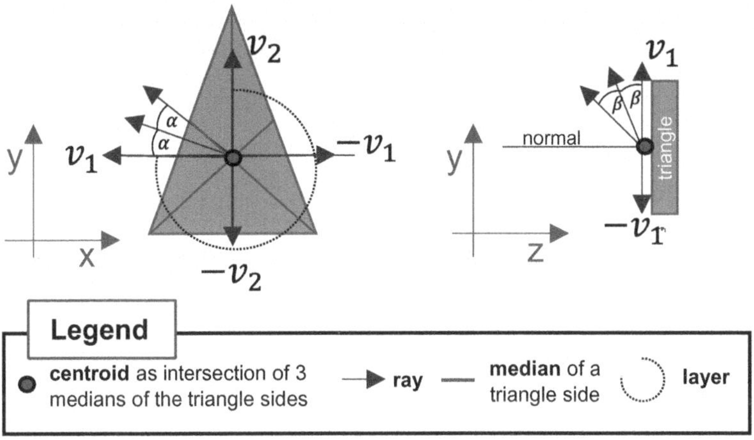

Fig. 3. The creation of the centroid of the triangle and rays in a hemisphere form

(0°, 90°] is chosen and for each subsequent layer, the vector v_1 is rotated upwards by β as shown in the right part of Fig. 3. Then, the layer is built by rotating around the normal. Again, to achieve an even distribution, the following must be ensured: $90° \equiv 0° \pmod{\beta}$. In every layer there are $\frac{360°}{\alpha}$ vectors, and there are $\frac{90°}{\beta}$ layers. Adding the normal, the total number of vectors and ultimately rays that are generated is $n_{Rays} = 1 + \frac{360°}{\alpha} \times \frac{90°}{\beta}$.

This method constructs an evenly distributed ray casting and allows the user to configure the algorithm for personal needs by setting α and β. However, so far this method only considers accessibility in linear direction. To support different kinds of support removal tools, new casts starting not in the triangle, but from previously cast rays are created. These are called nesting rays. This is done by using a recursive method for every ray.

Figure 4 illustrates an example of how the accessibility check in the AAA works and yields different results depending on the adjust parameters and the abstracted tool used.

In the first three steps, the parameters are set by the user depending on the tool. Depending on the tool, a so-called level is set, indicating how many times the rays should branch out during the accessibility examination. This is, for example, 0 for the chisel and 1 for the T-slot milling machine. If a nonlinear tool type is used, a branching distance x must be defined depending on the tool geometry (see Fig. 4). Then, the direction of the rays (emerging from the centroid) is followed until the branching distance x is reached, from which new rays are cast. Also, the hemisphere angles are adjusted by the user and affect the solution of the accessibility check as already described. Another aspect that is integrated using the ray tracing approach was proposed by Tran et al. [18]. By computing the distance between two surfaces, it can be determined whether tools can pass between them. This can be done by considering the normal of a triangle, a ray of which is already created in the previous procedure. Now, if the ray intersects with a surface, the distance d between the start of the ray and the intersection point can be determined. If the distance is smaller than a value set by the user, representing the tool diameter, a surface is deemed inaccessible, even if further investigation through nesting of rays deems it accessible.

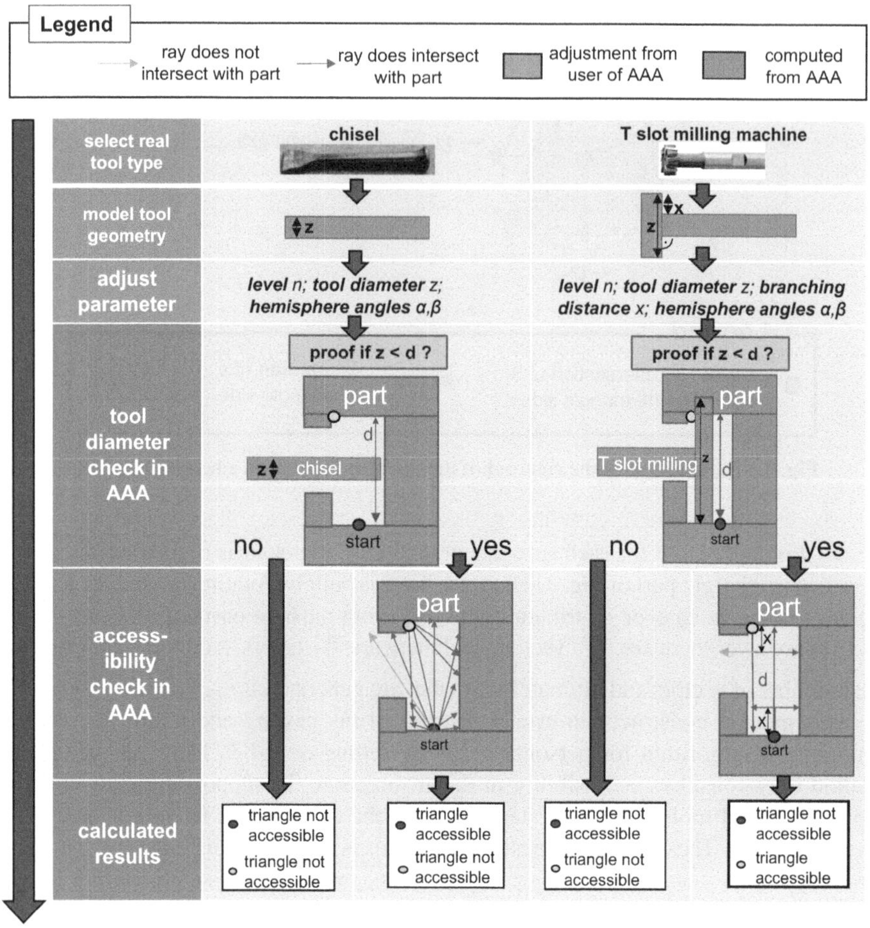

Fig. 4. The process of the accessibility check.

This procedure is shown in Fig. 4 under *Tool Diameter Check*. Depending on the result of the Tool Diameter Check, the Accessibility Check is then performed, and the calculation results are exemplarily presented for two triangles in Fig. 4. So, the AAA algorithm is designed to identify all areas of PBF-LB/M-manufactured parts where conventional tools may struggle to remove support structures due to accessibility issues. However, that the AAA algorithm does not consider the exception for round channels in PBF-LB/M parts. Therefore, overhanging areas for round channels up to a diameter of 8 mm can also be manufactured without support structures [21]. As a result, these part areas do not need to be checked for accessibility, as the support structures would not need to be removed during manufacturing. This exception is not yet covered by the ray tracing approach.

The presented algorithm was implemented in C# by using the Geometry3Sharp [20] library. In the next section, the AAA will be applied, and the validity will be assessed and discussed based on an initial sample access.

4 Case Study

The case study aims to conduct a proof of concept of the current AAA algorithm implementation based on the chosen ray tracing approach. Figure 5 outlines the conducted case study for performing this proof of concept. To this end, ten open source parts designed for AM were selected for this purpose [22].

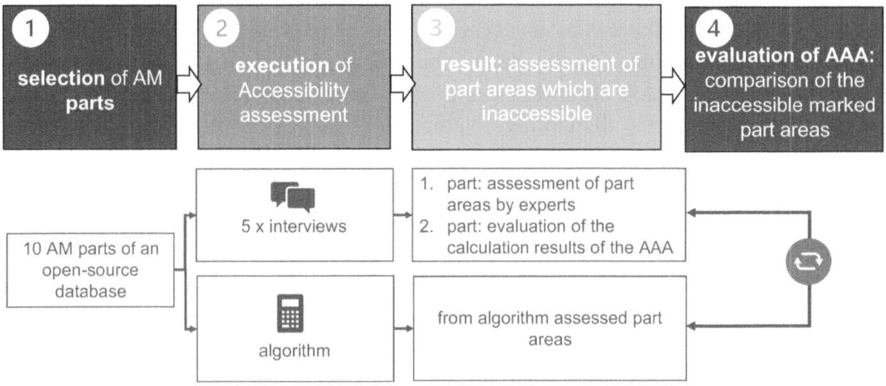

Fig. 5. Schematic depiction of the procedure of the case study.

With these parts five interviews were initially conducted with experts. Engineers who had supported at least 20 PBF-LB/M-printed parts in the past are categorized as experts. The interviews were conducted in 1.5-h sessions and were separated in two parts. In the **first part** the experts were tasked with identifying part areas where they believed supports could not be removed with tools or only with significant effort. Evaluation involved marking all part areas assessed as "inaccessible". Since all experts were accustomed to gauging distances through their experience with hammer and chisel, the same scenario was applied for evaluation purposes. Consequently, the experts shared their views on support removal solely using a hammer and chisel. Assessment with non-linear tools, therefore, remains to be conducted in future work. The same parts used for the interviews were evaluated for accessibility using the AAA algorithm. The aligned parameters for the AAA are shown in Table 1.

Table 1. Aligned parameters for AAA

name of parameter	parameter	value
tool diameter	z	5 mm
level	n	0
hemisphere angle	α	15°
	β	15°

The part areas marked by the algorithm were compared with those identified in the interviews. The analysis of these comparisons followed the logic shown in **Fig. 6** for one part, which was part of the case study as a representative example. Depending on the degree of overlap between the marked areas from the part identified in the interview and the part assessed by the AAA, the areas were classified as "missed", "correct markings", or "incorrect markings". In this context, "missed" denotes marked areas by the experts that were not detected by the AAA. "Correct markings" refers to the part areas where the marked areas matched between the experts and the AAA. "Incorrect markings" indicates that the AAA identified areas as inaccessible that the experts deemed accessible. On the left side, the assignment to the step of the case study from Fig. 5 was graphically indicated. Since the AAA algorithm cannot yet detect channels that can be printed without support structures, all markings evaluated as inaccessible by the AAA algorithm but not considered by the expert are manually excluded in the evaluation to allow for comparison. These results are also compared with the results without the exclusion of these part areas (see "incorrect markings (w/o channels)" in the "Results and Discussion" section).

In **the second part** of the interview, the experts were also presented with the algorithm's calculation results. Subsequently, they were asked to assess whether these results corresponded to their own markings, and if not, to indicate how many markings are missing or incorrectly placed by the algorithm.

5 Result and Discussion

Figure 7 presents the results of the first part of the interview. The diagram shows the percentage of the average of the part markings for the "missing", "incorrect markings" and "correct markings". The total count is derived from the number of marked part locations that needed to be compared between the parts evaluated by the AAA and experts, so it is the sum of the located part areas marked by both the engineer and the AAA. Each comparison of part markings is counted as one instance. The "incorrect markings (w/o channels)" refers to the percentage of marked part areas where the markings are located in channels of the part and therefore represent areas that, although assessed as inaccessible by the AAA, do not require support structures during the PBF-LB/M process regardless of the part orientation.

An analysis of the results already revealed a high degree of agreement of 61.6% between the part areas marked as inaccessible by experts and those identified by the AAA. In 10.3% of cases, the part areas marked as inaccessible by experts were not recognized by the AAA, and in 28.1% of cases, part areas were marked as inaccessible by the AAA, although the experts considered them accessible. To calculate the "standard deviation between the parts", the standard deviation of the average of each marking category across all individual parts (totaling 10 parts) was determined. Each of the five experts evaluated the 10 parts, resulting in a total of 50 evaluations compared with the AAA calculation results. Thus, this value serves as an indicator of the extent to which the assessment results fluctuate depending on the part geometry. Similarly, for calculating the "standard deviation between the experts", the average standard deviation among the engineers was determined. Thus, this value serves as an indicator of the variance

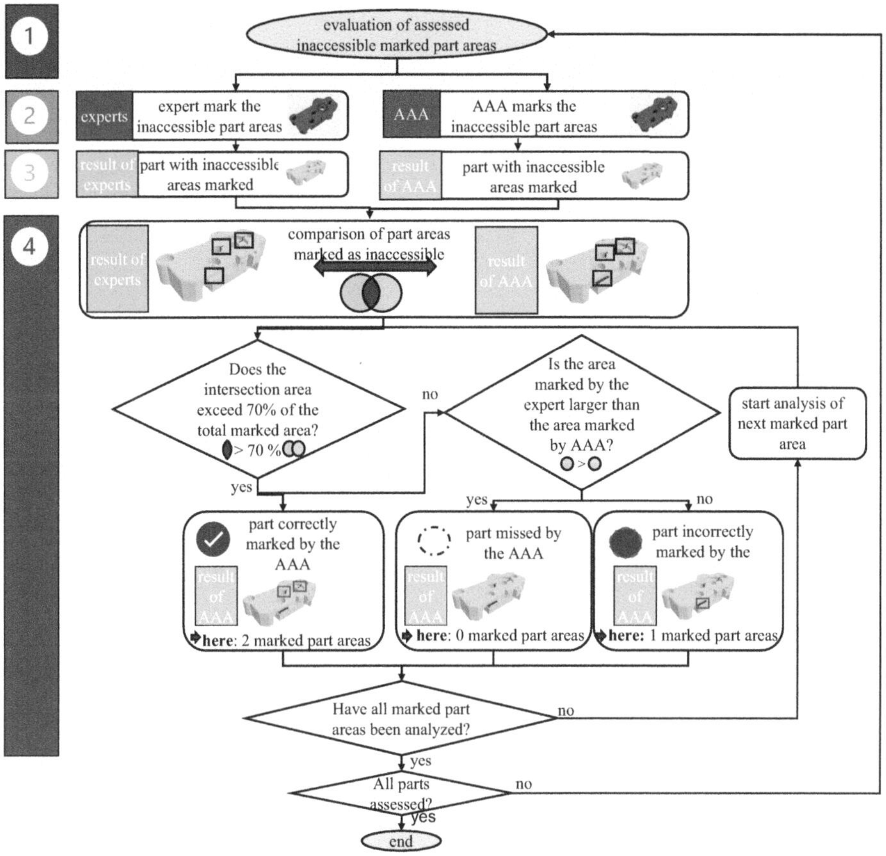

Fig. 6. Logic of analyzing marked areas between the AAA and expert, with reference to the case study.

in assessment among the engineers. The standard deviation among both the parts and the experts, particularly concerning the "correct markings" in contrast to the "average of part markings," is minimal. This observation suggests a lower degree of part dependency in these markings, as well as a higher consensus among the engineers regarding them. However, it should be noted that both standard deviations are high, highlighting the high level of uncertainty in the assessment of part accessibility through experts. Hence, it remains unclear whether the high standard deviation among the parts results from the algorithm making more accurate predictions for certain parts than for others, or from the lack of uniform agreement among the experts. Also due to the high standard deviation, the results should be further investigated in the by implementing practical experiments.

As previously mentioned, channels with a diameter of up to 8 mm can be printed without support structures, regardless of their orientation [21]. However, the current implementation of the AAA does not yet recognize part areas that could be additively manufactured without support structures. Therefore, the algorithm assesses the accessibility of these part areas, unlike the experts. The algorithm could be expanded to

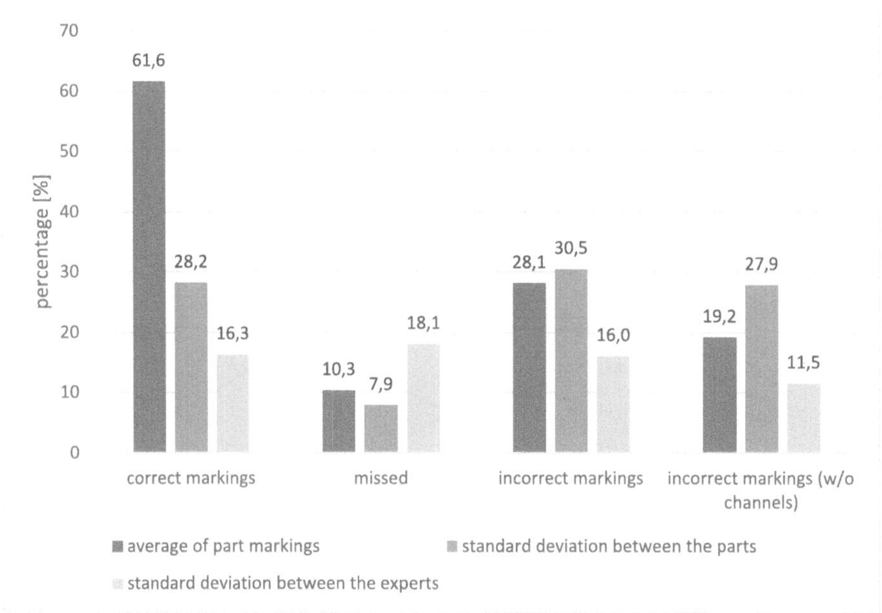

Fig. 7. Comparison between algorithm and interviewer marked parts.

recognize such channels and exclude these surface areas from the automatic accessibility assessment. If the part areas that can be manufactured without support structures are excluded (marked here as "incorrect markings w/o channels"), the "incorrect markings" would be reduced by an additional 8.9%, to 19.2%. This means that the total part areas classified as "correct markings" would increase to 70.5%. This demonstrates that the future expansion to include the automatic recognition of channels in PBF-LB/M-printed parts would be worthwhile to automatically identifying all inaccessible part areas.

Another aspect that should be considered is that part areas "missed" by the AAA were particularly those that can be reached by the tool, but where the tool's freedom of movement for processing is restricted by the part. This scenario should be considered in future improvements of the AAA by considering both the number and localization of rays emanating from the part during the assessment.

The results of the second part of the interview (see Fig. 8) also demonstrate a high level of agreement between the experts' own evaluation of the parts and the evaluation by the AAA. 64% of the parts marked by the algorithm were ultimately rated as correct by the experts, while 32% largely agreed with the results, except for one marking. Only 4% of the parts had engineers disagreeing with more than one marking by the algorithm. This also demonstrates an alignment between expert evaluations and the accessibility assessments provided by the AAA.

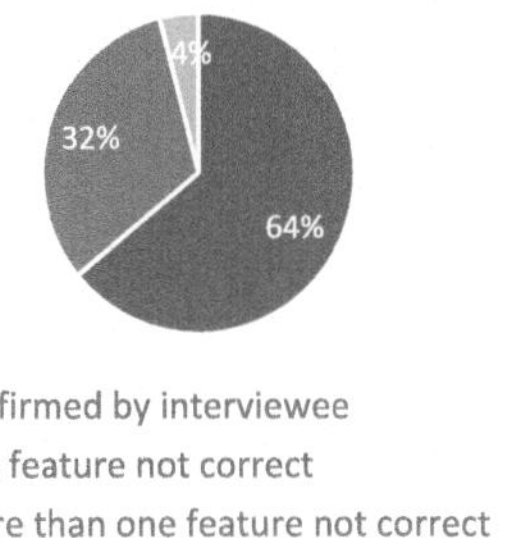

Fig. 8. Result of second part of interview.

6 Conclusion and Outlook

In this study, a successful first proof-of-concept was conducted for a selected ray tracing approach to automatically assess the accessibility of tools for removing support structures from PBF-LB/M manufactured parts (AAA). This was demonstrated by comparing the evaluations of inaccessible part areas of experts with the AAA algorithm, which already shows a high level of agreement. However, to further increase the matches in the future, the AAA should be enhanced by expanding its capabilities to identify part areas that can be additively manufactured without support structures regardless of part orientation, and by considering the freedom of movement of tools for support structure removal during the assessment.

Also, due to the inherent uncertainty associated with manual expert assessments, additional validation techniques are necessary to ensure the accuracy of the algorithm's results. Future efforts should focus on employing practical experiments, such as physically printing parts to support identified inaccessible and accessible areas, to verify the algorithm's predictions. Overall, while the AAA algorithm shows promise in identifying inaccessible part areas, ongoing research is needed to refine its functionalities and validate its accuracy. Upon completion of further investigations, this approach thus has the potential to be applied in orientation algorithms to automate the alignment of parts for the PBF-LB/M process.

Acknowledgments. The authors would like to thank the Federal Ministry of Economic Affairs and Climate Action for funding the research in the project "Ressourcenschonende Prozessroute für hochintegrierte Hydrauliksysteme am Beispiel einer elektrifizierten mobilen Ar-beitsmaschine" (HyRess, grant number 03LB3030G).

Disclosure of Interests. The authors have no competing interests to declare that are relevant to the content of this article.

References

1. Sakthivel, A.R., Kandasamy, J., Davim, J.P.: Managing Supply Chain Risk and Disruptions: Post COVID-19. Springer, Cham (2021)
2. Lee, H.: Synthesis report of the IPCC sixth assessment report (AR6). Longer report (2023). https://report.ipcc.ch/ar6syr/pdf/IPCC_AR6_SYR_LongerReport.pdf
3. Sweet, S., Swain, R.B. (eds.): Sustainable Consumption and Production, Volume I: Challenges and Development. Palgrave Macmillan Ltd, Cham (2021)
4. Brauckmann, O.: Smart Production: Wertschöpfung durch Geschäftsmodelle. Wertschöpfung durch Geschäftsmodelle. SpringerLink Bücher. Springer Science and Business Media, Berlin (2015)
5. Voigt, K.-I.M., Müller, J.: Digital Business Models in Industrial Ecosystems. Springer International Publishing, Cham (2021)
6. DIN Deutsches Institut für Normung e.V.: Additive Manufacturing - General principles - Fundamentals and vocabulary. Beuth, Berlin **01.040.25; 25.030** (2022)
7. Gibson, I., Rosen, D., Stucker, B.: Additive Manufacturing Technologies: 3D Printing, Rapid Prototyping, and Direct Digital Manufacturing. Springer, New York (2014)
8. Patel, C.D., Chen, C.-H.: Digital Manufacturing. The Industrialization of Art to Part 3D Additive Printing. Elsevier, San Diego (2022)
9. Wohlers, T., Campbell, R.I., Diegel, O., Kowen, J., Mostow, N.: Wohlers Report 2023 Trends. Forecasts. 3D Printing and Additive Manufacturing State of the Industry (2023)
10. DIN Deutsches Institut für Normung e. V.: DIN EN ISO/ASTM 52911–1:2020–05, Additive Fertigung_- Konstruktion_- Teil_1: Laserbasierte Pulverbettfusion von Metallen (ISO/ASTM 52911–1:2019); Deutsche Fassung EN_ISO/ASTM 52911–1:2019. Beuth Verlag GmbH, Berlin (2020)
11. Kumke, M.: Methodisches Konstruieren von additiv gefertigten Bauteilen. Dissertation (2018)
12. Kadir, A.Z.A., Yusof, Y., Wahab, M.S.: Additive manufacturing cost estimation models—a classification review. Int. J. Adv. Manuf. Technol. (2020). https://doi.org/10.1007/s00170-020-05262-5
13. Bi, Q.-Z., Wang, Y.-H., Ding, H.: A GPU-based algorithm for generating collision-free and orientation-smooth five-axis finishing tool paths of a ball-end cutter. Int. J. Prod. Res. (2010). https://doi.org/10.1080/00207540802570685
14. Ho, S., Sarma, S., Adachi, Y.: Real-time interference analysis between a tool and an environment. CAD Compute Aided Des (CAD Computer Aided Design) (2001). https://doi.org/10.1016/S0010-4485(00)00117-2
15. Chowdhury, S., Mhapsekar, K., Anand, S.: Part build orientation optimization and geometry compensations for additive manufacturing process. J. Manuf. Sci. Eng. (2017). https://doi.org/10.1115/1.4038293
16. Das, P., Mhapsekar, K., Chowdhury, S., Samant, R., Anand, S.: Selection of build orientation for optimal support structures and minimum part errors in additive manufacturing. Comput.-Aided Des. Appl. (2017). https://doi.org/10.1080/16864360.2017.1308074
17. Samant, R., Ranjan, R., Mhapsekar, K., Anand, S.: Octree data structure for support accessibility and removal analysis in additive manufacturing. Addit. Manuf. (2018). https://doi.org/10.1016/j.addma.2018.05.031
18. Zhang, B., Goel, A., Ghalsasi, O., Anand, S.: CAD-based design and pre-processing tools for additive manufacturing. J. Manuf. Syst. (2019). https://doi.org/10.1016/j.jmsy.2019.03.005
19. Tran, T.H.T., Mbow, M.M., Vignat, F., Nguyen, D.S.: Surface accessibility detection in CAD environment for additive manufacturing. In: Long, B.T., Kim, Y.-H., Ishizaki, K., Toan, N.D., Parinov, I.A., Vu, N.P. (eds.) Proceedings of the 2nd Annual International Conference on Material, Machines and Methods for Sustainable Development (MMMS2020), LNME, pp. 799–807. Springer, Cham (2021). https://doi.org/10.1007/978-3-030-69610-8_106

20. Allaire, G., Bihr, M., Bogosel, B.: Support optimization in additive manufacturing for geometric and thermo-mechanical constraints (2020). https://doi.org/10.1007/s00158-020-02551-1
21. DIN Deutsches Institut für Normung e.V.: DIN EN ISO/ASTM 52911-1, Additive Fertigung - Konstruktion - Teil 1: Laserbasierte Pulverbettfusion von Metallen: DE_307153562 (2019)
22. Alexander Kukan: Original-Prusa-i3. https://github.com/prusa3d/Original-Prusa-i3 (2019). Accessed 8 Feb 2024

Conceptualization of an Operational Capability Development Framework to Facilitate the Integration of Additive Manufacturing Based on a Use Case Perspective

Christopher Gustafsson(✉) and Christer Nygren

Mälardalen University, Hamngatan 15, 63105 Eskilstuna, Sweden
{christopher.gustafsson,christer.nygren}@mdu.se

Abstract. This paper proposes an operational capability development framework to facilitate additive manufacturing (AM) integration from an AM use case perspective. We conducted a single case study in collaboration with master students and AM actors. Data was collected through interviews with respondents from five AM actors and additional data collection methods. Data was analyzed thematically. The conceptualization of the proposed framework was performed iteratively by revisiting the literature and empirical data. The proposed framework provides insights on targeted operational performance and operational capabilities, including its constituents and development mechanisms, based on AM use cases which are evaluated regarding AM's potential and fit to fulfill a company's needs. We evaluated the framework based on selected AM use cases and additional insights from industry professionals and academics. This paper introduces a novel framework for operational capability developed to facilitate AM integration. Thereby, yielding another way of working with AM integration based on an AM use case perspective for engineers, managers, and other decision-makers. Therefore, we recommend that future research should extend the framework to include, evaluation using additional AM use cases, the development of dynamic capabilities, and evaluate other industrial contexts.

Keywords: 3D Printing · Adoption · Implementation · Production System Development · Advanced Manufacturing Technology

1 Introduction

In today's manufacturing industry, companies explore and exploit various advanced manufacturing technologies to sustain competitive advantage and improve business performance [1]. Additive manufacturing (AM), also known as 3D printing, is key with essential capabilities for manufacturing companies to unlock sustained competitive advantage and improved business performance [2, 3]. AM can be regarded as an advanced manufacturing technology [4], similar to artificial intelligence [5], drones [6], and mobile robots [7], among others, since AM uses computer-based technologies and capabilities

Published by Springer Nature Switzerland AG 2024
M. Thürer et al. (Eds.): APMS 2024, IFIP AICT 733, pp. 338–352, 2024.
https://doi.org/10.1007/978-3-031-71645-4_23

for different operational purposes intended for industrial contexts [8]. Manufacturing companies, especially well-established ones, that investigate the transition of extending (complementing and not necessarily replacing) existing operational capabilities with new AM-based operational capabilities still find it challenging [2, 9, 10]. Some examples of challenges are related to being embedded in traditional ways of working and culture over several decades, too high investment costs, uncertainty in value-adding opportunities, unwillingness to try again, AM being perceived as a complex technology, and not mature enough AM technology for specific industrial use cases, among others [see e.g., 11–16].

In the current advanced manufacturing technology and AM literature, there is a lack of research regarding operational capability development frameworks. Despite the lack, there exist a few in the operations and production management literature [17, 18] and information technology literature [19]. On the other hand, according to the business literature, there exist organizational capability development frameworks [20, 21]. However, these are focused more on developing a business structure (i.e., organizational) rather than on developing how you get things done (i.e., operational). Additionally, these frameworks do not focus on AM. In this research, the idea is to provide engineers, managers, and other decision-makers in manufacturing companies with an operational capability development framework as a guide to, not only implement and adopt AM, but to integrate AM with existing operational capabilities, resources, and ways of working, to sustain competitive advantage and improve business performance. Previous research has highlighted the importance of involving AM actors (e.g., service providers, machine providers, suppliers, research institutes, and universities) in operational capability development [22–25] and that AM integration might be use case dependent [26–28].

Based on this background, the purpose of this paper is to propose an operational capability development framework to facilitate AM integration from an AM use case perspective.

2 Frame of Reference

2.1 Additive Manufacturing Integration Process and Use Case Fit

AM is defined as the "*process of joining materials to make parts from 3D model data, usually layer upon layer, as opposed to subtractive manufacturing and formative manufacturing technologies*" [29]. AM integration is seen as the next step beyond AM implementation and AM adoption where the technology is an integral part of existing operational capabilities, resources, and ways of working. Previous research has identified value-adding areas in a generic company to facilitate AM integration with a use case perspective [27, 28]. A use case can be referred to as the "*application of a technology for a specific operational purpose*" [6, p. 568]. AM use case types [28] can be explained as follows: *prototyping* (i.e., physical artifacts to reproduce a portion or the whole of a final product for testing and validation during development), *incremental product launch* (i.e., uncertainty reduction and increased speed of a new product launch), *improved delivery* (i.e., improving and streamlining the delivery of goods and services to customers), *production tools* (i.e., all shape-giving devices and non-shape-giving devices used during the manufacturing of a product), *process concentration* (i.e., reducing the number of steps needed to manufacture a portion or the whole of a product), *custom products* (i.e., changing product characteristics, product model modification, or one-time design of a product meeting specific customer's needs), and *enhanced designs* (i.e., enhancing aesthetics and functionality of a product). To evaluate whether an AM use case is suitable for a company, the following *technology fit* model [6] could be used: (1) introduce new technology, (2) identify use cases, (3) pilot use case, (4) scale use case; and (5) restart from step 2. In between steps (2 & 3, 3 & 4, and 4 & 5), the use case should be evaluated based on its fit regarding factors such as operational, supply chain, economic, strategic, organizational, and behavioral.

Previous research has provided several frameworks for AM integration processes. One study focused on a company pull approach for the AM adoption process and highlighted its potential business impact [30]. The second study developed an AM implementation framework with several factors that managers and other decision-makers should consider [16]. On the other hand, previous research has provided simplified three-stage processes based on established literature [31]. The first study investigated the role of service providers in AM adoption [25] and the other one investigated AM implementation for mass customization [32]. The simplified three-stage integration process [31] includes *pre-integration* (i.e., the resources and practices relevant to the success or failure of AM integration), *integration* (i.e., the successful and consistent realization of a working order after achieving AM integration), and *post-integration* (i.e., continuous improvements of technical and business activities after achieving AM integration).

2.2 Operational Capability Development

Companies that are seeking a sustainable competitive advantage [33, 34] should investigate developing their existing resources and capabilities for long-term success in their respective markets [35]. In the strategic management literature, an interesting way to achieve this is to avoid extrapolating solely on the resource-based view [35] and combine the theoretical lens with capability theory. Capability theory is a way for companies to adapt, innovate, and reshape as a response to uncertain and changing environments. In general, a company has two kinds of capabilities, namely ordinary capabilities and dynamic capabilities [36]. Ordinary capabilities are operational in nature which explains why they also can be called operational capabilities [37]. Operational capabilities are interesting since they explain company-specific resources and capabilities used for different operational purposes and their role in creating and sustaining competitive advantage [36, 37]. In this research, operational capability refers to company-specific sets of skills, processes, routines, facilities, equipment, administration, and coordination developed in the operations management system used for problem-solving through the configuration of operational resources [36, 37]. Operational capabilities can be differentiated into the following constituents: *skills* (i.e., specific abilities, competence, knowledge, and expertise a person possesses to perform a task), *facilities and equipment* (i.e., physical and digital assets a company uses for daily operations), *processes and routines* (i.e., actions used to accomplish tasks or achieve goals), and *administration and coordination* (i.e., management of assets and coordination of efforts to achieve goals).

The following taxonomy [37] presents a way to determine which operational capabilities to develop: *operational improvement* (i.e., incrementally refining and reinforcing existing operational resources and practices), *operational innovation* (i.e., radically improving existing operational resources and practices or creating and implementing new and unique operational resources and practices), *operational customization* (i.e., creating knowledge by extending and customizing operational resources and practices), *operational cooperation* (i.e., creating healthy and stable relationships with people from different internal functional organizations and external actors), *operational responsiveness* (i.e., quick and easy reaction to changes in inputs and output requirements), and *operational reconfiguration* (i.e., accomplishing the necessary transformation to re-establish the fit between operational strategy and the market, when their equilibrium has been disturbed). Furthermore, the means of operational capabilities in sustaining competitive advantage is their relation to other factors and potential impact on operational performance such as *cost* (e.g., reduced manufacturing cost), *quality* (e.g., improved product performance), *delivery* (e.g., reduced time-to-market), and *flexibility* (e.g., enabled product customization) [37, 38].

Managers in companies often face the task of developing operational capabilities in the best way possible [17–19], especially if the company has certain requirements for operational performance. Therefore, a study presented the following mechanisms for operational capability development [17] and are presented as follows: *learning* (i.e., how companies develop operational capabilities in-house), *building* (i.e., how companies develop operational capabilities from other internal functional organizations, suppliers, and customers for extending the existing operational capability base), and *acquiring* (i.e., how companies utilize the operational capabilities of other actors such as machine providers, service providers, other companies, consultancy firms, universities, and research institutes).

3 Research Methodology

3.1 Research Design

This research proposes an operational capability development framework to facilitate AM integration from an AM use case perspective. This research was conducted in collaboration with 10 master's students from Mälardalen University who explored AM integration in a university course focusing on the value creation perspective. Additionally, the research was set up collaboratively with five AM actors (see Table 1) of different sizes according to European Legislation [39]. Both the 10 master's students and the five AM actors were selected based on opportunistic sampling [40]. The selection of the 10 master's students was based on the following criteria (1) they explored AM integration from a value creation perspective, (2) they have an interest in AM, and (3) they were allowed to contribute to ongoing AM research. The selection of the five AM actors was based on the following criteria (1) they have expert competence or capabilities in AM integration, (2) they are directly or indirectly involved in the product development and production of AM parts, and (3) they are involved or have been involved with customers and other stakeholders in the manufacturing industry directly or indirectly related to AM integration.

The research followed the single case study guidelines [41] since they are useful for exploring phenomena in depth compared to multiple case studies that explore phenomena from multiple perspectives. The single case study focused on the AM journey manufacturing companies should partake in toward AM integration based on insights from the AM actors' perspectives. This research took additional inspiration from longitudinal case study guidelines [42, 43] to help capture the understanding and knowledge of facilitating AM integration from the context and perspectives of the AM actors over time.

Before the initiation of this single case study, one of the authors had contact with the five AM actors, informal meetings with the respondents of the five AM actors and several of their colleagues, and managed to visit some of their facilities. One of the authors already had extensive experience and knowledge in AM and the other author gained experience and knowledge in 2021. The planning of the single case study was initiated and initially planned by the authors which had identified the five AM actors. Thereafter, one of the authors continued the planning together with the 10 master students who then together conducted the data collection. Then, the 10 master students conducted the initial data analysis. Lastly, both authors conducted additional data analysis, revisited the literature, collected initial insights from potential users from the industry, and summarized the findings in this paper. This single case study unofficially started in 2019 and officially started with proper research planning at the beginning of 2022 until 2024.

Table 1. Summary of the five AM actors

Designation	AM business	Turnover	No. of employees (size)	Location
Company A	Development, design, production, education, sales, consultancy	>$2M	10–49 (small)	Sweden
Company B	Prototyping, material development, material sales, machine elements, design, development	>$3M	10–49 (small)	Sweden
Company C	Technical consultancy	>$79M	>250 (large)	Sweden
Company D	Production, machine provider	>$5M	10–49 (small)	Sweden
Company E	Service provider, machine provider, production, education	>$0,8M	10–49 (small)	Sweden

3.2 Data Collection

Data collection was conducted through a total of five semi-structured interviews with the five Respondents A-E from Company A-E (see Table 2). The interviews were conducted from late April to early May 2022 and held online via Teams/Zoom. The interviews were not recorded at the request of the respondents. To mitigate this, the person leading the interview had one, and sometimes, two additional people assisting with taking notes. The people involved in each of the interviews were responsible for transcribing the material which was done manually and resulted in comprehensive summaries presented in Word documents. Additional data was collected from informal meetings, AM expert opinions, observations from factory and facility tours, online and offline seminars and workshops held specifically by the respondents or by the company, company documents, and extensive field notes. A total of three AM use cases were selected (see Table 3) to evaluate the operational capability development framework. They were selected based on intensity sampling [40] since they (1) provide examples of AM use case types, covering one example per AM use case type, and (2) contain rich and intense, but not extreme, descriptions of the specific use case type.

Table 2. Summary of the respondents from the five AM actors

Designation	Job role	Area(s) of expertise	Case company	Work experience (years)
Respondent A	Senior metallurgist	AM materials	Company A	5
Respondent B	Owner & managing director	General contractor in rapid prototyping	Company B	18
Respondent C	AM expert	AM in general, materials, process, mechanical engineering	Company C	2
Respondent D	Production technician	AM production planning and operations	Company D	8
Respondent E	Technical salesman	Sales, marketing, project management	Company E	2

3.3 Data Analysis

Data analysis was conducted thematically [44] to locate emerging themes and patterns in the collected data. The analysis process was performed in the following steps: (1) open-coding by grouping excerpts into first-order concepts, (2) axial-coding by grouping the first-order concepts into second-order themes, and (3) higher-level abstraction by grouping the second-order themes into third-order dimensions. This analysis process mitigated the identification of overlapping and emerging concepts contributing to potentially novel insights [45]. The analysis of the empirical findings provided a wide variety of constituents for an initial operational capability development framework. This was then supported by the frame of reference by providing theoretical concepts to the operational capability development framework. Additionally, the operational capability development framework was shown to two other AM experts (both with 5+ years of AM work experience), one university lecturer (with no AM work experience), and two managers (one with no AM work experience and the other with 1+ years of AM work experience) from a global heavy vehicle manufacturing company for additional feedback. Thereafter, the operational capability development framework was evaluated by providing information from the selected AM use cases. Lastly, support from the frame of reference provided additional theoretical concepts, thereby extending the operational capability development framework. The conceptualization of the operational capability development framework was done iteratively through triangulation [46] between (1) insights from the interviews, (2) insights from additional data collection, and (3) reviewing the literature. The data collection and data analysis were performed through several iterations until reaching data saturation and theoretical saturation. This meant that additionally collected data and data analysis did not add any significant novel findings and there were enough findings to fulfill the purpose [47, 48].

Table 3. Summary of the three AM use cases

Designation	Use case	Type	Sector	Country	Ref.
Case A	AM for prototyping	Prototyping	Production	United States	[49]
Case B	Production fixtures	Production tools	Automotive	Belgium	[50]
Case C	Heat exchanger	Enhanced designs	Automotive	Australia	[51]

4 Results

4.1 Emergence of an Operational Capability Development Framework

The empirical findings provided a wide variety of insights that companies are required to consider when integrating AM. Based on these findings, an initial framework emerged consisting of (1) *operational performances* (initial description: potential benefits of AM technology), (2) *constituents of operational capabilities* (initial description: new required capabilities), and (3) *development mechanisms* (initial description: development opportunities). In terms of operational performances, the respondents mentioned several benefits including reduced part cost, reduced time-to-market, improved part functionality, reduced production process steps, enhanced carbon footprint and handprint, and simplified supply chain, among others. Regarding constituents of operational capabilities, all respondents agreed that companies need certain design knowledge, material knowledge, economic understanding, hobby printers, safety procedures, AM versus traditional manufacturing, standardized way of working with AM, collaboration with part manufacturers, commitment from staff, and decide on make or buy, among others. Vis-á-vis development mechanisms, the respondents recommended less than a handful of approaches including practical learning, learning through physical examples, online is not the reality, and providing support when needed.

By revisiting the literature, as a starting point, the initial framework could be expanded by including the following steps (1) *select AM use case* [28], one of seven value-adding AM areas in a company, and pursue a function-driven design strategy or manufacturing-driven design strategy [27], (2) *identify operational performance* [37, 38], potential benefits and opportunities of the selected AM use case, (3) *identify constituents of operational capabilities* [36, 37], required for reaching the targeted operational performance, (4) *identify operational capabilities* [37], by combining various constituents of operational capabilities and determine which operational capabilities to pursue, and (5) *select development mechanism* [17], selecting suitable development strategies for the operational capabilities. Thereafter, previous research has suggested (6) *evaluate AM fit* [6] and deciding whether the selected AM use case is a suitable fit to solve the company's need(s). This evaluation could be used to further explore the AM use case as a pilot case, and if applicable, scale the use case and explore if it is suitable as a business case. Once finished with the evaluation, whether a positive or negative outcome was reached, additional AM use cases should be explored. The reasoning includes (7) *develop AM experience and knowledge* about a wide range of AM use cases, consequently building a library of experience and knowledge over time.

Insights from the two AM experts highlighted the importance of the proper selection of the AM use case, including AM material selection (e.g., polymer vs metal) and AM process selection (e.g., powder-based, filament-based, or liquid-based) as part of the AM use case selection process. Additionally, arguments pointed out that if the part was not designed for AM from the beginning, it probably would not generate any significant value for a company. Even this statement requires a sense of reflection since some areas in a company (e.g., spare parts and production tools) need minimal or no design changes at all [27, 28]. Insights from the two managers highlighted the need for additional steps before selecting AM and deciding if the technology is suitable based on the company's needs and brings enough value (e.g., lowest cost possible, fastest delivery possible, and highest quality possible). The starting point would then be (1) identifying the *need(s) in the company* and (2) *evaluate AM potential to solve the need(s)* to decide whether AM both solves the needs and brings enough value to the company. To avoid re-inventing the wheel, it is recommended to include internal cross-functional input and external expertise from AM actors throughout the process [18] from evaluating AM potential to evaluating AM fit. Thereby forming the final version of the operational capability development framework (Fig. 1) to facilitate AM integration based on a use case approach.

4.2 Evaluation of the Operational Capability Development Framework Based on the Selected AM Use Cases

The final version of the operational capability development framework (Fig. 1) to facilitate AM integration was evaluated based on the selected AM use cases (Table 3). However, due to the nature of the selected AM use cases such as being based on secondary data rather than primary data, only steps including *select AM use case* until *select development mechanism(s)* could be evaluated. The other steps required additional data which was not accessible in the selected AM use cases and should be included in future evaluations.

Case A: AM for Prototyping. In Case A [49], the challenge was to find a solution to *reduce cost and increase speed* during the prototyping process (need). The study

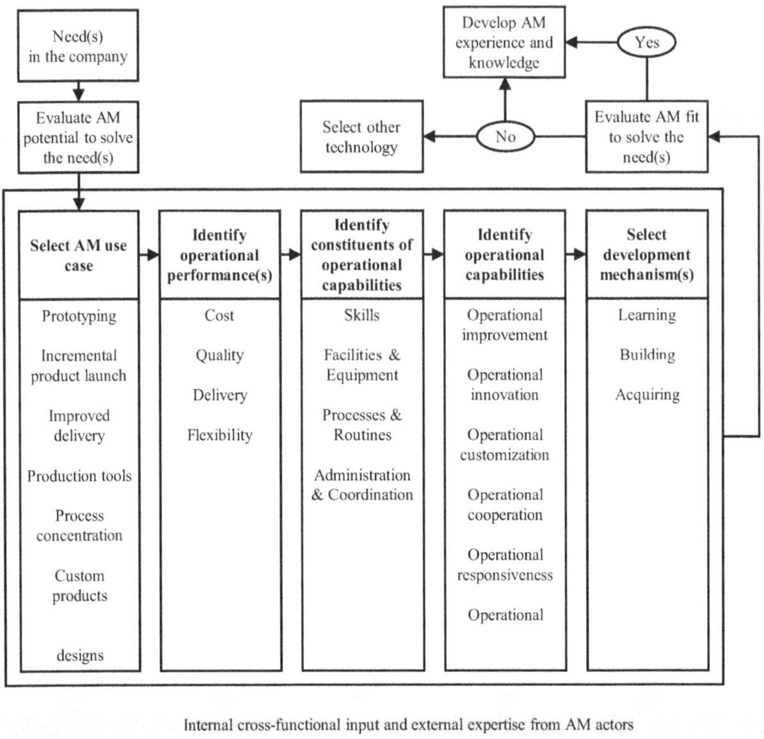

Fig. 1. Operational capability development framework to facilitate AM integration based on a use case perspective

focused on the *prototyping* (AM use case) of parts (e.g., oil caps and hood pin plates) for lightweight vehicles. AM provided a *faster new product introduction, saving weeks of additional work* and *saving $ 3,000 per product in production costs* (operational performance). Part concepts were created in *CAD software from Solidworks* and *a ProJet Color Jet Printer 360 AM machine* was selected for the production of physical models (constituents of operational capabilities). Such investments provided the engineering team to *make critical design decisions* and the sales team with *a preview of upcoming products* (operational capabilities). Investments were *acquired from different AM actors*, and later, the company's workforce and their suppliers *learned to adapt to the new ways of working with AM* (development mechanism).

Case B: Production Fixtures. In Case B [50], the challenge was to *increase efficiency* in the production line of lightweight vehicles (need). The study focused on *production tools* (AM use case), such as fixtures, used in the assembly line. AM provided a fixture with a *48% cost reduction, 64% weight reduction*, and a *faster assembly process* (operational performance). The company *collaborated with an AM actor* that designed and produced a fixture with an *AM machine* using *AM materials such as PA12* for the connection points, and *TPU* for the buffers, among others (constituents of operational capabilities). This investment allowed the company to *think additive* and allowed them to *continuously* *improve* the assembly line (operational capabilities). The investments into the new fixture were *acquired by an AM actor* (development mechanism).

Case C: Heat Exchanger. In Case C [51], the challenge was to *increase the performance* and *reduce the weight* of a heat exchanger to be applied across multiple industries such as aerospace, automotive, energy, chemical, and processing, among others (need). The study focused on *enhanced designs* (AM use case) of the next generation of heat exchangers. With AM, it became possible to realize a heat exchanger with *3 times higher performance of heat rejection, reducing the pressure drop by 2/3, 22% weight reduction, increased design flexibility* for optimum placement inside the vehicle, and *part consolidation* into a single part thus *removing assembly time and costs* (operational performance). After scanning the AM landscape, an *AM actor was selected for collaboration* that developed (using *computational fluid dynamics* and *non-linear thermos-mechanical finite element modeling*) and produced (using *AM software tools* for data preparation, process optimization, and quality assurance) the heat exchanger within 6 months (constituents of operational capabilities). The new heat exchanger highlights a *radical improvement with an AM heat exchanger*, and consequently, a *new way of working with AM* (operational capabilities). The investments led to the *expansion of existing AM business* with additional AM expertise through *collaboration with an AM actor* (development mechanism).

5 Discussion

The evaluation of the operational capability development framework based on the selected AM use cases revealed a systematic way of facilitating AM integration, especially in the *pre-integration stage* [31]. Because, at this point, significant information has been collected regarding the resources and practices relevant to the success or failure of AM integration. The evaluation further revealed preliminary indications that different AM use cases include different operational performance and operational capabilities including its constituents and development mechanisms. However, this evidence is not conclusive and needs to be quantified including all seven AM use case types and additional AM use cases per type. Additionally, the evaluation revealed that companies initially acquired operational capabilities (e.g., services in the design and production of AM parts) from AM actors, and then, began building and learning (e.g., think additive and continuously improve) over time. These indications are similar to previous research in operational capability development in digital servitization [17].

Thereby, the following step (evaluate AM fit to solve the needs), equals the go or no-go point toward the *integration stage* [31]. This evaluation should not be taken lightly [6] and needs to consider, for example, the characteristics of the selected AM use case [27, 28] and its business impact [30]. Then, a trial-and-error approach combined with either manufacturing-driven or function-driven design strategies [13, 27] is recommended in developing AM experience and knowledge for the successful or unsuccessful and consistent realization of a working order with AM integration. However, this framework is unlikely to be a "*one-size-fits-all*" framework that can explain operational capability development in manufacturing companies [18, p.53].

Concerning previous research, the proposed operational capability development framework facilitates AM integration with the following contributions: (1) reducing uncertainties in AM integration through a use case perspective [13, 27, 28], meaning that the AM use case dictates which operational capabilities should be developed, (2) providing resources to engineers, managers, and other decision-makers to systematically identify operational capabilities, including its constituents and development mechanisms, based on specific AM use cases [6, 13, 24], (3) providing a new way of working for engineers, managers, and other decision-makers with AM integration [13, 16, 30], (4) collaborating with AM actors throughout the AM integration process, if and when necessary [23–25], and (5) the notion of evolving the use case into a business case stems from evaluating AM fit of one or several pilot cases [6, 12, 30], and (6) allowing the operational capability development framework to evolve and be adapted to different industry-contexts, advanced manufacturing technologies, and use cases [6, 17–19].

Moreover, the empirical findings in the conceptualization of the operational capability development framework had to be complemented by literature. Another insight from this research revealed that the operational capability development framework should continuously evolve and be adapted according to the ways of working of industry professionals in manufacturing companies. Despite the initial evaluation from one university lecturer and two managers of a global heavy vehicle manufacturing company, additional evaluation from other roles such as engineering, purchasing, and sales, among others, as well as other manufacturing companies are recommended.

6 Conclusion

To conclude, this research proposed an operational capability development framework to facilitate AM integration from an AM use case perspective. To the best of our knowledge, this paper contributes to the AM literature with a novel conceptualization of an operational capability development framework to facilitate AM integration. The operational capability development framework allows engineers, managers, and other decision-makers a new way of working by systematically identifying targeted operational performances and operational capabilities including its constituents and development mechanisms, based on specific AM use cases which are then evaluated.

Research limitations include, but are not limited to, the small sample size of included respondents in the interviews, as well as included AM use cases and insights from industry professionals and academics to evaluate and further improve the operational capability development framework, thereby limiting generalizability. It is recommended to include more participants, not only additional AM actors, but industry professionals from specific industry sectors to improve validity. Additionally, it is recommended to include more AM use cases that are (1) based on primary data and (2) higher quantity per AM use case type, in the evaluation and further improving the operational capability development framework, thus reducing potential bias, and improving validity.

Some of the practical implications of the conceptualized operational capability development framework to facilitate AM integration include the lack of extensive empirical testing (e.g., providing this framework to industry professionals for them to test and evaluate). However, this framework is meant to be an evolving framework for future academics and industry professionals to use in AM integration.

Recommendations for future research include (1) improving the robustness of the operational capability development framework by evaluating additional AM use cases, (2) adding constituents for enabling dynamic capability development as part of the same framework, based on the notion that dynamic capabilities are developed based on operational capabilities, or develop a separate framework similar but focus on a dynamic capability development framework, (3) perform additional empirical evaluation and testing of the operational capability development framework in different industrial contexts such as the heavy vehicle manufacturing industry and other sectors, and (4) exploring similarities and differences between current education and training material and services provided by different AM actors such as service providers, machine providers, universities, and research institutes, among others.

7 Disclosure of Interests.

The authors have no competing interests to declare that are relevant to the content of this article.

Acknowledgments. Many thanks to the students of the university course INO213 in the spring of 2022 who participated in the data collection and initial data analysis as part of this research. This research project has been funded by the Knowledge Foundation within the framework of the ARRAY++Research School and the participating companies, and Mälardalen University, Sweden. The people who contributed with their knowledge are gratefully thanked for making these experiences available.

References

1. Sun, H., Hong, C.: The alignment between manufacturing and business strategies: its influence on business performance. Technovation **22**(11), 699–705 (2002). https://doi.org/10.1016/S0166-4972(01)00066-9
2. Niaki, M.K., Nonino, F., Tafakkori, K., Torabi, S.A., Kazemian, I.: The influence of manufacturing contexts on additive manufacturing-enabled competitive capabilities. J. Manuf. Technol. Manag. **33**(6), 1102–1123 (2022). https://doi.org/10.1108/JMTM-07-2021-0241
3. Culot, G., Orzes, G., Sartor, M., Nassimbeni, G.: The future of manufacturing: A Delphi-based scenario analysis on Industry 4.0. Technol. Forecast. Soc. Change **157**, 120092 (2020). https://doi.org/10.1016/j.techfore.2020.120092
4. Arvanitis, S., Hollenstein, H.: The Determinants of the adoption of advanced manufacturing technology: an empirical analysis based on firm-level data for swiss manufacturing. Econ. Innov. New Technol. **10**(5), 377–414 (2001). https://doi.org/10.1080/10438590100000015
5. Leberruyer, N., Bruch, J., Ahlskog, M., Afshar, S.: Toward Zero Defect Manufacturing with the support of Artificial Intelligence—Insights from an industrial application. Comput. Ind. **147**, 103877 (2023). https://doi.org/10.1016/j.compind.2023.103877
6. Maghazei, O., Lewis, M.A., Netland, T.H.: Emerging technologies and the use case: a multi-year study of drone adoption. J. Oper. Manag. **68**(6–7), 560–591 (2022). https://doi.org/10.1002/joom.1196

7. Fattouh, A., Chirumalla, K., Ahlskog, M., Behnam, M., Hatvani, L., Bruch, J.: Remote integration of advanced manufacturing technologies into production systems: integration processes, key challenges and mitigation actions. J. Manuf. Technol. Manag. **34**(4), 557–579 (2023). https://doi.org/10.1108/JMTM-02-2022-0087
8. Stentoft, J., Philipsen, K., Haug, A., Wickstrøm, K.A.: Motivations and challenges with the diffusion of additive manufacturing through a non-profit association. J. Manuf. Technol. Manag. **32**(4), 841–861 (2021). https://doi.org/10.1108/JMTM-04-2020-0125
9. Roscoe, S., Cousins, P.D., Handfield, R.: The microfoundations of an operational capability in digital manufacturing. J. Oper. Manag. **65**(8), 774–793 (2019). https://doi.org/10.1002/joom.1044
10. Holmström, J., Liotta, G., Chaudhuri, A.: Sustainability outcomes through direct digital manufacturing-based operational practices: a design theory approach. J. Clean. Prod. **167**, 951–961 (2017). https://doi.org/10.1016/j.jclepro.2017.03.092
11. Gustafsson, C., Sannö, A., Bruch, J., Chirumalla, K.: Exploring challenges in the integration of additive manufacturing. In: Kim, D.Y., von Cieminski, G., Romero, D. (eds.) APMS 2022. IFIP AICT, vol. 663, pp. 370–379. Springer, Cham (2022). https://doi.org/10.1007/978-3-031-16407-1_44
12. Yi, Y., Gläßner, C., Aurich, J.C.: How to integrate additive manufacturing technologies into manufacturing systems successfully: a perspective from the commercial vehicle industry. J. Manuf. Syst. **53**, 195–211 (2019). https://doi.org/10.1016/j.jmsy.2019.09.007
13. Martinsuo, M., Luomaranta, T.: Adopting additive manufacturing in SMEs: exploring the challenges and solutions. J. Manuf. Technol. Manag. **29**(6), 937–957 (2018). https://doi.org/10.1108/JMTM-02-2018-0030
14. Dwivedi, G., Srivastava, S.K., Srivastava, R.K.: Analysis of barriers to implement additive manufacturing technology in the Indian automotive sector. Int. J. Phys. Distrib. Logist. Manag. **47**(10), 972–991 (2017). https://doi.org/10.1108/IJPDLM-07-2017-0222
15. Ford, S., Despeisse, M.: Additive manufacturing and sustainability: an exploratory study of the advantages and challenges. J. Clean. Prod. **137**, 1573–1587 (2016). https://doi.org/10.1016/j.jclepro.2016.04.150
16. Mellor, S., Hao, L., Zhang, D.: Additive manufacturing: a framework for implementation. Int. J. Prod. Econ. **149**, 194–201 (2014). https://doi.org/10.1016/j.ijpe.2013.07.008
17. Momeni, K., Raddats, C., Martinsuo, M.: Mechanisms for developing operational capabilities in digital servitization. Int. J. Oper. Prod. Manag. **43**(13), 101–127 (2023). https://doi.org/10.1108/IJOPM-04-2022-0259
18. Wiengarten, F., Curach, C.F., Franke, H., Netland, T.H., Schmidt, F.K.: Impact pathways: towards an adapted understanding of the development of operational capabilities. Int. J. Oper. Prod. Manag. **43**(13), 50–67 (2023). https://doi.org/10.1108/IJOPM-08-2022-0475
19. Jarvenpaa, S.L., Mao, J.-Y.: Operational capabilities development in mediated offshore software services models. J. Inf. Technol. **23**(1), 3–17 (2008). https://doi.org/10.1057/palgrave.jit.2000125
20. Schriber, S., Löwstedt, J.: Tangible resources and the development of organizational capabilities. Scand. J. Manag. **31**(1), 54–68 (2015). https://doi.org/10.1016/j.scaman.2014.05.003
21. Saá-Pérez, P.D., García-Falcón, J.M.: A resource-based view of human resource management and organizational capabilities development. Int. J. Hum. Resour. Manag. **13**(1), 123–140 (2002). https://doi.org/10.1080/09585190110092848
22. Gustafsson, C., Sannö, A., Chirumalla, K., Bruch, J.: Integration of additive manufacturing in an industrial setting: the impact on operational capabilities. In: Alfnes, E., Romsdal, A., Strandhagen, J.O., von Cieminski, G., Romero, D. (eds) APMS 2023. IFIP AICT, vol. 690, pp. 590–604. Springer, Cham (2023) https://doi.org/10.1007/978-3-031-43666-6_40

23. Akmal, J.S., Salmi, M., Björkstrand, R., Partanen, J., Holmström, J.: Switchover to industrial additive manufacturing: dynamic decision-making for problematic spare parts. Int. J. Oper. Prod. Manag. **42**(13), 358–384 (2022). https://doi.org/10.1108/IJOPM-01-2022-0054
24. Friedrich, A., Lange, A., Elbert, R.: How additive manufacturing drives business model change: the perspective of logistics service providers. Int. J. Prod. Econ. **249**, 108521 (2022). https://doi.org/10.1016/j.ijpe.2022.108521
25. Chaudhuri, A., Rogers, H., Soberg, P., Pawar, K.S.: The role of service providers in 3D printing adoption. Ind. Manag. Data Syst. **119**(6), 1189–1205 (2019). https://doi.org/10.1108/IMDS-08-2018-0339
26. Gustafsson, C.: Exploring the integration of additive manufacturing: lessons learned and success factors of use cases. In: Klahn, C., Meboldt, M., Ferchow, J. (eds.) Industrializing Additive Manufacturing. AMPA 2023. Springer Tracts in Additive Manufacturing, pp. 423–439. Springer, Cham (2024) https://doi.org/10.1007/978-3-031-42983-5_28
27. Klahn, C., Fontana, F., Leutenecker-Twelsiek, B., Meboldt, M.: Mapping value clusters of additive manufacturing on design strategies to support part identification and selection. Rapid Prototyp. J. **26**(10), 1797–1807 (2020). https://doi.org/10.1108/RPJ-10-2019-0272
28. Fontana, F., Klahn, C., Meboldt, M.: Value-driven clustering of industrial additive manufacturing applications. J. Manuf. Technol. Manag. **30**(2), 366–390 (2019). https://doi.org/10.1108/JMTM-06-2018-0167
29. ISO/ASTM 52900:2021 Additive manufacturing—General principles—Terminology: Fabrication additive—Principes généraux—Terminologie. Switzerland: ISO/ASTM International (2021)
30. Rylands, B., Böhme, T., Gorkin, R., III., Fan, J., Birtchnell, T.: The adoption process and impact of additive manufacturing on manufacturing systems. J. Manuf. Technol. Manag. **27**(7), 969–989 (2016). https://doi.org/10.1108/JMTM-12-2015-0117
31. Voss, C.: Implementation: a key issue in manufacturing technology: the need for a field of study. Res. Policy **17**(2), 55–63 (1988). https://doi.org/10.1016/0048-7333(88)90021-2
32. Deradjat, D., Minshall, T.: Implementation of rapid manufacturing for mass customization. J. Manuf. Technol. Manag. **28**(1), 95–121 (2017). https://doi.org/10.1108/JMTM-01-2016-0007
33. Peteraf, M.A.: The cornerstones of competitive advantage: a resource-based view. Strateg. Manag. J. **14**(3), 179–191 (1993). https://doi.org/10.1002/smj.4250140303
34. Barney, J.: Firm resources and sustained competitive advantage. J. Manag. **17**(1), 99–120 (1991). https://doi.org/10.1177/014920639101700108
35. Wernerfelt, B.: From critical resources to corporate strategy. J. Gen. Manag. **14**(3), 4–12 (1989). https://doi.org/10.1177/030630708901400301
36. Teece, D.J.: A capability theory of the firm: an economics and (Strategic) management perspective. N. Z. Econ. Pap. **53**(1), 1–43 (2019). https://doi.org/10.1080/00779954.2017.1371208
37. Wu, S.J., Melnyk, S.A., Flynn, B.B.: Operational capabilities: the secret ingredient. Decis. Sci. **41**(4), 721–754 (2010). https://doi.org/10.1111/j.1540-5915.2010.00294.x
38. Miller, J.G., Roth, A.V.: A taxonomy of manufacturing strategies. Manage. Sci. **40**(3), 285–304 (1994). https://doi.org/10.1287/mnsc.40.3.285
39. EUR-Lex. https://eur-lex.europa.eu/SV/legal-content/summary/micro-small-and-medium-sized-enterprises-definition-and-scope.html. Accessed 12 Mar 202
40. Patton, M.Q.: Qualitative Research & Evaluation Methods, 3rd edn. SAGE, London (2002)
41. Yin, R.K.: Case Study Research: Design and Methods, 6th edn. Sage, Thousand Oaks (2018)
42. Pettigrew, A.M.: Contextualist research and the study of organizational change processes. Res. Methods Inf. Syst., 53–78 (1985)
43. Pettigrew, A.M.: Longitudinal field research on change: theory and practice. Organ. Sci. **1**(3), 267–292 (1990). https://doi.org/10.1287/orsc.1.3.267

44. Gioia, D.A., Corley, K.G., Hamilton, A.L.: Seeking qualitative rigor in inductive research: notes on the Gioia methodology. Organ. Res. Methods **16**(1), 15–31 (2012). https://doi.org/10.1177/1094428112452151
45. Eisenhardt, K.M.: Building theories from case study research. Acad. Manag. Rev. **14**(4), 532–550 (1989). https://doi.org/10.2307/258557
46. Flick, U.: Doing Triangulation and Mixed Methods. SAGE, London (2018)
47. Low, J.: A pragmatic definition of the concept of theoretical saturation. Sociol. Focus **52**(2), 131–139 (2019). https://doi.org/10.1080/00380237.2018.1544514
48. Glaser, B.G., Strauss, A.L.: The Discovery of Grounded Theory: Strategies for Qualitative Research. Aldine Publishing, Chicago (1967)
49. 3D Printing and Reverse Engineering Supercharge Steeda AutoSports' Production. https://www.3dsystems.com/learning-center/case-studies/projet-supercharges-steeda-autosports-production?ind=automotive. Accessed 13 Apr 2024
50. Think Additive: How Volvo Car Gent Has Reimagined Production Fixtures. https://www.materialise.com/en/inspiration/cases/volvo-car-gent-production-fixtures. Accessed 13 Apr 2024
51. Conflux Technology Leverage the Full Potential of a Heat Exchanger. https://www.eos.info/en-us/industries/customer-success-stories/conflux-heat-exchanger-produced-with-additive-manufacturing. Accessed 13 Apr 2024

Operations Management of Additive Manufacturing

Siavash Haghighat Khajavi[1](✉), Mika Salmi[2], and Jan Holmström[1]

[1] Department of Industrial Engineering and Management, Aalto University, Maarintie 8, 02150 Espoo, Finland
siavash.khajavi@aalto.fi

[2] Department of Mechanical Engineering, Aalto University, 02150 Espoo, Finland

Abstract. This article reviews the growing literature on additive manufacturing (AM) operations management and sheds light on the emerging research areas in this field. As the AM use cases of final parts rapidly expand, it is essential to focus on the operations management of this technology and determine the primary current and future research streams. A literature study method is utilized to select, review, and categorize articles in the field of AM. The 108 articles selected after the initial evaluation were carefully examined and categorized. The selected papers evaluate AM from an operations management perspective. This article categorizes the body of knowledge studying the application and operations management of additive manufacturing into three categories: studies concerned with the industry's current state, forward-looking studies with a conceptual approach, and forward-looking papers with empirical grounding. Different AM processes studied are also considered. Our categorization showed that the latter category is still under-researched and presents an opportunity for future investigations. Moreover, six emerging streams of research in the third category were recognized. In addition to pointing out the areas of research that require more attention, this article aims to assist the researchers in better positioning their research.

Keywords: Additive manufacturing · operations management · 3D printing · rapid manufacturing · logistics · manufacturing · management · supply chain

1 Introduction

Additive manufacturing (AM), also known as 3D printing, is a digital technology for producing physical objects layer by layer from a three-dimensional computer-aided design (CAD) file. A simple explanation of this production process is as follows. The process begins with generating a three-dimensional CAD model of the object with its details and dimensions. Next, a computer program slices the three-dimensional CAD file into thin two-dimensional (2D) cross sections (layers). Then, the 2D layers are sent to the three-dimensional printing machine one layer at a time. The machine produces the object by building each layer on top of the previous one, utilizing different raw material solidification methods in its production chamber [41, 67]. The process may take from a

M. Thürer et al. (Eds.): APMS 2024, IFIP AICT 733, pp. 353–377, 2024.
https://doi.org/10.1007/978-3-031-71645-4_24

few hours to a few days to produce an object, depending on its size, material and required production precision.

This technology, introduced initially as rapid prototyping (RP) and three-dimensional printing, was invented in the 1980s [29] to produce rough physical prototypes of final products. Since then, it has continued to evolve in different aspects. Currently, increasingly more parts produced with this method are reaching the suitable precision and quality necessary for final functional parts for special applications, such as air-cooling ducts for aircraft, hearing aid shells and medical prostheses [38]. The technical committee within ASTM (American Society for Testing and Materials) International recently adopted the term additive manufacturing for the technology, as it is no longer solely for prototype production. There are different AM processes available with different props and cons, but standards divide those into seven different processes. The advancements that have made AM possible are widening the machine's material range, improving precision and final quality, and reducing machine acquisition cost [67].

1.1 AM Application in Operations

With its swift advancements (e.g., precision, speed, affordability, and materials range) and inherent capabilities, AM technology has the potential to fundamentally revolutionize manufacturing operations and supply chains [85, 113]. What makes this production method a potentially disruptive technology for supply chain management is its characteristics. Holmström et al. [49] highlight the following benefits of AM methods over the conventional manufacturing methods:

- No need for tooling (economies of scale do not fully apply, which makes customization and design revisions possible).
- Feasibility of producing small production batches.
- Possibility for quickly changing the design.
- Product optimization for function (aka generative design).
- More economical custom product manufacturing (batch of one) plus the capability to produce complex geometries.
- The potential for simpler supply chains with shorter lead times and lower inventories.

Additionally, AM could significantly reduce material waste in the case of powder bed fusion of metals [85, 145] through design for performance (e.g., generative design), reduction of material scrap and increasing the material reuse. AM is still wasteful when it comes to polymer bed fusion [146]. These characteristics may enable the supply chain managers to manufacture any part (including customized parts) in various locations and batch sizes without the need to be concerned about massive tooling costs. Jeff DeGrange, former manager of Boeing Phantom Works' Direct Manufacturing Process, once said, "One day, we will be building parts on-demand in space, on aircraft carriers and at other points of use" [26]. Moreover, researchers [49, 101] who studied AM potential application in the spare parts supply chain, had indicated the possibility of introducing distributed production utilizing this technology.

From the beginning, the possibility to produce directly from a digital design file without tooling positioned AM as a method with radical implications on supply chains. However, technology limitations needed to be ironed out. AM was initially utilized for prototyping to significantly accelerate the pace while reducing the price of each prototype [42, 147]. However, AM technology at that stage had several severe limitations, such as low material variety, post-processing and process deficiencies which led to inaccuracies or structural weakness. In a later period, AM was used to produce master patterns (e.g., wax patterns for investment casting), molds, and soft tooling. Currently, the range of material, process accuracy, and printed parts structural integrity allow for the direct manufacturing of functional parts. In a growing number of instances, companies are replacing traditional manufacturing with AM in their operations to achieve functional and economic benefits, and some such companies include General Electric (GE), Boeing, and Phonak.

Among the early and well-publicized examples is Boeing using plastics powder bed fusion to produce air cooling ducts for the F-18 Super Hornet fighter jets. The reason for that was to meet the complex part geometries needed for the modified design [52] and the relatively small production batch size of only around 500 airplanes. GE's utilization of metal powder bed fusion for the manufacturing of fuel injectors in the new jet engine (LEAP) design has several objectives: to consolidate the parts, which would consequently reduce the cost of welding and testing and extend the product lifespan [148]. However, further potential uses were left untapped. These uses can become the source of competitiveness and cost-saving, such as broad spare parts production, product market launch, and the use of AM to enhance materials handling in an assembly line. A recent study by Jiang et al. [55] has analyzed the scenarios for the future of additive manufacturing applications. Their research recognizes the significant themes in the future of AM, and among them are the areas of production localization, spare parts provision, product development, and resource sharing. This literature review enables the gap spotting in the body of knowledge related to the substantial field of operations management and screens the research done with a forward-looking approach.

The research questions of this study are as follows;

1. What are the main categories of research in the field of AM which have an operations management perspective? What AM processes and materials are they studying?
2. Which themes are emerging at the cutting edge of research on the operations management of AM?

The remainder of this paper is structured as follows; Research method is Sect. 2 followed by the literature analysis on "forward-looking empirical research". Section 4 presents the discussion and Sect. 5 is conclusions.

2 Research Method

The methodology used for this research is the study and review of the relevant literature. A literature review utilizes a well-defined method to identify, analyze and interpret all available research related to a specific question, area, or phenomenon of interest, to provide a way to analyze the previous works on the topic comprehensively, without a bias and in a repeatable manner.

After planning the review and setting the research questions, the execution of the plan started when we used keywords to search for articles on Google Scholar and Scopus academic search engines for publications. In the next phase, the collected articles were analyzed and the relevant ones were selected for in-depth study. In the final phase of the literature review, we synthesized the findings and reported them.

The current body of knowledge which evaluates AM from an operations management perspective is still relatively limited but growing. A broad search with keywords, additive manufacturing in operations, rapid manufacturing in supply chain, 3D printing operations management and additive manufacturing operations improvement on the Google Scholar and the Web of Science resulted in 607 articles which then were analyzed by reviewing the abstracts and picking 108 closely relevant works for the categorization. Moreover, two recent literature reviews on the AM impact and AM management [53, 62] were utilized as complementary sources. Prevalent research themes in this field of study are classified into four categories; literature review studies, backward-looking empirical studies concerned with the feasibility of AM in operations management, forward-looking conceptual studies concerned with proposing new practices while AM is utilized for production and finally, forward-looking empirical studies concerned with improvement in AM operations while it is used for functional parts production. Similarly, AM processes and materials from the studies are listed. Based on this categorization and identifying the literature gap, all the relevant articles were reviewed and assigned to one or more categories. Results of this categorization are presented in Fig. 1.

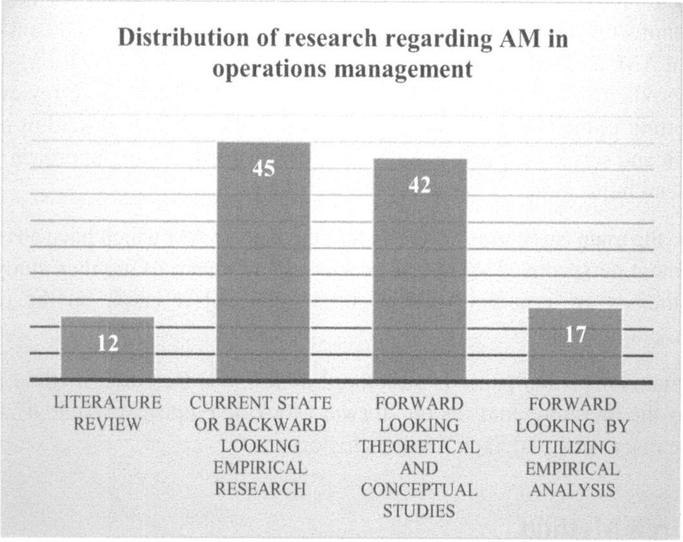

Fig. 1. Distribution of research regarding AM in operations management (more detail in Appendix)

The criteria used to classify the AM in operations management literature into the three classes of 1) Industry's current state, 2) Forward-looking studies with a conceptual approach, and 3) Forward-looking papers with empirical grounding were the following:

- Industry's Current State: Articles that study current use cases with empirical data and report the results of existing implementations are categorized under "Industry's current state".
- Forward-Looking Studies with a Conceptual Approach: Studies that do not include empirical use cases but propose future applications for AM in operations management conceptually are categorized as "Forward-looking studies with a conceptual approach".
- Forward-Looking Papers with Empirical Grounding: Articles that do not fall into the previous two classes and are not literature reviews but study a current use case and expand on that empirically through various methodologies, including simulations or scenario modeling, to empirically study the future implications of AM in operations management, are categorized as "Forward-looking papers with empirical grounding".

Articles such as Gao et al. [39], Kruth et al. [67] and Khorram Niaki, & Nonino [62], are categorized under the first column in Fig. 1 as they reviewed the literature regarding the AM. Articles such as Hopkinson & Dickens [50, 51], Mansour & Hague [84], Atzeni & Salmi [5], Baumers et al. [8, 9], Mellor et al. [89], are categorized under the empirical studies in AM with a current state or backward perspectives. Articles such as Holmström & Partanen [46], Holmström et al. [48, 49], Peres & Noyes [101], Sasson & Johnson [112], Tuck & Hague [127] and Tuck et al. [128], Lehmhus et al., [75], due to their forward-looking conceptual approach towards the study of the AM, are categorized under the third column in the Fig. 1.

To further clarify the forward-looking research perspective, the following example is depicted: In the spare parts supply chain, the backward-looking approach investigates the feasibility of AM utilization using the historical data to find out if the AM is feasible compared to conventional manufacturing. In contrast, the forward-looking approach goes one step further and investigates the characteristics and various operations management methods for AM implementation to achieve higher efficiency and effectiveness are. Forward-looking and empirically-based scenario research investigates a leading-edge context, considering actual supply chain settings and specific technological improvements in the short to medium term.

AM as a means for manufacturing functional parts is quite a young technology, and processes still require further improvements to make AM cost-competitive with conventional subtractive manufacturing. This consequently allows for broader applications for AM. The overall size of the AM industry in the year 2015, according to Wohlers report [139] was about \$5.1 billion. Compared to the estimated \$11.4 trillion total global manufacturing in the same year [124], this figure is just marginal. However, the overall size of AM industry grew to \$12.8 billion in the year 2020 [140].

Moreover, the AM field is advancing, and barriers are being addressed gradually. Such barriers include material cost and range, machine costs, process reliability and repeatability, parts structural resilience and surface finish, as well as labor intensiveness of pre- and post- processes. As the barriers are addressed, the utilization increases. The annual growth rate of the AM industry over the 27 years until 2016 has been around 26% [139]. However, it had slowed to 7.5% in 2020 [140]. Still, these are considered a relatively high figure, suggesting increasing adoption of this technology.

At this point, the bulk of the body of knowledge regarding AM is focused on research concerned with technology, process, material, and design—this is evident in journals such as the Rapid Prototyping Journal and the Additive Manufacturing journal from Elsevier. Another stream of research is to compare AM and conventional manufacturing from mechanical and economic perspectives [5, 37, 110]. This emphasis is understandable, as the AM for final parts production is still in its early stages. It is still not comparable with other conventional production methods regarding the application spectrum.

However, many studies have been conducted in the context of AM applications and its enabling of new operational capabilities (such as product customization, parts consolidation, and production decentralization). These are the studies aligned with this paper's direction and provide a forward-looking perspective for operations where AM is more widely used for the production of functional parts. Research by Walter et al. [134], Tuck et al. [128] and Holmström et al., [49] were among the early works to investigate the AM impact on the manufacturing supply chain. Tuck et al. [128], discussed how AM could influence the supply chains towards leaner, more agile systems that are used to serve the customized need of the consumers.

Walter et al. [134] and Holmström et al. [49], predicted the applicability of AM to replace the inventory in aerospace spare parts supply chains. Also, Berman, [11], identified spare parts and bridge manufacturing among promising AM applications. In their book, Gibson et al. [42] refer to "electronic spare parts" as one of the main AM drivers. They also take the example of F-18 fighter jet air cooling ducts produced by AM. The concept of electronic spare parts can benefit the user not just by eliminating the need to keep the molds or stamps but also through the decentralization of production. However, these studies remained at the conceptual level.

3 Forward-Looking and Empirical Research

Forward-looking studies that utilized empirical data to investigate AM are presented in Table 1. A detailed review is presented to pinpoint the literature gap and illustrate the areas for future research.

3.1 Customization

Tuck et al., [129], in a case study of fighter jet pilot ejector seat customization, using stereolithography AM method, investigated the capability of rapid manufacturing for functional custom parts production. They concluded that the material used by the vat photopolymerization method was not suitable for the demanding application of the ejector seat. However, reduction in production manual operation and workforce, which is needed in traditional customized manufacturing and lack of tooling, make rapid manufacturing a new type of agile manufacturing. They also found several improvement areas in rapid manufacturing technology, such as increasing the automation of pre-production processes (e.g.: 3D scanning and design optimization) and improving AM manufacturing speed. Reeves et al., [106], in a simple case review, also studied the AM as a tool for mass customization of products by investigating cases of AM use in the production of avatars and action figures (from online and offline video games) topographic data. They

Table 1. Articles with forward-looking perspective using an empirical approach.

Item	Article title
1	Production planning in additive manufacturing and 3D printing [76]
2	Predicting the future of additive manufacturing: A Delphi study on economic and societal implications of 3D printing for 2030 [55]
3	Quantifying the effects of additive manufacturing on supply networks by means of a facility location-allocation model [7]
4	Implications of additive manufacturing for spare parts inventory [120]
5	Additive manufacturing of biomedical implants: A feasibility assessment via supply-chain cost analysis [35]
6	Effects of combining product-centric control and direct digital manufacturing: The case of preparing customized hose assembly kits [81]
7	E-commerce channels for additive manufacturing: an exploratory study [36]
8	Additive manufacturing in the spare parts supply chain [141]
9	The impact of additive manufacturing in the aircraft spare parts supply chain: supply chain operation reference (SCOR) model-based analysis [80]
10	Analyzing Product Lifecycle Costs for a Better Understanding of Cost Drivers in Additive Manufacturing [79]
11	Additive manufacturing for mass customization [106]
12	Rapid Manufacturing Facilitated Customization [129]
13	Cost Estimation for Rapid Manufacturing Simultaneous Production of Mixed Components Using Laser Sintering [109]
14	A Web-based Manufacturing Service System for Rapid Product Development [71]
15	Additive manufacturing as a platform for introducing cyber-physical services [59]
16	Additive Manufacturing as an Enabler of Digital Spare Parts [61]
17	Economies of collaboration in build-to-model operations [45]

have pointed out the toollessness of the AM as the profound enabler of mass customization, which lowers the cost of producing individual units. Moreover, they discussed the potential savings in packaging and transshipment costs and lower lead-time between ordering and fulfillment as the benefits of a globally distributed AM supply chain.

3.2 Decentralized Production

Emelogu et al., [35] investigated the economic feasibility of AM in an in-situ production supply chain for the production of biomedical implants. They utilized a two-stage stochastic cost model to determine the number of AM facilities needed to be constructed as well as the number of implants to be ordered from AM or conventional manufacturing sites to minimize the total cost. An algorithm with an average sample approximation is used to find the solutions. Using primary data from a real-world case study, they pointed out the ratio between the unit production cost of AM and traditional manufacturing and item lead time and demand to be the key feasibility determinant parameters for AM implementation near the point of use for biomedical implant manufacturing. They recognized AM machine cost as a strong inhibitor for AM application in regional implant manufacturing at the current state of technology.

Jiang et al., [55] studied the future impact of additive manufacturing on society and the economy through scenarios based on a comprehensive Delphi survey. They conducted interviews and workshops with experts and performed a literature review to come up with 92 projections for the future of the AM in the year 2030. They narrowed down the number of projections to 18 by analyzing and combining. The 18 projections then were used in their Delphi survey. Based on their most probable scenario, in the year 2030, spare parts manufacturing of less critical parts will be done locally and by AM, while specialist conventional manufacturing hubs will produce critical parts with quality control capabilities. In the year 2030, they suggested that many AM-made parts are multi-material and/or contain embedded electronics. Also, according to their most probable scenario for the AM in the year 2030, the need to defend the intellectual property right for digital products will lead to novel practices. Moreover, they predicted that in the year 2030, critical regulatory measures would regulate AM file-sharing platforms. This study also discussed four extreme scenarios based on their Delphi survey findings.

Liu et al. [80] conducted a study investigating the AM impact on the aircraft spare parts supply chain. They utilized three scenarios to compare AM implementation in centralized and distributed settings with the current practices. They constructed the scenarios based on a case, using secondary data collected from the literature. Their findings highlighted the opportunities AM creates towards reducing spare parts safety stock in both centralized and distributed supply configurations. However, their study stopped short of providing a holistic impact projection for AM implementation in the spare parts supply chain and its feasibility.

Barz et al. [7], quantitatively compared the relationship between transportation cost and structure of a supply network for conventional subtractive manufacturing and AM. They modeled 700 two-stage capacitated facility location problem instances with seven node allocations, 25 geographical distributions, and four different buy-to-fly ratios and solved them with CPLEX mixed integer programming solver. They found that AM impact on resource efficiency through improved buy-to-fly ratios can significantly affect supply network structure while reducing raw material transportation. Moreover, they also realized AM would cause the production sites closer to the consumption locations, consequently reducing the overall transportation costs. Their study ignored all other operations costs related to machinery and material, production cycle time, and inventory.

Khajavi et al. [60], studied distributed spare parts provision as one of the three AM-enabled operations, was shown to provide additional flexibility through higher untapped production capacity compared to centralized production of spare parts. However, achieving a fully distributed implementation of AM is not feasible, even in the aerospace sector where downtime costs are exceedingly high, unless there is an increase in pre- and post-production autonomy and the ability to customize the production chamber size for the application. Additionally, AM machines need to become faster and cheaper.

3.3 Inventory Replacement

Wullms [141] studied the use of AM to replace the last time buy (to support the products in the last phase of the lifecycle when the manufacturing is to stop) for a selected number of parts in the Philips healthcare product portfolio. Wullms [141] found that in all cases, AM offers cost-saving over last time buy (which requires the storage of safety stock) due to the reduction of the parts inventory. Moreover, he pointed out the sensitivity of the results to the cost of additive manufacturing compared to traditional manufacturing. AM implementation will not lead to savings if AM is 10 times more expensive than traditional manufacturing.

Khajavi et al., [61], studies the replacement of physical spare part inventories by digital ones using the data collected from two large manufacturing companies. They conclude that a small percentage of spare parts are technically and economically viable for digitalization.

3.4 Product Lifecycle

In their paper, Lindemann et al. [79] proposed an analysis of AM cost drivers from a lifecycle perspective. They argued that since AM is a toolless method that enables design for performance (through complexity for free), it should be compared with conventional manufacturing methods based on product lifecycle and not just the build cost. They utilized an experimental production case by a metal AM machine to analyze the production cost breakdown. They concluded that AM machine cost is the biggest cost factor in the AM part production, followed by material cost and post-processing activities.

Khajavi et al., [59] investigated the benefits of integrating all the data required throughout a product lifecycle to the digital design of the part. They pointed out the benefits throughout the different phases of the lifecycle. In another large case study, Hedenstierna et al., [45] point out the economies of build-to-model production mode for products enabled by direct digital manufacturing.

3.5 Material Handling

Ruffo and Hague [109] performed a case study and showed the economic gains in lower production costs from mixing parts (in the production chamber) for laser sintering additive manufacturing. Moreover, they introduced a method for the cost assignment to various parts in mixed production. Li et al. [76] studied the production planning and scheduling of parts on different AM machines with different specifications and distinct part geometries to minimize the average production cost per volume. They mathematically modeled the problem and solved it using CPLEX. They also developed two different heuristics for the problem. They tested against the optimum solution using a numerical example to illustrate the high quality of results provided by the heuristic methods. They also illustrated the possibility of reducing production costs by better planning the parts in the AM machines' chambers.

Lyly-Yrjänäinen et al., [81] utilized a case study of customized industrial hydraulic hose production to investigate the impact of production digitalization and product-centric control on manufacturing operation. Their findings pointed to waste reduction, simplification of production planning and better customer responsiveness. The case company utilizes a direct digital manufacturing method where the digital model of the hose plays a significant role in labor productivity and delivery performance. Digital models of customized hoses are received from the clients, including the bill of material (BOM) specifying the hose length and diameter and the required straight or 90 degrees fittings. The case company keeps the hoses and fittings used by its clients. It does not initiate production until they receive an order. When they receive an order, the manufacturing cells that are equipped with digitally controlled cutting machines cut the hoses based on the order and kit number and print the work-order information on each hose, specifying the required fitting and the kit, order and other relevant information. Through the implementation of such direct digital manufacturing (DDM) process and integration of work-order data to the product (product-centric control), the case company improved the labor productivity by threefold in comparison with conventional methods of hose making used by its competitors while simplifying control and eliminating batch production, handling of batches and inventory management. Moreover, Lyly-Yrjänäinen et al. [81] suggested their research findings could be transferable to other cases, where DDM technology is generic for the production of required products and sharing the digital product model can replace product master data.

3.6 New Product Development

Lan et al. [71] explored the functionality of an Internet-based manufacturing service system with embedded AM for product development. They utilized a simple case study to illustrate the benefits of such a system. Findings suggested an online AM-based new product development system can cut the costs by 50% while also shortening the go-to-market time by 75%, compared to traditional methods. Eyers & Potter [36], utilized design science methodology in conjunction with interviews and secondary data to propose a framework of four distinct e-commerce channels for AM. Tele-marketing, collaborative manufacturing, localized manufacturing, and user manufacturing are the four e-commerce models discussed in their framework.

Khajavi et al. [149], studied hybrid manufacturing for the market launch phase of new product development. In hybrid production, the model-based direct digital manufacturing (e.g.: AM or incremental sheet forming) is utilized in the beginning when the demand volume is low and then a switch over to conventional manufacturing takes place as the volumes pick up. However, the right decision on the timing of the switchover is essential to keep the production cost and the risk of losses (in the event of product launch failure) low. A heuristic method to balance the risk and cost of new product launches while utilizing hybrid manufacturing was developed to consider different launch outcomes.

3.7 AM Processes Considered

ISO and ASTM categorize AM process into seven different classes: Material extrusion (MEX), Powder bed fusion (PBF), VAT photopolymerization (VPP), Directed energy deposition (DED), Material jetting (MJ), Binder jetting (BJ) and Sheet lamination (SL). Figure 2 illustrates the frequency of appearance of each AM process in the papers studied and related to the operations management literature. Some studies do not specify particular AM processes, and reviews and general descriptions of AM processes were excluded. The focus was on case studies and research specially addressing individual AM processes. It is evident that the most extensively studied processes are PBF, VPP and MEX. Studies on DED, MJ, BJ and SL are also available, albeit less frequently. This closely reflects the frequency of processes used in production. None of the studies examine composite materials recently developed using continuous fibers with the MEX method.

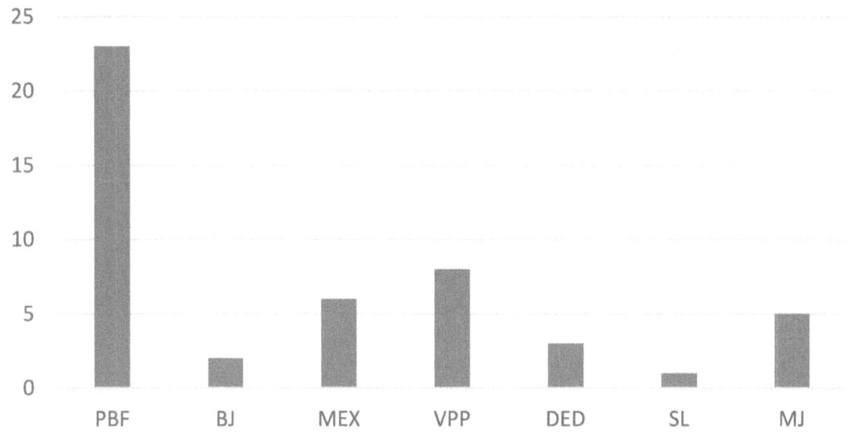

Fig. 2. Distribution of different AM processes in the operations management literature (more detail in Appendix)

4 Discussion

At this state, quite expectedly, the central part of the body of knowledge regarding AM is focused on research concerned with technology, process, material and design (Journals such as Rapid Prototyping Journal and Additive Manufacturing journal from Elsevier) as well as comparing AM and conventional manufacturing from mechanical and economic perspectives [5, 37, 110]. That is understandable, as the AM is an emerging technology and has not been yet fully established on a scale comparable to other conventional production methods. However, many studies have been conducted in the context of AM applications and their impact through enabling various capabilities (such as product customization, parts consolidation, and production decentralization). These are the studies aligned with this paper's direction and provide a forward-looking perspective for the AM while it is used for the production of functional parts.

One crucial thing in operations management research of AM is the lack of an actual AM process. AM studied as one common technique, even though there are significant differences inside same the category such as no supports required in powder bed fusion of plastics compared that metal with same processes requires heavy supports and support removal in the post-processing phase. In the future, more technology-related aspects should be taken into account and more focused case studies investigated.

5 Conclusions

This literature study contributed to the operations management of additive manufacturing by illustrating the under-researched areas. An in-depth categorization of the current AM literature related to operations management is presented.

As it stands, there are still a number of major technological challenges to be addressed in the field of AM. Such challenges include machine throughput, price, chamber size, process repeatability, reliability and standardization, material cost and requirements, intellectual property rights of digital design files, automation in the pre- and post-production processes, as well as digital model file formats. These areas need to be addressed in order to have AM prepared for wider mainstream applications [18, 150, 151]. However, this research was primarily motivated by an unfolding trend towards the utilization of AM to produce functional parts. As this utilization expands, practices aimed at enhancing AM-enabled operations are likely to become increasingly important. The aim and objective of this research are to facilitate the development of AM operations management.

Appendix

Categorization of AM literature;

Category 0: Literature review.
Category 1: Current state or backward-looking empirical research (concerned with the comparison of AM with conventional manufacturing form cost or engineering perspectives).
Category 2: Forward-looking theoretical and conceptual studies.
Category 3: Forward-looking (to the future when AM is used for functional parts) by utilizing empirical analysis (including the experimentation and simulations) (Table 2).

Table 2. The list and the classification of the reviewed articles

Item	Article	Category				Additive manufacturing	
		0	1	2	3	Processes	Materials
1	Achillas et al., (2015)		X				
2	Anthony et al., (2011)		X	X		PBF, BJ	Steel
3	Armillotta, (2008)		X				
4	Attaran, (2017)			X			
5	Atzeni, and Salmi, (2012)		X			PBF	Aluminium

(continued)

Table 2. (*continued*)

Item	Article	Category				Additive manufacturing	
		0	1	2	3	Processes	Materials
6	Sherman, (2009)		X				
7	Bals et al., (2015)			X			
8	Barz, et al., (2016)				X		
9	Baumers, et al., (2013)		X			PBF	Steel
10	Baumers, et al., (2016)		X			PBF	Titanium, Steel
11	Berman, (2012)			X			
12	Beyer, (2014)			X			
13	Bianchi, & Åhlström, (2014)			X			
14	Birtchnell, and Urry, (2013)			X		MEX	ABS, PLA
15	Bogers et al. (2016)			X			
16	Byun, & Lee, (2005)		X			PBF	Steel
17	Campbell, & Ivanova, (2013)			X			
18	Campbell et al., (2011)			X			
19	Chen et al., (2015)		X				
20	Christopher, & Ryals, (2014)			X			
21	Deradjat, & Minshall, (2015)		X				
22	Despeisse et al., (2017)			X		MEX	HIPS, PET, PLA
23	Di Angelo, and Di Stefano, (2011)		X				
24	Drizo, and Pegna, (2006)	X					
25	Emelogu et al., (2016)				X	PBF	
26	Eyers, & Potter, (2015)				X		

(*continued*)

Table 2. *(continued)*

Item	Article	Category				Additive manufacturing	
		0	1	2	3	Processes	Materials
27	Gao et al., (2015)	X					
28	Gibson, (2017)			X			
29	Hanumaiah et al., (2006)		X			PBF, VPP	Cu-Ni-Sn, photopolymer
30	Hasan et al., (2008)			X		PBF	Nylon
31	Holmström, & Partanen, (2014)			X		PBF	
32	Holmström et al., (2017)			X			
33	Holmström et al., (2010)			X			
34	Hopkinson, & Dickens, (2001)		X			VPP	
35	Hopkinson, & Dicknes, (2003)		X			VPP, MEX, PBF	Epoxy, ABS, Nylon
36	Huang et al., (2013)	X					
37	Jiang et al., (2017)			X	X		
38	Jonsson, & Holmström, (2016)			X			
39	Kellens et al., (2010)		X			PBF	Steel, PA2200
40	Kengpol, & O'Brien, (2001)		X				
41	Khorram Niaki, & Nonino, (2017)	X					
42	Khrais et al., (2011)		X				
43	Kietzmann et al., (2015)			X			
44	Kochan et al., (1999)			X			
45	Kohtala, (2015)	X					
46	Kruth et al., (1998)	X					
47	Lachmayer et al., (2017)		X			PBF	ABS
48	Lan, (2009)	X					
49	Lan et al., (2005)		X			PBF, VPP	

(continued)

Table 2. (*continued*)

Item	Article	Category				Additive manufacturing	
		0	1	2	3	Processes	Materials
50	Lan et al., (2004)				X		
51	Lan et al., (2008)		X			VPP	HXJ-971
52	Laplume et al., (2016)			X			
53	Le Bourhis et al., (2013)		X			DED	
54	Lehmhus et al., (2015)			X			
55	Li et al., (2017)				X	PBF	
56	Liao et al., (2014)		X				
57	Lindemann et al., (2012)				X	PBF	Steel
58	Lindemann et al., (2015)		X				
59	Liu et al., (2014)		X		X	PBF	
60	Lyly-Yrjänäinen et al., (2016)				X		
61	Mahapatra, and Panda, (2013)		X				
62	Manogharan et al., (2016)		X			PBF	Titanium
63	Mansour, and Hague, (2003)		X				
64	Mashhadi et al., (2015)			X			
65	Masood, and Soo, (2002)		X				
66	Masood, and Al-Alawi, (2002)		X				
67	Mellor et al., (2014)		X				
68	Mognol et al., (2006)		X			MJ, MEX, PBF	Wax, ABS
69	Mohajeri et al., (2014)			X			
70	Mohr, & Khan, (2015a)			X			

(*continued*)

Table 2. (*continued*)

Item	Article	Category				Additive manufacturing	
		0	1	2	3	Processes	Materials
71	Mohr, & Khan, (2015b)			X			
72	Moore et al., (2016)		X			MJ	RGD8430
73	Munguía et al., (2010)		X				
74	Nyman, & Sarlin, (2014)			X			
75	Oettmeier, & Hofmann, (2017)		X				
76	Pal et al., (2007)		X			MEX, MJ, VPP, SL	ABS, Polycarbonate, epoxy resin, wax, paper
77	Paul, and Anand, (2012)		X			PBF	
78	Pearce et al., (2010)			X			
79	Pérès, & Noyes, (2006)			X			
80	Petrick, & Simpson, (2013)			X			
81	Pour et al., (2016)	X		X			
82	Rao, and Padmanabhan, (2007)		X				
83	Reeves, (2008)			X			
84	Reeves et al., (2011)				X		
85	Rickenbacher et al., (2013)		X			PBF	
86	Roberson et al., (2013)		X			MEX, VPP, SL	ABS, PLA, PVC, photopolymer
87	Ruffo, and Hague, (2007)				X	PBF	Polyamide
88	Rylands et al., (2015)	X					
89	Sasson, & Johnson, (2016)			X			
90	Sealy, (2012)			X			

(*continued*)

Table 2. (*continued*)

Item	Article	Category				Additive manufacturing	
		0	1	2	3	Processes	Materials
91	Sharif Ullah et al., (2013)		X			VPP	Photopolymer
92	Sirichakwal, & Conner, (2016)				X		
93	Thomas, & Gilbert, (2014)	X					
94	Thompson et al., (2016)	X					
95	Tuck, and Hague, (2006)			X			
96	Tuck et al., (2006)			X			
97	Tuck et al., (2008)				X	VPP	
98	Tuomi, Karjalainen, (2006)		X			PBF	
99	Uriondo et al., (2015)	X					
100	Waller, & Fawcett, (2014)			X			
101	Walter et al., (2004)			X			
102	Weller et al., (2015)			X			
103	Vinodh et al., (2009)		X			SL	PVC
104	Witherell et al., (2017)			X			
105	Wullms et al., (2014)				X		
106	Xu et al., (2015)		X			BJ	
107	Yoon et al., (2014)		X			PBF, MEX, VPP	
108	Zanardini et al., (2016)		X	X			

References

1. Achillas, C., Aidonis, D., Iakovou, E., Thymianidis, M., Tzetzis, D.: A methodological framework for the inclusion of modern additive manufacturing into the production portfolio of a focused factory. J. Manuf. Syst. **37**(1), 328–339 (2015)

2. Anthony, R.W., Evans, M., Rennie, A.E.W., Kirkby, E.: Opportunities offered by additive manufacturing in creative businesses: informing designers. In: DS 69: Proceedings of E&PDE 2011, the 13th International Conference on Engineering and Product Design Education, London, UK, 08–09 September (2011)
3. Armillotta, A.: Selection of layered manufacturing techniques by an adaptive AHP decision model. Robot. Comput.-Integr. Manuf. **24**(3), 450–461 (2008)
4. Attaran, M.: Additive manufacturing: the most promising technology to alter the supply chain and logistics. J. Serv. Sci. Manag. **10**(3), 189 (2017)
5. Atzeni, E., Salmi, A.: Economics of additive manufacturing for end-usable metal parts. Int. J. Adv. Manuf. Technol. **62**(9–12), 1147–1155 (2012)
6. Bals, L., Daum, A., Tate, W.: From offshoring to rightshoring: focus on the backshoring phenomenon. AIB Insights **15**(4), 3 (2015)
7. Barz, A., Buer, T., Haasis, H.D.: Quantifying the effects of additive manufacturing on supply networks by means of a facility location-allocation model. Logist. Res. **9**(1), 1–14 (2016)
8. Baumers, M., Tuck, C., Wildman, R., Ashcroft, I., Hague, R.: Transparency built-in. J. Ind. Ecol. **17**(3), 418–431 (2013)
9. Baumers, M., Dickens, P., Tuck, C., Hague, R.: The cost of additive manufacturing: machine productivity, economies of scale and technology-push. Technol. Forecast. Soc. Change **102**, 193–201 (2016)
10. Baumers, M., Tuck, C., Wildman, R., Ashcroft, I., Hague, R.: Energy inputs to additive manufacturing: does capacity utilization matter? Eos **1000**(270), 30–40 (2011)
11. Berman, B.: 3-D printing: the new industrial revolution. Bus. Horiz. **55**(2), 155–162 (2012)
12. Beyer, C.: Strategic implications of current trends in additive manufacturing. J. Manuf. Sci. Eng. **136**(6), 064701 (2014)
13. Bianchi, M., Åhlström, P.: Additive manufacturing: towards a new operations management paradigm. In: 20th International EurOMA Conference, May 2014
14. Birtchnell, T., Urry, J.: Fabricating futures and the movement of objects. Mobilities **8**(3), 388–405 (2013)
15. Birtchnell, T., Urry, J., Cook, C., Curry, A.: Freight miles: the impact of 3D printing on transport and society. Mobilities **8**(3), 388–405 (2013)
16. Bogers, M., Hadar, R., Bilberg, A.: Additive manufacturing for consumer-centric business models: implications for supply chains in consumer goods manufacturing. Technol. Forecast. Soc. Change **102**, 225–239 (2016)
17. Byun, H.S., Lee, K.H.: A decision support system for the selection of a rapid prototyping process using the modified TOPSIS method. Int. J. Adv. Manuf. Technol. **26**(11–12), 1338–1347 (2005)
18. Ituarte, I.F., Coatanea, E., Salmi, M., Tuomi, J., Partanen, J.: Additive manufacturing in production: a study case applying technical requirements. Phys. Procedia **78**, 357–366 (2015)
19. Campbell, D.T., Stanley, J.C.: Experimental and quasi-experimental designs for research. In: Gage, N.L. (ed.) Handbook of Research on Teaching, pp. 171–246 (1966)
20. Campbell, T.A., Ivanova, O.S.: Additive manufacturing as a disruptive technology: implications of three-dimensional printing. Technol. Innov. **15**(1), 67–79 (2013)
21. Campbell, T., Williams, C., Ivanova, O., Garrett, B.: Could 3D Printing Change the World? Technologies, Potential, and Implications of Additive Manufacturing. Atlantic Council, Washington, DC (2011)
22. Castellion, G., Markham, S.K.: Perspective: new product failure rates: influence of argumentum ad populum and self-interest. J. Prod. Innov. Manag. **30**(5), 976–979 (2013)
23. Chen, D., Heyer, S., Ibbotson, S., Salonitis, K., Steingrímsson, J.G., Thiede, S.: Direct digital manufacturing: definition, evolution, and sustainability implications. J. Cleaner Prod. **107**, 615–625 (2015)

24. Christopher, M., Ryals, L.J.: The supply chain becomes the demand chain. J. Bus. Logist. **35**(1), 29–35 (2014)
25. Cohen, M.A., Agrawal, N., Agrawal, V.: Winning in the aftermarket. Harv. Bus. Rev. **84**(5), 129 (2006)
26. Cole, W.: Breaking the Mold: Boeing engineers and technologists are constantly developing better ways to design and make products, Frontiers. http://www.boeing.com/news/frontiers/archive/2004/december/ts_sf03.html. Accessed 10 Mar 2013
27. Crawford, C.M.: New product failure rates: a reprise. Res. Manage. **30**(4), 20–24 (1987)
28. Cruz Sanchez, F.A., Boudaoud, H., Muller, L., Camargo, M.: Towards a standard experimental protocol for open source additive manufacturing. Virt. Phys. Prototyp. **9**(3), 151–167 (2014)
29. Bourell, D.L., Beaman, J.B., Leu, M.C., Rosen, D.W.: A brief history of additive manufacturing and the 2009 roadmap for additive manufacturing: looking back and looking ahead. In: US-Turkey Workshop on Rapid Technologies (2009)
30. Deradjat, D., Minshall, T.: Implementation of additive manufacturing technologies for mass customisation. In: Proceedings of the 2015 International Association for Management of Technology Conference (IAMOT 2015), pp. 2079–2094 (2015)
31. Despeisse, M., et al.: Unlocking value for a circular economy through 3D printing: a research agenda. Technol. Forecast. Soc. Change **115**, 75–84 (2017)
32. Di Angelo, L., Di Stefano, P.: A neural network-based build time estimator for layer manufactured objects. Int. J. Adv. Manuf. Technol. **57**, 215–224 (2011)
33. Di Stefano, G., Peteraf, M., Verona, G.: Dynamic capabilities deconstructed: a bibliographic investigation into the origins, development, and future directions of the research domain. Ind. Corp. Change **19**(4), 1187–1204 (2010)
34. Drizo, A., Pegna, J.: Environmental impacts of rapid prototyping: an overview of research to date. Rapid Prototyping J. **12**(2), 64–71 (2006)
35. Emelogu, A., Marufuzzaman, M., Thompson, S.M., Shamsaei, N., Bian, L.: Additive manufacturing of biomedical implants: a feasibility assessment via supply-chain cost analysis. Addit. Manuf. **11**, 97–113 (2016)
36. Eyers, D.R., Potter, A.T.: E-commerce channels for additive manufacturing: an exploratory study. J. Manuf. Technol. Manage. **26**(3), 390–411 (2015)
37. Frazier, W.E.: Metal additive manufacturing: a review. J. Mater. Eng. Perform. **23**(6), 1917–1928 (2014)
38. Levy, G.N., Schindel, R., Kruth, J.P.: Rapid manufacturing and rapid tooling with layer manufacturing (LM) technologies, state of the art and future perspectives. CIRP Ann.-Manuf. Technol. **52**(2), 589–609 (2003)
39. Gao, W., et al.: The status, challenges, and future of additive manufacturing in engineering. Comput.-Aided Des. **69**, 65–89 (2015)
40. Gibson, I.: The changing face of additive manufacturing. J. Manuf. Technol. Manage. **28**(1), 10–17 (2017)
41. Gibson, I., Rosen, D.W., Stucker, B.: Additive Manufacturing Technologies: Rapid Prototyping to Direct Digital Manufacturing. Springer, New York (2009)
42. Gibson, I., Rosen, D.W., Stucker, B.: Additive Manufacturing Technologies, vol. 238. Springer, New York (2010)
43. Hanumaiah, N., Ravi, B., Mukherjee, N.P.: Rapid hard tooling process selection using QFD-AHP methodology. J. Manuf. Technol. Manage. **17**(3), 332–350 (2006)
44. Hasan, S., Rennie, A.E.W.: The application of rapid manufacturing technologies in the spare parts industry. In: Nineteenth Annual International Solid Freeform Fabrication (SFF) Symposium, 4–8, Austin, TX, August 2008

45. Hedenstierna, C.P.T., Disney, S.M., Eyers, D.R., Holmström, J., Syntetos, A.A., Wang, X.: Economies of collaboration in build-to-model operations. J. Oper. Manage. **65**(8), 753–773 (2019)
46. Holmström, J., Partanen, J.: Digital manufacturing-driven transformations of service supply chains for complex products. Supply Chain Manage. **19**(4), 421–430 (2014)
47. Holmström, J., Holweg, M., Khajavi, S.H., Partanen, J.: The direct digital manufacturing (r) evolution: definition of a research agenda. Oper. Manag. Res. **9**(1–2), 1–10 (2016)
48. Holmström, J., Liotta, G., Chaudhuri, A.: Sustainability outcomes through direct digital manufacturing-based operational practices: a design theory approach. J. Cleaner Prod. **167**, 951–961 (2017)
49. Holmström, J., Partanen, J., Tuomi, J., Walter, M.: Rapid manufacturing in the spare parts supply chain: alternative approaches to capacity deployment. J. Manuf. Technol. Manage. **21**(6), 687–697 (2010)
50. Hopkinson, N., Dickens, P.: Rapid prototyping for direct manufacture. Rapid Prototyping J. **7**(4), 197–202 (2001)
51. Hopkinson, N., Dickens, P.: Analysis of rapid manufacturing—using layer manufacturing processes for production. Proc. Inst. Mech. Eng. Part C **217**(1), 31–39 (2003)
52. Hopkinson, N., Hague, R., Dickens, P. (eds.): Rapid Manufacturing: An Industrial Revolution for the Digital Age. Wiley, Hoboken (2006)
53. Huang, S.H., Liu, P., Mokasdar, A., Hou, L.: Additive manufacturing and its societal impact: a literature review. Int. J. Adv. Manuf. Technol. **67**, 1–13 (2013)
54. Ilyas, I.P.: 3D machine vision and additive manufacturing: concurrent product and process development. In: IOP Conference Series: Materials Science and Engineering, vol. 46, no. 1, 012029. IOP Publishing, Bristol (2013)
55. Jiang, R., Kleer, R., Piller, F.T.: Predicting the future of additive manufacturing: a Delphi study on economic and societal implications of 3D printing for 2030. Technol. Forecast. Soc. Change **117**, 84–97 (2017)
56. Jonsson, P., Holmström, J.: Future of supply chain planning: closing the gaps between practice and promise. Int. J. Phys. Distrib. Logist. Manage. **46**(1), 62–81 (2016)
57. Kellens, K., Yasa, E., Dewulf, W., Duflou, J.: Environmental assessment of selective laser melting and selective laser sintering. In: Going Green–CARE INNOVATION 2010: From Legal Compliance to Energy-Efficient Products and Services, Vienna, 8–11 November 2010
58. Kengpol, A., O'Brien, C.: The development of a decision support tool for the selection of advanced technology to achieve rapid product development. Int. J. Prod. Econ. **69**(2), 177–191 (2001)
59. Khajavi, S.H., Holmström, J., Baumers, M.: Additive manufacturing as a platform for introducing cyber-physical services. In: 3rd International Conference on Progress in Additive Manufacturing (Pro-AM), Singapore (2018)
60. Khajavi, S.H., Partanen, J., Holmström, J.: Additive manufacturing in the spare parts supply chain. Comput. Ind. **65**(1), 50–63 (2014)
61. Khajavi, S.H., Salmi, M., Holmström, J.: Additive manufacturing as an enabler of digital spare parts. In: Managing 3D Printing, pp. 45–60. Palgrave Macmillan, Cham (2020)
62. Khorram Niaki, M., Nonino, F.: Additive manufacturing management: a review and future research agenda. Int. J. Prod. Res. **55**(5), 1419–1439 (2017)
63. Khrais, S., Al-Hawari, T., Al-Araidah, O.: A fuzzy logic application for selecting layered manufacturing techniques. Expert Syst. Appl. **38**(8), 10286–10291 (2011)
64. Kietzmann, J., Pitt, L., Berthon, P.: Disruptions, decisions, and destinations: enter the age of 3-D printing and additive manufacturing. Bus. Horiz. **58**(2), 209–215 (2015)
65. Kochan, D., Kai, C.C., Zhaohui, D.: Rapid prototyping issues in the 21st century. Comput. Ind. **39**(1), 3–10 (1999)

66. Kohtala, C.: Addressing sustainability in research on distributed production: an integrated literature review. J. Cleaner Prod. **106**, 654–668 (2015)
67. Kruth, J.P., Leu, M.C., Nakagawa, T.: Progress in additive manufacturing and rapid prototyping. CIRP Ann.-Manuf. Technol. **47**(2), 525–540 (1998)
68. Lachmayer, R., Gembarski, P.C., Gottwald, P., Lippert, R.B.: The potential of product customization using technologies of additive manufacturing. In: Bellemare, J., Carrier, S., Nielsen, K., Piller, F.T. (eds.) Managing Complexity, pp. 71–81. Springer, Cham (2017). https://doi.org/10.1007/978-3-319-29058-4_6
69. Lan, H.: Web-based rapid prototyping and manufacturing systems: a review. Comput. Ind. **60**(9), 643–656 (2009)
70. Lan, H., Ding, Y., Hong, J.: Decision support system for rapid prototyping process selection through integration of fuzzy synthetic evaluation and an expert system. Int. J. Prod. Res. **43**(1), 169–194 (2005)
71. Lan, H., Ding, Y., Hong, J., Huang, H., Lu, B.: A web-based manufacturing service system for rapid product development. Comput. Ind. **54**(1), 51–67 (2004)
72. Lan, H., Ding, Y., Hong, J., Huang, H., Lu, B.: Web-based quotation system for stereolithography parts. Comput. Ind. **59**(8), 777–785 (2008)
73. Laplume, A.O., Petersen, B., Pearce, J.M.: Global value chains from a 3D printing perspective. J. Int. Bus. Stud. **47**(5), 595–609 (2016)
74. Le Bourhis, F., Kerbrat, O., Hascoet, J.Y., Mognol, P.: Sustainable manufacturing: evaluation and modeling of environmental impacts in additive manufacturing. Int. J. Adv. Manuf. Technol. **69**(9–12), 1927–1939 (2013)
75. Lehmhus, D., et al.: Cloud-based automated design and additive manufacturing: a usage data-enabled paradigm shift. Sensors **15**(12), 32079–32122 (2015)
76. Li, Q., Kucukkoc, I., Zhang, D.Z.: Production planning in additive manufacturing and 3D printing. Comput. Oper. Res. **83**, 157–172 (2017)
77. Liao, S., Wu, M.J., Huang, C.Y., Kao, Y.S., Lee, T.H.: Evaluating and enhancing three-dimensional printing service providers for rapid prototyping using the DEMATEL based network process and VIKOR. Math. Probl. Eng. **2014**, 349348 (2014)
78. Lindemann, C., Reiher, T., Jahnke, U., Koch, R.: Towards a sustainable and economic selection of part candidates for additive manufacturing. Rapid Prototyping J. **21**, 216–227 (2015)
79. Lindemann, C., Jahnke, U., Moi, M., Koch, R.: Analyzing product lifecycle costs for a better understanding of cost drivers in additive manufacturing. In: 23rd Annual International Solid Freeform Fabrication Symposium, Austin, TX, USA, 6–8 August (2012)
80. Liu, P., Huang, S.H., Mokasdar, A., Zhou, H., Hou, L.: The impact of additive manufacturing in the aircraft spare parts supply chain: supply chain operation reference (SCOR) model-based analysis. Prod. Plann. Control **25**(13–14), 1169–1181 (2014)
81. Lyly-Yrjänäinen, J., Holmström, J., Johansson, M.I., Suomala, P.: Effects of combining product-centric control and direct digital manufacturing: the case of preparing customized hose assembly kits. Comput. Ind. **82**, 82–94 (2016)
82. Mahapatra, S.S., Panda, B.N.: Benchmarking of rapid prototyping systems using grey relational analysis. Int. J. Serv. Oper. Manage. **16**(4), 460–477 (2013)
83. Manogharan, G., Wysk, R.A., Harrysson, O.L.: Additive manufacturing-integrated hybrid manufacturing and subtractive processes: economic model and analysis. Int. J. Comput. Integr. Manuf. **29**(5), 473–488 (2016)
84. Mansour, S., Hague, R.: Impact of rapid manufacturing on design for manufacture for injection moulding. Proc. Inst. Mech. Eng. Part B J. Eng. Manuf. **217**(4), 453–461 (2003)
85. Markillie, P.: A third industrial revolution. Economist. http://www.economist.com/node/21552901. Accessed 05 Dec 2012

86. Mashhadi, A.R., Esmaeilian, B., Behdad, S.: Impact of additive manufacturing adoption on future of supply chains. In: ASME 2015 International Manufacturing Science and Engineering Conference, pp. V001T02A064-V001T02A064. American Society of Mechanical Engineers (2015)
87. Masood, S.H., Soo, A.: A rule based expert system for rapid prototyping system selection. Robot. Comput.-Integr. Manuf. **18**(3–4), 267–274 (2002)
88. Masood, S.H., Al-Alawi, M.A.Z.E.N.: The IRIS rapid prototyping system selector for educational and manufacturing users. Int. J. Eng. Educ. **18**(1), 66–77 (2002)
89. Mellor, S., Hao, L., Zhang, D.: Additive manufacturing: a framework for implementation. Int. J. Prod. Econ. **149**, 194–201 (2014)
90. Mognol, P., Lepicart, D., Perry, N.: Rapid prototyping: energy and environment in the spotlight. Rapid Prototyping J. **12**(1), 26–34 (2006)
91. Mohajeri, B., Khajavi, S.H., Nyberg, T., Khajavi, S.H.: Supply chain modifications to improve additive manufacturing cost-benefit balance. In: 25th Annual International Solid Freeform Fabrication Symposium, Austin, TX, USA (2014)
92. Mohr, S., Khan, O.: 3D printing and its disruptive impacts on supply chains of the future. Technol. Innov. Manage. Rev. **5**(11), 20 (2015)
93. Mohr, S., Khan, O.: 3D printing and supply chains of the future: innovations and strategic logistics and supply chain management, p. 147 (2015)
94. Moore, S.S., O'Sullivan, K.J., Verdecchia, F.: Shrinking the supply chain for implantable coronary stent devices. Ann. Biomed. Eng. **44**(2), 497–507 (2016)
95. Munguía, J., Lloveras, J., Llorens, S., Laoui, T.: Development of an AI-based rapid manufacturing advice system. Int. J. Prod. Res. **48**(8), 2261–2278 (2010)
96. Nyman, H.J., Sarlin, P.: From bits to atoms: 3D printing in the context of supply chain strategies. In: Proceedings of the 47th Hawaii International Conference on System Science, pp. 4190–4199. IEEE (2014)
97. Oettmeier, K., Hofmann, E.: Additive manufacturing technology adoption: an empirical analysis of general and supply chain-related determinants. J. Bus. Econ. **87**(1), 97–124 (2017)
98. Pal, D.K., Ravi, B., Bhargava, L.S.: Rapid tooling route selection for metal casting using QFD–ANP methodology. Int. J. Comput. Integr. Manuf. **20**(4), 338–354 (2007)
99. Paul, R., Anand, S.: Process energy analysis and optimization in selective laser sintering. J. Manuf. Syst. **31**(4), 429–437 (2012)
100. Pearce, J.M., Blair, C.M., Laciak, K.J., Andrews, R., Nosrat, A., Zelenika-Zovko, I.: 3-D printing of open source appropriate technologies for self-directed sustainable development. J. Sustainable Dev. **3**(4), 17 (2010)
101. Pérès, F., Noyes, D.: Envisioning e-logistics developments: making spare parts in situ and on demand: state of the art and guidelines for future developments. Comput. Ind. **57**(6), 490–503 (2006)
102. Petrick, I.J., Simpson, T.W.: 3D printing disrupts manufacturing: how economies of one create new rules of competition. Res. Technol. Manage. **56**(6), 12–16 (2013)
103. Pour, M.A., Zanardini, M., Bacchetti, A., Zanoni, S.: Additive manufacturing impacts on productions and logistics systems. IFAC-PapersOnLine **49**(12), 1679–1684 (2016)
104. Rao, R.V., Padmanabhan, K.K.: Rapid prototyping process selection using graph theory and matrix approach. J. Mater. Process. Technol. **194**(1–3), 81–88 (2007)
105. Reeves, P.: How rapid manufacturing could transform supply chains. Supply Chain Q. **2**(4), 32–36 (2008)
106. Reeves, P., Tuck, C., Hague, R.: Additive manufacturing for mass customization. In: Fogliatto, F., da Silveira, G. (eds.) Mass Customization, pp. 275–289. Springer, London (2011). https://doi.org/10.1007/978-1-84996-489-0_13

107. Rickenbacher, L., Spierings, A., Wegener, K.: An integrated cost-model for selective laser melting (SLM). Rapid Prototyping J. **19**(3), 208–214 (2013)
108. Roberson, D.A., Espalin, D., Wicker, R.B.: 3D printer selection: a decision-making evaluation and ranking model. Virtual Phys. Prototyp. **8**(3), 201–212 (2013)
109. Ruffo, M., Hague, R.: Cost estimation for rapid manufacturing—simultaneous production of mixed components using laser sintering. Proc. Inst. Mech. Eng. Part B J. Eng. Manuf. **221**(11), 1585–1591 (2007)
110. Ruffo, M., Tuck, C., Hague, R.: Cost estimation for rapid manufacturing—laser sintering production for low to medium volumes. Proc. Inst. Mech. Eng. Part B J. Eng. Manuf. **220**(9), 1417–1427 (2006)
111. Rylands, B., Bohme, T., Gorkin III, R.A., Fan, J.P., Birtchnell, T.: 3D printing—To print or not to print? Aspects to consider before adoption—A supply chain perspective (2015)
112. Sasson, A., Johnson, J.C.: The 3D printing order: variability, supercenters and supply chain reconfigurations. Int. J. Phys. Distrib. Logist. Manage. **46**(1), 82–94 (2016)
113. Schumpeter: Additive manufacturing: Print me a jet engine. The Economist (2012). http://www.economist.com/blogs/schumpeter/2012/11/additive-manufacturing. Accessed 05 Dec 2012
114. Schwartz, P.: The Art of the Long View: Paths to Strategic Insight for Yourself and Your Company. Doubleday (1991)
115. Sealy, W.: Additive manufacturing as a disruptive technology: how to avoid the pitfall. Am. J. Eng. Technol. Res. **11**(10), 86–93 (2012)
116. Sharif Ullah, A.M.M., Hashimoto, H., Kubo, A., Tamaki, J.I.: Sustainability analysis of rapid prototyping: Material/resource and process perspectives. Int. J. Sustainable Manuf. **3**(1), 20–36 (2013)
117. Shell: People and connections; Global scenarios to 2020; Public summary. Global Business Environment, Shell International (2002)
118. Sherman, L.M.: Additive manufacturing—New capabilities in materials and equipment advance rapid prototyping and production of short-run parts by "direct digital manufacturing." Plast. Technol. **55**(3), 35 (2009)
119. Simao, H., Powell, W.: Approximate dynamic programming for management of high-value spare parts. J. Manuf. Technol. Manage. **20**(2), 147–160 (2009)
120. Sirichakwal, I., Conner, B.: Implications of additive manufacturing for spare parts inventory. 3D Print. Addit. Manuf. **3**(1), 56–63 (2016)
121. Slack, N., Brandon-Jones, A., Johnston, R.: Essentials of Operations Management. Financial Times Prentice Hall, Upper Saddle River (2011)
122. Stevenson, W.J.: Operations Management. McGraw-Hill, New York (2005)
123. Takeuchi, H., Nonaka, I.: The new product development game. Harv. Bus. Rev. **64**(1), 137–146 (1986)
124. The World Bank: Manufacturing, value added (% of GDP) (2017). http://data.worldbank.org/indicator/NV.IND.MANF.ZS. Accessed 1 Sept 2017
125. Thomas, D.S., Gilbert, S.W.: Costs and cost effectiveness of additive manufacturing. NIST Spec. Publ. **1176**, 12 (2014)
126. Thompson, M.K., et al.: Design for additive manufacturing: trends, opportunities, considerations, and constraints. CIRP Ann.-Manuf. Technol. **65**(2), 737–760 (2016)
127. Tuck, C., Hague, R.: The pivotal role of rapid manufacturing in the production of cost-effective customised products. Int. J. Mass Customisation **1**(2/3), 360–373 (2006)
128. Tuck, C., Hague, R., Burns, N.: Rapid manufacturing: impact on supply chain methodologies and practice. Int. J. Serv. Oper. Manag. **3**(1), 1–22 (2006)
129. Tuck, C., Hague, R., Ruffo, M., Ransley, M., Adams, P.: Rapid manufacturing facilitated customization. Int. J. Comput. Integr. Manuf. **21**(3), 245–258 (2008)

130. Tuomi, J., Karjalainen, J.: Cost model for rapid manufacturing. J. New Gener. Sci. **4**(1), 110–117 (2006)
131. Uriondo, A., Esperon-Miguez, M., Perinpanayagam, S.: The present and future of additive manufacturing in the aerospace sector: a review of important aspects. Proc. Inst. Mech. Eng. Part G-J. Aerosp. Eng. **229**(11), 2132–2147 (2015)
132. Vinodh, S., Sundararaj, G., Devadasan, S.R., Kuttalingam, D., Rajanayagam, D.: Agility through rapid prototyping technology in a manufacturing environment using a 3D printer. J. Manuf. Technol. Manage. **20**(7), 1023–1041 (2009)
133. Waller, M.A., Fawcett, S.E.: Click here to print a maker movement supply chain: how invention and entrepreneurship will disrupt supply chain design. J. Bus. Logist. **35**(2), 99–102 (2014)
134. Walter, M., Holmström, J., Tuomi, H., Yrjölä, H.: Rapid manufacturing and its impact on supply chain management. In: Proceedings of the Logistics Research Network Annual Conference, pp. 9–10 (2004)
135. Waters, D.: Operations Management: Producing Goods and Services. Pearson Education, Upper Saddle River (2002)
136. Weller, C., Kleer, R., Piller, F.T.: Economic implications of 3D printing: market structure models in light of additive manufacturing revisited. Int. J. Prod. Econ. **164**, 43–56 (2015)
137. West, J., Kuk, G.: Proprietary benefits from open communities: how MakerBot leveraged Thingiverse in 3D printing. SSRN2544970 (2014)
138. Witherell, P., Lu, Y., Jones, A.: Additive manufacturing: a trans-disciplinary experience. In: Kahlen, J., Flumerfelt, S., Alves, A. (eds.) Transdisciplinary Perspectives on Complex Systems, pp. 145–175. Springer, Cham (2017). https://doi.org/10.1007/978-3-319-38756-7_6
139. Wohlers Associates (2016). https://wohlersassociates.com/press71.html. Accessed 1 Sept 2017
140. Wohlers report (2021). https://wohlersassociates.com/press83.html. Accessed 1 Aug 2021
141. Wullms, B., Arts, J.J., Topan, E.E., van der Schoot, H., Schindler-Chaloub, J.: Additive manufacturing in the spare parts supply chain. Eindhoven University of Technology (2014)
142. Xu, X., Meteyer, S., Perry, N., Zhao, Y.F.: Energy consumption model of binder-jetting additive manufacturing processes. Int. J. Prod. Res. **53**(23), 11-7005–11-7015 (2015)
143. Yoon, H.S., et al.: A comparison of energy consumption in bulk forming, subtractive, and additive processes: review and case study. Int. J. Precis. Eng. Manuf.-Green Technol. **1**(3), 261–279 (2014)
144. Zanardini, M., Bacchetti, A., Zanoni, S., Ashourpour, M.: Additive manufacturing applications in the domain of product service system: an empirical overview. Procedia CIRP **47**, 543–548 (2016)
145. Petrovic, V., et al.: Additive layered manufacturing: Sectors of industrial application shown through case studies. Int. J. Prod. Res. **49**(4), 1061–1079 (2011)
146. Faludi, J., Hu, Z., Alrashed, S., Braunholz, C., Kaul, S., Kassaye, L.: Does material choice drive sustainability of 3D printing? Int. J. Mech. Aerosp. Ind. Mech. Eng. (2015)
147. Gibson, I., Rosen, D.W., Stucker, B.: Design for additive manufacturing. In: Additive Manufacturing Technologies, pp. 299–332. Springer, Boston (2010). https://doi.org/10.1007/978-1-4419-1120-9_11
148. Kellner, T.: GE is building the world's largest 'additive' machine for 3D printing metals (2017). https://www.ge.com/news/taxonomy/term/6335. Accessed May 2024
149. Khajavi, S.H., Partanen, J., Holmström, J., Tuomi, J.: Risk reduction in new product launch: a hybrid approach combining direct digital and tool-based manufacturing. Comput. Ind. **74**, 29–42 (2015)

150. Chekurov, S., Salmi, M.: Additive manufacturing in offsite repair of consumer electronics. Phys. Procedia **89**, 23–30 (2017)
151. Spalt, P., Bauernhansl, T.: A framework for integration of additive manufacturing technologies in production networks. Procedia CIRP **57**, 716–721 (2016)

Advances in Production Management Systems

A Worker-Centric Order Release Method Based on Workload Control: An Assessment by Simulation

Lin Ma[1,2,3], Mingze Yuan[1,3], Ting Qu[1,3,4](✉), Lei Liu[5], Kai Zhang[1,3,6], Congdong Li[1,3,7], and Matthias Thürer[8]

[1] Institute of Physical Internet, Jinan University, Zhuhai 519070, China
qutingejnu.edu.cn

[2] Department of Industrial and Manufacturing Systems Engineering, The University of Hong Kong, Hong Kong, China

[3] Guangdong International Cooperation Base of Science and Technology for GBA Smart Logistics, Jinan University, Zhuhai 519070, China

[4] School of Intelligent Systems Science and Engineering, Jinan University, Zhuhai 519070, China

[5] Business School, Guilin University of Electronic Technology, Guilin 541004, China

[6] The Hong Kong Polytechnic University Shenzhen Research Institute, Shenzhen 518057, China

[7] Chool of Management, Jinan University, Guangzhou 510632, China

[8] Chair of Factory Planning and Intralogistics, Chemnitz University of Technology, Erfeschlager Straße 73, 09125 Chemnitz, Germany

Abstract. This study presents a novel paradigm of production planning and control by introducing a worker-centric workload control order release approach, which is particularly suitable for the manufacturing shops involving robots and workers, such as the human-machine collaborative production system. Specifically, a production order in the real-life production process may consist of multiple operations. Some of these operations are exclusively performed by machines and others necessitating collaboration between workers and machines. This will make it more difficult for managers to balance the workload on the shop floor, even deteriorating the available capacity of workers and machines. Therefore, this study first classifies the production orders into two kinds of tasks (i.e., independent tasks and collaborative tasks) based on their processing attributes before holding them in a pre-shop pool. Subsequently, we give priority to releasing the workload of collaborative tasks (i.e., triggered by the workers) and subsequently release independent tasks (i.e., triggered by the independent machines) to balance the workload across machines at the order release stage. The experimental results demonstrate that prioritizing worker workload balance can significantly improve the shop performance in terms of throughput time and tardiness time. Furthermore, we observe that prioritizing workers' workload also mitigates tardiness typically associated with high load periods when expediting the processing of urgent orders. In other words, prioritizing workers' workload to some extent diminishes the disparity between high load and low load periods due to this is conducive to avoiding premature idle or overload of small capacity resources, therefore leading to a more balanced workload on the manufacturing shops.

M. Thürer et al. (Eds.): APMS 2024, IFIP AICT 733, pp. 381–393, 2024.
https://doi.org/10.1007/978-3-031-71645-4_25

Keywords: Worker-Centric · Order Release · Workload Control · Production Planning and Control

1 Introduction

Through on-site investigations of a leading air conditioning manufacturing company in the Guangdong-Hong Kong-Macao Greater Bay Area of P. R. China. It has been determined that workers remain indispensable production resources in this industrial era, since they are particularly suited for handling tasks that demand creativity and autonomous decision-making. Workers strongly complement machine capabilities by performing flexible and highly adaptable tasks, thereby significantly enhancing system throughput performance [1]. Instead of directly releasing the orders onto the shop floor, they are retained in a pre-order pool in front of the shop floor to meet a specific demand criterion. Orders can be generally classified into two types of tasks: those that machines can independently complete (i.e., independent task) and those that necessitate collaboration between machines and workers (i.e., collaborative task). When preparing to release an order onto the shop floor, it is imperative to have available machine capacity for independent tasks, as well as both available machine and worker capacity for collaborative tasks. Any unavailability of machine or worker capacity can cause disruptions in the entire production process, even leading to production blocking.

WorkLoad Control (WLC), a concept that is conducive to improving the production system throughput performance and reducing product tardiness by minimizing Work-In-Progress (WIP) level and balancing the workload across production resources, including both workers and machines. It has found widespread application in small and medium-sized make-to-order manufacturing companies. Previous approaches of WLC primarily focused on the shops that utilized diverse machine resources, such as flow shops, job shops, and assembly shops. However, there has been limited research considering workers as the production resources in the WLC. The existing studies that did consider workers as production resources often treated them as auxiliary machines assigned to specific production tasks. Factors such as human flexibility, efficiency, and assignment were taken into account in these studies [3]. Some studies also explored the integration of humans as a means to augment machine capacity and enhance shop throughput performance [4]. Nevertheless, our investigation revealed that the few existing studies assume that workers possess the same high level of availability as machines, treating workers as mere auxiliaries to machines and assuming that they need to be available whenever machines require assistance. Obviously, worker availability and efficiency are influenced by various factors, including mind state, physical state, age, skill level, and learning rate [5]. These factors can significantly impact worker availability and result in a misalignment between workers and machines when both of them work together on the same tasks. Consequently, this not only undermines machine efficiency, but also deteriorates the overall system throughput performance.

As mentioned above, the production process not only considers the impact of human dynamics on the throughput performance of the production system, but also considers the issue of worker role transitions during the production process. Specifically, this refers to the shift from workers assisting machines to machines assisting workers in decision-making. This reversal of roles empowers workers to finish complex production tasks efficiently and flexibly with the aid of machines. Additionally, it is able to ensure that when dynamic disturbances arise from workers, the likelihood of these disturbances negatively affecting system throughput performance is minimized. By elevating the importance of workers, a certain level of flexibility is guaranteed for workers, thereby preventing them from being the primary cause of performance degradation in the production system. Consequently, a collaborative human-machine manufacturing ecosystem can be established. In response, this study will start by asking the following question:

How to establish a novel worker-centric production planning and control approaches to fit the future manufacturing paradigm?

This study aims to transform a novel insight that emphasizes a human-centric WLC approach and proposes a worker-centric order release approach based on WLC (i.e., machines assist workers). Specifically, it takes into account the existence of multiple operations within an order in the actual production process, some of which are produced by a machine independently, while others involve collaboration between workers and machines. Firstly, orders are classified based on their processing attributes, such as independent tasks and collaborative tasks [6], and then consolidated in a pre-shop pool. Secondly, we give priority to releasing the workload of collaborative tasks (i.e., triggered by the workers) and subsequently release independent tasks (i.e., triggered by the independent machines) to balance the workload across machines at the order release stage. This consideration is influenced by various fluctuation factors, which helps to overcome the system performance loss caused by the unstable efficiency of workers. By using simulation, the experiment results demonstrate that the human-centric production system not only enables workers to fully utilize their capabilities but also significantly enhances shop floor performance. Finally, the conclusions obtained in this study are not only conducive to leading the researchers in this field to think about how to design various human-focused PPC solutions in future manufacturing paradigm such as Industry 5.0, and can inspire the companies that utilize the human workers as production resources to make more appropriate PPC solutions.

The remainder of this paper is structured as follows. Section 2 offers a comprehensive review of relevant literature concerning human-machine scheduling, worker assignment, and WLC order release. In Sect. 3, we introduce an innovative production planning and control paradigm and provide a detailed explanation of the methodology employed in constructing a simulation model. Section 4 shows the description and discussion of the experiment results. Finally, Sect. 5 concludes our study and reveals the managerial implications derived from the findings, and proposes future research works.

2 Literature Review

This section provides a comprehensive review of research in the field of WLC that incorporates worker-related resources. From the perspective of shop floor operation management, there are three elements that require coordination within the shop: human-machine scheduling, worker assignment, and WLC order release. Many studies assume that shop capacity is fixed and static, representing stable resources, while workers possess flexibility and dynamism. The uncertainty associated with human can impact machine status. Section 2.1 will review the literature on the human-machine scheduling, while Sect. 2.2 will provide an overview of the WLC order release by considering workers.

2.1 Human-Machine Scheduling

The continuous execution of tasks by both humans and machines offers advantages for enhancing the productivity of production systems [7]. Production tasks can be categorized and assigned to machines or humans based on their respective expertise in different process types, with the aim of maximizing system performance. However, humans, in comparison to machines, are highly susceptible to their states and the production environment, which makes it challenging to analyze and model their behavior [5]. Consequently, a significant amount of research has been conducted in this area. From a static perspective, considering that machines are better suited for repetitive tasks, Ghosh [8] initially developed a scheduling model using discrete event simulation, based on the balanced timetable method, to determine the availability and task qualification of workers. Nahavandi [9] assigned such tasks to machines while allocating more complex, high-value-added, or creative tasks to humans. However, taking into account subjective factors such as fatigue levels, Tan et al. [10] proposed a method for joint scheduling of humans and machines to mitigate the impact of fatigue on system performance. From a dynamic perspective, humans are highly dynamic production resources, and operator errors account for approximately 60% to 90% of system failures [11]. Therefore, considering the dynamism of humans during task execution is also a focal point in this field of research. Pupa et al. [7] addressed the uncertainty in human production scheduling by introducing a two-tier dynamic scheduling architecture for flexible job shops, taking into consideration the variability and constraints associated with humans during task execution. The integration of dynamic scheduling models with human factors plays a crucial role in production decision-making. Wang et al. [12] proposed an active learning-based human-machine interaction approach that enables more efficient training of workers in production scheduling models and significantly enhances the learning capability of such models. However, due to external and internal factors affecting humans, the inevitable decrease in task execution efficiency poses a significant challenge that needs to be addressed through post-solving methods to mitigate these adverse impacts. Human and machine collaboration in executing order tasks is advantageous for improving the performance of production systems in many situations [7]. Order tasks can be categorized and assigned to machines and/or humans based on their expertise in different process types, with the goal of maximizing system throughput performance. However, humans are highly susceptible to their states and the production environment, making it challenging to analyze and model their behavior [5].

2.2 Workload Control Order Release

Existing studies on WLC often involve integrating work flexibility into the order review and release stages of a DCR shop [13]. Fredendall et al. [14] developed a worker assignment problem based on the Order Review/Release (ORR) method for highly diverse make-to-order production environments. After determining the appropriate job types for the workers, this approach tackles essential issues regarding worker assignment, including their movement, timing, and selection [15]. In response to this challenge, Portioli-Staudacher et al. [16] introduced a novel approach to output control, which enhances system capacity by reallocating idle workers to suitable machines, without requiring additional capacity. This research effectively addresses the question of worker assignment. Given that worker flexibility significantly impacts work efficiency, the issue of selecting the appropriate worker arises. In real production processes, it is more common for multiple machines to wait for a worker rather than having one machine with multiple workers to choose from. Therefore, Thürer et al. [3] concentrated on the problem of assigning workers to the subsequent machine. Another research aspect centers on addressing worker characteristics. Franchini et al. [17] initially suggested enhancing WLC performance by improving the flexibility of human resources, particularly by augmenting workers' proficiency in critical skills. Thürer et al. [13] examined the interchangeability of worker flexibility within the task shop and devised rules for worker assignment. The researchers discovered that WLC order release functions are beneficial in enhancing DRC shop performance across varying staffing levels. Moreover, to enhance WLC performance, some researchers have incorporated worker assignment into multiple critical stages of WLC. For instance, Costa et al. [18] integrated order review and release, as well as worker assignment, as capacity adjustment strategies to integrate input control into a pure flow shop. Worker assignment aims to adjust worker idle time rather than increase the overall system capacity. Renna et al. [19] used game theory to propose new worker assignment rules applicable to DRC shops. The authors constructed a Gale-Shapley model using a periodic review strategy and formed "marriages" consisting of workers and machines. Bergenhenegouwen et al. [20] introduced available workers during the WLC order release stage to assist machines in production, avoiding premature machine idleness. However, it should be noted that researchers still consider humans as an auxiliary production resource alongside machines and assume that when a worker is allocated to a machine that needs them, the worker is available and exhibits no dynamic behavior.

3 Simulation Model

Simulation is a powerful tool for experimenting with different system designs [21]. In this study, two types of formalized standard shop models, a General Flow Shop (GFS) and a Pure Job Shop (PJS), were utilized to avoid potential confounding interactions that could hinder our understanding of the primary experimental factors. Section 3.1 provides an initial summary of the modeled floor shop and job characteristics in the simulation. Before taking out the shop floor dispatching rules used to sequence jobs in Sect. 3.3, the WLC order release procedure will be described in Sect. 3.2. Finally, an overview of the experimental design is provided, and the measures used to evaluate performance are introduced in Sect. 3.4.

3.1 Overview of Modeled Shop and Job Characteristics

The simulation model comprises a PJS with random routings and a GFS with directed routings, implemented in Python programming language using the SimPy module. To avoid introducing additional confounding factors during this experiment, we adopted a standardized setup of a small-sized DRC shop configuration from existing literature (see as [3]). Consequently, we restricted the number of machines to 6 and workers to 2, enabling us to determine the cumulative workload of workers at various machines. In practice, each workstation always comprises a robot and a worker, where the robots are responsible for processing precise tasks, while the worker handles decision-making tasks. Each machine serves as a distinct and dedicated capacity resource. Consistent with prior literature on DRC shops, each machine possesses a constant capacity, and the worker's capacity is also fixed [3]. The availability of workers is limited by their capacity since each worker can handle only one task with a machine concurrently. The routing of a job, including the required machines, workers, and their sequence, is established once order arrives. All machines are equally likely to be visited, and at any given point in the production line, a machine can process at most one order, preventing reentrant flow. In other words, all workers have an equal chance to be visited, and each order is encountered only once in the processing routing of a specific worker. The processing times of operations adhere to a truncated 2-Erlang distribution, with a maximum value of 4 time units and an average value of 1 time unit. The timing settings are assumed to be both sequentially independent and partially operation-dependent. Finally, due dates are established by incorporating a random allowance factor (uniformly distributed between 20 and 35 time units) into the job processing times.

3.2 Order Release Procedures

There are many order release methods in the workload control literature. The LUMS COR (Lancaster University Management School Corrected Order Release) has been demonstrated as the most effective approach in the majority of shop scenarios [2]. It is also the method employed in this study. LUMS COR uses a periodic release procedure to regulate and balance the workload within a shop at regular intervals. The procedure is outlined below:

1. All jobs in the set of jobs J in the pre-shop pool are sorted according to planned release dates.
2. The job $j \in J$ with the highest priority is considered for release first.
3. Take R_j to be the ordered set of operations in the routing of job j. If job j's processing time p_{ij} at the i th operation in its routing – corrected for machine position i – together with the workload W_s released to machine s (corresponding to operation i) and yet to be completed fits within the workload norm N_s at this machine, that is $\frac{p_{ij}}{i} + W_s \leq N_s \forall i \in R_j$, then the job is selected for release. That means it is removed from J and its load contribution is included, i.e., $W_s = \frac{p_{ij}}{i} + W_s \forall i \in R_j$. Otherwise, the job remains in the pool and its processing time does not contribute to the station load.
4. If the set of jobs J in the pool contains any jobs that have not yet been considered for release, then return to Step 2 and consider the job with the next highest priority. Otherwise, the release procedure is completed, and the selected jobs are released to the shop floor.

To ensure stable workload levels of workers and machines, a worker-centric WLC order release approach is proposed by using two-stage procedure can be adopted when the inter-arrival times of jobs are exponentially distributed:

1. Priority the workload of workers, involves preventing the workload of each worker from exceeding a predefined norm;
2. Balance the workload of machines, aims to prevent the workload of machines from exceeding a predefined norm under the premise of maintaining a balanced worker workload.

Specifically, the production orders are categorized into two types of tasks: independent tasks and collaborative tasks, based on the processing attributes in their routings. A two-step order release strategy is then adopted. In the first step, priority is given to the release of orders at machines involving workers. In the second step, orders processed by independent machines are released. The worker-centric order release approach aims to prioritize the workload balance of workers before considering the overall system workload.

3.3 Shop Floor Dispatching Rules

In the manufacturing process, determining the processing priority of jobs in the queue in front of each machine is essential by using shop floor priority dispatching rules. This study utilized two well-established and effective rules commonly employed in DCR job shops: (i) the Operation Due Date (ODD) rule, which prioritizes the job with the most urgent operation due date; (ii) the Modified Operation Due Date (MODD) rule (see as [22]). It prioritizes jobs based on the lowest priority number, determined by the maximum value between the job's due date δ_{ij} and the earliest completion time. Specifically, the completion time for an operation with processing time p_{ij} is determined by $\max(\delta_{ij}, t + p_{ij})$ where t denotes the dispatching decision-making time. The MODD shifts between a focus on ODDs to complete work orders on time and a focus on speeding up work orders during periods of high workload.

3.4 Experimental Design and Performance Measures

The experiment factor settings are shown in Table 1. A full factorial design was employed, consisting of a total of 68 scenarios ($2 \times 2 \times 2 \times 8 + 2 \times 2$). Each scenario was repeated 100 times. After a warm-up period of 3000 time units, results were collected for over 10,000 time units. These parameter settings allowed us to obtain stable results while maintaining a reasonable level of simulation runtime. Four main performance measures are used to evaluate the shop performance: (i) the gross throughput time; (ii) the shop floor throughput time; (iii) the percentage tardy, and; (iv) the mean tardiness time.

4 Results

The detailed performance results to further explore shop floor differences are presented in Sect. 4.1, and Sect. 4.2 shows the results of different dispatching rules. The main results for the different dispatching rules are illustrated in Fig. 1 and Fig. 2 for GFS

Table 1. Experiment parameters.

Factors	Levels
Order release methods	LUMS COR and IMM
Order release strategies	Machine-centric and worker-centric
Dispatching rules	ODD and MODD
Norms	5,6,7,8,9,10,11,12
Shop types	GFS and PJS
Number of machines	6
Number of workers	2

and PJS, respectively. Results are presented in form of performance curves, where each performance curve represents one dispatching rule, and each data point on a curve represents a specific norm level. Moving from right to left along the curve, the workload norms get tighter.

4.1 Performance Assessment: GFS with Directed Routings

By comparing the performance curves depicted in the respective figures, the performance of different dispatching rules in GFS with directed processing routings can be assessed. Figure 2 provides a comparison of the gross throughput time, mean tardiness time, and percentage tardy performance with respect to the shop floor throughput time results.

We can observe from Fig. 2 that there is little difference in the shop floor throughput time and performance between the ODD dispatching rule and the MODD dispatching rule. Moreover, as the workload norm gradually increases, the difference between the two rules becomes smaller. This indicates that the increase in the suitability of orders has little impact on the workloads of workers and machines, regardless of whether order conversion is accelerated during periods of high load.

In terms of gross throughput time, prioritizing the balance of worker workloads can significantly reduce the gross throughput time of the shop as the workload norm increases step-wise. Furthermore, as the workload norm continues to increase, the performance of gross throughput time improves. This may be because the capacity of workers is smaller than that of machines. Prioritizing the balance of worker workloads can reduce their idle times and overloads, avoiding imbalances where worker workloads are prioritized. This is beneficial for balancing the workloads of workers and improving system performance, resulting in a decrease in overall throughput time.

Regarding tardiness performance, we collected data on the percentage tardy and mean tardiness time. It was observed that as the workload norm gradually increases, prioritizing the balance of worker workloads leads to a gradual decrease in mean tardiness time, which serves as an indicator of overall tardiness performance. Conversely, if worker workload balance is not taken into account and only the balance of workloads on machines (i.e., workers and machines) is emphasized, negative effects may arise.

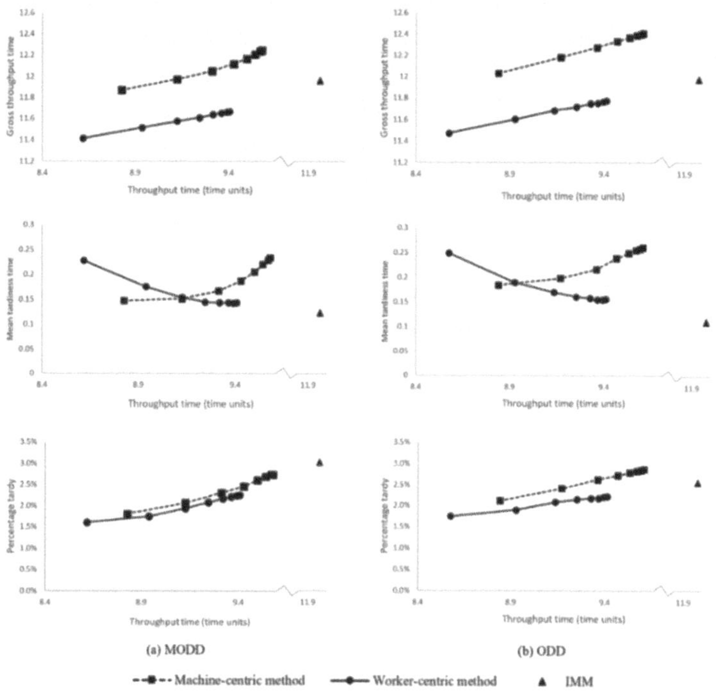

Fig. 1. The results for two order release strategies in GFS with: (a) MODD dispatching rule, and (b) ODD dispatching rule.

In other words, imbalances in worker workloads can result in more tardiness in larger orders. However, the use of the MODD dispatching rule slightly mitigates this situation.

In terms of the percentage tardy, when prioritizing worker workload balance, it can slightly reduce the percentage tardy compared to the performance in mean tardiness time. As previously mentioned, this could be attributed to the increased occurrence of tardiness for larger orders. While the proportion of such orders remains constant, their mean tardiness time is rising.

4.2 Performance Assessment: PJS with Random Routings

The results for the PJS are given in Fig. 2 for the gross throughput time, mean tardiness time, percentage tardy, respectively.

The shop floor throughput time performance in PJS with random routing is significantly enhanced by considering the prioritization of worker workload, as shown in Fig. 3 compared to Fig. 2. Moreover, most evaluation metrics exhibit slight improvements in PJS compared to GFS, which uses directed routing. PJS shows a more gradual enhancement in gross throughput time compared to GFS as the workload increases. This pattern is observed for both the ODD and MODD dispatching methods, with the improvement being more significant when the workload norm is larger. Notably, the ODD dispatching rule achieves the lowest gross throughput time.

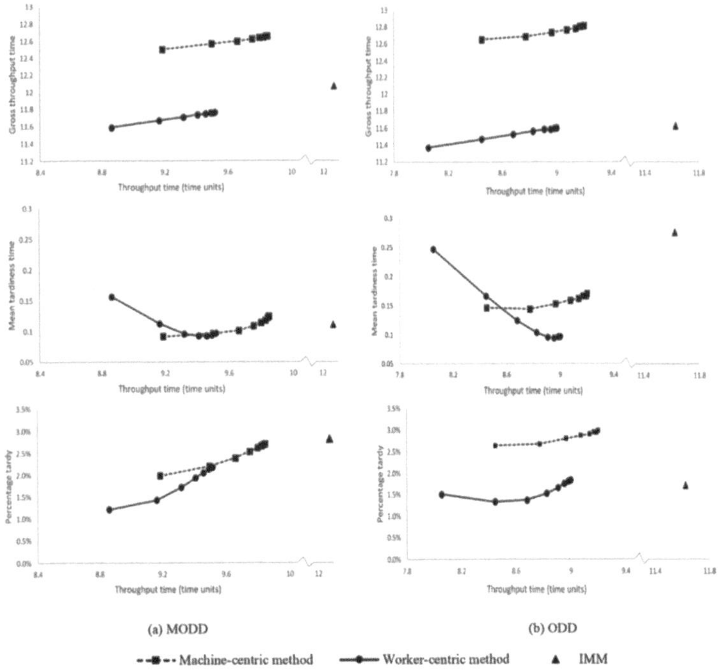

Fig.2. The results for two order release strategies in PJS with: (a) MODD dispatching rule, and (b) ODD dispatching rule.

In terms of mean tardiness time, there is a notable performance disparity between the MODD and ODD dispatching rules. The use of the MODD dispatching rule leads to a consistent improvement trend in the workshop's performance as the workload norm is eased and the prioritization of worker workload is taken into account, leading to a reduction in mean tardiness time. However, when the workload norm is stringent, prioritizing the balance of workers' workload is less effective compared to balancing the workload on machines. This could be attributed to the fact that in tight workload norm conditions, both the worker workload norm and the corresponding machine workload norm are relatively small, resulting in increased tardiness in larger orders and an overall increase in terms of mean tardiness time.

When using the MODD dispatching rule, the efficiency improvement in terms of percentage tardy is not as pronounced as when using the ODD dispatching rule. However, considering the prioritization of worker workload, the percentage tardy performance with the MODD dispatching rule improves as the workload norm gradually tightens. This could be attributed to the accelerated conversion of orders during high load periods, which enhances the overall system performance in terms of both workshop throughput and tardiness.

4.3 Performance Analysis

This study shows that balancing the workload of workers can improve the performance of the production system in terms of the shop floor throughput time, the gross throughput time, the percentage tardy and the mean tardiness time, regardless of whether it is in GFS or PJS. This is because balancing the relatively lower capacity resources (i.e., workers) is conducive to improving their utilization, reducing the idle time or overload of the lower capacity resources, and thereby reducing shop throughput time and tardiness. Additionally, using the MODD and ODD shop floor priority dispatching rules, we discovered that prioritizing the balance of worker's workload is particularly beneficial when combined with the MODD dispatching rule, which involves expedited processing of urgent jobs. Prioritizing worker's workload can better respond to the urgent demands of high load period orders and significantly reduce tardiness, especially the mean tardiness time.

In conclusion, giving priority to balancing the workload of workers not only improves the shop performance in terms of the throughput time and tardiness time but also avoids high load periods at machines. This provides a distinct research direction for exploring more granular workload balancing of production resources in the field of workload control and offers a new opportunity for research in worker-centric production planning and control methods in human-machine collaborative production systems.

5 Conclusions and Future Works

This study presents a novel paradigm of human-centric production planning and control by introducing a worker-centric workload control order release approach. Specifically, we consider that a production order in the actual production process may involve multiple operations, with some being exclusively performed by machines and others necessitating collaboration between workers and machines. Firstly, orders are classified based on their processing attributes, such as independent tasks and collaborative tasks [6], and then consolidated in a pre-shop pool. During order release stage, we prioritize the workload of collaborative tasks and subsequently release independent tasks to achieve workload balance across machine machines. Our experimental results demonstrate that prioritizing the balance of work's workload significantly enhances the shop performance in terms of throughput time and tardiness time. Furthermore, we observe that giving priority to balancing the workload of workers also mitigates delays typically associated with high load periods when expediting the processing of urgent orders. In other words, prioritizing workers' workload to some extent diminishes the disparity between high load and low load periods due to this is conducive to avoiding premature idle or overload of small capacity resources, therefore leading to a more consistent workload in the manufacturing shops. The primary contribution of our research is to propose a novel WLC order release method within a specific paradigm, and we substantiate the performance advantages of this method through simulation.

This study a novel approach of order release in the field of WLC and even PPC by shifting the perspective from a machine-centric to a human-centric paradigm. The findings of this study can provide valuable insights for the evolving manufacturing paradigm, particularly the recently proposed Industry 5.0 paradigm, which emphasizes the importance of human beings in production and manufacturing systems. Furthermore, this study highlights that in real-world shop floor environments, where human workers are utilized as production resources responsible for manufacturing products, human variability should be accounted for in PPC solutions, as various human factors can directly impact the performance of production systems. Additionally, managers should pay close attention to human behavior processes in the production process, as this will help them develop more appropriate PPC solutions and provide the necessary methodological support.

Future works should fully leverage the significant potential of the WLC concept in mitigating production process variability, and explore the impact mechanisms of varying degrees of human variability on WLC order release methods. Furthermore, it is necessary to integrate production process data from the real-life factories in the context of Industry 3.0 or Industry 4.0 to assess the reliability of the proposed human-centric order release approach. This will also be conducive to validating the applicability of the proposed order release approach to the actual production processes of the majority of manufacturing companies. In addition, with the introduction and rapid development of Industry 5.0, it is necessary to propose specific PPC solutions for human-robot collaborative shop floors based on the foundation of numerous conceptual studies from multiple perspectives.

References

1. Autor, D.H.: Why are there still so many jobs? The history and future of workplace automation. J. Econ. Perspect. **29**(3), 3–30 (2015)
2. Thürer, M., Stevenson, M., Silva, C., Land, M.J., Fredendall, L.D.: Workload control and order release: a lean solution for make-to-order companies. Prod. Oper. Manag. **21**(5), 939–953 (2012)
3. Thürer, M., Zhang, H., Stevenson, M., Costa, F., Ma, L.: Worker assignment in dual resource constrained assembly job shops with worker heterogeneity: an assessment by simulation. Int. J. Prod. Res. **58**(20), 6336–6349 (2020)
4. Costa, F., Thürer, M., Portioli-Staudacher, A.: Heterogeneous worker multi-functionality and efficiency in dual resource constrained manufacturing lines: an assessment by simulation. Oper. Manag. Res. **16**(3), 1476–1489 (2023)
5. Destouet, C., Tlahig, H., Bettayeb, B., Mazari, B.: Flexible job shop scheduling problem under Industry 5.0: a survey on human reintegration, environmental consideration and resilience improvement. J. Manuf. Syst. **67**, 155–173 (2023)
6. Ferreira, C., Figueira, G., Amorim, P.: Scheduling human-robot teams in collaborative working cells. Int. J. Prod. Econ. **235**, 108094 (2021)
7. Pupa, A., Van Dijk, W., Secchi, C.: A human-centered dynamic scheduling architecture for collaborative application. IEEE Robot. Autom. Lett. **6**(3), 4736–4743 (2021)
8. Ghosh, J.: Distributed task scheduling and human resource distribution in industrial service solution production: a simulation application. Int. J. Prod. Res. **51**(10), 2900–2914 (2013)
9. Nahavandi, S.: Industry 5.0—a human-centric solution. Sustainability 2019 **11**(16) (2020)
10. Tan, Q., Tong, Y., Wu, S., Li, D.: Anthropocentric approach for smart assembly integration and collaboration. J. Robot. (2019)
11. Kumar, P., Gupta, S., Gunda, Y.R.: Estimation of human error rate in underground coal mines through retrospective analysis of mining accident reports and some error reduction strategies. Saf. Sci. **123**, 104555 (2020)

12. Wang, D., Guan, L., Liu, J., Ding, C., Qiao, F.: Human–machine interactive learning method based on active learning for smart workshop dynamic scheduling. IEEE Trans. Hum.-Mach. Syst. (2023)
13. Thürer, M., Stevenson, M., Renna, P.: Workload control in dual-resource constrained high-variety shops: an assessment by simulation. Int. J. Prod. Res. **57**(3), 931–947 (2019)
14. Fredendall, L.D., Melnyk, S.A., Ragatz, G.: Information and scheduling in a dual resource constrained job shop. Int. J. Prod. Res. **34**(10), 2783–2802 (1996)
15. Salum, L., Araz, Ö.U.: Using the when/where rules in dual resource constrained systems for a hybrid push-pull control. Int. J. Prod. Res. **47**(6), 1661–1677 (2009)
16. Portioli-Staudacher, A., Costa, F., Thürer, M.: The use of labour flexibility for output control in workload controlled flow shops: a simulation analysis. Int. J. Ind. Eng. Comput. **11**(3), 429–442 (2020)
17. Franchini, L., Caillaud, E., Nguyen, P., Lacoste, G.: Workload control of human resources to improve production management. Int. J. Prod. Res. **39**(7), 1385–1403 (2001)
18. Costa, F., Portioli-Staudacher, A., Nisi, D., Rossini, M.: Integration of order review and release and output control with worker's allocation in a pure flow shop. IFAC-PapersOnLine **52**(13), 2632–2637 (2019)
19. Renna, P., Thürer, M., Stevenson, M.: A game theory model based on Gale-Shapley for dual-resource constrained (DRC) flexible job shop scheduling. Int. J. Ind. Eng. Comput. **11**(2), 173–184 (2020)
20. Bergenhenegouwen, T., Kasper, T.A., Bokhorst, J.A., Land, M.J.: Managing premature idleness in high-variety manufacturing. Flexible Serv. Manuf. J. 1–23 (2023)
21. Mourtzis, D.: Simulation in the design and operation of manufacturing systems: state of the art and new trends. Int. J. Prod. Res. **58**(7), 1927–1949 (2019)
22. Baker, K.R., Kanet, J.J.: Job shop scheduling with modified due dates. J. Oper. Manag. **4**(1), 11–22 (1983)

Understanding Part Complexity: A Novel Approach for the Identification of Complexity-Influencing Part Characteristics

Greta Tjaden[1(✉)], Nick Große[2], and Anne Meyer[3]

[1] TRUMPF SE + Co. KG, Ditzingen, Germany
greta.tjaden@trumpf.com
[2] TU Dortmund University, Dortmund, Germany
[3] Karslruhe Institute of Technology, Karlsruhe, Germany

Abstract. Machine tool manufacturers must understand the requirements of their customers to survive in challenging markets, e.g. by understanding their customers' order characteristics. We want to explore part complexity as a measure to fill this gap of data-based customer information. This paper presents the procedure that we developed for part complexity assessment and is centered around expert consultation and the results from implementing this procedure at our industry partner, a machine tool manufacturer for sheet metal processing. Our contributions are (1) a literature review on part complexity and assessment approaches, (2) a definition of part complexity, (3) a tried-and-tested procedure for the assessment of part complexity including an online tool, (4) 80 sheet metal processing geometries, (5) codebooks containing the relevant part characteristics to assess part complexity of an exemplary production unit, (6) detailed explanation of the labeling results for three geometries, and (7) refinements for enabling transferring the labeling results to new geometries using a classification algorithm.

Keywords: Part Complexity · Machine Tool Manufacturing · Sheet Metal Processing · Random Forest Classifier

1 Introduction

To survive in challenging markets, machine tool manufacturers must understand the requirements of their customers. We found use cases along the product life cycle that would benefit from such data-driven customer information [1], like data-driven customer segmentation for product and portfolio management [2] or a KPI set for customer assessment for sales and consulting [3]. By understanding part complexity, machine tool manufacturers gain more knowledge about their customers and can more effectively identify and meet customer needs. The part complexity approaches found in the literature lack a common definition, while

Published by Springer Nature Switzerland AG 2024
M. Thürer et al. (Eds.): APMS 2024, IFIP AICT 733, pp. 394–409, 2024.
https://doi.org/10.1007/978-3-031-71645-4_26

the part complexity assessment approaches are missing evaluation mechanisms. We want to answer the questions "How can we assess part complexity?" and "What characteristics make parts complex?".

The contributions of our paper are (1) a literature review on part complexity and assessment approaches, (2) the development of a definition for part complexity in sheet metal processing, (3) a mixed-methods approach for assessing part complexity including (4) 80 laser cutting geometries used for the complexity labeling [4]. We put our complexity assessment procedure into practice, using the TruLaser Center 7030 as an exemplary production unit. Furthermore, we (5) give an excerpt of the codebooks describing part complexity influencing characteristics, and (6) explain three labeling geometries and their labeling results in detail. Lastly, we (7) present the results of transferring our results to further geometries using a classification algorithm.

Our paper is structured as follows: Sect. 2 presents the state of the art of part complexity. Then, we present our methodology and its adaption to our exemplary production unit in detail in Sect. 3. Section 4 contains a critical reflection of our method, explains the labeling results of five geometries in detail, and shows our approach to transferring our labeling results to other parts.

2 Related Work

2.1 Definition of Part Complexity

Part complexity is mentioned without being explained any further in [5–9]. The authors of [10] provide a rough definition of complex surfaces as "surfaces that do not represent a classical geometry (circle, arc, etc.)", while the authors of [11] state that complex products are characterized by being rare. These definitions seem insufficient to assess the part complexity for a machine tool manufacturer for sheet metal processing. The influence of the part's manufacturability is brought up in [12] by stating that "complex geometries [...] are extremely difficult to form with conventional stamping". Other authors seem to agree with the perspective of complexity as an indicator of manufacturability [8,12–17]. One must keep in mind that three of these eight papers stem from the same research group, namely [13–15].

We derive our definition of part complexity in sheet metal processing based on the results of Sect. 2. The researchers of [8,12–17] agree that the part's manufacturability is an indicator of the part's complexity. We conclude that the more complex a geometry is, the harder it is to manufacture. Since the part complexity reflects the manufacturability, the part complexity will change when the manufacturing process changes. We derive the following definition:

We define sheet metal part complexity as the degree of manufacturability for geometry and process step. The harder a geometry is to manufacture, the more complex the geometry is. Moreover, the sheet metal part complexity will underly change, as the manufacturability may change over time, e.g. due to changes in the production process.

Now, we must develop a procedure to assess the Sheet Metal Part Complexity and identify the complexity influencing part characteristics.

2.2 Previous Work of Part Complexity Assessment

In addition to part complexity definitions, we found prior researchers determining part complexity. The authors of [8] developed a system for automatic feature extraction of sheet metal CAD files since none of the approaches they identified in the literature provided their required level of detail. In [18], they extract geometrical features, such as holes and edges, aiming to generate a triangle mesh of the geometries' surface. The authors of [19] elaborate on the complexity of CAD models instead of the geometries, e.g. further considering the internal structure of CAD files. They also state that the geometry complexity influences the CAD complexity. They analyze a database containing 370 CAD models and analyze them—among others—regarding their geometry features. The authors of [17] state that "shape complexity of a part is usually described in qualitative terms like low, medium, high, and very high". They analyze the complexity of a casting process with the use case design for manufacturability. They want to identify drivers for high costs within the geometries to alter the responsible geometrical features as early in the design process as possible to lower the cost. The authors of [17] define equations for the calculations for each criterion for the complexity, such that it returns values between 0 (low complexity) and 1 (high complexity). They put their approach into practice, using 40 CAD data files from a tooling manufacturing company. They accessed the cost data from the senior tool maker, covering a period of two years and ignoring inflation. To derive the weights for their six criteria, they perform a multiple regression analysis. In the evaluation of their equation, they calculate the complexity of not yet considered parts, reaching an accuracy of 97.7%, satisfying their accuracy requirement of at least 95%. The authors of [20] aim to assess the manufacturability of castings to optimize production planning. They derived characteristics from interviews with casting engineers, product designers, and purchase managers. These characteristics include the area, number of cores, casting size and weight, and section thickness. They determined the weights of the calculation parameters using the analytic hierarchy process (AHP). To do so, they created a hierarchical structure of the problem to assess the priorities of the attributes relative to each other, evaluated the consistency of the relative priorities, and calculated the weights based on the verified relative priorities of the attributes. Moreover, they demonstrate their framework on an industrial example. The authors of [21] analyze production efficiency—consisting of structural complexity comparable to part complexity, production complexity, and management complexity—on the example of China's offsite construction industry. The structural complexity covers surface-to-volume ratio, number of openings (holes), reinforcement ratio (hinting at production time), number of embedded parts, and weight of exposed rebar (hinting at the structural complexity of the products). In addition to the literature research, the authors of [21] conduct a field search, visiting three factories to observe the production processes and to semi-structured interview experts.

To verify the impact of the complexity features and the production efficiency, they let 182 experts from the factories and researchers rank their influence, using the Likert Scale. The authors of [22] assess the complexity of a part for CAD modeling by the number of surfaces. They let CAD trainers assess the complexity of ten parts, resulting in a part being considered complex when it has more than 250 surfaces. They state that their definition could be refined by taking further characteristics into account, such as the size of the surfaces or the ratio of planar and non-planar surfaces. The authors of [23] also work with expert surveys to assess the part complexity and observe a relation between the number of triangles of a geometry and its complexity: The more triangles a geometry has, the more complex it is. They further define that when the geometry decreases in size, the complexity increases. They calculate complexity as the ratio of the product of the surface and the number of triangles, and the volume.

In conclusion, researchers in previous work used between 40 and 370 parts to analyze part complexity influencing characteristics. Many approaches relied on experts, e.g. due to labeling [21–23] or interviews [20]. The authors of [21] used a Likert scale from 1 to 5, while [17] used a scale from 0 to 1. However, none of the approaches found in the literature can be adapted as is to our machine tool manufacturer for sheet metal processing: They lack (1) a common definition of part complexity. The most prominent missing points in the assessment approaches identified in the literature are (2) evaluation mechanisms to assess the quality of the experts' answers and (3) distribution across areas of expertise and geography. We want to build upon these previous researchers and develop our procedure to identify part complexity influencing part characteristics, using a machine tool manufacturer for sheet metal processing as an example.

3 Methodology

3.1 Scientific Approach to Assess Part Complexity in Sheet Metal Processing

Ideally, we would utilize field data from the production units about the geometrical characteristics of the part and their production. Until we can access such field data, we rely on human labeling technology. We follow a mixed methods approach combining both empirical and qualitative research elements. Subsequently, we conducted an individual depth analysis, investigating geometry characteristics that influence the complexity, and decided to collect the data with an online survey. Both the development of the online survey tool and the decisions on experiment parameters underwent feedback loops from both scientific and industry perspectives. Before the final data collection, we furthermore conducted a pilot test and incorporated the findings into the online survey tool and the decision on parameters.

Surveys are highly structured interviews that aim to collect comparable data to identify similarities and dissimilarities [24]. A sub-form of surveys are computer-delivered self-administered questionnaires, also labeled computer-assisted self-interviews (CASIs) [24]. An advantage of this labeling tool is that errors stemming from the presence of the interviewer or other survey participants are highly unlikely to occur since there will be no interaction between the interviewer and the participants during the labeling [24]. We develop a labeling tool to allow for independent operation from the participants' schedule and their locations. Since we will distribute our labeling tool via the Internet, we can include employees on different sites. The participants can interrupt the survey after each geometry to cater to their roles in the organization.

3.2 The Exemplary Production Unit

The industry partner of this paper is a German machine tool manufacturer for sheet metal processing that offers machine tools for the most relevant sheet metal processing steps, like the highly automated laser cutting production unit TruLaser Center 7030 which serves as our exemplary production unit. The TruLaser Center 7030 offers three process steps in total, the process step laser cutting and two options for part handling, depending on the part size: Parts between 30 mm × 30 mm and 160 mm × 160 mm are ejected through the SmartGate, a hole under the laser cutting head. Parts between 90 mm × 60 mm and 2000 mm × 1500 mm are unloaded with the SortMaster Speed, a robot for automated unloading of parts using suction cups [25].

3.3 The Labeling Tool for the CASIs

The labeling tool interface is depicted in Fig. 1. The labeling participants label the complexity of each geometry for each process step of the TruLaser Center 7030: Laser cutting, part handling with the SortMaster Speed, and part handling with the Smart Gate. The example shows a geometry that is suitable for both part handling with the SortMaster Speed and the Smart Gate. At the top, the participants see how many geometries they have labeled out of the total number of geometries. Below, the geometry is visualized with scales in both x- and y-direction to give the participants the geometrical dimensions. The labeling participants can additionally choose to visualize the part's bounding box (red), its convex hull (blue), and its centroid (green) for further information in the lower right corner. Next to it, we provide additional geometric information like the part weight, area, dimensions of the bounding box in x- and y-direction, the ratio of the part area to the area of the bounding box, the ratio of the part area to the area of the convex hull, and the part handling type. We chose mild steel as the material since this was the most produced material in the production unit in the 12 months before our labeling. For this material, the most used sheet thickness is 3 mm, so we pre-defined the sheet thickness. As we are looking at two-dimensional geometries, we believe that a two-dimensional labeling interface provides sufficient detail of information.

At the lower left corner are the scales for the labeling of the complexity of the three process steps. According to [24], rating scales usually provide between 3 to 7 answer alternatives. We use a balanced scale that provides an "equal number of categories above and below the midpoint" [24]. We exclude the possibility for

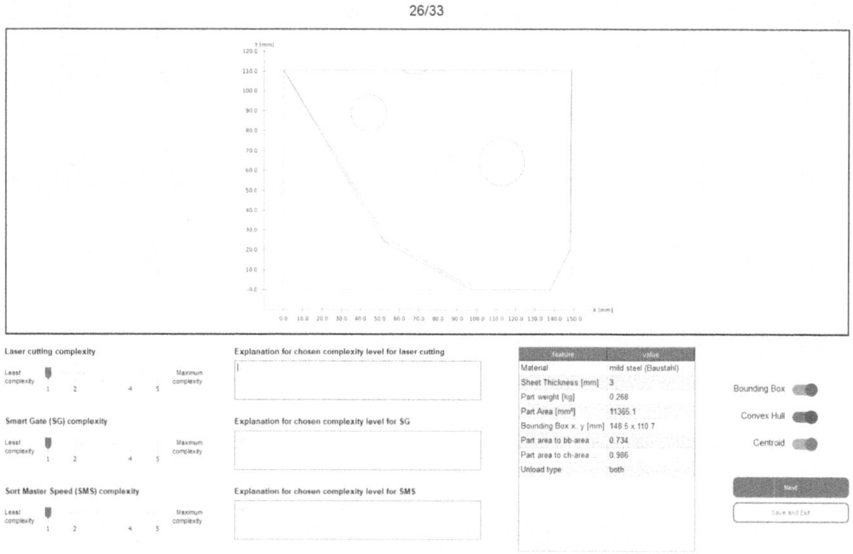

Fig. 1. Interface of the Labeling Tool. Own visualization. (Color figure online)

the participants to remain neutral in their rating by excluding the average score, making our scale a forced-choice rating scale instead of an unforced-choice rating scale. Forced-choice rating scales inhibit the risk of producing a bias when many participants want to select medium complexity but are not given this opportunity [24]. However, we strongly assume that in our case, the majority if not all participants will have an opinion on a part's complexity. From our point of view, the risk of producing a bias is smaller than the risk of indecisive participants. Previous work used a Likert scale ranging from 1 to 5 [21,26]. Since we assume that a distinction between not complex (1), a little bit complex (2), more complex (4), and highly complex (5) is sufficient, we choose a Likert scale without the average response option. By only providing this scale as an answer option, we designed our survey to have structured responses, sometimes also named closed responses [24]. To conclude, our scale will be a multiple-choice, single-response scale so that participants cannot answer with more than one value per question. The scales range from 1 (least complexity) to 5 (maximum complexity) without 3 as the middle option. On the left side of the scales, the users document an explanation for their labeling. To analyze the participants' explanations in English and German, we create categories and assign fitting explanations to these categories. We create these categories exploratively and inductive [27] while categorizing the participants' explanations for the chosen part complexity for the process step. To ensure the quality of these assignments, one researcher assigns the categories, and another evaluates the assignments for accuracy and consistency. To evaluate the resulting codebook, we select a labeling result with the most different explanations and ask a third person to annotate these explanations based on our codebook. Then, we evaluate the differences in annotation.

3.4 The Labeling Participants

The labeling participants should be experts on the selected machine tool. Hence, we perform purposive sampling [24] to select the labeling participants purposefully based on their expertise in the sheet metal processing production processes. To ensure that a variety of perspectives on complexity is covered, we want to include at least two people per area to label the geometries, covering product management, development, remote operation, machine operation, and machine training.

3.5 The Sample Size

Since limitless geometries exist, we cannot rely on established sampling methods. Hence, we define the sample size for the labeling based on the research presented in Sect. 2. While [19] used 370 geometries, [17] used 40, and [22] used 10. The two latter sample sizes seem rather low, so we define 80 geometries as our sample size to balance the contradicting requirements of generating a large database and keeping the labeling effort to a minimum. Furthermore, we will check for inconsistency in the participants' answers by repeating a subset of 10 geometries in each of the three weeks of the labeling.

3.6 Exploratory Data Analysis of Geometries

The majority of the geometries is handled exclusively with the SortMaster Speed. In contrast to this, only a small portion of geometries is exclusively unloaded with the SmartGate, or apply to both unloading options. While some geometries have maximum lengths and widths for our exemplary production unit, we see many geometries having smaller lengths and widths. 59 geometries have inner contours, while 21 geometries do not. The majority of the geometries having inner contours have between 1 and 15 inner contours and some geometries have between 16 and 31 inner contours. In addition, two geometries have 52 and 219 inner contours, which makes them the geometries with an extraordinarily high amount of inner contours. Although the majority of the area of the inner contours is rather small, we have a significant number of outliers. The cutting length is unevenly distributed, with the median below 2,500 mm, the majority of the parts having a cutting length below 10,000 mm, and outliers having more than 25,000 mm cutting length. In contrast to the cutting length, the geometric area has many outliers. The greater the area of the geometry is, the higher is also the cutting length. The majority of the parts have a cutting length of under 10,000 mm. Approximately half of the geometries are rectangular and five geometries are circular. The majority of the part has little to no undercuts.

4 Results

4.1 Data Collection

We collected the data over the course of three weeks in 2023. After a participant completed the labeling of one week, he or she got the next set of geometries. Due to the scheduling of the introduction meetings, not all participants started on the same data, and some participants took more time than one week.

We initially started with 40 participants in the labeling, stemming from the areas of product management, remote and machine operation, development, and customer training. Out of these 40 participants, half completed the first week, with a number of 20. Subsequently, four more participants dropped out, with 16 participants completing the second week. 15 participants completed all three weeks of the labeling, making up 37.5% of the starting group. Product management was not covered anymore.

4.2 Analysis of Repetition Test

Ten geometries were randomly selected and labeled each week to analyze the consistency of the participants' answers. These geometries include eight geometries exclusively applicable to part handling with the SortMaster Speed, one geometry applicable for part handling with both the SortMaster Speed and the Smart Gate, and one geometry exclusively applicable for part handling with the SmartGate. First, the repeating geometries are analyzed for each participant and process step. Unfortunately, we could not find limits of deviation in the literature that are suitable for our repetition test. This is why we defined the limits of deviation by ourselves: Deviations smaller or equal to 1 are fine, while deviations greater than 1 are considered large deviations. Especially with our forced-choice scale design, where the middle value 3 is not an option, this means that the participant switched between low and high complexity. We decided that occurring critical deviations of under 30% were okay and to exclude participants from further considerations for process steps if critical deviations were 30% or above. This threshold works for the process steps of laser cutting and part handling with the SortMaster Speed, which were represented in the repeating geometries with 10 and 9 geometries, respectively. For part handling with the SmartGate, we only have two repeating geometries. Here, a critical deviation in one of the two parts already leads to more than 30% deviation and excludes the participant from this process step. This results because the majority of the geometries are applicable for part handling with the SortMaster Speed and not with the SmartGate.

Figure 2 exemplarily shows the analysis of the repeating geometries for participant P1 and the process step laser cutting. The geometries are named from A to J, while the numbers 0, 1, and 2 indicate the week of the labeling. For clarity, each geometry has been assigned a color. Here, we see a high consistency in the participant's answers: Eight of ten geometries have been labeled the same over the course of three weeks. Geometry B has a little deviation since it has been labeled with complexity 2 in the first week and complexity 1 in the two subsequent weeks. There is only one critical deviation for geometry C, which has been labeled with complexity 2 in the first week and complexity 4 in the two subsequent weeks. This results in 10% critical deviation for the process step laser cutting, which lies below the defined limit of 30%. Hence, participant P1 passed the repetition test for laser cutting, and his or her labeling results for the process step laser cutting will be considered in further analysis.

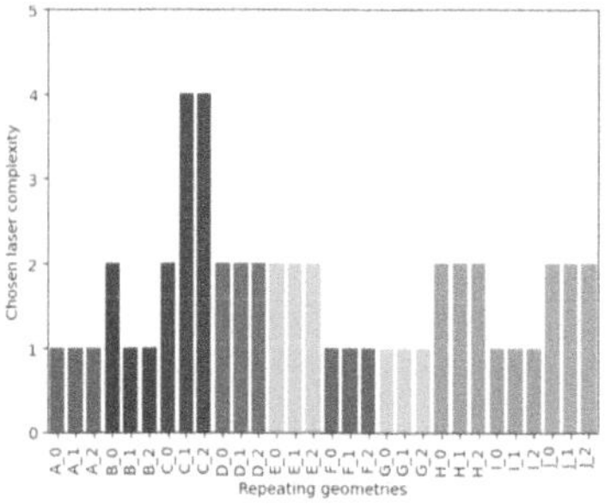

Fig. 2. Analysis of repetition test for participant P1 for the process step laser cutting. Own visualization.

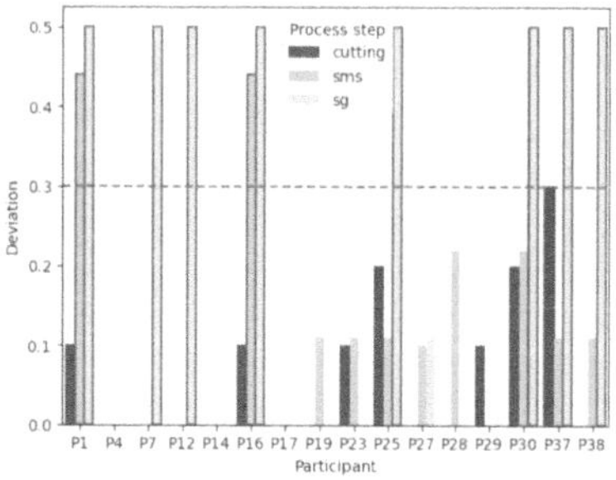

Fig. 3. Analysis of repeating geometries. Own visualization.

We followed the same procedure for all participants and process steps. The resulting labeling deviations for all participants are visualized in Fig. 3. On the x-axis, we see the participants, while the y-axis indicates the critical deviation. The process steps of laser cutting (cutting), part handling with SortMaster Speed (sms), and part handling with SmartGate (sg) are abbreviated in the legend. The limit in critical deviation of 30% is highlighted with the red dotted line. The three participants P4, P14, and P17 have labeled the geometries highly consistent and without any critical deviation. The five participants P19, P23, P27, P28, and P29 have labeled the geometries with critical deviation, but without reaching the limit for any of the process steps. P37 is the only participant who reached the limit of 30% deviation for the process step laser cutting, while P1 and P16 reached the limit for the process step part handling with the SortMaster Speed. The participants P1 and P16 also reached the limit for part handling with the SmartGate, together with the six other participants P7, P12, P25, P30, P37, and P38. Concluding, one participant will not be considered for the process step laser cutting, and two participants will not be considered for the process step part handling with the SortMaster Speed. Eight and hence, half of the participants will not be considered for the process step part handling with the SmartGate.

4.3 The Codebook

The codebooks contain the categories of the experts' explanations for the chosen part complexity for each process step. We followed the scientific approach described in Subsect. 3.3. One researcher categorized the explanations and created the codebook, while another researcher checked the categorizations and the codebooks. We created three codebooks in total, one for each process step. To evaluate the codebooks, we first selected the labeling result with the most diverse explanations for the chosen complexity. Then, we asked a third person to categorize the explanations using the codebook.

Table 1. Excerpt of the codebook for laser cutting.

Category	Explanation
no_detailed_explanation	The participant did not choose part complexity 1 and did not provide any explanation.
no_issues	The participant chose part complexity 1.
notches_problematic	This category summarizes potential manufacturing problems due to part characteristics such as notches, slots, taps, and flaps.

We improved the codebooks, where indicated. The resulting deviations between our categorization and the categorization by the third person were small, satisfying our requirements. For reasons of scope, we show an excerpt of the codebook for laser cutting in Table 1.

4.4 3 Repeating Geometries as an Example

In the further analysis of the labeled geometries, we exclude the participants who labeled too inconsistently to pass the repetition test. We compute the average process step complexity by calculating the average value of the complexity labels of the remaining participants. For the analysis of the categorized explanations from the codebooks, we make lists.

Fig. 4. Repeating geometry A. Own visualization.

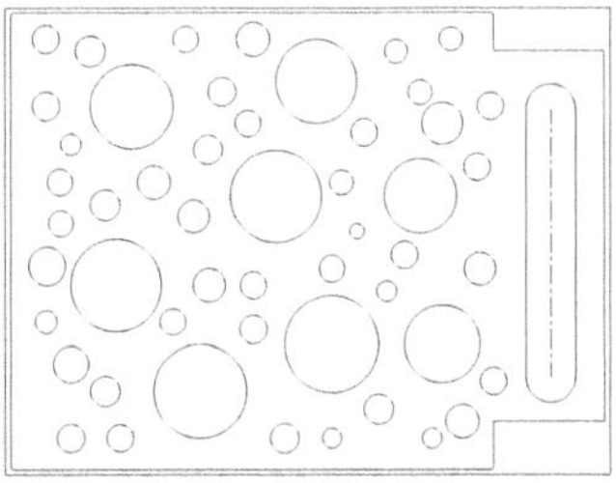

Fig. 5. Repeating Geometry C. Own visualization

Repeating geometry A is visualized in Fig. 4 and is a rectangle. After being laser cut, it will be sorted out with the SortMaster Speed. The labeling participants who passed the repetition test all labeled this geometry with complexity 1 for laser cutting. The average complexity for laser cutting for geometry A is 1. We only have "no_issues" as the categorized explanation. Geometry A is perceived as very simple to laser cut. We get a similar result for part handling with the SortMaster Speed. We have one outlier with one participant labeling it with complexity 4 in one week, stating that the size of the geometry was making it difficult to be handled with the SortMaster Speed. However, before and after labeling it with complexity 4, this participant labeled the geometry with 1, like every other labeling participant. The average complexity for part handling with the SortMaster Speed for geometry A is 1.1.

Fig. 6. Repeating geometry B. Own visualization.

Geometry B is visualized in Fig. 6. In contrast to geometry A, geometry B appears to be longer in shape and has indents in the outer contour. Compared to geometry A, we see a higher variance in the labeling, especially for part handling with the SortMaster Speed. We calculated an average laser cutting complexity of 1.2 and an average complexity for part handling with the SortMaster Speed of 1.6. The explanations for the higher laser cutting complexity are the undercut with one mention and two mentions of the scrap skeleton being problematic. These explanations are related to the indents in the outer contour of the geometry. We have seven mentions that the bending may be problematic, followed by three mentions of the length. We have each two mentions of placement of the pins of the SortMaster Speed, the clamping being problematic, the size, and the undercuts. Furthermore, we have one mention that the placement of the suction cups of the SortMaster Speed may be problematic. Here, we can link the categorized answers to the indents in the outer contour of geometry B as well as the narrowness of the geometry. The bending, the size, and the length seen as problematic by the participants are related to the narrowness, while the clamping and the undercut are related to the indents in the outer contour. We do not see a link between the placement of both the suction cups and the pins of the SortMaster Speed and the geometry.

Figure 5 shows repeating geometry C, a rectangular geometry with many inner contours, consisting of circles of various sizes and one elongated hole. We see a high variance in the labeling for the laser cutting, with complexity values ranging from 1 to 4. The calculated mean value for the laser cutting complexity is 1.5 The inner contour being problematic for the manufacturing process has been mentioned eleven times. The slugs being problematic was mentioned twice. The heat, degree of cutting, possible tilting, scrap cutting being required, falling detection, placement of the ejection cylinder, and size were all mentioned once. Here, we can see the link between the inner contours being the mostly mentioned characteristics making the part hard to manufacture and the geometry itself, having many inner contours and a rather large area of inner contours. For part handling with the SortMaster Speed, We see a similar variance in complexity, where the complexity values also range from 1 to 4. We calculated a mean complexity value for part handling with the SortMaster Speed of 1.8. The placement of the suction cups being problematic was mentioned seven times, and the inner contours and the placement of the pins being problematic were mentioned five times each. That the scrap must be cut was mentioned once. Again, we can see the link between the categorized explanations and the geometry itself. The inner contours—both the amount and the combined inner contour area—provide less space for the placement of the suction cups and the pins of the SortMaster Speed shuttle. The mention of scrap cutting is also related to the inner contours since some of the inner contours are prone to clamping due to their size. This can be prevented by cutting the scrap produced by the inner contours.

4.5 Transferring the Results to Other Geometries Using Random Forest Classifier

Using algorithms, we analyzed the transferability of our insights into what characteristics make parts complex for the process steps of our exemplary production unit. Our goal was to train an algorithm on our labeling results and then use this algorithm to classify new geometries that were not labeled before. We achieved the best classification accuracies with a random forest classifier. We defined complexity values smaller than 1.2 as small, between 1.2 and 1.8 as medium, and greater than 1.8 as high. The missing values were assigned "not_applicable". For the random forest classifier, we tested leave one out cross-validation, stratified k-fold cross-validation, and repeated k-fold cross-validation. Leave one out cross-validation had the best results. We performed a hyperparameter optimization using GridSearch for leave one out cross-validation, leading to a best mean cross-validated score of 0.7625 for laser cutting, 0.7 for part handling with the SortMaster Speed, and 0.925 for part handling with the SmartGate. One must keep in mind that for the latter, we only have 16 data points in our data set, which relativizes the rather high accuracy. The accuracy of 0.7625 and 0.7 is not good enough to use the classifications made by the algorithm for decision-making processes. We assume the limited number of geometries and insufficient part characteristics to be the reasons. For instance, the analysis of the three repeating geometries has shown that large inner contours appear to highly influence the complexity of part handling with the SortMaster Speed but unfortunately, we can only calculate the sum of all inner contours. We believe that with more fine-granular part characteristics, the part complexity will be reflected more accurately, leading to better classification results.

5 Discussion

Due to the endless possibilities of geometries, the statistical representativity of our labeling geometries could not be ensured. Instead, we opted to randomly select 80 geometries and have ten repeating geometries, since that seemed to be a good trade-off between having a high enough number to get trustworthy results and not having time-consuming labeling.

We ensured the quality of our data collection by implementing the repeating geometries as an evaluation mechanism to observe the participants' labeling consistency over time. Since we did not find suitable limits in the literature, we subjectively chose 30% of high deviations for eliminating inconsistent labeling. However, one must keep in mind that we did not provide the middle option "3" in the rating scale, which might provoke a bigger deviation. Furthermore, the participants' labeling of the repeating geometries may be influenced by the prior geometries. Although the repeating geometries' order did not change, the participants gained more experience in complexity labeling each week, which may have influenced their labeling.

As explained in Subsect. 4.4, the relation between the part and its geometrical characteristics on the one hand, and the participants' labeling and explanations, on the other hand, are visible. This proves the internal validity of our part complexity assessment procedure.

For the external validity, we evaluated both the generalizability of our part complexity assessment procedure and the results. We demonstrated the generalizability of the method by adapting it to our exemplary production unit. Hence, our part complexity assessment method can be adapted to further production units and technologies. The supporting labeling tool, of course, must be adapted to the production processes of the chosen production units. The transferability of our results regarding the part complexity influencing part characteristics to other geometries is limited: Due to the small dataset of 80 unique geometries and its imbalancedness with a majority of parts with rather low complexity, the chosen algorithms did not achieve satisfying results. However, we are confident that by labeling more geometries and incorporating more complex geometries, the performance of the Random Forest algorithm can be significantly improved.

6 Conclusion and Outlook

We answered the initial research questions of "How can we assess part complexity?" and "What characteristics make parts complex?" by developing our own part complexity assessment procedure based on the literature findings and by implementing it on an exemplary production unit. To better understand part complexity, we have screened definitions of part complexity and assessment approaches in the literature. Based on these results, we derived a definition of part complexity for sheet metal processing and developed a mixed-methods procedure to assess part complexity. This procedure covers the selection of experts for complexity labeling, an online labeling tool, and a subset of repeating geometries to evaluate the experts' labeling consistency. We demonstrated the applicability of our procedure by putting it into practice, using a highly automated laser cutting production unit of a machine tool manufacturer for sheet metal processing as an example. The labeling results include categorized explanations for the chosen complexity and a mean complexity value for each geometry and applicable process steps.

Our results inhibit several limitations. First, we only accessed experts from one company. In the future, including customers using the selected production units may be beneficial to ensure a broadened perspective. Second, we only labeled 80 geometries. We expect to get more valuable insights into what characteristics make geometries complex and improve the accuracy of the classification algorithm by incorporating more geometries. Third, we only covered one production step of the process chain in sheet metal processing. This was beneficial for evaluating our part complexity assessment and demonstrating its applicability. However, to employ part complexity throughout the process chain, we must cover further production units and process steps, e.g. bending.

In future work, we want to define additional part characteristics that capture the relevant geometrical information for part complexity, such as the inner contour area. To better train an algorithm for transferring our results to new geometries, we want to expand the amount of labeling geometries. Ideally, we would get access to geometrical information from machine tools in the field. Furthermore, we want to add to the use cases for part complexity known in the literature of production optimization and production technology selection to explore the contribution of part complexity to the demand for data-driven information about the customers along the product life cycle.

References

1. Tjaden, G., Meyer, A., Rinciog, A.: Towards adaptable customer segments and reference geometries. In: Kim, D.Y., von Cieminski, G., Romero, D. (eds.) APMS 2022. IFIP AICT, vol. 663, pp. 321–328. Springer, Cham (2022). https://doi.org/10.1007/978-3-031-16407-1_38
2. Tjaden, G., Feyhl, L., Meyer, A.: Prototyping a data-driven customer segmentation utilizing machine usage data for product portfolio management. Procedia CIRP **120**, 572–577 (2023)
3. Tjaden, G., Baier, A., Strache, M., Regelmann, C., Meyer, A.: Maximizing customer satisfaction in sheet metal processing: a strategic application of the customer health score. In: Alfnes, E., Romsdal, A., Strandhagen, J.O., von Cieminski, G., Romero, D. (eds.) APMS 2023. IFIP AICT, vol. 690, pp. 136–150. Springer, Cham (2023). https://doi.org/10.1007/978-3-031-43666-6_10
4. Tjaden, G.: 80 2D laser cutting geometries. https://doi.org/10.5281/zenodo.10991360
5. Buff, B., Magnus, C., Zhu, J.H., Meier, H.: Robot-based incremental sheet metal forming–increasing the geometrical complexity and accuracy. In: Key Engineering Materials, vol. 549, pp. 149–155. Trans. Tech. Publ. (2013)

6. Claus, F., Hagen, H., Hamann, B.: Calculating the gravity-free shape of sheet metal parts. Int. J. Adv. Manuf. Technol. **113**(11), 3401–3417 (2021)
7. Salunkhe, S., Kumar, S., Teraiya, S.: Automatic feature recognition of sheet metal parts (2015)
8. Kumar, S., Singh, R., Panghal, D., Salunkhe, S., Hussein, H.M.A.: Feature extraction and manufacturability assessment of sheet metal parts. In: Kumar, S., Hussein, H.M.A. (eds.) AI Applications in Sheet Metal Forming. TMMME, pp. 41–66. Springer, Singapore (2017). https://doi.org/10.1007/978-981-10-2251-7_3
9. Lam, T.F., Sze, W.S., Tan, S.T.: Nesting of complex sheet metal parts. Comput.-Aided Des. Appl. **4**(1–4), 169–179 (2007)
10. Esener, E., et al.: Accurate localization technique for stamping sheet metal parts with complex surfaces. Indian J. Eng. Mater. Sci. (IJEMS) **28**(4), 409–419 (2021)
11. Turco, A.L., Maggioni, D.: The knowledge and skill content of production complexity. Res. Policy, 104059 (2020)
12. Schuh, G., Bergweiler, G., Fiedler, F., Bickendorf, P., Colag, C.: A review on flexible forming of sheet metal parts. In: 2019 IEEE International Conference on Industrial Engineering and Engineering Management (IEEM), pp. 1221–1225. IEEE (2019)
13. Lohtander, M., Lanz, M., Varis, J., Ollikainen, M.: Breaking down the manufacturing process of sheet metal products into features. Mechanics **64**(2), 40–48 (2007)
14. Lohtander, M., Varis, J.: Manufacturability of large sheet metal part. In: 13th International Conference on Proceedings of 13th International Conference. Mechanika (2008)
15. Lohtander, M., Varis, J.: Manufacturing features in cutting shapes and punching holes in sheet metal. In: 19th International Conference on Production Research ICPR-19. Valpariso, Chile. CD-Proceedings (2007)
16. Qamar, S.Z., Chekotu, J.C., Al-Maharbi, M., Alam, K.: Shape complexity in metal extrusion: definitions, classification, and applications. Arab. J. Sci. Eng. **44**, 7371–7384 (2019)
17. Joshi, D., Ravi, B.: Quantifying the shape complexity of cast parts. Comput.-Aided Des. Appl. **7**(5), 685–700 (2010)
18. Radvar-Esfahlan, H., Tahan, S.-A.: Robust generalized numerical inspection fixture for the metrology of compliant mechanical parts. Int. J. Adv. Manuf. Technol. **70**(5), 1101–1112 (2014)
19. Camba, J.D., Contero, M., Pérez-López, D., Company, P.: A database framework for the characterization and classification of parametric models based on complexity metrics to support data analytics. In: International Manufacturing Science and Engineering Conference, vol. 58745, p. V001T02A007. American Society of Mechanical Engineers (2019)
20. Chougule, R.G., Ravi, B.: Variant process planning of castings using AHP-based nearest neighbour algorithm for case retrieval. Int. J. Prod. Res. **43**(6), 1255–1273 (2005)
21. Liu, Y., et al.: Study on complexity of precast concrete components and its influence on production efficiency. Adv. Civil Eng. **2022** (2022)
22. Bodein, Y., Rose, B., Caillaud, E.: Explicit reference modeling methodology in parametric cad system. Comput. Ind. **65**(1), 136–147 (2014)
23. Valentan, B., Brajlih, T., Drstvenšek, I., Balič, J.: Development of a part-complexity evaluation model for application in additive fabrication technologies. Stroj. Vestnik J. Mech. Eng **10**, 709–718 (2011)
24. Cooper, D.R., Schindler, P.S., Cooper, D.R., Schindler, P.S.: Business research methods (2003)

25. TRUMPF. TruLaser Center 7030 (2024). https://www.trumpf.com/de_DE/produkte/maschinen-systeme/2d-laserschneidmaschinen/trulaser-center-7030/. Accessed 18 Mar 2024
26. Greco, A., Manco, P., Gerbino, S.: On the geometrical complexity index as a driver for selecting the production technology. In: Rizzi, C., Campana, F., Bici, M., Gherardini, F., Ingrassia, T., Cicconi, P. (eds.) ADM 2021. LNME, pp. 3–12. Springer, Cham (2022). https://doi.org/10.1007/978-3-030-91234-5_1
27. Döring, N., Bortz, J.: Forschungsmethoden und evaluation. Springer, Wiesbaden (2016). https://doi.org/10.1007/978-3-642-41089-5

Covariance Structure Analysis of the Collaboration Between Local Enterprises and Local Governments for the Creation and Communication of Value in Local Brands

Yu Fujii[1](✉), Tomomi Nonaka[2], Fumihito Oura[1], and Seiko Shirasaka[1]

[1] Graduate School of System Design and Management, Keio University, Collaboration Complex, 4-1-1 Hiyoshi, Kohoku-Ku, Yokohama 223-8526, Kanagawa, Japan
Y.Fujii8313@keio.jp

[2] Department of Industrial and Management Systems Engineering, School of Creative Science and Engineering, Faculty of Science and Engineering, Waseda University, Shinjuku, Japan

Abstract. Establishing regional brands is indispensable for revitalizing local areas; however, advancing such initiatives requires collaboration with a diverse array of stakeholders. This study focuses on local enterprises, which are pivotal stakeholders in regional branding strategies, and aims to elucidate the impact of their collaboration with local governments and degree of brand penetration on their willingness to engage in regional branding. We define regional branding as comprising two types: "comprehensive regional branding," which involves branding the region, and "resource-based regional branding," which pertains to branding specific local products and services. We propose a hypothetical model suggesting that the depth of collaboration with local governments influences local enterprises' engagement in regional branding. We conducted a questionnaire survey among local businesses in Fukuyama City, Hiroshima Prefecture, Japan, and verified the hypothetical model through covariance structure analysis using the survey data. The results revealed a moderate correlation between the willingness to participate in regional branding and the extent of collaboration with local governments. Furthermore, we identified factors that influence local enterprises' deepening relationships with local governments and those that enhance brand penetration among local enterprises, specifically the impact of the degree of understanding of the local culture and the attractiveness of the brand.

Keywords: Local Brands · Covariance Structure Analysis · Questionnaire Survey · Value Creation · Service Systems Design · Value Creation

Published by Springer Nature Switzerland AG 2024
M. Thürer et al. (Eds.): APMS 2024, IFIP AICT 733, pp. 410–424, 2024.
https://doi.org/10.1007/978-3-031-71645-4_27

1 Introduction

1.1 Background

According to the Statistics Bureau of the Ministry of Internal Affairs and Communications, Japan's population in 2022 was 124.947 million, a decrease of 556,000 from the previous year, marking the 12th consecutive year of decline. The only region experiencing population growth is Tokyo, while the remaining 46 prefectures continue to experience a population decrease [1]. Owing to the declining birthrate and aging population, the working-age population (ages 15–64) has been decreasing since its peak in 1995 and is expected to drop to 57.25 million by 2050, exacerbating labor shortages and shrinking the economy, with particularly severe impacts in rural areas [2].

As the population continues to shift to the Tokyo metropolitan area, rural areas face increasing difficulties, and regional and local revitalization has gained attention. Regional revitalization is a policy by the Cabinet Office that supports initiatives such as promoting Society 5.0 in local areas and urban regeneration [3]. Local revitalization generally refers to activities aimed at stimulating local economic, cultural, and industrial activities, and enhancing residents' motivation to contribute to their communities. In 2014, the government formulated a "comprehensive strategy for towns, people, and work creation," promoting regional revitalization efforts [4].

Efforts to establish regional brands have increased. In 2006, the Japan Patent Office established the "Regional Collective Trademark System," allowing trademarks in the format of "region + product (service) name" to be registered. This change has enabled many local governments and organizations to actively pursue the creation of regional brands using this new system. Research on regional branding surged between 2002 and 2007 and has continued to be a significant area of study.

With the intensification of regional-branding efforts, the number of scholarly articles on "regional brands" increased between 2002 and 2007. Although publications have gradually declined since then, research on regional brands has continued to be prolific. The spike in research activity during this period can be attributed to several legislative developments aimed at regional promotion, including the introduction of the Regional Collective Trademark System in 2006, the Act on Promotion of Utilization of Regional Resources by Small and Medium Enterprises, the Act for Promotion of Regional Collaboration among Agriculture, Commerce, and Industry, and the Act on Promotion of Local Production for Local Consumption.

These legislative frameworks have facilitated the emergence of regional brands across various localities and municipalities. However, the formation of regional brands ideally involves the participation of stakeholders throughout the value-creation process. Traditional approaches, as discussed by Sadahira (2017), Ibe (2011), and Tsuboi (2006), predominantly focused on consumer relations, with less emphasis on engaging other stakeholders [5–7]. Not all local enterprises, municipal bodies, and residents are enthusiastic about participating in regional branding. Therefore, this study focuses on local enterprises, which are pivotal stakeholders in regional branding strategies, and aims to elucidate the impact of their collaboration with local governments and the degree of brand penetration on their willingness to engage in regional branding. We propose a hypothetical model suggesting that the depth of collaboration with local governments influences local enterprises' engagement in regional branding. We conducted a questionnaire survey among local businesses in Fukuyama City, Hiroshima Prefecture, Japan, and verified the hypothetical model through covariance structure analysis using the survey data. From the perspective of service and system design, this analysis contributes to the development of effective methods for local governments to convey regional brand value.

The value-co-creation process focuses on the value customers experience in the unique context of using services or products, thereby forming customer satisfaction and brand perceptions. For example, in the context of regional branding, consumers not only purchase products but also participate in the cultural and value backgrounds of the products, thus co-creating brand value.

In this study, we analyze collaborative cases between businesses and local governments in urban areas to examine how these stakeholders contribute to value co-creation. Specifically, we analyze how regional resources are utilized and how branding strategies are formulated to appeal to the market and deepen consumer relationships.

In this study, we adopt the service-dominant Logic (S-D logic) perspective to explore the concept of 'value co-creation' in regional brand creation. Value co-creation is a process in which product or service providers collaborate with customers to create value, transcending traditional exchange value [8]. This theory views customers not merely as recipients of value, but also as active participants in the value-creation process.

The value co-creation process focuses on the value customers experience within the unique context of using services or products, thereby shaping customer satisfaction and brand recognition. For example, in the context of regional branding, consumers not only purchase products but also engage in the cultural and value backgrounds of the products, co-creating brand value with local businesses and other consumers.

1.2 Previous Research

The early 2000s witnessed a significant increase in scholarly articles on regional brands, prompted by the establishment of legal frameworks aimed at regional promotion. Research conducted by scholars such as Aoki (2004) and Akutsu and Amano (2007) [9, 10] extensively explored themes such as "the concept and definition of regional brands," "the development and nurturing of regional brands," and "the assessment and effectiveness of regional brands." This body of work has laid the groundwork for defining regional brands as essential tools for regional revitalization. A growing consensus suggests that the value of regional brands should be assessed not only in the short term but also over the long term. This perspective is reflected in the studies by Harada (2010) and Haneda and Watanabe (2013), which focus on the "sustainability of regional brands" [11, 12]. Moreover, the widespread adoption of the internet and social media highlights the critical role of digital marketing in enhancing the recognition and evaluation of regional brands, as discussed by Sadahira and Saito (2017) [5].

Strategic efforts in regional branding involve communicating the brand's value to consumers and encouraging their purchasing decisions. Ibe (2011) emphasized the necessity of differentiating regional brands from competitors and strategically promoting brand recognition to boost consumer purchases [6]. Similarly, Tsuboi (2006) noted that branding tourist destinations and shopping districts should establish a distinct identity from other regions, thereby solidifying a specific image in consumers' minds and increasing the likelihood of visits [7].

However, challenges and strategic considerations in the regional branding process have been highlighted. Hatuzawa (2015) pointed out that misalignments between strategic initiatives and the realities of production areas present significant challenges, underscoring the importance of developing branding strategies that reflect the actual production structures and industrial characteristics of the region [13]. Saito (2011) argues that forming food brands requires building value and supply chains, deepening relationship marketing, systematizing brand management techniques, and developing regional revitalization strategies linked to food industry clusters [14].

Ishihara (2009) stated the imperative of clearly articulating and communicating the value of a regional brand to consumers and business partners after it has been established and shared. He asserted that active engagement in these processes is crucial for the establishment of a regional brand. Furthermore, mechanisms need development that encompass 'value articulation,' 'value communication,' and 'value realization' to effectively convey the regional brand's value [15]. This strategy highlights the importance not only of communicating value to consumers but also to small and medium-sized enterprises and sole proprietors who are in cooperative relationships with the entities managing the regional brands. Despite extensive research on communicating value to consumers [5–7], a noticeable gap remains in studies focusing on the methods of conveying value to stakeholders involved in value creation.

1.3 Research Objectives

This study focuses on the involvement of stakeholders other than consumers in a regional brand strategy. It aims to elucidate the impact of the depth of collaboration between local businesses and governments, as well as the degree of regional brand penetration among local businesses, on their willingness to participate in regional branding.

A covariance structure analysis was performed using survey data collected from local enterprises. By analyzing how the penetration of regional brands and the depth of collaboration with local governments relate to the engagement levels of local enterprises in regional branding, this study seeks to identify the key factors necessary for formulating new strategies to enhance the recognition of regional brands. By clarifying the impact of the depth of collaboration with local governments and the degree of regional brand penetration on willingness to engage in regional branding, this study aims to assist in developing methods for local governments to convey regional brand value to local businesses.

This study focused on Fukuyama City, Hiroshima Prefecture, Japan. Located in the easternmost part of Hiroshima, Fukuyama is the second-most populous city in the prefecture after Hiroshima, the prefectural capital. Additionally, Fukuyama City is an ancient castle town with a history spanning over 700 years. It serves as a leading industrial city in the Chugoku region, noted for its concentration of steel- and automotive-related industries [16].

2 Regional Brand Studies in Brand Theory Research

Although research on regional branding is scarce, the theory of branding has been debated for more than half a century. A significant factor that drew the attention of both researchers and practitioners to branding theory in Japan was the emergence of the concept of "Brand Equity" in the early 1990s [6]. Furthermore, the concepts most applied in regional branding discussions are "Brand Equity" and subsequently developed "Brand Identity." This chapter provides an overview of foundational branding theory research that also informs regional branding studies, followed by a review of the regional branding literature.

2.1 Brand Theory Studies

The emergence of "Brand Equity" in the 1980s in the United States marked a pivotal shift in brand management, driven by the dynamics of mergers and acquisitions and a growing emphasis on the long-term value of brands [6]. Aaker conceptualized "Brand Equity" as a set of assets linked to a brand's name and symbols, including brand loyalty, name awareness, perceived quality, and other proprietary brand assets, fundamentally altering the strategic approach to branding [17]. Further elaborating on this, Aaker introduced "Brand Identity," which outlines a brand's unique characteristics and value propositions that connect emotionally and functionally with consumers [17]. Aoki (2006) noted that prior to the widespread acceptance of "Brand Equity," branding was often merely a component of marketing strategies. However, recognizing the intangible asset value of brands has become essential for sustaining competitive advantage [12].

2.2 Summary of Research on Regional Brand

This section explores the constituents of a regional brand by reviewing various studies of regional branding. In Japan, under the 1959 trademark law, trademarks composed of regional and product names were not eligible for registration, with a few exceptions, such as if they achieved national recognition [18]. However, the revised law enacted in June 2005 (implemented in April 2006) introduced the "Regional Collective Trademark System." This system allows for the registration of trademarks composed solely of "region name + product/service name" under certain conditions, facilitating the early stages of regional brand development [18]. Given this background, research on regional branding has noticeably increased since the early 2000s, revealing diverse interpretations of regional brand concepts.

Ito (2009) defines the concept of a regional brand as "a product (including services) that is produced or offered in a specific region and is well known not only in that region but also in surrounding areas and nationally, as well as its name [19]." Based on the "Regional Brand Manual" created by the Organization for Small & Medium Enterprises and Regional Innovation in 2005, the Brand Research Institute (2008) views a regional brand as "consumer evaluations of a region, constituting one of the intangible assets the region possesses, composed of both the brand of the region itself (Regional Brand) and the brands of products that capitalize on the region's characteristics (Products Brand) [20]". This perspective likely builds on the notions associated with "Brand Equity [12]".

Furthermore, comparing regional and corporate brands, Aoki (2004) stated, "A comprehensive regional brand is similar to a corporate brand, while a regional resource brand resembles a product brand." However, he also notes, "The concept of 'region' serves both as the venue for individual regional resource branding and as the object of branding itself, which distinguishes it from typical corporate branding and underscores the uniqueness of regional brands [10]."

From this discussion, the definition of a Regional Brand in this study encompasses both the brand of the region itself (regional brand) and regional resource brands (product brand).

3 Hypothetical Model of Regional Brand Entry Process Considering the Depth of Collaboration Between Local Governments and Local Enterprises

Building upon the discussions in the previous chapters, this section proposes a hypothetical internal structural model for regional branding, considering the depth of collaboration between municipalities and the region.

Figure 1 displays the hypothetical model, where observed variables are represented by squares and latent variables introduced by the analyst to explain the relationships between unobservable variables are represented by ellipses. The model first introduces observed variables such as the degree of alignment between the regional brand and the local culture or strengths, awareness of the regional brand, its appeal, and pride in the region. These are linked to a latent variable representing "awareness of the regional brand." The latent variable "depth of collaboration with the region" is connected to observed variables about the level of cooperation with local governments and the quality and frequency of communication between these authorities and local businesses. We posit that these two latent variables influence the intention to engage with a regional brand. Furthermore, a hypothesis was formulated proposing a reciprocal interaction between "depth of collaboration with the municipality" and "awareness of the regional brand."

Next, we shall elaborate on the hypothesis concerning the relationship between the latent variable "awareness of the regional brand" and the observed variables. Tanimoto (2008) asserts that the primary agents responsible for constructing a regional brand need to not only enhance brand recognition and associations through advertising and other information dissemination activities but also cultivate and strengthen a positive and intimate relationship between the regional brand and local residents [21], underscoring the significant role of brand awareness and recognition in the development of a regional brand. Additionally, Aotani (2010) remarks that mutual understanding not only enhances the functional aspects of a brand but also enables the provision of emotional value to consumers [22], suggesting that measuring the favorable impressions—or the allure—of a regional brand can facilitate an understanding of the brand's perception. Moreover, Sakamoto (2016) comments that the inherent "localness" rooted in a region's nature, history, culture, and traditions generates expectations towards the regional brand [23], indicating a correlation between the recognition of the regional brand and the understanding of the region's culture.

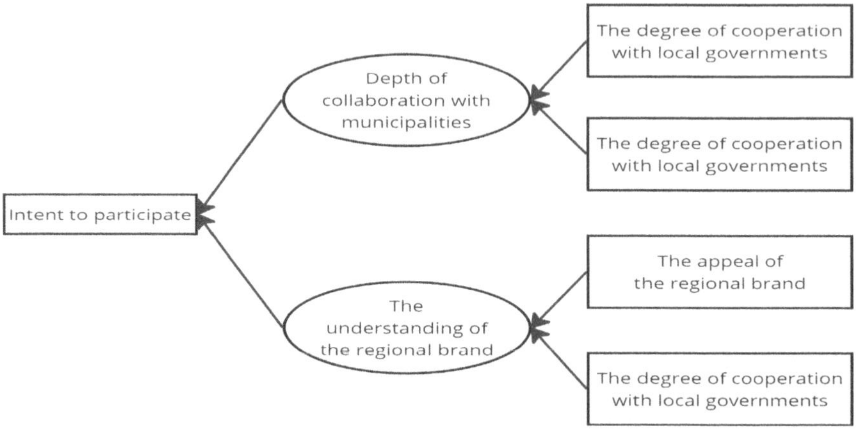

Fig. 1. Hypothetical model of collaboration with local governments and interest in regional branding.

Next, we discuss the hypothesis between the latent variable "depth of collaboration with municipalities" and the observed variables. Aotani (2010) asserts that "enhancing regional relationships through dialogue among businesses (regions), consumers (regions), and retailers can provide greater empirical value [22]." Additionally, Saito (2011) notes that "if consumers, local residents, municipalities, and other relevant parties can share a brand image through branding efforts, a strengthened identity is expected, leading to regional revitalization using a variety of products and services [13]."

These considerations underscore the importance of collaborations between municipalities and local businesses in regional branding strategies. The quality and frequency of communication between municipalities and local enterprises are critical elements in revitalizing regional brands. Additionally, much prior research highlights the significance of cooperation with local businesses and residents when municipalities are the main actors in regional branding initiatives [14, 21, 22]. However, quantitative analyses of these aspects are limited. Therefore, this study uniquely posits that the degree of cooperation and frequency of communication are essential observed variables in measuring the "depth of collaboration with municipalities" and incorporates them into the hypothetical model.

Furthermore, the hypothesized posited that the two aforementioned latent variables interacted with each other. For instance, an increase in interactions with local governments is believed to enhance understanding of their activities. Consequently, this deepened understanding could lead to a greater comprehension of regional branding strategies. Consequently, the effectiveness of the regional brand is recognized, prompting increased participation in regional branding. Positing through this sequence of developments, the number of enterprises engaging with regional brands gradually increases.

4 Hypothetical Model of the Regional Brand Entry Process Considering the Depth of Collaboration Between Local Governments and Local Enterprises

4.1 Survey Objectives and Overview for Local Business Stakeholders

An inquiry has been designed to assess whether deepening collaboration with local governments increases interest in regional brands among local business stakeholders. Through surveying their awareness and interest in collaboration with municipalities and regional branding, and analyzing the results, this survey aims to validate the hypothetical model proposed in Sect. 3.

4.2 Survey Design

To analyze the depth of collaboration with local governments and interest in regional branding, a questionnaire survey was conducted targeting individuals who owned or were employed by local businesses in Fukuyama City, Hiroshima Prefecture.

The survey adhered to the "Research Ethics Guidelines" established by the university, ensuring that participants were fully informed about their involvement in the research and its lack of impact on personnel evaluations before obtaining their consent. The survey administration method involved contacting local business stakeholders via social media to provide explanations and secure their prior consent, followed by an online questionnaire. The measurement technique used a Likert scale. Furthermore, to minimize the psychological burden on the participants, the questionnaire was structured with a neutral scale across five levels.

Furthermore, Shinohara (2020) noted that when survey respondents were asked about personal matters, approximately one in four exhibited vanity bias, desiring to present themselves in a more favorable light, which can lead to biased responses [24]. Accordingly, we proceed with the analysis while being cognizant of the possibility that the responses may be overly positive. The survey consisted of demographic and multiple-choice items. It includes a total of 25 questions using a five-point Likert scale ranging from "strongly disagree" to "strongly agree," focusing on understanding the "Fukuyama Brand"—a regional brand officially recognized and registered by Fukuyama City—and the depth of collaboration with local authorities.

4.3 Questionnaire Survey Items

Table 1 presents the questionnaire survey items, formulated by adapting prior research on regional branding strategies directed at consumers rather than focusing on local businesses.

Questions Q1 to Q1-7 were designed to assess the understanding and penetration of regional brands among local business stakeholders. Questions Q2-1 to Q2-14 were designed to investigate the extent of collaboration between these local business stakeholders and local governments, who are the principal agents of the regional brand.

Table 1. Questionnaire survey items.

Demographic Information: Gender, Age, Occupation	
Q1	Are you familiar with the "Fukuyama Brand"?
Q1-1	Are you aware that the "Fukuyama Brand" is officially recognized and registered by Fukuyama City?
Q1-2	Are you satisfied with the initiatives of the "Fukuyama Brand"?
Q1-3	Do you like the "Fukuyama Brand"?
Q1-4	Do you find the "Fukuyama Brand" appealing?
Q1-5	How well do you think the "Fukuyama Brand" harmonizes with the local values and culture?
Q1-6	To what extent do you think the "Fukuyama Brand" represents the region's characteristics and strengths?
Q1-7	How closely do you think the "Fukuyama Brand" aligns with the regional identity?
Q2-1	Is your company certified and registered under the "Fukuyama Brand" that is spearheaded by Fukuyama City?
Q2-2	Do you feel that your company collaborates with the local governments?
Q2-3	How actively do you think your company participates in municipal projects and events?
Q2-4	To what extent do you believe your company contributes to the development of the municipality?
Q2-5	How much do you think your company contributes to the local community?
Q2-6	How actively is your company involved in local events and projects?
Q2-7	To what extent do you think your company responds to social needs?
Q2-8	How frequently does your company initiate communication with local governments?
Q2-9	How frequently do the local governments initiate communication with your company?
Q2-10	How satisfied are you with the quality of communication between your company and the local governments?
Q2-11	Do you feel that your company's activities enhance local pride?
Q2-12	To what extent do you think your company's success affects the region's reputation?
Q2-13	Do you believe that your company's contributions to the region contribute to residents' pride?
Q2-14	Are you currently participating in the "Fukuyama Brand", or do you intend to participate in the future?

5 Covariance Structure Analysis of a Survey to Validate the Hypothetical Model of the Regional Brand Entry Process Considering the Depth of Collaboration Between Local Governments and Local Enterprises

5.1 Analysis Objectives

To analyze the internal structure that promotes the intention to participate in regional branding, this study tests a hypothetical model of the regional brand entry process, which incorporates the depth of collaboration between municipalities and regions, as described in Sect. 3. The method of analysis involved validating the hypothetical model through covariance structure analysis using the survey results presented in Sect. 4.

5.2 Survey Subjects

A survey was conducted among local business stakeholders in Fukuyama City, Hiroshima Prefecture, yielding valid responses from 21 participants. The respondents were categorized as local business managers, employees, or self-employed industrial coordinators. The survey was conducted from November 15 to November 18, 2023.

5.3 Result of Covariance Structure Analysis

5.3.1 Hypothetical Model of Regional Branding Based on Collaboration Between on Local Government and Local Enterprises

When municipalities manage regional brands, collaboration with a wide range of stakeholders is essential, including but not limited to leading municipal bodies. Among these, partnerships with local businesses that sustain the regional economy are considered particularly crucial in regional branding strategies. As Aotani (2010) pointed out, "enhancing relationships between regions through dialogue among businesses (region), consumers (region), and retailers is needed to provide more experiential value, [21]" emphasizing that bidirectional communication between municipalities and local enterprises is indispensable for the development of regional brands. This section focuses on the quality and frequency of communication between municipalities and local enterprises, as indicated by the survey questions (Table 1). We examined the relationship between the depth of collaboration with municipalities, understanding of the regional brand, and intention to engage with the regional brand using covariance structure analysis. The content of each observed variable survey question was detailed.

The internal structural hypothesis model proposed in Sect. 3 (Fig. 1) suggests a potential relationship between the depth of collaboration with municipalities and understanding of the regional brand. Thus, in this section, using the survey questions designed in Sect. 4 as observed variables, we designate "depth of collaboration with municipalities" and "awareness of the regional brand" as the corresponding latent variables and analyze the relationships between these variables through covariance structure analysis.

5.3.2 Result of Analysis

Figure 2 presents the results of the covariance structure analysis of the hypothetical model, using survey data from local business stakeholders in Fukuyama City, Hiroshima Prefecture, Japan. The Comparative Fit Index (CFI) and Root Mean Square Error of Approximation (RMSEA) served as fit indices. The CFI value was 1.000, and the RMSEA value was 0. These figures (Fig. 3) fall within an acceptable range, indicating that the results support the model depicted in the diagram.

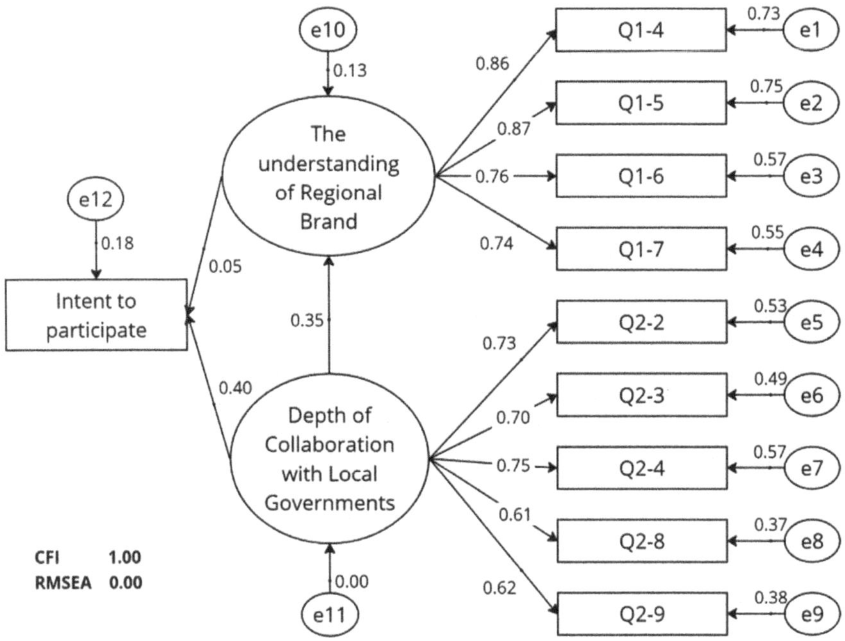

Fig. 2. Result of covariance structure analysis.

In the model depicted in Fig. 3, a weak correlation exists between the latent variables of regional brand awareness and depth of collaboration with local governments, with a path coefficient of 0.35. The paths between the latent and observed variables show high values, indicating a strong association between regional brand penetration and each observed variable.

Specifically, the path coefficient with question Q1-4, assessing the appeal of the brand, was 0.86, indicating a strong association. Similarly, question Q1-5, examining the extent to which a brand harmonized with regional values and culture, also showed a high path coefficient of 0.87.

Additionally, high values were observed for the observed variables associated with the depth of collaboration with local governments. For instance, Q2-2, assessing the extent of cooperation with local governments from the perspective of local business stakeholders, had a path coefficient of 0.73. Question Q2-3, examining the frequency of local business stakeholders' participation in municipality-sponsored events, had a path coefficient of 0.70.

The path coefficients between the latent and observed variables representing the intention to participate in the regional brand showed that regional brand penetration had virtually no correlation, with a path coefficient of 0.05. In contrast, the depth of collaboration with local governments exhibited a moderate correlation, with a path coefficient of 0.40.

5.4 Discussion

The path coefficient between the latent variables of awareness of the regional brand and depth of collaboration with local governments was low at 0.35. This indicates a weak connection between local business stakeholders' awareness of the regional brand and their depth of collaboration with local governments.

Relationships with local governments undeniably affect the penetration of regional brands among local business stakeholders. However, influences from other elements not included in this model could enhance the penetration of regional brands. One such factor is the depth of collaboration among local businesses.

Specifically, the utilization of methods for sharing information and images related to regional brands during networking events should be considered.

Next, we consider the paths between the observed and latent variables.

The path between the latent variable representing awareness of the regional brand and Q1-4 had a high value of 0.86. This suggests that accurately conveying the appeal of a regional brand to local businesses is crucial for its penetration.

Furthermore, high path coefficients were confirmed among the three questions (Q1-5, Q1-6, and Q1-7), indicating an understanding of the regional brand. The understanding of a regional brand is shaped by its harmony with the local culture, its expression, and alignment, and the awareness of the brand is influenced by these aspects of understanding and its appeal. Additionally, the three observed variables that comprise this understanding highlight the importance of showcasing regional characteristics to enhance brand awareness among consumers, echoing previous observations in the literature [23].

This suggests that methods effective in enhancing brand awareness among consumers can be similarly applied to local businesses. As indicated in the literature, understanding the inherent characteristics of a region is crucial to increasing the penetration of a regional brand. This is significant not only in communication with consumers but also in transmission from local governments to local businesses. Conveying the cultural and characteristic backgrounds of the region in the construction of a brand is anticipated to be effective.

High values of path coefficients among the three questions (Q2-2, Q2-3, and Q2-4) that indicate the degree of cooperation with local governments suggest that proactive involvement in municipally sponsored events and the consciousness of contributing to municipal development are crucial for building relationships between local enterprises and local governments. Active engagement with municipal initiatives is necessary to enhance collaboration between local businesses and municipalities. Furthermore, a moderate correlation was observed between the quality and frequency of communication (Q2-8, Q2-9) and the depth of collaboration with municipalities.

This indicates that in the relationship between local governments and businesses, communication should not be one-sided but mutual, as this contributes to the depth of collaboration.

Finally, the path coefficient between the observed variable of intent to participate (Q2-14) and the latent variable of awareness of the regional brand was very low at 0.05, indicating no significant relationship. However, a moderate correlation (0.40 between intent to participate (Q2-14) and depth of collaboration with local governments suggests that local businesses' willingness to engage in regional brands is influenced by the extent of their collaboration with local governments.

6 Conclusion

This study reveals a low path coefficient (0.35) between the latent variables of regional brand awareness and the depth of collaboration with local governments, indicating a weak connection. This suggests that, while local government relationships impact regional brand penetration among businesses, other unexplored factors may also play a significant role. For instance, the depth of interbusiness collaboration can potentially enhance brand penetration.

The high path coefficients (0.86) between the latent variable of brand awareness and the observed variables (Q1-4, Q1-5, Q1-6, and Q1-7) underscore the importance of effectively communicating a regional brand's appeal to local businesses and its alignment with local culture. This alignment significantly influences the comprehension and attractiveness of a brand, thereby facilitating its broader acceptance.

Moreover, active participation in municipal initiatives and the quality and frequency of communication with local governments (Q2-2, Q2-3, Q2-4, Q2-8, and Q2-9) were critical for fostering strong relationships between local enterprises and government bodies, as indicated by the significant path coefficients. Effective communication enhances the depth of collaboration in two ways:

Notably, the intent to participate in regional branding (Q2-14) showed only a moderate correlation (0.40) with collaboration depth, suggesting that, while government engagement is influential, other factors also impact businesses' commitment to regional branding initiatives.

This analysis highlights the intricate dynamics between local businesses and governments in regional branding efforts and underscores potential areas for further investigation to enhance the effectiveness of such collaborations.

Acknowledgments. This work was partially supported by JSPS KAKENHI Grant Number JP23K22158.

References

1. Statistics Bureau, Ministry of Internal Affairs and Communications. https://www.stat.go.jp/data/jinsui/2022np/index.html#a05k01-a. Accessed 28 Mar 2024

2. Ministry of Internal Affairs and Communications: Information and Communications in Japan, White Paper 2022 - Decline in the Productive Age Population. https://www.soumu.go.jp/johotsusintokei/whitepaper/ja/r04/html/nd121110.html. Accessed 28 Mar 2024
3. Ministry of Internal Affairs and Communications: Regional Power Creation Group, Special Measures Law for Sustainable Development in Depopulated Areas. https://www.soumu.go.jp/main_content/000752615.pdf. Accessed 28 Mar 2024
4. Cabinet Office, Regional Revitalization Promotion Office, Regional Revitalization. https://www.chisou.go.jp/sousei/index.html. Accessed 28 Mar 2024
5. Sadahira, M., Saito, S., Matsuura, K.: Image strategy and branding of local contents through social network in Japanese. J. Inform. Arts Shobi Univ. **21**, 1–16 (2017)
6. Ibe, Y.: The role of regional branding in regional activation in Japanese. Bull. Niigata Univ. Manag. **17**, 63–75 (2011)
7. Tsuboi, A.: Trends and issues in constructing regional brands in Japanese. Region. Policy Res. Takasaki City Univ. Econ. **8**(3), 188–199 (2006)
8. Stephen, L., Vargo, R., Lush, F.: Evolving to new dominant logic for marketing. J. Mark. **68**, 1–17 (2004)
9. Akutsu, S., Amano, M.: Regional brands and their management challenges in Japanese. Jpn. Mark. J. **27**(1), 4–19 (2007)
10. Aoki, Y.: Perspectives and frameworks in building regional brands in Japanese. J. Shoko Chukin Bank Inst. Commer. Ind. (2004)
11. Harada, T.: Context design of area branding in Japanese. In: Proceedings of the Japan Association for Management Diagnosis, pp. 40–43 (2010)
12. Haneda, H., Watanabe, A.: What is competitive advantage model for region? Branding "MINAMIISE in Japanese. Bull. Faculty of Urban Manag. Fukuyama City Univ. **4**, 29–42 (2013)
13. Hatsuzawa, T.: Local industries' branding strategy and its problems: case studies of Japan Brand Business on Naruko Industrial Region and personal training business on Tajimi Ceramic Industrial Region in Japanese. J. Center Region. Aff. Fukushima Univ. **27**(1), 4–12 (2015)
14. Saito, O.: New direction and strategic problem about local brand in Japanese. J. Food Syst. Res. **18**(2) (2011)
15. Ishihara, S.: The research about the creating of regional brand for the agricultural and marine products in regional society: the study about the value and the creating method of regional brand for the activation of regional industry in Japanese. Doctoral thesis, Graduate School of Regional Society, Hirosaki University, pp. 1–188 (2009)
16. Hiroshima Prefecture, Fukuyama City Official Website, Profile of Fukuyama City (2024). city.fukuyama.hiroshima.jp. Accessed 29 Mar 2024
17. Aaker, D.A.: Managing Brand Equity: Capitalizing on the Value of a Brand Name. The Free Press, A Division of Macmillan, Inc. Translated by Toyama, K., Nakata, Y., Ozaki, K., Kobayashi, T.: Brand Equity Strategy - Creating Competitive Advantage with Names, Symbols, Slogans. Diamond Inc. (1994)
18. Cabinet Office. Economic Analysis of Regional Brands. In: Regional Economy 2017, Chapter 2, Section 2. https://www5.cao.go.jp/j-j./cr/cr17/chr17_02-02.html. Accessed 28 Mar 2024
19. Ito, T.: Effective brand management for the regional collective trademarks: how should a strong brand for the development of the regional economy be built? Japanese. Bull. Miyagi Univ. Sch. Project Des. **12**, 15–27 (2009)
20. Brand Research Institute, Inc.: Methodology of Building Regional Brands, First Edition: What is a Regional Brand? Definition of Regional Brand. Accessed 28 Mar 2024
21. Tanimoto, T.: A consideration on place marketing and brand management in Japanese. Econ. J. Ehime Univ. **27**(2/3), 57–82 (2008)

22. Aotani, M.: Consumer behavior in the local brand and problems in Japanese. J. Rural Probl. **177**, 343–357 (2010)
23. Sakamoto, K.: Design marketing research-approach for composite of local brands in Japanese. Spec. Issue Japan. Soc. Sci. Des. **17**(1), 62–71 (2010)
24. Shinohara, M.: Biased replies in questionnaire surveys in Japanese. J. Faculty Educ. Buddhist Univ. **31**, 15–31 (2020)

Developing 3D Production Simulation Models in Industrial Production Systems

Akshay Goyal(✉), Yuji Yamamoto, and Ioanna Aslanidou

Mälardalen University, 631 05 Eskilstuna, Sweden
{akshay.goyal,yuji.yamamoto,ioanna.aslanidou}@mdu.se

Abstract. This research presents a comprehensive framework for the development, integration, and formulation of 3D simulation models within industrial production systems. It aims to provide the required guidelines for utilizing 3D simulation technology, improving system productivity, assisting with decision-making, and furthering system optimization. The research highlights the considerable influence of 3D simulation technology on production systems, starting with developing and evaluating the system current state and its potential in manufacturing industries. The framework includes elements related to collecting data, the creation of 3D production simulation models, validation, and verification phases; setting objectives for the integration of 3D simulation; and the choice and advancement of customized simulation technologies that meets specific industry demands. The framework is developed with the help of a case study where the authors present the steps and information necessary to facilitate the modelling of the current state version within 3D production simulation model. Through the case study presented, this research illustrates the practical application of this framework, aiming to serve as an exemplary guide for academic and industrial practitioners.

Keywords: 3D Production Simulation · Modelling · Discrete Event Simulation. · Conceptual Framework · Industrial Production Systems

1 Introduction

Discrete Event Simulation (DES), focuses on the simulation of the operation of a system as a discrete series of events over time, making it ideal for studying complex production systems [1]. It is becoming increasingly important in both industries and academics because of its ability to generate detailed, realistic virtual models of manufacturing processes, allowing for the optimization of production lines, reducing production time and costs, and in its related decision-making [1]. It is also an essential tool for research and education, enabling a hands-on approach to analyzing complex industrial systems and experimenting with new ideas and configurations without the risks and expenses associated with physical shop-floor testing [2]. With the more extensive use of simulation and the development of more advanced simulation software, simulation models are increasingly being developed in 3D compared to the schematics-based versions used in early simulation tools.

Published by Springer Nature Switzerland AG 2024
M. Thürer et al. (Eds.): APMS 2024, IFIP AICT 733, pp. 425–436, 2024.
https://doi.org/10.1007/978-3-031-71645-4_28

Within Industry 4.0, the integration of 3D simulation models within manufacturing systems has become an important component for improving production design, decision-making processes, and system efficiency [6]. As companies progress down the complex path of digitalization and automation, a methodical, well-defined framework for developing and implementing 3D production simulation models becomes increasingly important [3]. This need arises from the complexities involved in accurately simulating and optimizing production environments to align with the dynamic requirements of industrial operations [4]. The existing environment exposes a considerable gap in the methodology used to generate and integrate 3D production simulation models. Industries are facing the issue of not just embracing 3D simulation technology, but also efficiently using these tools to improve production efficiency, optimize workflows, and rethink production design [3]. In order to realize the full potential of the technology, an approach for developing and implementing simulation models that is further building on the framework for discrete event simulation [8], is needed. The barrier to utilizing 3D simulation technology to its fullest extent may include the lack of a comprehensive framework for developing, evaluating, and putting into practice these models. This disparity highlights the necessity of a methodical strategy that includes assessing industry-specific requirements, picking appropriate simulation tools, and carefully integrating simulation outcomes into processes that include making decisions.

To address this challenge, this research focuses on developing a systematic conceptual framework to help navigate the process of developing and adopting 3D production simulation models. This research furthers on the research done [8], in regards to the steps in a simulation study, thereby providing the 3D production simulation outlook presented through the developed conceptual framework. This framework aims to close the existing gap by providing an approach that includes evaluating current production processes, identifying optimization opportunities, selecting, and customizing 3D simulation tools, and aligning simulation results with strategic business objectives. The process of a production simulation project uses production data to create a "Current State Simulation" that mirrors a real production system. This is followed by an "Optimization" phase to refine processes. The resulting data then feeds into a "Future State Simulation" to model and forecast potential improvements and the performance of the system post-optimization.

The research question that will be answered through this research is, "*What essential data should be considered for developing a 3D production simulation model?*". Thus, a framework has been designed to address this research question. Furthermore, the presented case study serves as an exemplary practical application and verification, for facilitating framework implementation.

2 Literature Review

The development of virtual current state 3D production simulation models necessitates a comprehensive understanding of various manufacturing processes, including assembly, machining, and quality control operations, to enhance production efficiency and product quality [3]. Essential data for these models include detailed descriptions of internal logistics, highlighting the flow of materials and the integration of supply chain dynamics within the production environment [6]. Furthermore, accurate modeling requires accurate process machine parameters, such as cycle times and setup times, which are critical for optimizing production schedules and minimizing downtime [8]. Furthermore, real-time operational metrics and environmental data are also incorporated into the simulation, which helps to forecast production results under different situations. The accuracy of simulation models is increased by combining these layers of precise data with 3D scanning technologies [5] along with the CAD (Computer Aided Design) data [7], which ensures the spatial context and geometrical precision of the manufacturing setup. The following sub-sections further elaborate on the different factors essential for developing simulation models.

2.1 Shop-Floor 2D Layout

In order to develop an accurate 3D model and optimize it, it is necessary to have a foundational blueprint that provides a simplified yet detailed representation of the manufacturing environment [7]. This is why 2D layout is important for developing a model in 3D production simulation. Before implementing the 2D simulation model, a thorough analysis and validation of workflow, space utilization, and process efficiency can be achieved by transforming the spatial and operational data from the 2D layout into a dynamic 3D simulation environment [2]. A 2D shop-floor layout can lower costs and increase planning accuracy by spotting inefficiencies early in the design process. It also allows for the testing of different layout scenarios without the need for physical adjustments, which can result in optimized manufacturing processes through better decision-making [7]. The utility of the 2D layout in the context of robot assembly cell layout optimization extends to facilitating decision-making in layout design, improving material flow, and enhancing overall production efficiency by enabling detailed analysis and scenario testing in a virtual environment [2].

2.2 Process Data for Simulation

In the field of designing industrial production systems, simulation models stand out as a crucial tool for improving operational efficiency and layout optimization [1]. The accurate and complete collection and integration of process data is an important component from simulation effectiveness standpoint. The literature demonstrates a consensus on the importance of including precise process data into simulation models in order to achieve realistic and actionable results [2, 8].

Process data includes a variety of operational information, such as machine cycle times, operator movements, material handling times, and storage capacity [8]. This data serves as the foundation for simulation models, allowing for virtual replication of the production environment [6]. Simulation models can serve as an environment for testing various layout and operational assumptions without incurring the expenses and disruptions associated with physical reconfigurations. Recent advances in simulation technologies have highlighted the value of high-fidelity process data. Several authors [6, 7] have presented that the precision and accurateness of input data have a direct impact on the reliability of simulation results. For instance, thorough time-motion studies and real-time data collecting using IoT sensors, have been recognized as viable ways for capturing the intricacies of the manufacturing process [7]. This data not only enriches the simulation model, but it also improves the system's predictive power, allowing for more exact layout optimization.

Furthermore, integrating process data into simulation models allows for a more detailed understanding of system bottlenecks and inefficiencies [8]. Therefore, by simulating the interdependencies and interactions between various components of the manufacturing system, stakeholders can identify and address possible problems before they occur in the actual production environment. This proactive approach to system design and optimization has been complimented for its ability to significantly reduce downtime while increasing overall system throughput.

2.3 Incorporating Process CAD Data

Incorporating process CAD data into virtual 3D production simulation environments is an important step towards improving accuracy and realism in manufacturing system design and optimization [7]. Initially, this procedure requires the collection of CAD files for a variety of components, including machines, robots, conveyor belts, and other important assets identified during the process mapping step. Following the acquisition of these CAD files, the next stage is to integrate them into the simulation program, which brings digital representations of physical components to life inside a virtual 3D environment [7].

Precisely mirroring the actual configuration in the virtual environment requires careful scaling and positioning of these components to their precise dimensions and relative places [6]. The process involves several decision-making aspects. Decisions include determining the level of detail necessary for each component, deciding how to handle components with moving parts or variable configurations, and resolving how interactions between different components will be simulated [6, 7]. In addition, concerns regarding the computational load of the simulation must be balanced against the integrity of the CAD models in order to maintain functional simulation speed. Each decision has a direct impact on the simulation's utility, accuracy, and feasibility, highlighting the complexities and necessity of efficiently merging process CAD data into 3D production simulation environments.

2.4 3D Scanning of Production Systems

3D scanning is a novel approach to investigating possibilities to enhance manufacturing systems through technology innovation. 3D scanning of industrial manufacturing processes has several advantages [5]. Primarily, it provides exceptional precision in capturing the current state of production system environment, which is critical for incorporating real-world data into virtual simulation models. This precision is especially useful when CAD files are unavailable, such as with older machines or custom-built components, as it ensures that no asset is unaccounted for during the simulation. Once captured, these 3D scans can be integrated into virtual 3D production simulation models [9]. This integration not only facilitates a thorough awareness of the physical workspace, but it also enables rigorous design of production processes, considering every spatial element.

The research [5] presents how differences between the simulated and actual manufacturing setups can be identified, enabling adjustments to be made before they impact production. This capability highlights the significant benefit of 3D scanning in situations where adaptability and precision are critical, ensuring that the virtual models used for planning and optimization closely resemble the real-world arrangement [5]. This connection allows manufacturers to predict and correct potential inefficiencies, dramatically improving operational performance and reducing downtime.

2.5 3D Production Simulation Tool

Commercially available 3D production simulation software, Visual Components® [9], has become important in the field of continuously developing manufacturing industries. Furthermore, this technology enables the accurate modelling, simulation, and optimization of production lines and manufacturing processes with hitherto unachievable precision and detail. This software package is useful as it enables foreseeing of operational issues, to visualize complex production scenarios, and to assess the effects of adjustments without interfering with actual production flow. Simulation tools give decision-makers the ability to test various configurations and scenarios in a risk-free virtual environment, which helps manufacturers increase productivity, cut costs, and shorten time-to-market.

3 Methodology

3.1 Simulation Methodology

The methodological process of facilitating a production simulation project involves using production data to develop a virtual "Current State Simulation" that models an existing real production system, then feeding the results into an "Optimization" phase where processes are analyzed and refined (see Fig. 1). The "Future State Simulation" then uses the updated data from the optimization to model potential changes and improvements, providing a forecast of how the system might perform after implementing the optimizations.

The loop between optimization and the future state represents an iterative process in which the results of the future state simulation inform further optimization attempts, resulting in a cycle of continual process improvement. Hence, this research presents the framework developed for identifying the data required for creating the current state simulation model. The developed framework focuses on the environment of 3D production simulation for identifying essential data that is needed to construct a simulation model of the current state.

3.2 Case Study

The case study presented in this paper focuses on the construction of a 3D production simulation model for the automated FESTO Mobile Phone Assembly system, which was principally undertaken by the main author. The model focuses on an actual manufacturing system located at the Mälardalen Industrial Technology Centre (MITC AB), which

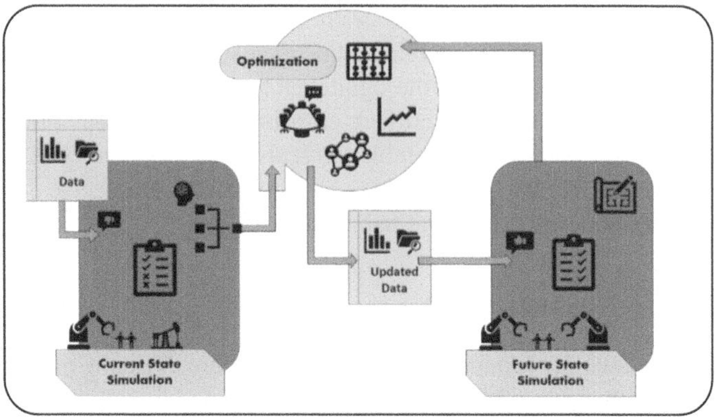

Fig. 1. Integrated simulation-optimization loop for production system enhancement

demonstrates the integration of theoretical concepts into practical implementation. The goal of this simulation case was to arrive at the "current state simulation" of the FESTO assembly line in a virtual environment.

The FESTO assembly system [10–12], regarded for its precision and efficiency in assembly operations, served as an ideal platform for integrating and demonstrating 3D production simulation capability as part of an industrial production system. The next section describes the conceptual framework, which describes the detailed process of simulation model development. This framework serves not only as the model's structural backbone, but also as a strategic method for simulating the complex dynamics of the FESTO assembly line. Using this methodology, the research intends to demonstrate how the simulation matches the assembly station's processes, paving the door for potential improvements and optimization in an industrial context.

4 Conceptual Framework

Developing a conceptual framework for a 3D production simulation model is a complex process that aims to closely replicate a real-world production system. According to the frameworks and methods presented in the literature [5, 7, 8], this necessitates a comprehensive combination of data gathering, model creation, simulation, and validation to ensure that the simulation accurately and effectively depicts actual processes. The purpose of this conceptual framework, the goal is to achieve the "current state simulation", for an industrial production system. This is limited to processes which can be simulated using discrete event simulation, wherein the simulation time progresses based on a pre-determined time schedule.

The process begins with a thorough data collecting phase in which many different kinds of data are acquired (see Fig. 2). This consists of CAD data from process equipment, robots, conveyors, and production cells, as well as inventory data. It also includes process-related information like cycle times, production quantities, and rates.

Data on operators, Automated Guided Vehicles (AGVs), forklifts, and shop-floor layout pathways are also important at this stage. Each element of data, from raw materials

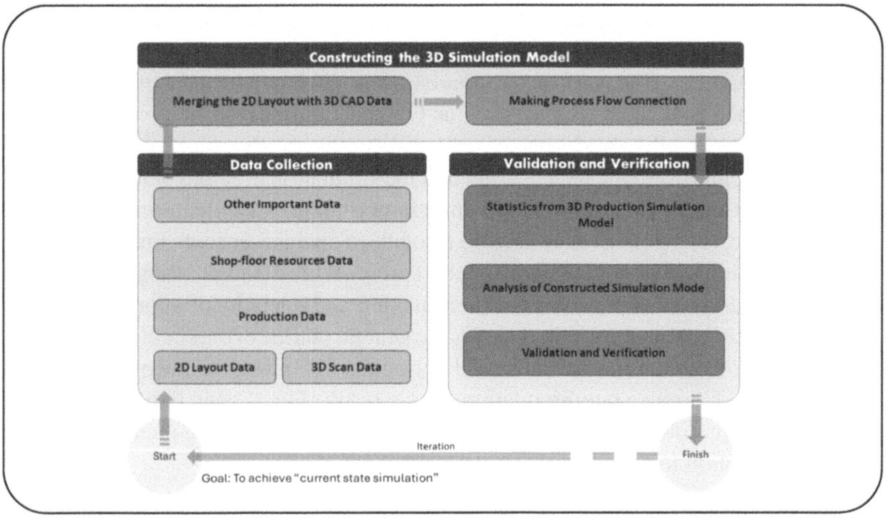

Fig. 2. Conceptual framework for facilitating the development of 3D production simulation model.

to operator characteristics, helps to build a detailed foundation for the simulation model. Following data collection, the next stage is to merge the 2D shop-floor layout with 3D CAD data in the simulation environment. This includes importing the 2D layout and properly matching it with real-world locations and dimensions. CAD files for machines and components are then loaded and positioned correctly on top of the 2D shop-floor layout. Adjustments are made to ensure that all components represent real-world traits and features, paving the way for precise simulation.

For environments where a 2D layout is not readily available, 3D scanning technology can be employed to capture the physical setup of the production system. This process generates point cloud data, which can then be imported into the simulation environment. This capability ensures that even in the absence of pre-existing layout data, a detailed and accurate model can be constructed by converting the point cloud data into a usable 3D model.

Making process flow connections is another critical step. It involves defining how materials move between processes—from their raw material state to their final product state—using various transportation modes like manual handling, AGVs, robots, or forklifts. This step also includes programming robots and conveyor systems, as well as creating assembly and disassembly methods to handle various batch sizes, which reflects the complexity of real-world operations.

Once the simulation model has been developed, it goes through a series of validation and verification steps. The initial validation is comparing simulation run data (such as machine utilization, process statuses, and production rates) to actual production output data to ensure accuracy. The model is further validated by real-world assessments, in which simulated results are compared to acquired data to determine the simulation's dependability and usefulness. Furthermore, the model is validated to confirm that it was properly implemented and performs as intended, reflecting the known features and behaviors of the concerned production system.

The final analysis of the constructed simulation model is conducted using established Key Performance Indicators (KPIs), such as Overall Equipment Effectiveness (OEE), which includes availability, performance, and quality metrics. This analysis helps in identifying areas of improvement and ensuring that the simulation provides valuable insights into the production process, ultimately aiding in optimizing manufacturing operations. This detailed process from data collection through to analysis ensures that the 3D production simulation model not only accurately represents the manufacturing environment but also provides a reliable tool for decision-making and operational improvement.

Consequently, to sum up, as Fig. 1 shows the general outline of the simulation process, it is crucial to construct a simulation model during the first phases of the process, which are "Data" and "Current State Simulation." As a result, Fig. 2 goes into greater detail about these two preliminary phases that are shown in Fig. 1, stressing their significance and their connection to the general simulation framework.

5 Exemplary Case Study: Developing 3D Production Simulation Model of FESTO Mobile Phone Assembly System

5.1 FESTO Assembly Station Description

The FESTO line, a core component of the 3D production simulation model, comprises of two primary stations linked by an Automated Guided Vehicle (AGV) known as Robotino [10, 11]. This AGV plays a crucial role in the transfer of mobile phones, which are securely positioned on carriers, between the stations [11]. Each station is specialized and automated; the first station includes storage or inventory, inspection, and robot (assembly and mounting of circuit board and two fuse components) while the second station focuses on processes including vision inspection, drilling, cover mounting, pressing, labelling, and finally placing of finished products [11, 12]. Robotino's design function ensures that it navigates the floor efficiently, minimizing transfer times and optimizing the flow between stations [10]. The agility and precision of the AGV is essential for maintaining a seamless production process and is thus accurately represented within the simulation model.

5.2 Developed 3D Production Simulation Model

Data Collection

Developing the 3D production simulation model required integrating several sources of data in order to produce a reliable and useful depiction of the manufacturing process. Acquiring Computer-Aided Design (CAD) data was the fundamental building block that followed the conceptual framework that was presented; this provided precise geometry and physical attributes of the manufacturing assets. To capture the essence of the actual FESTO line, Visual Components® software was used to digitally reconstruct machines, production lines, robots, and AGVs (see Fig. 3). To provide a dynamic and responsive simulation environment, precise data on process cycle timings, production volumes, setup times, and tool changeover times were gathered. Shop-floor resource data was also integrated, such as timetables for resource availability and operator and AGV access points.

Model Creation

The durability of the simulation model depended heavily on the smooth integration of 3D CAD models with 2D layout data. The virtual environment's work cell layouts and real space constraints could be precisely simulated thanks to this integration. To improve the realism of the model, dynamic interactions were added, such as Robotino's mobility, the flow of goods through the assembly line, and interactions between workers and machinery.

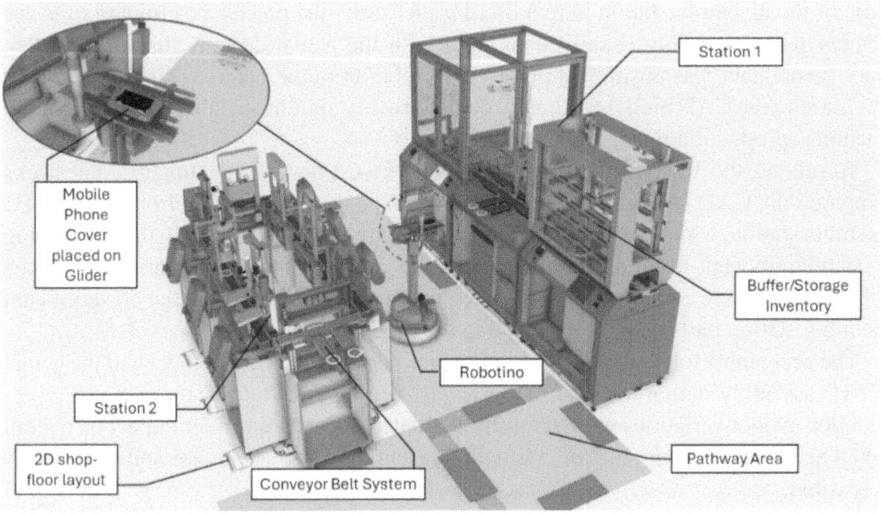

Fig. 3. 3D Production Simulation Model of automated FESTO Mobile Phone Assembly system developed in Visual Components® software.

Integration with 2D Layouts

Accurate representation of work cell configurations and spatial restrictions in the simulation was assured by integrating 2D layout data with 3D CAD models. The assessment of various layout scenarios and their effects on production efficiency was made easier by this integration.

Dynamic Interactions

To simulate shop floor processes, the model included dynamic aspects like AGV mobility and its defined pathway area (see Fig. 3). A more thorough examination of the production flow was made possible by the addition of dynamic interactions.

Application and Outcomes

A simulation model that accurately reflected the intricacy and workflow of a modern manufacturing floor was created by combining operational data with elaborate CAD representations. This model can be used to facilitate production analysis, streamline procedures, and detect possible process constraints in a risk-free virtual sandbox. The simulation also provided information on safety precautions and ergonomics, which helped to create a more comprehensive picture of the manufacturing environment.

5.3 Verification and Validation

A thorough Validation and Verification (V&V) phase was carried out to guarantee the quality and dependability of the model. Its main goal was to validate the 3D CAD data to accurately reflect the FESTO assembly station in real life. This phase's main objective was to confirm that every component was made with exact dimensions that matched those of the assembly station in real life. Furthermore, the precise positioning of every element was thoroughly examined to align with the actual design, guaranteeing that every component was positioned precisely in relation to the others. This also involved making sure the CAD model accurately depicted every significant component, including machines, production lines, robots, and AGVs.

To validate the 3D CAD data, several scenarios were developed. Dimensional checks compared the CAD components' dimensions to actual measurements from the FESTO assembly station, while positional verification required cross-referencing the positioning of each component in the CAD model with the real-world arrangement. Another key stage was to ensure component inclusion, which involved confirming that all significant assembly station parts were represented in the model.

The acceptable tolerance for inconsistencies between the CAD model and the actual FESTO assembly station was set to a low level, assuring great fidelity in the model representation. Minor variations were carefully evaluated to determine their impact on overall model accuracy, ensuring that the virtual representation was as precise and trustworthy as feasible.

6 Discussion

The proposed conceptual framework served as a practical extension of the presented literature, operationalizing its theoretical insights within the tangible context of the FESTO case study. This juxtaposition demonstrates that, while the literature explains the fundamental criteria for simulation [6–8], the framework provides a methodical technique for putting these notions into action, as evidenced by the detailed implementation in the case study. The FESTO case study demonstrates the framework's practical utility, particularly in its depiction of the AGV 'Robotino' and displays a specific implementation that goes beyond the generalized advice of current literature.

The case study emphasizes the benefits of 3D production simulation. This tool enables a thorough study and optimization of production processes without the accompanying expenses or disruptions of physical prototypes. It gives a proactive technique to foresee and remedy production inefficiencies, therefore highlighting its value as a decision-making aid in manufacturing.

The synthesis of the literature review, conceptual framework, and case study implicitly addresses the research question about the necessary data for creating a 3D manufacturing simulation model. The successful deployment of the FESTO simulation demonstrates the importance of incorporating complete data, such as CAD details and operational parameters. As a result, this not only emphasizes the importance of a wide range of reliable data for simulation accuracy, but it also provides a nuanced solution to the research question, confirming the significance of 3D production simulation in the context of industrial systems.

In response to the research question— "What essential data should be considered for developing a 3D production simulation model?"—the study emphasized on the need for collecting thorough and varied dataset. This included precise geometric CAD models of manufacturing components, detailed process statistics such as cycle times and throughput times. The findings indicate that the fidelity of a 3D production simulation model is directly proportional to the breadth and correctness of the underlying data, highlighting the importance of rigorous data integration in developing an effective and trustworthy 3D production simulation model.

7 Conclusion

The need for a conceptual framework in the industrial sector is driven by the requirement to facilitate the development of simulation models within virtual environments. This framework is applicable to a diverse array of industrial systems that benefit from discrete event simulation, such as automotive, pharmaceutical, and hardware manufacturing. The versatility of this tool makes it valuable across both practical industry applications and academic research, meeting key industrial demands.

Moreover, this research addresses the question, "What essential data should be considered for developing a 3D production simulation model?" by emphasizing the importance of comprehensive data collection. CAD models for accurate geometric representation, comprehensive process data (cycle times, setup times, production quantities), shop-floor layouts (two and three dimensions), resource data on machine and human availability, and 3D scanning data for setups that currently exist but lack CAD models are some of the important data, which has been presented in the proposed framework (see Fig. 2). High-fidelity 3D simulation models that accurately represent the current production states and enable robust optimization and decision-making are ensured by integrating these different types of data.

Furthermore, the research presented herein explores deeply into the requirements and application of 3D production simulation in industrial systems. By combining scientific theories with a structured conceptual framework and applying it to the FESTO case study, the study validates 3D simulation's potential in improving manufacturing processes. The case study demonstrates that extensive data acquisition—including CAD data, operational parameters, and 3D scanning precision—is critical to building a realistic simulation model. This confluence of data and technology allows for a virtual yet realistic analysis of production systems, promoting informed decision-making and process optimization.

Acknowledgments. This research work has been partially funded by the Knowledge Foundation (KKS) within the framework of INDTECH + Graduate Research School project (Grant Number 20220132), Mälardalen Industrial Technology Centre AB and Mälardalen University and partially funded by XPRES (Excellence in Production Research).

Disclosure of Interests. The authors have no competing interests to declare that are relevant to the content of this article.

References

1. Bandyopadhyay, S., Bhattacharya, R.: Discrete and Continuous Simulation: Theory and Practice. CRC Press, Boca Raton (2014)
2. Beck, J.: Towards an automated system for robot assembly cell layout optimization. In: Kiefl, N., Wulle, F., Ackermann, C., Holder, D. (eds) ARENA2036. LNCS, pp. 108–117. Springer, Cham (2022). https://doi.org/10.1007/978-3-031-27933-1_11
3. Law, A.M.: How to build valid and credible simulation models. In: 2022 Winter Simulation Conference (WSC). LNCS, pp. 1283–1295 (2022)
4. Malega, P., Gazda, V., Rudy, V.: Optimization of production system in plant simulation. Simulation **98**(4), 295–306 (2021)
5. McAtee, S., Dunn, M., Nagarajah, R.: Simulation scan comparison for process monitoring using 3D scanning in manufacturing environments. Int. J. Adv. Manuf. Technol. **74**, 823–834 (2014)
6. Delbrügger, T., et al.: Multi-level simulation concept for multidisciplinary analysis and optimization of production systems. Int. J. Adv. Manuf. Technol. **103**, 3993–4012 (2019)
7. Strahilov, A., Hämmerle, H.: Engineering workflow and software tool chains of automated production systems. In: Multi-disciplinary Engineering for Cyber-Physical Production Systems: Data Models and Software Solutions for Handling Complex Engineering Projects, pp. 207–234 (2017)
8. Banks, J. (ed.): Handbook of Simulation: Principles, Methodology, Advances, Applications, and Practice. Wiley, Hoboken (1998)
9. Visual Components 4.8 Homepage. https://www.visualcomponents.com/. Accessed 19 Apr 2024
10. Fattouh, A., Chirumalla, K., Ahlskog, M., Behnam, M., Hatvani, L., Bruch, J.: Remote integration of advanced manufacturing technologies into production systems: integration processes, key challenges and mitigation actions. J. Manuf. Technol. Manag. (2023)
11. da Cunha, C., Cardin, O., Gallot, G., Viaud, J.: Designing the digital twins of reconfigurable manufacturing systems: application on a smart factory. IFAC-PapersOnLine **54**(1), 874–879 (2021)
12. Mortensen, S.T., Chrysostomou, D., Madsen, O.: A novel framework for virtual recommissioning in reconfigurable manufacturing systems. In: 2017 22nd IEEE International Conference on Emerging Technologies and Factory Automation (ETFA), pp. 1–4. IEEE (2017)

From Product-Service Systems to Smart Product-Service Systems. A Survey About German Companies and Practical Implications

Christian Schiller[1], Mike Freitag[1](✉), and Oliver Hämmerle[2]

[1] Fraunhofer Institute for Industrial Engineering IAO, Nobelstr. 12, 70569 Stuttgart, Germany
mike.freitag@iao.fraunhofer.de

[2] Institute of Human Factors and Technology Management IAT, , Universität Stuttgart, Nobelstrasse 12, 70569 Stuttgart, Germany

Abstract. This paper discusses a study conducted in Germany on smart product-service systems and their adoption among companies. The study examines the current state of the implementation of smart product-service systems, the importance of digital technologies, motivations for adoption, challenges faced, and the need for support. The findings show that 40% of surveyed companies are preparing to use or already utilize applications for smart product-service systems. Large companies and those in the industrial sector are more likely to have implemented smart product-service systems. Companies with experience in smart product-service systems prioritize digital technologies such as the Internet of Things. Motivations for adoption include leveraging data, increasing competitiveness, meeting customer requirements, and creating new value. Challenges vary between experienced and inexperienced companies, with the former facing issues related to data protection, security, and costs, while the latter struggle with strategies, concepts, and human resources. System and solution providers have the highest adoption rate, followed by product providers, with service providers lagging behind. The study provides valuable insights for companies considering the implementation of smart product-service systems and offers guidance on technology integration, motivations, challenges, and the need for support.

Keywords: product-service systems · smart service · service business model · cyber-physical production system · service lifecycle management · service engineering · SME

1 Introduction

Since at least the early 1990s, the phenomenon of increasing complementarity between services and products has been the subject of discussion in science and business [1]. Physical products have increasingly been complemented by various services to enhance customer satisfaction and remain competitive against other companies.

Presently, services are often created on an ad-hoc basis and are only incorporated into a company's overall offering after the product has been developed. In some companies,

Published by Springer Nature Switzerland AG 2024
M. Thürer et al. (Eds.): APMS 2024, IFIP AICT 733, pp. 437–448, 2024.
https://doi.org/10.1007/978-3-031-71645-4_29

products and services are developed simultaneously. This typically involves more or less structured coordination loops between the two development processes. In both scenarios, these are bundles of products and services rather than integrated systems.

Since the late 2000s, however, there has also been increasing research activity on the integration of product and service development processes [2]. Initial activities on methodological integration were also taking place at that time [3]. The end result was intended to be a solution that only delivers added value to both customers and providers when used together as a system.

For some years now, advancing digitalization has also increasingly presented the opportunity to further develop traditional products into cyber-physical systems that can communicate with other products and the internet [4]. This creates completely new potential for the provision of data-based and intelligent ("smart") services, enabling companies to further boost their competitiveness and generate additional added value, in some cases with completely new business models [5–7]. The resulting systems of networked products and smart services are also referred to as smart product-service systems. The networked ("smart") products serve as boundary objects to enables costumers and providers mutual benefits [8]. Smart product-service systems are distinguished by their customer-oriented approach to generate concrete added value for users [9, 10]. They are also characterized by a high degree of complexity, dynamism, and multi-layeredness [11]. The high amount of interdisciplinarity of PSS in particular leads to tensions that must be taken into account during development [12].

Following this introduction, the paper is structured into four parts:

- Section 2 describes the components of a smart product-service system and provides an example from the electroplating industry to illustrate this.
- The framework of the study is presented in Sect. 3, which includes distinguishing between successful and less successful companies.
- Section 4 presents selected results of the study. The focus here is on the acceptance of smart product-service systems in Germany today and in the future.
- Section 5 then follows with a brief summary.

2 Reference Model for the Product-Service System

To better contextualize the questions of this study, this section provides a brief description of what the authors consider to be a smart product-service system. Both of the components are described in detail along with the connections between them.

To support the development and configuration of product-service systems, a six-level reference framework was devised, which was subsequently delineated into five distinct layers (see Fig. 1). The inner four layers delineate the cyber-physical production system (CPPS) of the electroplating plant (as introduced in [13] and [14]), while the outer layer elaborates on the surrounding service through service engineering [15–17]. These layers also serve to indicate the extent or status of the digital transformation within a product-service system. The dots depicted here represent the layers affected in the current stage of development. This reference framework amalgamates components of smart service life cycle management and the CRISP DM model [18], thereby combining the realms of service and CPPS within a singular reference framework. For a comprehensive explanation of this reference framework, refer to Leiden et al. [19]. This paper presents a concise overview. Detailed information can be found in [20–22].

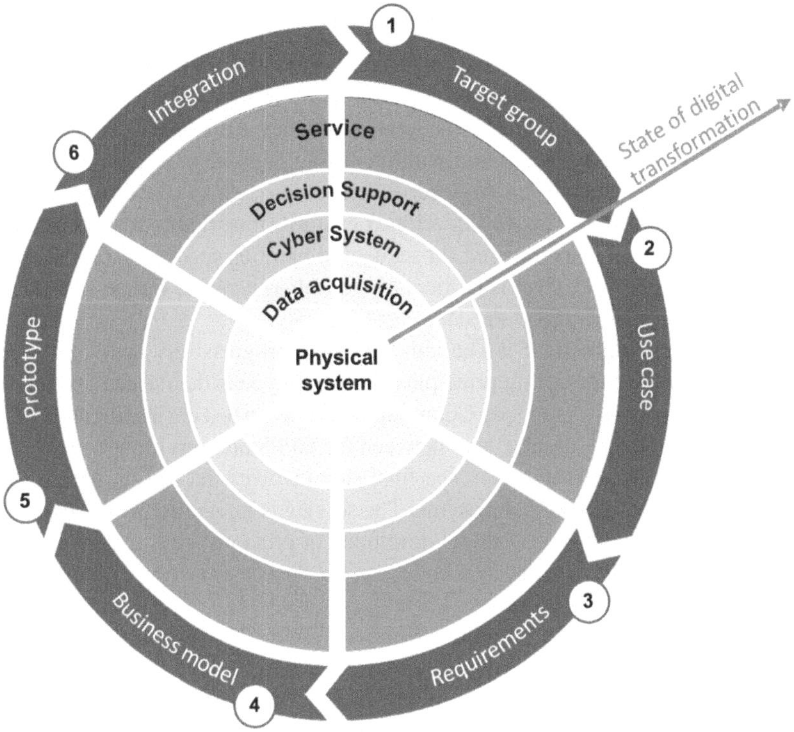

Fig. 1. Reference framework for the development of product-service systems (based on [20]).

Moreover, the reference framework not only offers methodological support for smart service development but also gauges the current extent of their digital transformation. The more layers a user can address with their systems, the more advanced their digital transformation status.

The subsequent sections elaborate on the individual steps of the reference framework from both a service and CPPS perspective, highlighting their distinct features. They also describe the additional layers of a smart service production system in comparison to a product-service system. The reference framework is divided into the following six phases:

Target Group Definition (1): During this initial step, the customer target group is identified, determining the audience for the new smart service and corresponding business model. Differentiation among target groups may be necessary based on various factors such as distribution channels, relationship types, profitability levels, or willingness to pay. In the electroplating context, distinctions can be made between in-house and contract electroplating as well as among companies of different sizes.

Use Case Development (2): The second step involves establishing a specific use case tailored to the defined target group. This includes outlining the existing service provision, identifying unique selling points of the business model, and defining the value proposition. The use case and value proposition drive the requirements for decision support and interface with the CPPS, especially relevant when considering mixed reality approaches.

Requirement Definition (3): Requirements from both a service and CPPS perspective are finalized based on previous steps to develop the prototype. Methods for the cyber system are determined, taking into consideration whether a data-based or model-based approach is more suitable. Integration within existing IT systems is crucial, especially in view of the variability in technology across industries such as electroplating.

Business Modeling (4): A well-developed business model serves as a cornerstone for complete service engineering and subsequent integration processes. Osterwalder & Pigneur's business model canvas is particularly useful, as it can be further refined into a service business model canvas if needed.

Prototype Development (5): The aim here is to progressively develop the smart service using service engineering principles. The prototype is then tested at a technical readiness level (TRL) of four to five. Compatibility between test environments is crucial, ensuring intensive testing of interfaces between the CPPS and service sphere.

Smart Service Integration (6): The final step involves integrating the developed smart services within operational systems. The service is integrated within the process landscape, utilizing the business model developed in previous steps. Achieving a high TRL between 8 and 9 ensures the system is ready for use, and involves rigorous testing of hardware and software systems.

This brief overview of smart product-service systems is also intended to facilitate better understanding of the content of this study.

3 General Information About the Study

The study was published in 2022 in Germany and involved a quantitative survey of companies [23]. The questionnaire was developed on the basis of extensive research and analysis on the topic of smart product-service systems. Furthermore, the researchers incorporated the experience from previous, comparable project work.

The survey focused on various aspects of development, such as current and future priorities, existing development priorities, and the time required for the development process. It also explored the development of smart product-service systems, including the experience of the surveyed companies, the importance of digital technologies, motivations for working on these systems, and challenges faced. General questions about the company, such as the number of employees and industry, were also part of the survey [23].

The target groups for the survey were companies primarily focused on products with accompanying services, as well as service providers heavily reliant on technology. The majority of respondents were from the manufacturing industry, while the remaining companies were from the services sector. The companies were classified based on their number of employees, with small, medium, and large enterprises represented [23].

A total of 71 companies completed the online questionnaire. At around three-fifths of the total, most of the responding companies are in the secondary sector (manufacturing industry). The remaining companies come from the tertiary sector (services) [23].

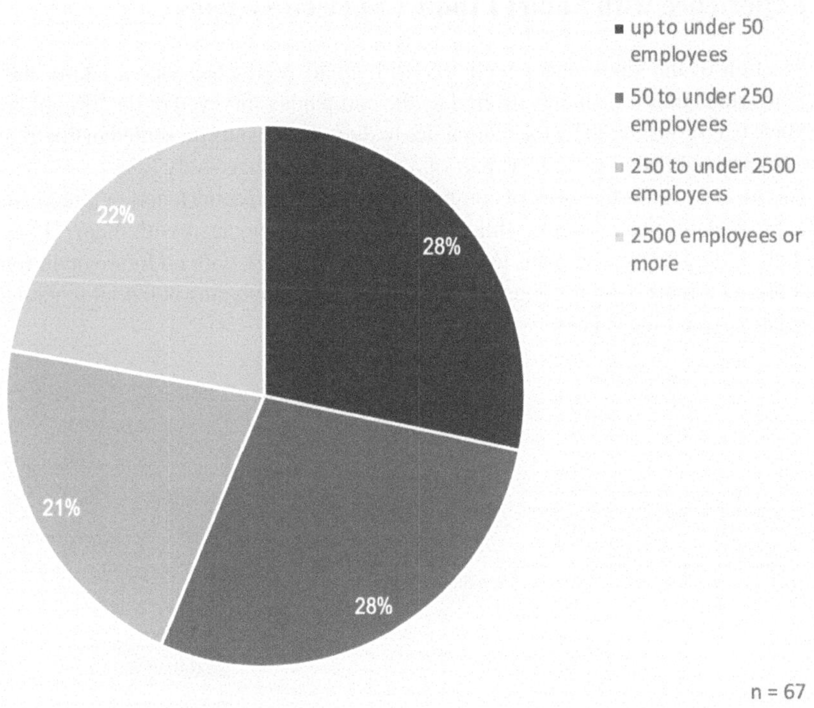

Fig. 2. Distribution by company size [23].

To assess company size, the participating companies were classified according to their number of employees (see Fig. 2). 28 percent of the companies surveyed are small businesses with up to 49 employees. A further 28 percent of the companies are medium-sized enterprises with up to 249 employees. Large companies with 250 to 2,499 employees account for 21 percent, and companies with more than 2,500 employees make up 22 percent [23].

An important subject of the study was the question regarding specific differences between successful and less successful companies. To filter out the successful companies from the pool of responding companies, the question of the development of important company key figures—number of employees, turnover, and profit—was taken as the starting point. In this regard, the companies surveyed were asked to indicate how they had developed over the last three years in comparison to their own sector [23].

A cluster analysis was used to derive three groups of companies from the answers to this question: The first group has largely positive key figures ("very successful companies"), the second group has average key figures ("successful companies"), and the third group of companies has somewhat negative key figures ("less successful companies"). In total, the first group comprises 16 companies, the second group 30 companies, and the third group 19 companies [23].

4 Experience with Smart Product-Service Systems

This section of the study deals explicitly with smart product-service systems and the extent to which they are already offered by the companies surveyed or the form in which they have been considered by the companies to date. The results presented in this chapter were already published in 2022 as part of a german-language study [23].

The survey reveals that 40% of companies are either preparing to use or already using smart product-service systems, while 36% have not yet engaged with them. However, over half of the latter group plans to do so, leaving only 17% with no future intentions to adopt these systems. About a quarter are in the phase of analyzing potential or gathering information (see Fig. 3) [23].

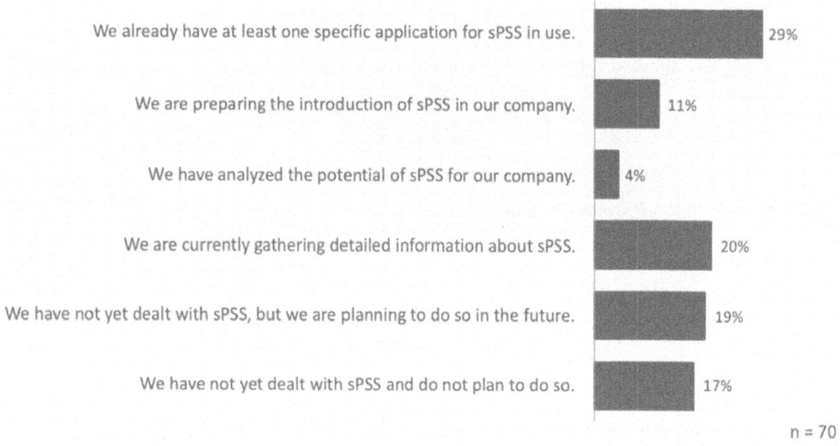

Fig. 3. Experience with smart product-service systems (sPSS) [23].

Large companies and those in the industrial sector are more likely to use smart product-service systems compared to small and medium-sized enterprises (SMEs) and companies in the services sector. A notable difference also exists based on company success [23].

The very successful companies are by far the most likely to already have smart product-service systems in place. As already mentioned, the relatively small number of cases in this breakdown must be taken into account (see Fig. 4) [23].

The study shows that companies with experience in smart product-service systems attach greater importance on digital technologies than those that have only analyzed the potential, gathered information, or have not yet dealt with such systems at all. The most significant differences between these two groups are seen in their views on the Internet

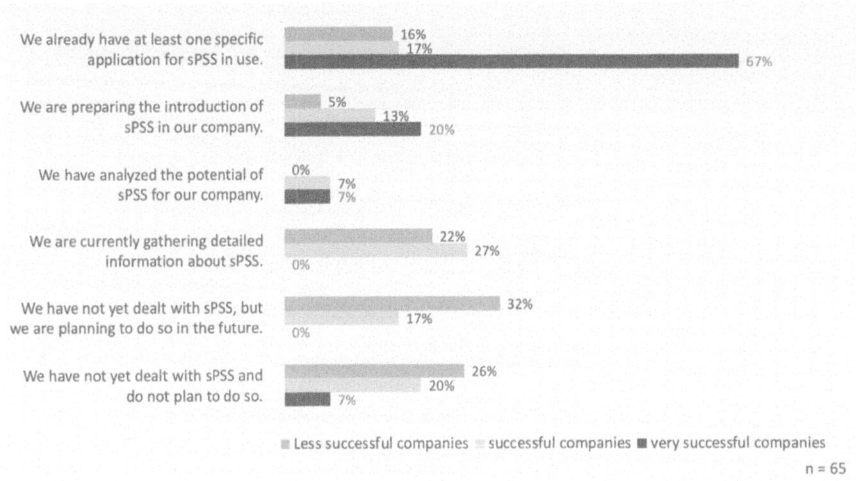

Fig. 4. Experience with smart product-service systems broken down by company success [23].

of Things and sensors and actuators. Conversely, there is little difference between the two groups regarding technologies considered least important, such as Blockchain and robotics [23].

There are also significant differences in the motivation to engage with smart product-service systems between the companies that already use smart product-service systems or are preparing to use them and the other companies (see Fig. 5). The former group cites motivational reasons considerably more frequently than the latter group [23].

Companies with experience in smart product-service systems are more motivated to engage with these systems. They aim to use existing data for new service offers, boost their competitiveness, meet customer requirements more precisely, and create new value systems. On the other hand, companies without practical experience in smart product-service systems perceive more and greater challenges. They particularly highlight a lack of strategies, concepts, and human resources as significant challenges, unlike experienced companies who are more concerned with data protection, data security, and costs [23].

Experienced companies also express a slightly greater need for support, though not uniformly across all areas. This may be because their experience allows them to assess their support needs more accurately. Specifically, these companies indicate a higher need for support in integrating product and service development, implementing intelligent components, and customer integration. Conversely, less experienced companies show a slightly higher need for support in developing business models [23].

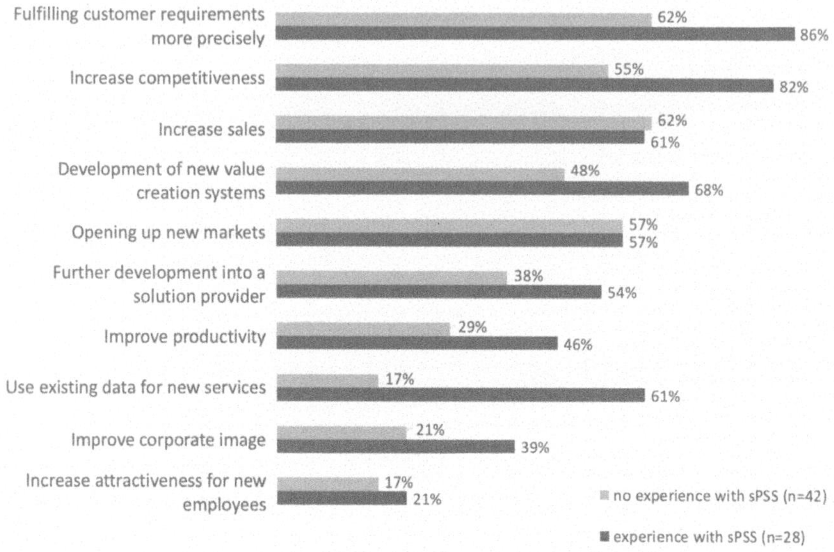

Fig. 5. Motivation to deal with smart product-service systems divided into companies with and without experience with smart product-service systems [23].

5 Practical Implications

The study has yielded interesting findings on smart product-service systems and their prevalence in the industry. It serves as a basis for recommending appropriate actions to companies, in particular small and medium-sized enterprises. Users in the field are now faced with the question of how to meet the organizational challenges of developing smart product-service systems and how to define suitable development processes. They also need to identify how best to gain initial experience in using smart product-service systems and how to develop interdisciplinary skills in the process.

Companies with experience in smart product-service systems prioritize digital technologies such as the Internet of Things. Motivations for adoption include leveraging data, increasing competitiveness, meeting customer requirements, and creating new value. Challenges vary between experienced and inexperienced companies, with the former facing issues related to data protection, security, and costs, while the latter struggle with strategies, concepts, and human resources. System and solution providers have the highest adoption rate, followed by product providers, with service providers lagging behind. An example how to help this companies is show on Fig. 6.

5.1 Defining Development Processes for Smart Product-Service Systems

Defining development processes is also key to the successful development and introduction of smart product-service systems. The survey revealed that very successful companies and successful companies are far more likely to have somewhat defined development processes than less successful companies. These companies also exhibit a greater motivation in tackling this issue. Furthermore, they considerably fewer challenges and estimate a lower support requirement than companies that currently do not have any defined development processes in place.

There are many ways to define development processes within a company. On the one hand, new hires may be able to provide the company with the appropriate methodological skills and content-related impetus on this topic. Together with experienced employees working in the area of product and service development (if available), a company-specific development process for smart product-service systems can be established.

One example to do is to use the reference framework – see Fig. 6.

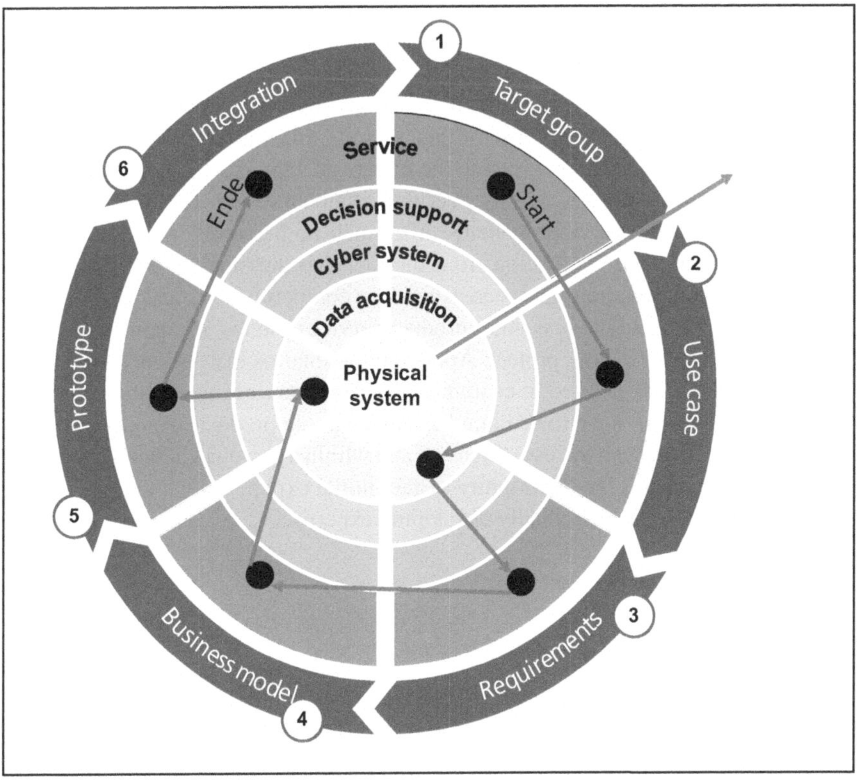

Fig. 6. Use Case: A company from the plating industry

The physical system forms the nucleus of the system. In this example, it is a company that builds air cleaning components for the electroplating industry. Instead of selling components the company will sell maintenance service over the lifetime of the air cleaning components. The company used the reference framework in Fig. 6.

First step a start point is to fix the target group. Second step is to identify use cases. Step is to collecting requirement from business side and also from the technical data. Data are then collected in the second layer through sensors. These data undergo either a data-driven or model-based analysis in the subsequent layer of the cyber system. The decision support layer marks the transition to service, initiating interaction between individuals and IT systems. In this step 5 a prototype has to show that the interactions are working. These prototype systems must be made user-friendly through suitable visualization methods in order to ensure high acceptance levels. Finally, the service layer builds upon preceding layers, incorporating operational requirements into the Integration in step 7.

5.2 Implementing Smart Product-Service Systems to Gain Practical Experience

The study confirmed that there is a direct link between corporate success and experience with the use of product-service systems. However, there is no clear information about the exact direction of this correlation. It is therefore possible, on the one hand, that more successful companies are more likely to engage in smart product-service systems and, on the other hand, that companies that engage in smart product-service systems are more successful precisely because of this. Despite this, this correlation should still be taken into consideration, in part due to the high expectations attached to introducing smart product-service systems, and the fact that companies with previous experience also value smart product-service systems at a much higher level. As a result, these companies can draw their motivation from the fact that they have a better chance of meeting customer expectations and increasing their competitiveness and sales.

In order to implement the first smart product-service systems, it is advisable to start with a small, manageable pilot project. Application scenarios that are only used within the company or are only offered to customers after internal testing are often suited for this. These offers do not have to be complex. Even simple products can provide added value for customers through the use of sensors and within a narrowly defined application framework. This presents the opportunity to gain initial experience with minimal effort and manageable risk, while gradually developing expertise.

6 Summary

The study focuses on smart product-service systems and examines the extent to which the surveyed companies offer them or are considering them. Of the surveyed companies, 40% are preparing to use smart product-service systems or already have applications for them. However, 36% of the companies have not yet dealt with smart product-service systems, although more than half of them plan to do so in the future. Only 17% of the respondents stated that they do not intend to engage with smart product-service systems. The large companies are more likely to already use smart product-service systems compared to the SMEs surveyed. Companies in the industrial sector are more likely to use smart product-service systems compared to those in the services sector. The most successful companies are the most likely to have already implemented smart product-service systems.

The companies with experience in smart product-service systems place greater importance on digital technologies, particularly the Internet of Things and sensors and actuators. The motivation to engage with smart product-service systems is significantly higher for the companies with experience, as they aim to use existing data to offer new services, increase competitiveness, meet customer requirements more precisely, and build new value creation systems. Those companies without practical experience in smart product-service systems perceive challenges such as lack of strategies, concepts, and human resources to be greater. In contrast, the companies with experience see challenges related to data protection, data security, and costs as being greater. The experienced companies also have a higher need for support in areas such as product-service development, implementation of intelligent components, and customer integration.

In terms of the types of providers, system and solution providers are most likely to use or introduce smart product-service systems, followed by product providers, and service providers have the lowest adoption rate.

Overall, the study highlights the varying levels of engagement with smart product-service systems among companies, the importance of digital technologies, and the motivation and challenges associated with implementing smart product-service systems.

Acknowledgements. This article was written as part of two projects:

The "Integriertes Design smarter Produkt-Service-Systeme in KMU (bi.smart) research and development project is funded by the German Federal Ministry of Education and Research (BMBF) within the "Innovations for Tomorrow's Production, Services, and Work" program (funding number 02J19B042) and implemented by the Project Management Agency Karlsruhe (PTKA).

The "AgileAssembly" research and development project is funded by the German Federal Ministry of Education and Research (BMBF) within the "Innovations for Tomorrow's Production, Services, and Work" program (funding number 02J21C124) and implemented by the Project Management Agency Karlsruhe (PTKA).

The authors are responsible for the content of this publication.

References

1. Engelhardt, W.H., Kleinaltenkamp, M., Reckenfelderbäumer, M.: Leistungsbündel als Absatzobjekte. Zeitschrift für Betriebswirtschaftliche Forschung **45**(5), 394–426 (1993)
2. Spath, D., Meiren, T., Münster, M.: F&E-Management für Lösungsanbieter. In: Fleig, G.: Integration von Produkt und Service. Auf dem Weg zum Lösungsanbieter, pp. 73–87. Handelsblatt Verlagsgruppe, Düsseldorf (2012)
3. Stark, R., Müller, P.: HLB-Entwicklungsmethodik – generischer entwicklungsprozess, generierung von anforderungen und absicherung hybrider leistungsbündel. In: Meier, H., Uhlmann, E. (Ed.), Integrierte Sach- und Dienstleistungen: Vermarktung, Entwicklung und Erbringung hybrider Leistungsbündel. Springer-Verlag, Berlin, Heidelberg (2012). https://doi.org/10.1007/978-3-642-25269-3_3
4. Abramovici, M. (Ed.): Engineering smarter Produkte und Services, Plattform Industrie 4.0 Studie, acatech – Deutsche Akademie der Technikwissenschaften, Munich (2018)
5. Freitag, M., Hämmerle, O.: Die Entwicklung von Smart Services im Maschinen- und Anlagenbau: Ein Leitfaden. Fraunhofer IAO, Stuttgart (2021)
6. Chowdhury, S., Haftor, D. and Pashkevich, N.: Smart product-service systems (Smart PSS) in industrial firms: a literature review. In: Procedia CIPR, Special Issue: 11th CIRP Conference on Industrial Product- Service Systems, vol. 73, pp. 26–31 (2018). https://doi.org/10.1016/j.procir.2018.03.333

7. Rizvi, M.A.K., Chew, E.: Towards systematic design of cyber-physical product-service systems. In: Proceedings of the DESIGN 2018 15th International Design Conference, pp. 2961–2974 (2018). https://doi.org/10.21278/idc.2018.0248
8. Beverungen, D., Müller, O., Matzner, M., Mendling, J., vom Brocke, J.: Conceptualizing smart service systems. Electron Markets **29**, 7–18 (2019). https://doi.org/10.1007/s12525-017-0270-5
9. Maleki, E., Belkadi, F., Bernard, A.: A meta-model for product-service system based on systems engineering approach. Procedia CIRP **73**, 39–44 (2018). https://doi.org/10.1016/j.procir.2018.04.016
10. Meier, H., Uhlmann, E.: Integrierte Industrielle Sach- und Dienstleistungen, Springer-Verlag Berlin, Heidelberg (2012). https://doi.org/10.1007/978-3-642-25269-3
11. Kuhlenkötter, B., et al.: New Perspectives for Generating Smart PSS Solutions – Life Cycle. Methodologies and Transformation. Procedia CIRP **64**, 217–222 (2017). https://doi.org/10.1016/j.procir.2017.03.036
12. Paliyenko, Y., Heinz, D., Schiller, C., Tüzün, G.-J., Roth, D., Kreimeyer, M.: Requirements for a smart product-service system development framework. Proc. Design Soc. **3**, 3085–3094 (2023). https://doi.org/10.1017/pds.2023.309
13. Leiden, A., Herrmann, C., Thiede, S.: Cyber-physical production system approach for energy and resource efficient planning and operation of plating process chains. J. Cleaner Prod., 125160 (2020)
14. Leiden, A., Kölle, S., Thiede, S., Schmid, K., Metzner, M., Herrmann, C.: Model-based analysis, control and dosing of electroplating electrolytes. Int. J. Adv. Manuf. Technol. (2020)
15. Qu M., Yu S., Chen D., Chu J., Tian B.: State-of-the-art of design, evaluation, and operation methodologies in product service systems. Comput. Ind. **77**, 1–4 (2016)
16. Pezzotta, G., Pirola, F., Pinto, R., Akasaka, F., Shimomura, Y.: A service engineering framework to design and assess an integrated product-service. Mechatronics. **31**, 169–179 (2015)
17. Ojasalo, J., Ojasalo, K.: Using service logic business model canvas in lean service development. In: Gummesson, E, Mele, C, Polese, F., (Eds.) Service Dominant Logic, Network and Systems Theory and Service Science: Integrating Three Perspectives for a New Service Agenda. Naples: Conference Proceedings 5th Naples Forum on Service (2016)
18. Shearer, C.: The CRISP-DM Model: the new blueprint for data mining. J. Data Warehousing **5**, 13–22 (2000)
19. Leiden, A., et al.: Entwicklung von Smart Services – Ein Bezugsrahmen und Anwendung in der Galvanotechnik. In: Sörgel, T., (Ed.), Jahrbuch Oberflächentechnik. Leuze, Bad Saulgau (2020)
20. Schiller, C., et al.: Reference model for product-service systems with an use case from the plating industry. In: IFIP International Conference on Advances in Production Management Systems, pp. 335–342. Springer Nature Switzerland, Cham (2022)
21. Freitag, M., Schiller, C.: Approach to test a product-service system during service engineering. In: Procedia CIRP 64, pp. 336–339 (2017). https://doi.org/10.1016/j.procir.2017.03.059
22. Freitag, M., Schiller, C., Hämmerle, O.: Guideline to develop smart service business models for small and medium sized enterprises. In: Dolgui, A., International Federation for Information Processing (IFIP): Advances in Production Management Systems. Artificial Intelligence for Sustainable and Resilient Production Systems. Proceedings. Pt. V: IFIP WG 5.7 International Conference, APMS 2021, Nantes, France, September 5–9 (2021)
23. Friedrich, M., Schiller, C.: Smarte Produkt-Service-Systeme – Eine Bestandsaufnahme unter deutschen Unternehmen. Study. Fraunhofer IAO, Stuttgart (2022)

The Impact of Blockchain Implementations on Supply Chain Collaboration

Yutong Bai(✉) and Ming K. Lim

University of Glasgow, Glasgow G12 8QQ, UK
y.bai.1@research.gla.ac.uk

Abstract. Blockchain technology is increasingly gaining attention from practitioners and academics in supply chain management due to its potential to enhance transparency, efficiency, and trust among trading partners. This study explores the impact of blockchain deployment on supply chain collaboration (SCC), with a focus on information sharing, trust, and commitment. Through a multiple case study of six companies, we found that while blockchain enhances information sharing by facilitating digital transformation in the supply chain, concerns about data privacy pose significant challenges. Moreover, blockchain technology can foster trust through immutable record keeping and automated smart contracts, alleviating concerns related to data tampering and enhancing the reliability of transactions. Commitment within SCC is also influenced by blockchain, as it requires joint investment in specific assets, thereby strengthening partnerships. However, asymmetric willingness to adopt blockchain, especially among smaller supply chain participants, may undermine these benefits. Our findings provide valuable insights into the strategic implementation of blockchain in supply chains and highlight the crucial balance needed between technology adoption and stakeholder engagement.

Keywords: Blockchain technology · supply chain collaboration · supply chain management

1 Introduction

Blockchain technology has gained significant attention from enterprises and researchers for its potential in data security, transparency, and smart contracts [1]. In the field of supply chain management, blockchain-based applications such as traceability platform have been widely discussed as game-changers in supply chain [2]. The immutability of transaction information recorded on the blockchain, due to its unique chain structure, has led researchers to believe that blockchain might have a revolutionary impact on organisational relationships within the supply chain [3]. Researchers argued that blockchain can promote information exchange and benefit supply chain collaboration (SCC) by enhancing information security[4]. Moreover, the impact of blockchain technology on trust in collaboration has been discussed in the literature, with some researchers suggesting that it could even reconstruct the emotional-based trust in traditional collaborations [5].

Published by Springer Nature Switzerland AG 2024
M. Thürer et al. (Eds.): APMS 2024, IFIP AICT 733, pp. 449–460, 2024.
https://doi.org/10.1007/978-3-031-71645-4_30

Current research primarily focuses on the technological aspects of blockchain, such as its ability to protect transactions and reduce fraud through decentralization and immutability[6]. However, there is still a significant gap in understanding the specific mechanisms under blockchain's impacts on SCC. Therefore, this study aims to fill this research gap by exploring how blockchain technology impacts information sharing, trust, and commitment among supply chain partners. We conduct a qualitative investigation through semi-structured interviews with supply chain stakeholders from six case companies to gain a deeper understanding of the actual effects of blockchain implementation.

2 Literature Review

2.1 Blockchain Technology and Its Impacts on Supply Chain Management

As the underlying technology of Bitcoin, blockchain was first proposed in Satoshi Nakamoto's white paper in 2008. It is defined as a decentralized distributed ledger where transaction records are added in chronological order and has an immutable characteristic [7]. The unique consensus mechanism and chain structure provide the foundation for its tamper-proof characteristic. Before adding any transaction record, it needs to be verified through consensus algorithms such as proof of work and proof of stake. A certain number or size of transactions constitutes a block, which is linked together as a chain. Figure 1 illustrates the architecture of blockchain. The block header of each block contains the hash value of the previous block, and if the transaction record in the block is modified, the hash value will also change accordingly. Therefore, the close relationship between blocks forms and further creates the resistance of blockchain to data tampering. The longer the chain structure, the higher the trust in the blockchain [8].

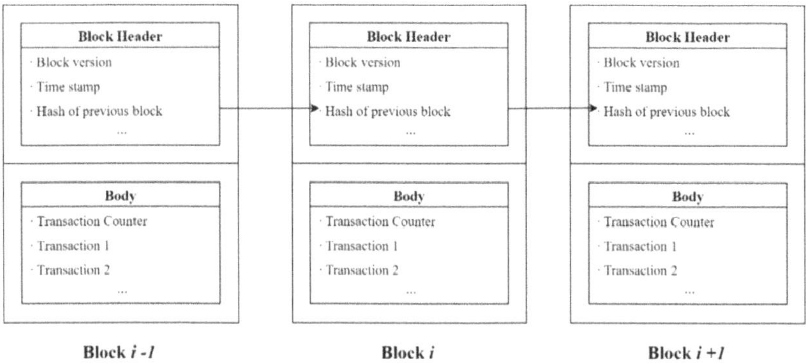

Fig. 1. Blockchain architecture

As a trusted decentralized distributed database technology, blockchain's potential in operations and supply chain management (OSCM) has been extensively discussed in the literature. In terms of data and information management, the tamper-proof characteristic can improve information transparency in the supply chain and facilitate traceability.

A recent research has proposed a blockchain traceability framework for a multi-layer textile and apparel supply chain and demonstrated its good performance in simulation [9]. Immutability also enables blockchain to leverage its advantages in digitization. Another study proposed that blockchain can benefit risk management in the supply chain by promoting the construction of digital platforms and combining with IoT technology [10]. The application of blockchain in supply chain finance (SCF) regarding the digitization of core documents has also been discussed by the literature [11].

2.2 Supply Chain Collaboration and Its Key Dimensions

Discussions about SCC have been continuous since the 1990s. As a driving force behind effective supply chain management [12], SCC is usually defined as a relationship where supply chain members work together to obtain collaborative benefits [13]. In the literature under the topic of SCC, the importance of three dimensions—information sharing, trust, and commitment—is widely recognized. These elements are considered crucial for the effective functioning and success of collaborative relationships within supply chains [14].

Information Sharing

In supply chain management, information sharing is often defined as the extent to which key or proprietary information is shared with supply chain partners [15]. However, information sharing in traditional supply chains faces issues such as lack of visibility [16] and data reliability [17]. As an emerging information technology, blockchain's positive role in information and resource sharing has been widely discussed [18]. Researchers suggest that blockchain can address common problems in data management, such as product tampering and fraud, through smart contracts and consensus mechanisms, thus securely sharing information in medical supply chains [19]. From the perspective of Transaction Cost Economics (TCE), the immutability of blockchain can obviously increase the credibility of shared information, thereby constraining opportunistic behaviour. High-quality information sharing is important in SCC [20], which explains why the literature widely anticipates the technology's role in facilitating collaboration [21].

Trust

Trust is defined as the degree to which members of an organization have a trust orientation towards their collective partners [22]. As a significant factor for the success of supply chain relationships, this structure is considered more important than other forms of trust, such as interpersonal trust [23]. In inter-organizational exchanges, trust creates an environment that encourages enterprises to go beyond the minimum requirements of the relationship, increasing the likelihood of mutual benefits [24].

Commitment

Commitment reflects the long-term exchange beliefs of partners, where the committing party values the continuation of the relationship and is willing to make necessary sacrifices for the stability of the relationship [25]. Commitment to a relationship can improve communication between organizations and facilitate the coordination of buyer-supplier relationships [26]. Researchers also indicated that relationship commitment can achieve effective cooperation by strengthening the relationship between supply chain partners [27]. Furthermore, some studies believe that commitment is necessary for any long-term organizational relationship, as a lack of commitment can lead to vulnerability in supply chain relationships and make it difficult to achieve common goals [28].

3 Methodology

Considering the early stage of blockchain technology implementations in supply chain management and the scarcity of empirical research in the field of research, we have chosen to use qualitative methods for our study. Multiple case study is widely recognized as an excellent tool for investigating new phenomena and addressing "how" questions. Additionally, case study could facilitate the exploration of the underlying mechanism behind a new phenomenon rather than merely testing the correlations between known variables and concepts. This approach aligns more closely with our research question and research aim.

Semi-structured interview is chosen as our primary data source, while we also employing secondary data such as archives and news for triangulation. We endeavoured to choose six case companies from various industries. In the process of case selection, we first identified potential target enterprises through an initial search and review. Subsequently, we conducted further screening through an investigation of secondary data. Considering that the research objective is to investigate the impact of blockchain technology application on SCC, we only selected enterprises that have either successfully run pilot projects or have fully implemented blockchain technology for further contact. This focused approach ensures that our study is grounded in practical, real-world applications of blockchain, allowing us to gather insights from those who have firsthand experience with the technology's deployment and its effects on SCC. When selecting interviewees, we aimed to cover a range of occupations, from top management to technical experts, to ensure the validity of the data.

All interviews were recorded with the consent of the interviewees, with each interview lasting approximately 45 to 60 min. The recordings were transcribed and organized into manuscripts, which were then analysed and coded using NVivo. Through multiple iterations, we developed and refined the coding structure. From the generated codes, we identified patterns and themes, which were documented in case templates for further analysis and discussion.

4 Results and Discussion

4.1 Information Sharing

Enhanced Performance Attributed to Technological Characteristics

In our investigation, several interviewees mentioned the positive role of blockchain technology in information sharing. Case company A is committed to implementing blockchain technology as a foundation for the digitization of key documents in the supply chain, and they also emphasized in interviews that this technology could significantly improve the efficiency of information sharing and greatly reduce the transmission costs of critical documents. The CEO of company A likened the impact of implementing blockchain on the supply chain to the significance of 5G technology for autonomous driving, highlighting the revolutionary transformation brought about by improved efficiency in information sharing. At the same time, this analogy also well explains the significance of blockchain for information sharing: the concept of information sharing was not proposed after the advent of blockchain technology; however, blockchain provides a new way and architecture for sharing, solving the pain points in traditional information sharing and thus greatly enhancing sharing security and efficiency.

Meanwhile, some respondents expressed their views on the impact of blockchain on information sharing from a long-term perspective when asked. For example, an employee of company F noted that blockchain technology is currently officially implemented only in a few of their business segments, but they have a positive outlook on its prospects:

> *"We are still interfacing with more companies... This (blockchain-based) platform will have a strong flow of information, forming an active ecosystem."*

Reluctant Due to Opportunism

Interviewees also expressed their concerns regarding the use of blockchain technology in information sharing. A manager from company E mentioned that some of their clients lack trust in the technology, even though they have heard about blockchain as a prominent information technology with excellent performance in data security.

> *"...but now we may have to integrate it with the Internet of Things, and then customers' information will be more exposed throughout the entire supply chain."*

This concern arises partly because the implementation of blockchain in the supply chain is still at a pilot or initial stage, and worries about opportunistic behaviour are often particularly emphasized by stakeholders in the early stages of many technologies' development. For blockchain, from the perspective of the enterprises implementing this technology, they indeed share or expose more information on the platform, even though they have received assurances that this information will only be made public to the supply chain partners they have agreed to share with. However, many businesses express concerns and hesitation about the potential risks associated with transitioning to new technology.

It is evident, therefore, that although blockchain is expected to be a tool to limit opportunistic behaviour, and one of the motivations for some companies to deploy blockchain technology is to eliminate such behaviour, concerns related to information leakage and speculative use of information, in turn, become obstacles to the deployment of blockchain technology.

Willingness Asymmetry

When considering the impact of blockchain on information sharing, the consistency of willingness to share among parties in the supply chain is also an indelible dimension. In case E, a manager mentioned that when the willingness to share information among supply chain partners is inconsistent, it could conversely have a negative impact on cooperation. Especially for the weaker parties in the supply chain, such as small and medium-sized enterprises (SMEs) or even micro-enterprises in upper-tier of supply chain, this asymmetry in willingness can enhance their insecurity in cooperation. Some SMEs believe that being exposed to information in front of more dominant enterprises makes them more passive in cooperation and even enhances the bargaining power of the already dominant enterprises.

4.2 Trust

Reduced Apprehensions Regarding Data Tampering

From our interviews, we discovered that the implementation of blockchain technology can enhance trust among supply chain partners by reducing concerns about data tampering. In traditional supply chains, data storage is often centralized at single or few control points, which might be managed by some organizations within the supply chain, thus posing risks of alteration or misuse [6]. In contrast, blockchain provides a decentralized information storage and sharing system, where each transaction is recorded across multiple nodes in the network, rather than in a single database. These records are protected by complex encryption algorithms; once a transaction is verified and added to the blockchain, it is permanently recorded and becomes nearly impossible to change. Altering any information would require consensus from the majority of nodes on the network, which is practically infeasible, especially in large and dispersed networks. This immutability significantly reduces the possibility of data being maliciously modified, as any unauthorized changes would be detected and rejected by other parts of the network.

Based on this property, case companies D and F have extended the implementation of blockchain to their supply chain finance operations. For example, in a project by company D, by generating digital certificates and using blockchain to encrypt share and consensus validation, the credit transmission channels between financial institutions, suppliers, and core enterprises can be connected, alleviating to some extent the financing difficulties faced by small and medium-sized enterprises.

Moreover, the action of recording itself can also enhance trust in collaboration. A technical expert from company D mentioned,

> *"...accountability based on records is the greatest significance of (blockchain) at present."*

When asked for further explanation, the interviewees added that because transactions recorded on the blockchain cannot be tampered with, this act of recording itself can also "build" trust, making individuals in the collaboration feel more secure. At the same time, this willingness to record can also demonstrate one's sincerity and enthusiasm in the collaboration to supply chain partners, as it equates to "volunteering for supervision."

Furthermore, smart contracts on the blockchain can automatically execute terms in business agreements, further ensuring the transparency and reliability of transactions, reducing trust issues caused by misunderstandings or opportunistic behaviour. Through smart contracts, the parties involved in a transaction can preset conditions, and once these conditions are met, the corresponding transactions will automatically be executed without the need for third-party verification.

Therefore, these features of blockchain—decentralized data management, immutable records, and the automatic execution of smart contracts—collectively reduce distrust factors in supply chain collaboration. It provides a more transparent and secure environment where every participant in the supply chain can see the complete transaction history and trust that this history is accurate. This enhanced trust helps to promote tighter and more efficient supply chain collaboration.

Increased Trust Among Existing Supply Chain Partners
In fact, among the vast majority of case companies (A, B, C, and E), blockchain technology is often used to strengthen existing cooperation rather than to support the establishment of new partnerships. As the CEO of company A mentioned,

"At the current stage, we are mainly focusing on post-trade applications."

This means that the implementation of blockchain in these case companies is primarily deployed in existing partnerships, i.e., there is first a collaboration intention (even collaborative behaviour), and then blockchain technology is deployed.

When asked about the reasons, the interviewee explained, "*This is due to the needs of our customers.*" A manager from company C also pointed out that their current business is more about utilizing the implementation of blockchain technology within existing collaborations to achieve supply chain transparency. After actual application, many interviewees noted that based on information sharing and transparency, trust in existing collaborative relationships can be further enhanced. Because a highly transparent supply chain can reduce the constant concerns about behavioural and environmental uncertainties in cooperation.

In the supply chain, trust is one of the basic requirements for long-term cooperative relationships, as it often encompasses the trust that both sides have in obtaining additional benefits within the partnership [29]. The implementation of blockchain technology in already established relationships can enhance trust, thereby making long-term cooperative relationships tighter.

Scepticism Among Stakeholders
As an emerging technology, although the potential of blockchain implementation is anticipated by many researchers and practitioners [30], many interviewees still expressed their concerns or those from their supply chain partners when actually deploying the technology. For instance, an employee from company D noted that some small and medium-sized enterprises (SMEs) in the supply chain remain sceptical about the technology because they do not understand blockchain on a technical level, and thus cannot trust the claimed superior features of the technology.

Additionally, multiple interviewees expressed concerns about the possibility of opportunistic behaviour during the transition from traditional supply chains to those based on blockchain technology. For example, an interviewee from company E mentioned that some of their clients, upon hearing about blockchain projects, first react with scepticism, wondering if it is another scam, "like the once-popular P2P financing," as they described. In such scenarios, it is challenging to advance blockchain projects, as deploying them appears to some supply chain partners as opportunistic behaviour itself, raising concerns that introducing blockchain technology could lead to a loss of their own interests.

However, some interviewees hold a positive outlook. For example, a senior manager from case company C mentioned that although at the current stage many supply chain partners are only willing to try deploying blockchain in very basic applications, such as logistics traceability, they believe that in the next five years, blockchain applications will become "common sense" and more heavily relied upon by clients.

"Once they get used to the technology and the conveniences it brings, they'll be open to other applications, like smart contracts and automatic execution."

4.3 Commitment

Joint Investment in Relationship-Specific Assets

When asked about the impact of blockchain on commitment in SCC, some interviewees highlighted it as a form of "joint investment," emphasizing its significance for commitment in collaboration. Relational Specific Assets are defined as assets that are created or acquired in a specific business relationship, holding special value for the parties involved in that relationship but potentially less valuable in other uses or relationships [31]. A technical expert from company B mentioned,

"When we and our partners decide to use blockchain, it's essentially an investment in this specific technological asset together."

In fact, many interviewees noted that this investment includes not only financial expenditure but also the investment of time and resources, as well as adaptation and adjustments in technology and processes. From the perspective of Transaction Cost Economics (TCE), it will significantly increase transaction costs. An employee from company A's marketing department used the word "sincerity" to describe this joint investment in specific assets,

"...to show each other sincerity, collaborating in this project is an important step..."

Therefore, it is evident that the deployment of blockchain implementation is driven not merely by its technological attributes or superior performance. The act of joint investment itself also represents a confidence in the success and durability of the collaboration. By investing in relational specific assets, supply chain partners demonstrate to each other their valuation of the collaborative relationship and their trust for the future of relationship. This expression of trust aids in strengthening the willingness to collaborate, establishing more robust partnership ties. Since these investments are often made specifically for the collaborative relationship, they constitute a form of commitment that is difficult to transfer, indicating that both parties are interested in and positively estimate the long-term maintenance of the partnership.

Perceived Coercive Commitment

Although joint implementation of blockchain is often considered by researchers and practitioners as a commitment to supply chain collaboration, discussions on whether such commitment is voluntary have been almost entirely overlooked. In our survey, some interviewees mentioned the concepts of "forced commitment" or "coerced commitment". For instance, an interviewee in the operations department from company E noted,

"For groups that are in a weaker position within the supply chain, they feel that the dominant parties are forcing them to commit and join."

Similarly, an employee from company D mentioned,

"When dealing with small and micro enterprises upstream in the supply chain (to facilitate and promote the implementation of blockchain technology), sometimes resistance is encountered... They feel coerced by the core companies. They believe that core companies are trying to facilitate blockchain (project implementation) for their own benefit."

In discussions related to commitment, the conditions of 'strong' and 'weak' groups are often distinctively discussed by the interviewees. When asked what constitutes a "weak group" in the supply chain, interviewees typically described such roles using terms like smaller scale (e.g., small and medium-sized enterprises or micro-enterprises) or not dominant in negotiations. Indeed, most interviewees viewed the deployment of blockchain technology as a form of commitment among partners. However, there were varying degrees of concern from interviewees with different backgrounds or professions regarding the willingness of different groups to make such commitments.

A marketing manager from company F used the term "big picture" when describing their blockchain project,

"Although there will definitely be some resistance in the process, it's imperative for all enterprises along the supply chain to jointly commit to using (blockchain), and it reflects our big picture thinking and long-term mindset."

Similarly, the general manager of company E mentioned the holistic considerations of core companies in implementing blockchain technology,

"...(Strong) companies are pushing this (blockchain technology) not just out of self-interest but for the benefit of the entire supply chain and even extending to the industry chain."

Power is often considered one of the antecedents of commitment behaviour in the supply chain, yet discussions on how it interacts with the concept of willingness to commit remain scant [32]. From our interviews, it is evident that in the context of power asymmetry, more dominant enterprises in the collaboration often believe their commitment to, or push for, partners' commitment to deploying blockchain technology benefits the entire supply chain; however, for enterprises in weaker positions within the supply chain, such commitment is sometimes made under pressure from core companies or customers.

5 Discussion and Directions for Future Research

Through multi-case analysis and consideration of the three dimensions of SCC, we have identified the impact of blockchain technology implementation on collaboration. As an emerging decentralized technology, blockchain was considered by our interviewees to significantly enhance the trustworthiness of information shared within the supply chain and reduce concerns about data tampering and double-spending issues in financing processes.

However, like any emerging information technology, the adoption of blockchain in supply chains has initially experienced confusion and challenges. Many interviewees expressed concerns about potential opportunistic behaviours during the transformation process, echoing the literature that blockchain does not guarantee absolute transaction security. Vulnerabilities in smart contracts could potentially have devastating consequences for ongoing projects [33]. Additionally, during the digital transformation process driven by blockchain, doubts about whether all supply chain partners have equal willingness to share information and make commitments were also raised by interviewees. As indicated in a study [34], these enforced commitments might be driven by power relationships.

Therefore, future research could delve deeper into how to overcome potential technical challenges in the implementation process of blockchain technology and avoid introducing new information security issues. How to protect weaker parties in supply chain collaboration and prevent them from falling into extremely passive situations is also worth further consideration by researchers and practitioners. This approach will not only address the technological and strategic challenges but also foster a more equitable and transparent environment within the digital transformations enabled by blockchain in supply chain.

6 Conclusion

This study explores the implementation of blockchain technology in six case companies and its impact on supply chain collaboration and its key dimensions: information sharing, trust, and commitment. The findings indicate that this technology enhances information sharing and trust within SCC by facilitating the digitalization of key documents and ensuring their secure and cost-effective transmission. Moreover, the adoption of blockchain technology is also viewed by partners as a commitment to maintaining long-term relationships. However, concerns about data privacy, readiness for technology integration, and the asymmetric willingness to adopt new technology during the deployment of blockchain cannot be overlooked, especially for upstream enterprises in a passive position within the supply chain. Considering its crucial role in digital transformation, the deployment of blockchain in supply chains is set to become more widespread. Protecting the interests of upstream suppliers and maintaining a balance of power within cooperative relationships will be a focal point for future researchers and practitioners.

This research is significant both theoretically and managerially. From a theoretical perspective, it explores how SCC is affected during the process of deploying blockchain, initially filling a gap in this area of research and providing direction and guidance for subsequent studies under this topic. From managerial perspectives, the findings reveal the impacts of deploying blockchain technology on collaboration, offering insights for managers in relevant fields and serving as a reference for practitioners who are in the process of deploying or preparing to deploy blockchain technology.

Disclosure of Interests. The authors declare that they have no known competing financial interests or personal relationships that could have appeared to influence the work reported in this paper.

References

1. Kshetri, N.: Blockchain's roles in meeting key supply chain management objectives. Int. J. Inf. Manage. **39**, 80–89 (2018)
2. Schmidt, C.G., Wagner, S.M.: Blockchain and supply chain relations: a transaction cost theory perspective. J. Purchasing Supply Manage. **25**, 100552 (2019)
3. Brookbanks, M., Parry, G.: The impact of a blockchain platform on trust in established relationships: a case study of wine supply chains. Supply Chain Manage. **27** (2022)
4. Rejeb, A., Keogh, J.G., Simske, S.J., Stafford, T., Treiblmaier, H.: Potentials of blockchain technologies for supply chain collaboration: a conceptual framework. Int. J. Logist. Manage. **32** (2021)
5. Qian, X. (Alice), Papadonikolaki, E.: Shifting trust in construction supply chains through blockchain technology. Eng. Construct. Archit. Manage. **28** (2021)
6. Cole, R., Stevenson, M., Aitken, J.: Blockchain technology: implications for operations and supply chain management. Supply Chain Manage. **24** (2019)
7. Treiblmaier, H.: The impact of the blockchain on the supply chain: a theory-based research framework and a call for action. Supply Chain Manage. **23** (2018)
8. Narayanan, A., Bonneau, J., Felten, E., Miller, A., Goldfeder, S., Clark, J.: Bitcoin and Cryptocurrency Technologies Introduction to the book (2016)
9. Agrawal, T.K., Kumar, V., Pal, R., Wang, L., Chen, Y.: Blockchain-based framework for supply chain traceability: A case example of textile and clothing industry. Comput. Ind. Eng. **154**, 107130 (2021)
10. Rauniyar, K., Wu, X., Gupta, S., Modgil, S., Lopes de Sousa Jabbour, A.B.: Risk management of supply chains in the digital transformation era: contribution and challenges of blockchain technology. Ind. Manage. Data Syst. (2022)
11. Bai, Y., Liu, Y., Yeo, W.M.: Supply chain finance: what are the challenges in the adoption of blockchain technology? J. Digit. Econ. **1** (2022)
12. Min, S., et al.: Supply chain collaboration: What's happening? Int. J. Logist. Manage. **16**, 237–256 (2005)
13. Whipple, J.M., Lynch, D.F., Nyaga, G.N.: A buyer's perspective on collaborative versus transactional relationships. Ind. Market. Manage. **39**, 507–518 (2010)
14. Simatupang, T.M., Sridharan, R.: Design for supply chain collaboration. Bus. Process Manage. J. **14**, 401–418 (2008)
15. Kocoglu, I., Imamoglu, S.Z., Ince, H.: Inter-organizational Relationships in Enhancing Information Sharing: The Role of Trust and Commitment. Bus. Rev. Cambridge. **18** (2011)
16. Madhavika, N., Jayasinghe, N., Ehalapitiya, S., Wickramage, T., Fernando, D., Jayasinghe, V.: Operationalizing resilience through collaboration: the case of Sri Lankan tea supply chain during Covid-19. Qual. Quant. **57**, 2981–3018 (2023). https://doi.org/10.1007/s11135-022-01493-8

17. Korepin, V., Dzenzeliuk, N., Seryshev, R., Rogulin, R.: Improving supply chain reliability with blockchain technology. Maritime Econ. Logist. (2021)
18. Wang, Z., Zheng, Z., Jiang, W., Tang, S.: Blockchain-enabled data sharing in supply chains: model, operationalization, and tutorial. Prod. Oper. Manag. **30**(5) (2021)
19. Dwivedi, S.K., Amin, R., Vollala, S.: Blockchain based secured information sharing protocol in supply chain management system with key distribution mechanism. J. Inf. Secur. Appl. **54**, 102554 (2020)
20. Youn, S.H., Hong, P., Nahm, A.: Supply chain partnerships and supply chain integration: the mediating role of information quality and sharing. Int. J. Logist. Syst. Manage. **4**(4) (2008)
21. Sheu, C., Yen, H.J.R., Chae, B.: Determinants of supplier-retailer collaboration: evidence from an international study. Int. J. Oper. Prod. Manage. **26**, 24–49 (2006)
22. Zaheer, N., Trkman, P.: An information sharing theory perspective on willingness to share information in supply chains (2017)
23. Fawcett, S.E., Osterhaus, P., Magnan, G.M., Brau, J.C., McCarter, M.W.: Information sharing and supply chain performance: the role of connectivity and willingness. Supply Chain Manage. **12**, 358–368 (2007)
24. Panayides, P.M., Venus Lun, Y.H.: The impact of trust on innovativeness and supply chain performance. Int. J. Prod. Econ. **122**, 35–46 (2009)
25. Wu, N.C., Nystrom, M.A., Lin, T.R., Yu, H.C.: Challenges to global RFID adoption. Technovation **26**, 1317–1323 (2006)
26. Narayandas, D., Rangan, V.K.: Building and sustaining buyer-seller relationships in mature industrial markets. J. Mark. **68**, 63–77 (2004)
27. Afshan, N., Motwani, J.: An investigation of antecedents and consequences of supplier integration: a study in Indian context. Measur. Bus. Excell. **25**, 138–151 (2020)
28. Krause, D.R., Handfield, R.B., Tyler, B.B.: The relationships between supplier development, commitment, social capital accumulation and performance improvement. J. Oper. Manage. **25** (2007)
29. Sahay, B.S.: Understanding trust in supply chain relationships. Ind. Manage. Data Syst. **103**, 553–563 (2003)
30. Queiroz, M.M., Telles, R., Bonilla, S.H.: Blockchain and supply chain management integration: a systematic review of the literature (2020)
31. Sambasivan, M., Siew-Phaik, L., Abidin Mohamed, Z., Leong, Y.C.: Factors influencing strategic alliance outcomes in a manufacturing supply chain: role of alliance motives, interdependence, asset specificity and relational capital. Int. J. Prod. Econ. **141**, 339–359 (2013)
32. Paluri, R.A., Mishal, A.: Trust and commitment in supply chain management: a systematic review of literature. Benchmarking **27** (2020)
33. Zhao, X., Chen, Z., Chen, X., Wang, Y., Tang, C.: The DAO attack paradoxes in propositional logic. In: 2017 4th International Conference on Systems and Informatics, ICSAI 2017 (2017)
34. Zhao, X., Huo, B., Flynn, B.B., Yeung, J.H.Y.: The impact of power and relationship commitment on the integration between manufacturers and customers in a supply chain. J. Oper. Manage. **26**, 368–388 (2008)

An Overview of Cloud-Based Services for Smart Production Plants

Auday Al-Dulaimy[1,2](✉), Leo Hatvani[1], Moris Behnam[1], Anas Fattouh[1], and Koteshwar Chirumalla[1]

[1] Mälardalen University, Västerås, Sweden
{auday.aldulaimy,leo.hatvani,moris.behnam, anas.fattouh,koteshwar.chirumalla}@mdu.se
[2] Dalarna University, Falun, Sweden
aay@du.se

Abstract. Cloud computing is a game-changer model that opens new directions for modern manufacturing. It enables services and solutions that help improve the productivity and efficiency of smart production plants. The main objective of the paper is to provide a summary of the various cloud-based manufacturing services currently being offered to manufacturers or that could be offered in the future. Additionally, the paper aims to discuss the various enabling technologies used to support the integration of cloud manufacturing in the manufacturing industry. Furthermore, the paper categorizes the different services based on their functionalities and maps them to four levels of production such as plant level, production line level, machine level, and process level. The categorization of services and mapping them to appropriate levels in production can enhance efficiency and productivity in the manufacturing industry. The study advances the discussion on cloud-based manufacturing from the types of services and enabling technologies perspective.

Keywords: Cloud computing · manufacturing · cloud manufacturing services · digital transformation · digital servitization

1 Introduction

In manufacturing, the production system refers to the operations and resources that are set in different layers to facilitate the production of specific product(s). To improve productivity, such industrial systems need to be shaped in a model that can utilize all manufacturing services centrally. This model is called Cloud Manufacturing (CMfg). CMfg expands the principles of cloud computing (Everything-as-a-Service, XaaS) to cover the manufacturing environment, resulting in new types of services called Manufacturing-as-a-Service (MaaS) [2]. In this way, CMfg encapsulates manufacturing resources, network capacity, and Internet of Things (IoT) features into services, and then offers these services to the manufacturers upon request so that they can improve the productions [23].

Published by Springer Nature Switzerland AG 2024
M. Thürer et al. (Eds.): APMS 2024, IFIP AICT 733, pp. 461–475, 2024.
https://doi.org/10.1007/978-3-031-71645-4_31

By using these services, manufacturers can avoid the complications related to infrastructure management, software updates, and security concerns. This results in secure, cost-efficient, reliable, and on-demand manufacturing services for the whole life cycle of manufacturing. Manufacturers can redirect their resources that were previously allocated to maintaining on-premises servers toward other tasks like innovation, strategic initiatives, and core competencies. Moreover, cloud computing enables manufacturers to scale their operations up or down quickly, depending on market demands, without incurring significant capital expenditures.

Cloud services, along with other Industry 4.0 technologies and concepts, are the keys to digital transformation in manufacturing, which is currently unlocking new possibilities in terms of automation, process improvement, and sustainability [3]. To reach their full potential, these technologies need to be added to the manufacturing facility while keeping in mind very important aspects such as maintainability, cybersecurity, interoperability, etc.

The paper aims to give an overview of the cloud-based manufacturing services offered, or that can be offered, to manufacturers to support the digital transformation, and also epitomize the enabling technologies that support cloud manufacturing integration. Moreover, the paper categorizes the services and maps them to different levels in manufacturing production. The main contributions of this paper are providing answers to the following research questions:

RQ1: What are cloud-based manufacturing services, and what are their main characteristics?

RQ2: What are the concepts and technologies that are discussed in the literature to enable cloud-based services in manufacturing?

RQ3: Which cloud-based services, discussed in the literature, can be used across the hierarchic levels of production systems?

The rest of this paper is organized as follows: Sect. 2 gives a background about the levels of production systems. Section 3 defines cloud-based manufacturing services and discusses their characteristics. Section 4 summarizes the existing cloud-based manufacturing services. Section 5 lists the concepts and technologies that enable cloud services in the manufacturing industry. In Sect. 6, we discuss and show how cloud services can be utilized at the levels of production systems. Section 7 concludes the paper.

2 Hierarchic Levels of a Production System

In this section, we discuss the hierarchic levels of a production system. Production can be categorized into levels that reflect manufacturing operations based on increasing levels of detail [25]. These levels are: (1) Plant Level, which refers to the overall production capability and performance of an entire manufacturing facility, (2) Production Line Level, which focuses on a specific production line within the plant, (3) Machine Level, which refers to the machines that form a production line, and (4) Process Level, which is the most granular level that focuses on optimizing each process for efficiency, quality, and minimizing waste.

Categorizing the production into these four levels enables manufacturers to understand their operations at different scales, helping them identify bottlenecks, allocate resources efficiently, and make data-driven decisions to improve production. However, the cloud computing model offers services that help improve efficiency at each level.

3 Cloud-Based Manufacturing Services

Cloud computing model provides manufacturers with efficient, flexible, and innovative services that improve competitiveness. It offers secure, scalable, and cost-effective Infrastructure, Platform, and Software Services. Through Infrastructure as a Service (IaaS), manufacturers can scale their computing resources on demand, avoiding the need for infrastructure investments. With Platform as a Service (PaaS), manufacturers can facilitate custom application development and accelerate the deployment of manufacturing-specific software solutions. Using Software as a Service (SaaS), manufacturers can access cutting-edge applications for different tasks. In this section, we define cloud-based manufacturing services, discuss their characteristics, and summarize the widely used existing ones.

3.1 Definition

There is no standard and comprehensive definition for cloud-based manufacturing services in the literature, so in this paper, we define them as ***"a range of interoperable services provided by cloud service providers that encapsulate infrastructure resources, platforms, software, network capacity, and IT capabilities. These services are offered to manufacturers on demand and enable them to scale their manufacturing needs flexibly, improve production processes and product designs, and simplify factory automation"***. This definition helps manufacturers, managers, and engineers understand the functionalities of the services to be mapped into a specific level of production, which results in enhancing productivity.

3.2 Characteristics

The main characteristics of cloud manufacturing services are:

- ***Diverse:*** Manufacturers require diverse manufacturing services to efficiently provide smart solutions based on their product types, production volumes, and market demands. Such services utilize a wide range of resources, concepts, and technologies.
- ***On-demand:*** Manufacturers can get cloud manufacturing services, possibly automatically, as needed from the different service providers.
- ***Customer-Centric:*** Manufacturers' needs are at the center of cloud manufacturing services' design and operations. Many cloud service providers work closely with the manufacturers to understand their specific needs, which helps in designing and offering customized manufacturing services.

- ***Flexible and Scalable:*** The cloud resources needed to run manufacturing services are elastically provisioned and released to scale in/out (possibly up/down) rapidly based on manufacturers' demands. Cloud resources are available for provisioning to manufacturers at any time, with no restrictions on the quantity. Manufacturing services need cloud resources that can be provisioned and released elastically to meet the manufacturers' demands. Manufacturers can scale in or out (possibly up or down) rapidly as required. There are no restrictions on the quantity of cloud resources that can be provisioned/released, and they are available at any time.
- ***Continuous Deployment:*** With modern cybersecurity considerations and agile software development models, easy software updates become a paramount issue. Cloud-based software services benefit from continuous deployment and continuous development. This, in turn, reduces the initial software cost, increases reliability, and reduces downtime.
- ***Interoperable:*** Cloud-based software services commonly share the platform that they are running on, thus making interoperability a matter of choice rather than an extensive technical investment. Within a single cloud platform, interoperability can be reliable, efficient, and secure. Even across different cloud platforms, interoperability can be reached more easily than in other situations using advanced networking architectures, high-grade encryption, and standardized protocols.

4 Categories of Cloud Manufacturing Services

Cloud computing services can be utilized in different areas of manufacturing processes and operations, most of which are summarized in this section.

4.1 The Methodology of Extracting Cloud Manufacturing Services

The cloud manufacturing services summarized in this paper are extracted from scientific publications including journal articles, conference papers, book chapters, reviews, and surveys. The services are extracted after following these steps:

- Retrieve the list of publications that includes all existing and potential cloud manufacturing services from the Web of Science database. An advanced document search with specific query keywords/terms is employed with these keywords/terms: *"cloud"*, *"manufacturing"*, *"services"*, and *"digital transformation"*. The retrieved list results in a total of 54 publications.
- Filter non-related publications from the list by reading the abstract, introduction, and conclusion. We removed irrelevant publications that did not discuss the cloud-based manufacturing services or their enabling technologies from further review. In total, 40 publications are selected after initial review.
- Read the entire content of the remaining relevant publications.
- Extract the service(s) presented in the resulting relevant publications.

4.2 Categories

Based on the services presented in the resulting relevant publications, the diverse services for the manufacturing industry are categorized and discussed in this section (refer to Fig. 1). However, these categories may not be mutually exclusive, as many manufacturing services overlap in their functions. Also, it is worth mentioning that the services presented here are mainly utilized in the production plants (i.e., level of production), but however, some of them can be used within and outside the plants (e.g., supply chain services).

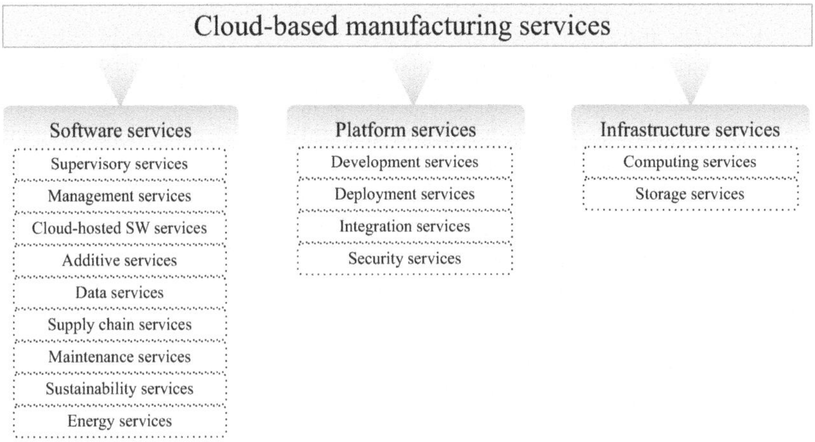

Fig. 1. Categories of cloud-based manufacturing services.

1. **Software services**: This category includes:
   - *Supervisory services:*
     - Monitoring services: Monitoring is a process that entails the systematic collection of data from various manufacturers, followed by the analysis of this data using cloud-based resources. This analysis aims to evaluate the information collected and subsequently oversee the condition of manufacturing processes, operations, or systems. The data collected can include anything from performance metrics to environmental factors, and the analysis itself can range from simple statistical summaries to complex learning algorithms. In general, the primary objective of monitoring different entities is to ensure that manufacturing processes are functioning effectively, efficiently, and safely. An example of monitoring services is Machine Downtime Monitor on AWS (MDM) [1].
     - Quality control services: These services enable real-time monitoring of sensor and IoT device data on the production floor, helping quality control teams promptly identify and address deviations from quality

standards. Also, a cloud-based quality control service provides centralized quality-related data such as product specifications, inspection results, and compliance documentation. Using such services, different quality control teams can collaborate seamlessly in a shared environment using cloud platforms to access, review, and update quality-related information. This ensures that all relevant manufacturers have access to the most up-to-date information to maintain consistency in quality control. Additionally, these services analyze historical data to predict potential quality issues. So, quality control teams can implement preventive measures and continuously improve processes by identifying patterns and trends in production data.
  - Safety and risk assessment services: These services provide risk assessments using different assessment techniques (for example via quantitative or qualitative analysis approaches) to diagnose and formulate improvement actions aiming at mitigating potential risks associated with specific activities [13].
  - Anomaly detection: Consistent functioning of manufacturing plant results in consistent data with predictable patterns. Thus, the use of industry grade and secure pattern detection algorithm enables anomaly detection [22] that is often correlated with the current or upcoming anomalies in the manufacturing process. Anomaly detection as a service has the advantages of collating multiple datasets, using cutting edge machine learning hardware and the latest algorithms.
- *Management Services:* These services help organize and plan the entire production process to maximize productivity, minimize costs, and ensure the production of high quality and meet customer demand. Cloud providers offer on-demand management services and tools to manage production resources in manufacturing systems. Mainly, this category includes:
  - Controlling: Such services allow production managers to access the management systems remotely to monitor and control their manufacturing processes from any location, providing greater flexibility and agility in responding to changing circumstances. It is a significant benefit for companies that have multiple production facilities or global operations.
  - Planning: Cloud-based planning services assist production managers in organizing and optimizing the entire production process to maximize productivity, minimize costs, and ensure high-quality output. These services encompass tasks such as production scheduling, inventory management, maintenance planning, logistics planning, and resource allocation. By harnessing cloud-based planning tools, production managers can develop optimized production schedules, manage inventory levels effectively, schedule maintenance activities proactively, optimize logistics operations, and allocate resources efficiently. Cloud-based planning services offer greater flexibility and agility,

allowing production managers to adapt quickly to changing market demands and operational conditions [17].

- *Cloud-hosted software services:*
  - Cloud-based Manufacturing Execution Systems (MES): Cloud-based MES software can be delivered as SaaS in public, private, or hybrid clouds. It is specifically designed to run in cloud data centers, aiming at providing real-time monitoring, scheduling, and execution capabilities to enhance visibility and control over the entire manufacturing process. For example: MES virtualization (vMES) by migration of MES workloads that were traditionally executed locally on physical machines to a cloud data center [19].
  - Cloud-based Enterprise Resource Planning (ERP): Cloud-based ERP systems provide a central platform for managing all aspects of a manufacturing business, including finance, inventory, human resources, and customer relationships. This fosters improved collaboration and data sharing across different departments.
- *Additive Services:* The cloud-based additive manufacturing services make it easier for manufacturers to access 3D printing resources by sending their designs to cloud service providers via the Internet, and the printed parts will be delivered to them (or to their customers). Moreover, some applications are designed to perform some tasks for the manufacturers, for example: applications for 3D printing job scheduling. With such services, manufacturers will not invest in expensive printing machines, and in training users who only need printing occasionally [27].
- *Data Services:* Big data in the manufacturing industry refers to the huge amount of data generated by a variety of heterogeneous sources during different production stages. Cloud service providers offer services by utilizing such data, including:
  - Data analysis: Analysing data is very important in many industrial applications, including product lifecycle management, process re-design, supply chain management, and production systems data analysis. In smart manufacturing, data analysis supports data-driven decision-making [14].
  - Data integration: In modern manufacturing, a variety of data is generated from operations at different levels of production. For example [6]: Manufacturing equipment can provide data that describes its operating state, which can be used to drive improvements and increase production output. Additionally, a Manufacturing Execution System (MES) can be used to collect data from production personnel to monitor process output and operator productivity. Data integration provides a unified view of all the data, and it is offered as a service that ensures the consistency and accuracy of data generated from different systems. The service also ensures that manufacturers have real-time access to the unified data [21].
- *Supply Chain Services:* In manufacturing, cloud computing can greatly benefit supply chain management. Supply chain partners can collaborate

through cloud platforms to improve communication, increase traceability, and reduce lead times. In addition, the cloud can help in utilizing cloud-based analytic tools to analyze historical data and enhance demand forecasting accuracy. According to Oracle, modern manufacturing involves various supply chain activities, including procurement, managing the product lifecycle, supply chain planning, logistics, and order management. Such activities may also extend to global trade activities, such as managing global suppliers and multinational production processes. an example of Supply chain services is the Cloud- and Digitally-based Supply Chain Management (SCM) systems offered by Oracle [20].

- *Maintenance Services:* The maintenance services could be [4,5]:
  - Breakdown maintenance, which is a strategy that usually performs maintenance when failures occur.
  - Time-based maintenance, which as the name suggests, is a strategy that usually performs maintenance at specific times.
  - Predictive maintenance, which is a proactive strategy that usually uses data analysis to detect anomalies and predict equipment failures before they occur. Using different enabling technologies and concepts (e.g., IoT and AI), it monitors equipment conditions, analyzing data trends to forecast potential breakdowns. This approach enables manufacturers to schedule maintenance tasks more efficiently, minimizing downtime, extending the life of the equipment, and reducing maintenance costs. Predictive maintenance represents a shift from traditional schedule-based maintenance to a more data-driven, condition-based strategy, enhancing operational efficiency and productivity in manufacturing environments.
- *Sustainability Services:* The use of cloud computing can enhance the sustainable innovation capability of manufacturers as part of their digital transformation development [11].
- *Energy Services:* Energy management in manufacturing involves the strategic monitoring, control, and conservation of energy usage across operations. By leveraging IoT devices, cloud-based platforms, and artificial intelligence, it provides real-time insights into energy consumption patterns and identifies inefficiencies. This approach enables manufacturers to optimize their energy use, reduce waste, and reduce costs by implementing smarter energy practices, such as adjusting production schedules, optimizing equipment operation, and automating energy systems based on demand. Beyond cost savings, effective energy management contributes to sustainability goals by reducing the environmental impact of manufacturing processes, making it a critical component of modern and environmentally friendly manufacturing strategies.

2. **Platform Services:** The platform services integrate development and runtime environments, offering manufacturers a point of control for the development and utilization of cross-platform application components [12]. Manufacturers can design and deploy customized applications to facilitate their manufacturing processes using these services. This category includes:

- *Development Services:* Cloud-based services benefit from a number of services aimed at making development smoother and more efficient, thus providing the basis for continuous development. This includes a variety of services for prototyping, building, testing, and recently even hosting full development environments in the cloud.
- *Deployment Services:* Software developed in the cloud can rely on cloud-provided features that control the deployment processes and thus enable continuous deployment. Deployment can be fine-grain controlled, from deploying a specific current or previous version of the service to a full update of the service.
- *Integration Services:* An integrated factory floor architecture enables seamless communication and connection between automated and non-automated systems, creating new automation functionalities. For example: The platform provided by Crosser for factory floor integration [7].
- *Security Services:* These services cover four levels: (1) ensuring an organization's core IT resources security at the network, host, and application levels, (2) identity and access management to improve operational efficiency and compliance with privacy and data protection requirements, (3) data protection and security to evaluate data security scenarios and make informed judgments regarding risk, and (4) cloud security as a service to extract security functions and capacities to assemble software as a service, which includes web security service, storage security service, and Identity and Access Management (IAM) service [26].

3. **Infrastructure Services:** Infrastructure services provide the required computing power and storage capacity. These services can be offered in different forms, for example, virtual machines and storage for hosting manufacturing applications, databases, etc. Mainly, these services are:
   - *Computing services:* Computing services offer computing power on demand, allowing manufacturers to scale computational resources at any time, as needed.
   - *Storage Services:* Storage services offer storage capacity to store manufacturers' data offsite in a secure and reliable way. This is especially beneficial for manufacturers who need to store a considerable amount of data, at the same time, they can ensure the data is regularly backed up.

5 Enabling Concepts and Technologies

The services in Sect. 4 are enabled by various concepts and technologies (refer to Fig. 2), including:

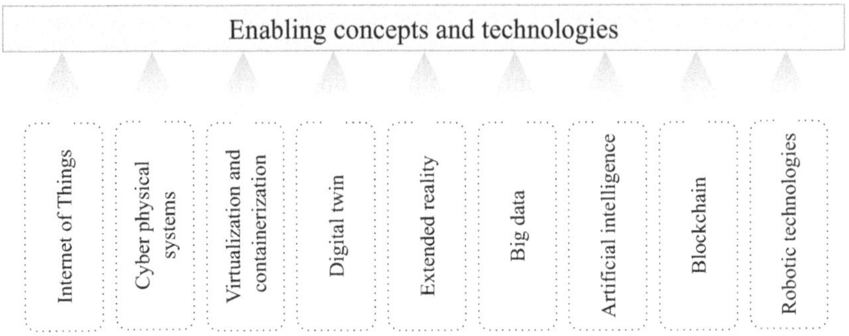

Fig. 2. Enabling concepts and technologies.

5.1 Internet of Things (IoT)

IoT is defined in [16] as: *"The Internet of Things (IoT) consists of physical entities (things) that were not necessarily intended for communication with each other and with the environment. In IoT, these things are able to identify themselves, communicate, and interact via computer networks, based on Internet technologies. They can act depending on external triggers or local logic"*. In the manufacturing industry, IoT plays a crucial role by connecting machines, devices, and systems to create smart and interconnected systems that facilitate efficient and effective production processes. This technology enables the real-time monitoring of machines and systems, helping manufacturers to identify potential problems and take necessary actions before they escalate into more significant issues.

5.2 Cyber Physical Systems (CPS)

In [16], CPS is defined as: *"CPS are systems consisting of tightly integrated physical and cyber components interconnected through one or more networks. The cyber components consist of computing and communication facilities (local or remote, e.g., embedded systems or cloud services) used for monitoring, automating and controlling physical systems and processes. CPS are normally based on complex feedback and control loops, where the physical components affect the cyber components and vice versa."*. Manufacturing systems have various operations and decision-making processes at all levels that need to be optimized. CPS utilizes fault-adaptive control systems and creates diagnostic models of the system's components and operations to achieve self-optimization [8].

5.3 Virtualization and Containerization

Virtualization is creating a virtual entity, like a Virtual Machine (VM), from a physical one to allow computer resource sharing, aiming at increasing resource utilization. Containerization, on the other hand, further enhances resource utilization by limiting the amount of computer resources needed to run different services. The container is a lightweight virtual entity compared to VM. Manufacturers have shown an increasing interest in the adoption of virtualization and containerization as these technologies offer benefits in terms of resource utilization, risk mitigation, cost efficiency, and return on investment [18,24].

5.4 Digital Twin

A "digital twin" is a virtual model replicating a physical object, system, or process in digital form. It is a realistic digital representation that closely mimics the real-world entity's qualities, functions, and efficiency. In the manufacturing industry, digital twins are increasingly utilized to simulate, monitor, and analyze physical assets and processes throughout their lifespan. This technology helps manufacturers optimize their operations, improve product quality, and reduce costs by identifying potential issues and testing solutions before implementing them in the real world.

5.5 Extended Reality Technologies

Extended Reality (XR) is the term that covers Augmented reality (AR), Virtual reality (VR), and Mixed reality (MR) technologies. These technologies can employ digital, cyber, virtual, and physical components to visualize complex procedures or operations in different areas of manufacturing. These technologies help simulate a manufacturing environment such that decision-makers, operators, and workers can find and rectify errors in the machines and consistently reduce human errors. This drives advances in maintenance efficiency [9].

5.6 Artificial Intelligence

Cloud computing infrastructure provides the storage and computing power required for the efficient use of intelligence. AI algorithms can be used for different applications in manufacturing, such as predictive maintenance and quality control. The intelligence can help manufacturers identify potential faults and prevent them from happening, ensure the quality of their products, and optimize their production processes to reduce costs. With the help of the cloud, manufacturers can access these technologies easily and at a lower cost compared with traditional solutions.

5.7 Big Data

In manufacturing, big data refers to the huge amount of data generated by various sources in the production environment. These sources include machine sensors, production processes, and other systems. Data analysis helps manufacturers improve productivity, reduce costs, and optimize their operations. In order to effectively use big data in the process of digital transformation of manufacturing, there are four stages that need to be followed [3]: (1) Gather data from a variety of sources, (2) Connect to create the knowledge context by extracting features and create metadata from diverse data sources, (3) Reason to make intelligent decisions by analyzing data in context and use context to exploit information, and (4) Adapt by considering the context, and by learning from history interaction patterns to design for the unexpected and predict service degradation and resource failures allowing for preventive maintenance.

5.8 Blockchain

In manufacturing, blockchain technology provides real-time transparency by storing data and making it available to all entities, and by ensuring that no changes can be made to the data without notifying the entities within the system [15]. Manufacturers are attracted to blockchain technology due to its ability to manage storing and accessing data in different processes and operations, like supply chain, security, and sustainability.

5.9 Robotic Technologies

In smart manufacturing systems, robotics are not likely to be preconfigured, but they need to cope with unexpected environmental and operational changes. Thus, advanced robotic-oriented technologies emerged to get intelligent and automatic levels of industrial robots. Examples of these technologies are [10]: Sensing, Motion planning, AI, and Human-inspired technology.

6 Mapping Services to the Production Levels

To optimize efficiency, manufacturers utilize various cloud-based services at the levels of production. In this section, the services obtained from the literature and discussed in Sect. 4 are mapped to different production levels within a factory that contributes mostly. In other words, we will specify the cloud manufacturing services that are needed at each level. Figure 3 illustrates the service-to-production level mapping.

As discussed in Sect. 2, the levels of production can be classified as *Plant Level*, *Production Line Level*, *Machine Level*, and *Process Level*. Utilizing different cloud-based services helps each level operate efficiently. Therefore, we map different services to each level as follows:

The cloud services that contribute to the Plant Level are Management, Cloud-hosted SW, Data, Supply chain, and Sustainability Services to ensure the smooth operation of the entire plant. Regarding Production Line Level, cloud services including Supervisory, Management, Cloud-hosted SW, Data, Sustainability, and Energy Services are mapped to this level to increase efficiency and maintain quality standards for production. For Machine Level, Data, Maintenance, Additive, Sustainability, and Energy Services enable real-time monitoring of machine performance to improve overall efficiency. Finally, the Process Level utilizes Data and Energy Services to enhance different processes at this level so that production and energy usage are managed efficiently. Looking at Fig. 3, certain services might fit multiple levels, and some fit only one specific level. For instance, Data services will contribute to all levels since data collected at all levels provide opportunities for optimization and smart decision making. In contrast, the Maintenance Service is only valid at the machine level. In general, using cloud-based services at specific production levels enables the adaptation to changes, resulting in a cost-effective, flexible, and responsive production process.

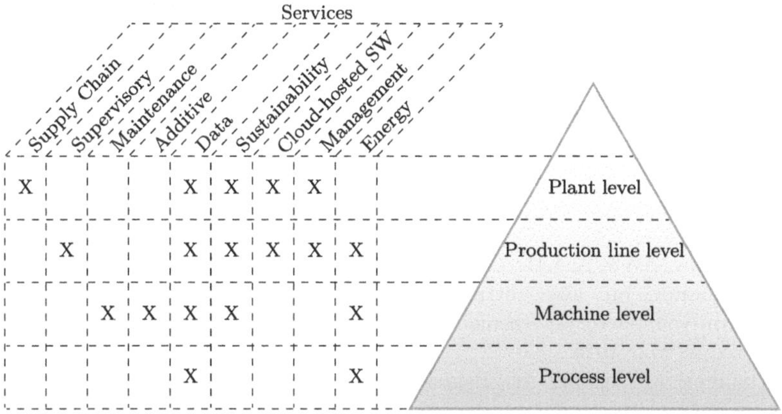

Fig. 3. Mapping of cloud-based services to the levels of a production system.

7 Conclusion and Future Work

The era of cloud computing has revolutionized manufacturing processes and operations. Cloud-based services can be utilized in various manufacturing areas. Manufacturers can access powerful tools, applications, platforms, and even the infrastructure of cloud service providers to facilitate collaboration, data analysis, decision-making, etc. In this paper, we discuss cloud-based manufacturing services and categorize them into three main categories: Software services (e.g., supervisory services), Platform services (e.g., deployments services), and Infrastructure services (e.g., computing services). Each category includes a range of services with different functionalities. Moreover, we discussed different concepts and technologies as enablers for cloud services. Then, we show how the services can be used across different stages of production systems. It is important to highlight that cloud-based services may overlap in their functions, even when they belong to different categories.

Paper Implications: The paper advances the understanding of modern manufacturing and presents a range of diverse services, aiming at helping manufacturers overcome or solve problems. Practically, the paper provides a simple framework for plant managers, automation engineers, and technology managers to identify and map relevant cloud-based services to their specific company or factory contexts, which results in enhancing productivity.

Future Studies: We (1) plan to make a detailed examination of the most feasible and potential edge-cloud manufacturing services that manufacturers can use in the future from multiple perspectives, such as technological, social, and organizational dimensions, (2) aim to show how enabling concepts and technologies can be utilized by specific services, and (3) target collaborating with different manufacturers via case study research to discuss services in different production levels and expand the content of this work based on the manufacturers' feedback.

Acknowledgments. This work was supported by the Excellence in Production Research (XPRES), a government funded Strategic Research Area (SRA) within manufacturing engineering in Sweden.

References

1. Ajay Swamy, AWS, A.A.B.: Digitally transform your factory with machine downtime monitor on aws. https://aws.amazon.com/blogs/architecture/digitally-transform-your-factory-with-machine-downtime-monitor-on-aws/ (2021). Accessed 26 Mar 2024
2. Al-Dulaimy, A., Ashjaei, M., Behnam, M., Nolte, T., Papadopoulos, A.V.: Fault tolerance in cloud manufacturing: an overview. In: International Conference on Mobile Computing, Applications, and Services, pp. 89–101. Springer (2022). https://doi.org/10.1007/978-3-031-31891-7_7
3. Borangiu, T., Trentesaux, D., Thomas, A., Leitão, P., Barata, J.: Digital transformation of manufacturing through cloud services and resource virtualization. Comput. Ind. **108**, 150–162 (2019)
4. Bousdekis, A., Mentzas, G.: Condition-based predictive maintenance in the frame of industry 4.0. In: Lödding, H., Riedel, R., Thoben, K.-D., von Cieminski, G., Kiritsis, D. (eds.) APMS 2017. IAICT, vol. 513, pp. 399–406. Springer, Cham (2017). https://doi.org/10.1007/978-3-319-66923-6_47
5. Bousdekis, A., Mentzas, G., Hribernik, K., Lewandowski, M., von Stietencron, M., Thoben, K.-D.: A unified architecture for proactive maintenance in manufacturing enterprises. In: Popplewell, K., Thoben, K.-D., Knothe, T., Poler, R. (eds.) Enterprise Interoperability VIII. PIC, vol. 9, pp. 307–317. Springer, Cham (2019). https://doi.org/10.1007/978-3-030-13693-2_26
6. Coronado, P.D.U., Lynn, R., Louhichi, W., Parto, M., Wescoat, E., Kurfess, T.: Part data integration in the shop floor digital twin: mobile and cloud technologies to enable a manufacturing execution system. J. Manuf. Syst. **48**, 25–33 (2018)
7. Crosser: Crosser factory floor integration. https://www.crosser.io/ (2024). Accessed 10 Apr 2024
8. Dafflon, B., Moalla, N., Ouzrout, Y.: The challenges, approaches, and used techniques of CPS for manufacturing in industry 4.0: a literature review. Int. J. Adv. Manuf. Technol. **113**, 2395–2412 (2021)
9. Dhanalakshmi, R., Mai, C.D., Latha, B., Vijayaraghavan, N.: AR and VR in manufacturing. In: Palanikumar, K., Natarajan, E., Sengottuvelu, R., Davim, J.P. (eds.) Futuristic Trends in Intelligent Manufacturing. MFMT, pp. 171–183. Springer, Cham (2021). https://doi.org/10.1007/978-3-030-70009-6_11

10. Du, Y., Wang, J., Wang, Z., Yu, F., Zheng, C.: Robotic manufacturing systems: a survey on technologies to improve the cognitive level in HRI. Procedia CIRP **107**, 1497–1502 (2022)
11. Fan, X., Wang, Y., Lu, X.: Digital transformation drives sustainable innovation capability improvement in manufacturing enterprises: based on FsQCA and NCA approaches. Sustainability **15**(1), 542 (2022)
12. Fuchs, J., Oks, S.J., Franke, J.: Platform-based service composition for manufacturing: a conceptualization. Procedia CIRP **81**, 541–546 (2019)
13. Gul, M., Lo, H.W., Yucesan, M.: Fermatean fuzzy topsis-based approach for occupational risk assessment in manufacturing. Complex Intell. Syst. **7**(5), 2635–2653 (2021)
14. Ismail, A., Truong, H.L., Kastner, W.: Manufacturing process data analysis pipelines: a requirements analysis and survey. J. Big Data **6**(1), 1–26 (2019)
15. Kasten, J.E.: Engineering and manufacturing on the blockchain: a systematic review. IEEE Eng. Manage. Rev. **48**(1), 31–47 (2020)
16. Lesch, V., Züfle, M., Bauer, A., Iffländer, L., Krupitzer, C., Kounev, S.: A literature review of IoT and CPS-what they are, and what they are not. J. Syst. Softw. **200**, 111631 (2023)
17. Liu, Y., Wang, L., Wang, X.V., Xu, X., Zhang, L.: Scheduling in cloud manufacturing: state-of-the-art and research challenges. Int. J. Prod. Res. **57**(15–16), 4854–4879 (2019)
18. Manufacturing.net: virtualization in manufacturing. https://www.manufacturing.net/operations/article/13240762/virtualization-in-manufacturing (2013). Accessed 16 Apr 2024
19. Morariu, O., Borangiu, T., Raileanu, S.: vMES: virtualization aware manufacturing execution system. Comput. Ind. **67**, 27–37 (2015)
20. Oracle: What is SCM (supply chain management)? https://www.oracle.com/se/scm/what-is-supply-chain-management/ (2024). Accessed 28 Mar 2024
21. Praxie: Data integration in manufacturing: unraveling the complexities and challenges. https://praxie.com/manufacturing-data-integration-challenges/ (2024). Accessed 18 Apr 2024
22. Stojanovic, L., Dinic, M., Stojanovic, N., Stojadinovic, A.: Big-data-driven anomaly detection in industry (4.0): an approach and a case study. In: 2016 IEEE International Conference on Big Data (Big Data), pp. 1647–1652 (2016). https://doi.org/10.1109/BigData.2016.7840777
23. Tao, F., Cheng, Y., Da Xu, L., Zhang, L., Li, B.H.: CCIoT-CMfg: cloud computing and internet of things-based cloud manufacturing service system. IEEE Trans. Industr. Inf. **10**(2), 1435–1442 (2014)
24. Technologies, A.: Virtualization benefits in manufacturing. https://appliedtech.pro/virtualization/ (2019). Accessed 16 Apr 2024
25. Viharos, Z.J., Monostori, L., Kemény, Z.: Automatic partitioning of problems through submodel decomposition a promising technique of digital enterprise technology. In: Digital Enterprise Technology: Perspectives and Future Challenges, pp. 165–173. Springer (2007). https://doi.org/10.1007/978-0-387-49864-5_19
26. Wang, Y., Ma, S., Ren, L.: A security framework for cloud manufacturing. In: International Manufacturing Science and Engineering Conference, vol. 45806, p. V001T04A022. American Society of Mechanical Engineers (2014)
27. Wang, Y., Zheng, P., Xu, X., Yang, H., Zou, J.: Production planning for cloud-based additive manufacturing-a computer vision-based approach. Robot. Comput.-Integr. Manuf. **58**, 145–157 (2019)

Sustainability Evaluation of Electronic Components: A Case Study of a Swedish Temperature-sensing Solutions Manufacturer

Trupti Borade[1](✉), Zuhara Zemke Chavez[1], and Emanuel Bengtsson[2]

[1] KTH Royal Institute of Technology, Stockholm, Sweden
borade@kth.se

[2] RISE Research Institutes of Sweden AB, Stockholm, Sweden

Abstract. The electronic industry faces increasing pressure to enhance product sustainability across the supply chain. Understanding the environmental impact of electronic devices is vital for informed decision-making and meeting customer demands for eco-friendly products among growing environmental concerns and regulatory pressures. This paper evaluates the sustainability impact of a Swedish electronic component manufacturer, TSS AB, specializing in temperature-sensing solutions, considering both the company's and their client's perspectives. Employing a life cycle assessment (LCA) and LCA software tools, we comprehensively analyze environmental impacts. Through a cradle-to-grave assessment, we evaluate production processes, reverse logistics, refurbishment, reuse, and disposal, identifying environmental hotspots for improvement. Additionally, we explore the influence of regulatory frameworks on incentivizing sustainable product development. Our findings contribute to the electronics sector's growing emphasis on sustainability, offering insights for companies aiming for net-zero emissions and circular product lifecycles.

Keywords: Sustainability Analysis · Electronic Devices · Life Cycle Assessment

1 Introduction

Climate change represents an unprecedented global challenge, impacting ecosystems, economies, and communities worldwide. Triggered by human activities such as the extensive burning of fossil fuels and rampant deforestation, there has been a marked increase in atmospheric greenhouse gas concentrations, which in turn has led to significant shifts in the Earth's climate. The Intergovernmental Panel on Climate Change emphasizes the urgent need for widespread emission reductions to limit global warming to 1.5 °C or 2 °C [1]. Collaborative action from governments, the private sector, and individuals is essential for effective mitigation and adaptation strategies to combat climate change. Failure to act decisively risks irreversible damage to climate systems, threatening a sustainable and equitable future.

Published by Springer Nature Switzerland AG 2024
M. Thürer et al. (Eds.): APMS 2024, IFIP AICT 733, pp. 476–491, 2024.
https://doi.org/10.1007/978-3-031-71645-4_32

Within this context, the electronics industry plays a crucial role as both a contributor to and a potential mitigator of environmental degradation. Despite its exponential growth driven by innovation and consumer demand, the industry faces significant environmental costs, including resource depletion, e-waste accumulation, and high energy consumption. According to the United Nations Environment Programme, global e-waste production is estimated at around 50 million tons annually, of which only a fraction is officially recycled [2]. This situation poses severe environmental and health risks, given the hazardous substances often contained in electronic waste [3].

The environmental impact of electronic components, such as temperature data loggers, requires thorough analysis given their contribution to greenhouse gas emissions and resource depletion throughout their lifecycle. Understanding these consequences is crucial for developing sustainable alternatives. However, despite increasing research on e-waste and electronics sustainability, specific knowledge gaps persist regarding the lifecycle impacts of niche electronic products like temperature data loggers. This underscores the necessity for targeted studies to fill these gaps in the existing literature.

Addressing these challenges, this paper explores the environmental impact of electronic components within the context of TSS AB, a small and medium sized enterprise (SME) specializing in temperature-sensing solutions. The adoption of a life cycle assessment (LCA) methodology in this study serves a crucial purpose in providing a systematic framework for evaluating the environmental implications of electronic products throughout their entire lifecycle. LCA enables not only the identification of environmental hotspots and areas for improvement but also facilitates the quantification of environmental impacts, allowing for informed decision-making. By examining the complete lifecycle from raw material extraction to end-of-life, LCA offers insights into the environmental consequences associated with each stage, thereby guiding strategies for sustainable product development and management. With the use of the OpenLCA software, this research enhances its analytical capabilities by efficiently constructing LCA models and integrating relevant databases, ensuring a comprehensive assessment of TSS AB's operations.

By employing LCA, this study aims to provide actionable insights that enable TSS AB to mitigate environmental impacts, optimize resource utilization, and foster a more sustainable approach to product lifecycle management. We aim to conduct a comprehensive cradle-to-grave assessment to evaluate the environmental impact of TSS AB's temperature data loggers, namely the TSS WebLogger II (WL II) and TempTracer BLE (TT BLE). Specifically, our objectives include investigating the efficiency of reverse logistics and product lifecycle extension, identifying opportunities for sustainability enhancement, and aligning our findings with broader sustainability goals such as the EU Sustainable Product Initiative. Through this research, we seek to provide electronic component manufacturers insights to foster a greener economy and contribute to the transition towards more sustainable industry practices.

1.1 Research Gaps and Objectives

Despite the increasing awareness and body of work on the environmental impacts of electronic waste, there is a paucity of detailed LCA studies focusing specifically on temperature data loggers. This gap in the literature underscores the necessity of our research. By conducting this LCA, we aim to fill these knowledge gaps, providing a detailed assessment of the environmental footprint of TSS AB's temperature data loggers. The objectives of this paper include:

- Conducting a comprehensive cradle-to-grave assessment to evaluate the environmental footprint of TSS AB's temperature data loggers, namely the WL II and TT BLE.
- Identifying critical hotspots in their lifecycle and proposing actionable strategies to mitigate adverse impacts.
- Investigating the efficiency of reverse logistics and product lifecycle extension.

1.2 Research Question

To better contextualize our study and facilitate interpretation, we propose the following research question: "How can the environmental impacts of temperature data loggers be minimized to enhance sustainability in the electronics industry?".

By employing LCA, this study aims to provide actionable insights that enable TSS AB to mitigate environmental impacts, optimize resource utilization, and foster a more sustainable approach to product lifecycle management. Through this research, we seek to provide electronic component manufacturers with insights to foster a greener economy and contribute to the transition towards more sustainable industry practices.

2 Theoretical Background

2.1 Electronic Components

The distinction of electronic components into active and passive segments reveals a broad spectrum of applications, from diminutive gadgets to substantial industrial machinery. While epitomizing technological progress, this diversity simultaneously propels environmental impacts, notably in resource depletion and e-waste proliferation. Innovative materials like graphene and carbon nanotubes symbolize the strides in enhancing electronic componentry's efficiency and sustainability [4]. However, the journey from raw materials to end-of-life, covered in the LCA, remains fraught with challenges, notably in emissions and hazardous waste.

Researchers such as Reyes et al. [5] explore recycling policies for end-of-life electronics in developing countries, highlighting that enhancing refurbishment and remanufacturing capabilities leads to increased profitability compared to merely focusing on collection. This approach aligns with the principles of Green Supply Chain Management and underscores the benefits of integrating eco-friendly practices throughout the product lifecycle [6]. Moreover, studies by Murphy [7] emphasize the importance of circular economy principles, such as material recycling and resource optimization, in reducing the environmental footprint of electronic products. However, gaps remain in how effectively these practices are implemented across different segments of the electronics industry, especially for components like temperature data loggers. Specific gaps include the lack of standardized practices for refurbishment, challenges in material recovery, and limited adoption of eco-design principles. Addressing these gaps through targeted LCAs can provide deeper insights and drive improvements in sustainability practices.

Printed Circuit Boards (PCBs) are crucial in electronics, providing connections between components across various applications. The research by Liu et al. [8] shows that paper-based PCBs offer a sustainable alternative to traditional materials, significantly reducing environmental impacts and supporting easier recycling processes, emphasizing the importance of sustainable end-of-life management and eco-friendly design in electronic manufacturing, which are key to reduce the environmental impacts of electronic products.

2.2 Temperature Data Loggers

A temperature data logger is a device designed for monitoring and recording temperature over time and is suitable for environments like refrigerators and ovens. They play a critical role in sectors such as life science, including pharmaceuticals and biotechnology, where precise temperature monitoring is essential for preserving the integrity and efficacy of sensitive products [9]. In these industries, temperature fluctuations can compromise the quality and effectiveness of medications, vaccines, and biological materials. Temperature data loggers provide continuous monitoring and recording of temperature variations, ensuring compliance with regulatory requirements and safeguarding the quality of temperature-sensitive products throughout storage and transportation.

2.3 Life Cycle Assessment Methodology

LCA as a tool provides a systematic approach for evaluating the environmental impacts associated with all stages of a product's lifecycle, including raw material extraction, manufacturing, distribution, use, and end-of-life [10]. OpenLCA, a widely-used software tool, supports the implementation of LCA studies by facilitating data collection, system analysis, and result calculation. The integration of databases like Ecoinvent enhances the depth and accuracy of the analysis. Previous studies, such as the one conducted by Andersen et al. [11], demonstrate the effectiveness of LCA in evaluating the environmental performance of consumer electronic devices, analyzing cradle-to-grave. In a parallel manner, the research conducted by Rigamonti et al. [12] underscores the significance of LCA in evaluating environmental impacts and directing eco-innovation initiatives toward fostering a sustainable circular economy through methodical assessment of various assumptions and models. While LCA offers valuable insights, challenges such as data availability and limitations in capturing all environmental impacts exist. Nonetheless, applying LCA enables companies to make informed decisions, optimize processes, and reduce their environmental footprint, contributing to overall sustainability objectives [13].

The LCA framework, described in ISO 14040:2006 and ISO 14044:2006 standards, comprises four key stages: goal and scope definition, inventory analysis, impact assessment, and interpretation. This systematic approach ensures rigorous analysis and meaningful insights into enhancing the sustainability of electronic devices.

ReCiPe 2016 is one of the scientific endpoint assessment method due to its scientific robustness and comprehensive coverage of 18 impact categories, including climate change and ecotoxicity, aligning with the EU Taxonomy's environmental objectives

[14]. This method quantifies greenhouse gas emissions alongside other critical impact categories highlighting the environmental footprint of the loggers.

3 Methodology

This study employs a case study approach within an SME manufacturer, TSS AB, utilizing LCA to evaluate the environmental impacts of their temperature data loggers systematically. The system boundary delineates the span from raw material acquisition to end-of-life treatment, considering phases such as raw material extraction, energy consumption, fabrication, distribution, utilization, and final disposal. Encapsulating potential environmental repercussions such as climate change, resource depletion, and toxicological impacts on health and ecosystems, the framework adheres to ISO 14040 and ISO 14044 standards, ensuring a structured and globally recognized approach to lifecycle assessments [10]. The study employs OpenLCA 1.11.0 software, integrated with the Ecoinvent 3.7.1 database, to facilitate the detailed sustainability analysis, providing access to a broad range of data covering various processes and materials. This approach enables a holistic assessment of the environmental footprints of the electronic devices in question, with an emphasis on identifying key environmental hotspots and opportunities for improvement, particularly focusing on energy consumption, climate impact, and waste generation.

4 Life Cycle Assessment of Temperature Loggers

4.1 Case Background

The case company, TSS AB is a Swedish SME, manufacturing temperature loggers. For this study the focus was on two of their products WL II and TT BLE. These products enhance pharmaceutical logistics by providing reliable temperature monitoring, ensuring product safety and regulatory compliance. Their advanced features and cloud-based analytics offer sophisticated temperature management in the supply chain. The WL II is a flexible, precise temperature monitoring device for pharmaceutical transportation, meeting the standards. It has USB connectivity, supports a wide temperature range (−30 °C to 70 °C), and offers a large memory capacity of 65,000 readings, ensuring IATA compliance [15]. The TT BLE is a wireless single-use logger that transmits data to the TSS Cloud for real-time monitoring across transport modes. It accommodates 50,000 data points and operates between −30 °C and 70 °C [16]. Both loggers provide large data storage, real-time monitoring, and regulatory compliance crucial for pharmaceuticals, but challenges like limited storage capacity and potential data loss highlight the need for robust data management strategies. The market for these loggers is global, with the UK and USA leading the list. Therefore, the paper further develops scenarios focused on these regions.

4.2 Life Cycle Assessment: Goal, Scope Definition and Scenario Development

The objective of this study is a Comparative LCA, evaluating the environmental impacts of two TSS products from a "cradle-to-grave" perspective, following ISO 14040/44 standards [10]. This method examines impacts from raw material acquisition to end-of-life, helping to pinpoint primary drivers of environmental impact across the lifecycle stages and guiding potential improvements.

The system boundaries include five key stages: raw material acquisition, manufacturing and assembly, distribution, usage, and end-of-life management. Processes are divided into foreground processes, directly managed by the company, and background processes, which lie outside direct control. This distinction helps focus our analysis on areas where TSS can most effectively reduce environmental impacts while acknowledging the broader context of the product's lifecycle.

The functional unit for this study is 'the use of a temperature logger for one complete mission,' which entails its delivery, utilization for temperature monitoring, and subsequent disposal or return for recalibration. The operational life of a logger, powered by a lithium battery and housed in a plastic casing, is primarily dictated by the durability of its printed circuit board assembly (PCBA), assumed to last approximately 50,000 h or 64 months based on typical usage conditions. This definition standardizes the assessment, allowing for the comparison of environmental impacts across different usage scenarios and lifecycle stages of the logger.

To assess the environmental footprint of TSS's temperature loggers two key scenarios were developed: the Britain Bubble Scenario and the Stateside Shuffle Scenario. These scenarios examine the impact of transporting loggers from Sweden to customer locations in the UK and the USA and explore outcomes based on single-use versus reuse after refurbishment.

The Britain Bubble Scenario

The Britain bubble scenario assesses the lifecycle within the UK (see Fig. 1), from manufacturing in Sweden to either end-of-life disposal or refurbishment for reuse. It assumes the loggers are shipped to the UK for use and subsequently transported 200 km to a disposal facility or alternatively back to Sweden for refurbishment.

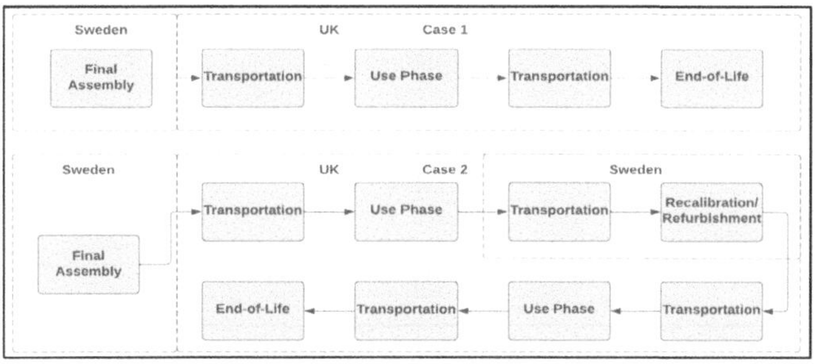

Fig. 1. Geographical boundary of the Britain Bubble Scenario Case 1 and Case 2

The Stateside Shuffle Scenario

The Stateside shuffle scenario focusing on the US, mirrors the Britain Bubble's structure but considers the longer transportation distances to the US (see Fig. 2). It evaluates the environmental impacts of manufacturing, delivery, usage, and end-of-life processes, including the potential for loggers to be returned to Sweden for recalibration and reuse.

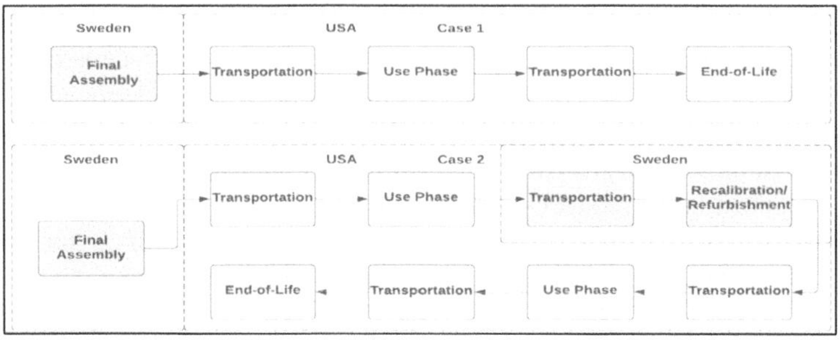

Fig. 2. Geographical boundary of the Stateside Shuffle Scenario Case 1 and Case 2

4.3 Life Cycle Inventory

The Life Cycle Inventory (LCI) section is pivotal in understanding the environmental implications of TSS's temperature data loggers, providing a foundation for our study's analysis. This inventory meticulously documents all relevant processes, inputs, and outputs across the products' lifecycle, highlighting the critical areas of energy consumption, material usage, and emissions.

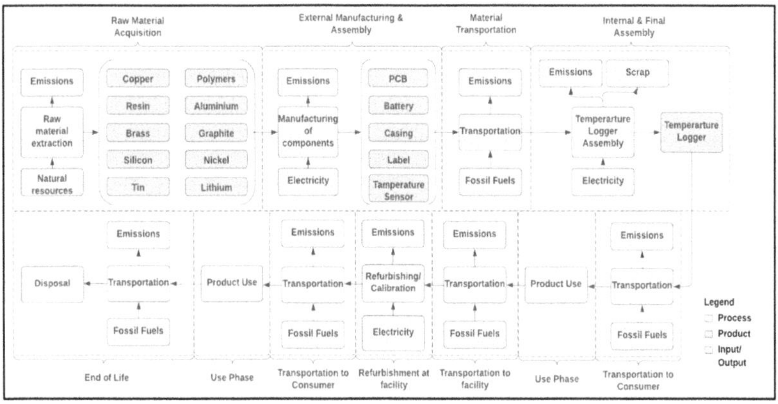

Fig. 3. Product System Overview

Figure 3 offers a schematic representation of the product system, marking the included processes and setting the boundaries of our study. This visual tool aids in comprehensively understanding the scope of our analysis. The inventory details the core components involved in the manufacture and assembly of both analyzed devices. This includes PCBAs, lithium batteries, casings, and labeling, each quantified to reflect their usage across different scenarios envisaged in the lifecycle assessment.

Table 1. Components of TSS products

Sr. No	Product	Process	Components	Weight (g)	Quantity for Case 1	Quantity for Case 2
1	WebLogger II	External Manufacturing (Input)	PCBA	12	1	1
			Lithium Battery	2,9	1	8
			Casing	21	1	8
			Label	0,01	1	16
		Internal Assembly (Output)	WL II	35	1	1
2	TempTracer BLE	External Manufacturing (Input)	PCBA	7	1	1
			Lithium Battery	3	1	16
			Casing	5	1	16
			Label	0,01	1	16
		Internal Assembly (Output)	TT BLE	15	1	1

Table 2. Lifecycle stage

Life cycle stage	Process	Location	Transport method	Distance in km
Cradle-to-gate	PCBA	Sweden	Truck	120
	Battery WB II	Indonesia	Air	8500
	Battery TT BLE	Japan	Air	8200
	Casing	Sweden	Truck	200
	Label	Sweden	-	-
	Internal Assembly	Sweden	-	-
Use Phase	Transportation	UK	Air	1500
		USA	Air	8500
End-of-Life	Transportation	Multiple Locations	Truck	-

Table 1 presents a detailed breakdown of these components, their respective weights, and quantities required per case scenario. This table serves as a crucial reference for understanding the material basis of the environmental impacts assessed.

The inventory incorporates both primary data from TSS and secondary data from the Ecoinvent database. The goal is to have a robust and accountable foundation for our environmental impact assessment. Table 2 outlines the lifecycle stages alongside the data sources and transportation methods, for transparency in the methodologies employed and the assumptions made during the LCA process.

4.4 Lifecycle Assessment Phases

Cradle-to-Gate. We begin with a detailed inventory that leverages both primary and Ecoinvent database sources to map out the logger's journey. Critical components such as the PCBA, battery, casing, and label are assembled in Stockholm, with a global supply chain for subcomponents. To ensure a thorough analysis, materials and transportation methods were precisely selected and modeled to reflect real-world scenarios. Assembly and testing at TSS require about 0.139 kWh per unit. This precise measurement of electricity use—derived from both TSS's records and the Ecoinvent database—facilitates a granular analysis of the energy footprint.

Use Phase. Upon assembly, loggers are transported internationally, with assumed distances to ensure realism. This phase's analysis emphasizes the operational efficiency of the loggers, noting the absence of direct emissions due to their battery-operated design, exploring single-use versus multiple-use scenarios facilitated by reverse logistics, emphasizing potential environmental impact reduction through reuse.

End-of-Life Management. This phase considers the disposal or recycling of loggers, with assumptions about transportation to disposal facilities. The analysis includes the separate processing of electronic and plastic components, considering the lifespan of the PCB and its impact on the logger's total operational life.

Allocation Procedures. To simplify the life cycle assessment (LCA) model, an allocation procedure is implemented. Particularly for components with multiple lifecycles like batteries and casings, an allocation strategy is adopted, reflecting Causal Allocation. This method distributes environmental impacts based on specific processes and their contributions over multiple missions of the components. For example, the battery and casing for WL II last up to two missions, each being 120 days long, totaling 16 missions during its lifetime. This implies that the WL II will return to the facility 16 times for recalibration and refurbishment before final disposal. During these intervals, the battery and casing will be replaced eight times since they can be reused once after the initial use. Similarly, for TT BLE, the battery and casing are replaced after every mission, resulting in replacement 16 times before final disposal.

This approach ensures realistic use patterns, considering the potential for components to be reused across multiple missions before final disposal. Table 3 encapsulates this allocation strategy, detailing component reuse and the total operational life of the loggers. By adopting this approach, the study provides a comprehensive framework for understanding the environmental lifecycle of TSS's temperature loggers, enabling an

Table 3. Calculation of components for each scenario

Lifetime of PCB		50 000 h ~ 64 months			
Duration of 1 mission		120 days = 4 months			
Logger	Parameter	UK		USA	
		Case 1	Case 2	Case 1	Case 2
WL II	Number of Missions per PCB	1	16	1	16
	Number of Batteries	1	8	1	8
	Number of Casings	1	8	1	8
	Number of Labels	1	16	1	16
	Total Distance to Customer in km	1 500	48 000	8 500	272 000
TT BLE	Number of Missions per PCB	1	16	1	16
	Number of Batteries	1	16	1	16
	Number of Casings	1	16	1	16
	Number of Labels	1	16	1	16
	Total Distance to Customer in km	1 500	48 000	8 500	272 000

accurate comparison and offering insights into potential areas for improvement in their lifecycle management.

4.5 Scenarios Results

Britain Bubble Scenario. Figure 4 visually summarizes impacts, highlighting production's significant impact. In the cradle-to-gate phase, from the raw material acquisition to the customer, dominates environmental impacts, notably in metal depletion and climate change, accounting for over 80% of total impact. While less significant, transportation impacts fossil depletion and climate change, comprising about 10–15% of the total. Although minor overall, the end-of-life phase notably affects terrestrial ecotoxicity and climate change, emphasizing the importance of effective end-of-life management strategies.

Stateside Shuffle Scenario. Figure 5 illustrates the breakdown of impacts across cradle-to-gate, transportation, and end-of-life stages. The graphical representation underscores the significant role of the production phase, contributing over 60% of total impacts, particularly in metal depletion and climate change. Transportation also plays a notable role, impacting fossil depletion and climate change by about 40%. Additionally, the end-of-life phase affects terrestrial ecotoxicity, albeit to a smaller extent. These findings highlight the imperative for interventions in production and disposal practices to bolster the environmental sustainability of temperature loggers.

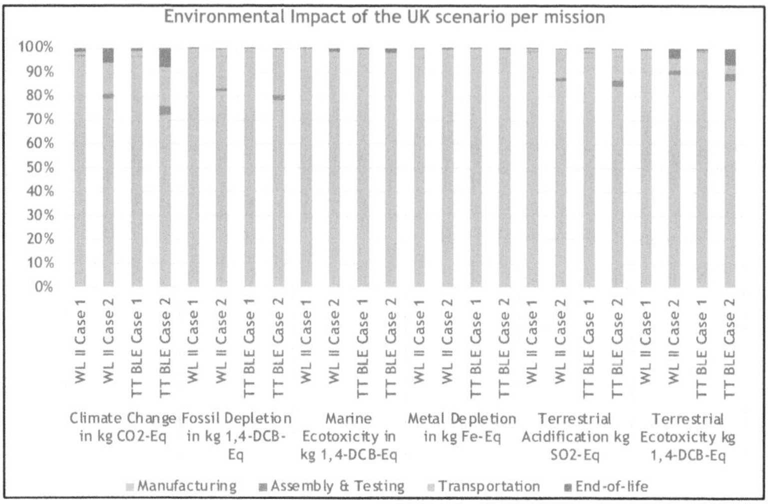

Fig. 4. Cradle-to-grave environmental impacts for the Britain Bubble Scenario

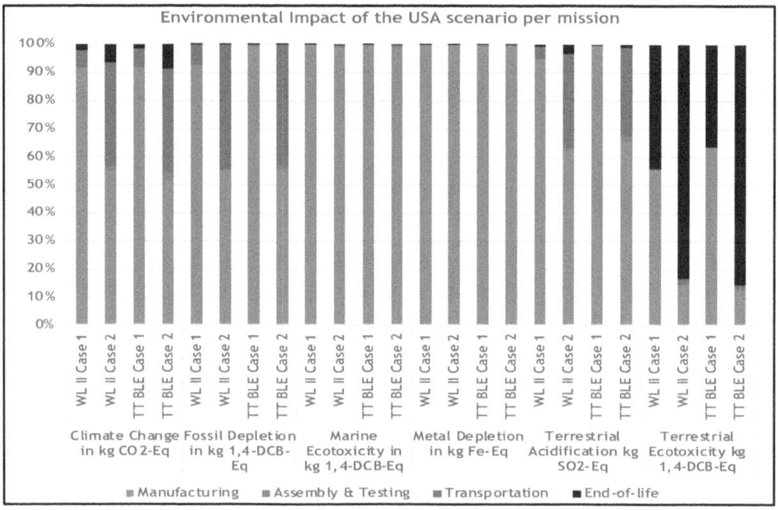

Fig. 5. Cradle-to-grave environmental impacts for the Stateside Shuffle Scenario

4.6 Comparison of Scenarios

This section distills insights on the environmental impacts of WL II and TT BLE loggers across different scenarios, with an emphasis on climate change (see Fig. 6). In Case 1, the temperature loggers are discarded after the first use, leading to a higher environmental impact due to the need for new materials for each deployment. Conversely, in Case 2, the same product is reused multiple times, thereby reducing the need for new materials and lowering the overall environmental impact associated with production and disposal.

This reuse scenario in Case 2 results in lower final impacts compared to Case 1, where the product is discarded after single use.

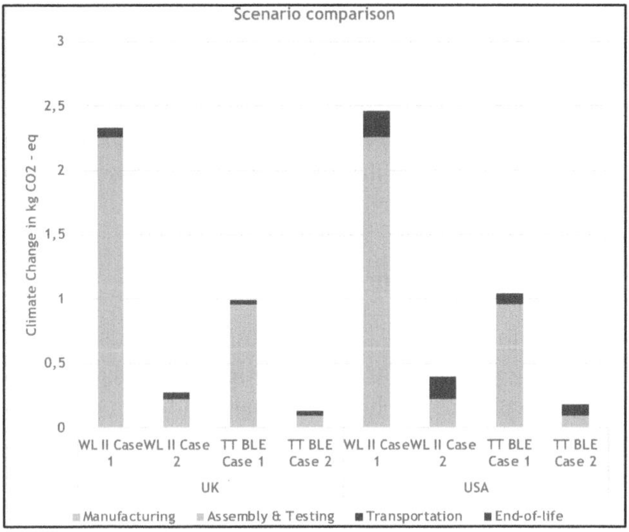

Fig. 6. Scenario Comparison

Across all scenarios, the production phase significantly shapes the environmental footprint, contributing 70% to 90% of the total impact. This underscores the critical need for sustainable manufacturing processes. Notably, in the USA case 2 for both loggers, transportation claims a larger impact share, nearly 50% of the total, due to extensive back-and-forth shipments for recalibration and refurbishment. This highlights the environmental cost of long-distance logistics. While less impactful than production and transportation, the end-of-life phase still influences certain environmental categories, especially terrestrial ecotoxicity, indicating the importance of effective disposal strategies.

A further analysis identified the PCBA as a significant environmental hotspot due to its material and metal usage in manufacturing, with transportation also crucial, especially in reverse logistics scenarios. The casing, particularly for WL II, significantly impacts the environment due to its weight and frequent replacements. While other subcomponents contribute less, the PCBA and transportation are critical hotspots in the cradle-to-gate phase, driven by material use and the impacts of recalibration and refurbishment. Within the PCBA, the temperature sensor is the primary contributor for both WL II and TT BLE devices, indicating a need for optimization. Other components like the PCB, voltage regulator for WL II, and Bluetooth module for TT BLE also significantly impact overall sustainability. These findings highlight the need for targeted eco-design and sustainability improvements focusing on the most impactful components.

4.7 Sensitivity Analysis

A sensitivity analysis was performed focusing on electricity consumption and mix during the manufacturing of temperature loggers. The analysis explored how variations in these factors affect the system, especially given their significant impact on the manufacturing phase's contribution to climate change. The analysis adjusted electricity consumption by $\pm$ 25% for both loggers, revealing its influence on overall energy requirements and environmental impact. Sweden's green energy sources minimize the climate change impact, whereas the US and China, relying more on fossil fuels, exhibit a higher environmental footprint. These findings emphasize the importance of considering energy sources in manufacturing decisions to reduce environmental impact.

5 Discussion and Recommendations

Given the increasing focus on sustainability regulations like the EU's Sustainable Product Initiative, conducting proactive LCAs is imperative for companies like TSS AB to ensure compliance and enhance supply chain sustainability. In our case study, the production phase was the primary contributor to the environmental footprint, emphasizing the need to optimize manufacturing processes and source low-impact materials. Recommendations include using eco-friendly PCBAs, reducing aluminum dependence, and developing comprehensive disposal plans to prioritize recycling and reusability. These measures align with the EU Taxonomy objectives of mitigating climate change and preventing pollution.

The adoption of refurbishment, recalibration, and reuse strategies is crucial for enhancing the sustainability of temperature loggers. By extending product lifespan and minimizing the need for new manufacturing, these strategies reduce environmental impacts and promote circular economy principles. While they may pose challenges, such as logistical complexities, the benefits in terms of sustainability and resource efficiency outweigh these concerns. Companies like TSS AB should actively explore and implement these strategies to align with sustainability objectives and regulatory frameworks.

Transparency and clarity in defining the study's scope and functional unit are crucial for understanding environmental impacts. More accurate primary data, particularly for end-of-life analysis, is needed to enhance precision. Optimizing production and responsible end-of-life disposal practices are key to reducing emissions and enhancing sustainability. Material and design optimizations, along with reducing transportation emissions, will significantly contribute to environmental goals. Continuous monitoring and reporting progress towards these goals are essential for ongoing improvement and compliance with regulations like the EU Sustainable Product Initiative. Manufacturing in greener electricity contexts, such as Sweden, can substantially reduce environmental footprints. Combining LCA expertise with product knowledge is essential, and companies can benefit from having LCA experts on staff, training existing employees, or a hybrid approach. Proactively conducting LCAs and implementing recommendations can enhance environmental performance, meet regulatory requirements, and position companies as leaders in sustainable manufacturing, contributing to a more environmentally responsible future.

6 Conclusion

This study conducted a comprehensive LCA of temperature loggers used in the pharmaceutical industry, focusing on the environmental impacts across their lifecycle. Adopting a methodological approach in line with ISO 14040/-44 standards, the study defined the assessment's goal, scope, and functional unit, encapsulating the loggers' utilization per mission until end-of-life. Various scenarios were explored to account for differences in materials, energy sources, and transportation, providing an detailed view of the environmental footprint.

Key findings from the study reveal that the production phase significantly influences the loggers' environmental footprint, accounting for over 90% of the total impact. Transportation emerged as the second-largest contributor, emphasizing the need for optimizing logistics and production processes. The end-of-life phase, while less impactful, still significantly affects categories like climate change and ecotoxicity, underscoring the importance of effective disposal strategies. The PCBA and transportation were identified as critical hotspots, with the casing of the WL II logger also making a notable impact due to its weight.

The sensitivity analysis highlighted Sweden's greener electricity mix as less impactful on climate change compared to the USA and China, suggesting manufacturing in Sweden or similar contexts could significantly reduce environmental footprints. Moreover, adopting a refurbishment, recalibration, and reuse strategy markedly lowers impacts compared to a single-use approach, advocating for the implementation of reverse logistics to enhance sustainability. Overall, the LCA provided the case company with valuable insights for improving their products' environmental performance, aligning with the EU Taxonomy and Sustainable Product Initiative's goals. By addressing identified issues and implementing recommended improvements, the company can reduce its environmental impact and contribute to a more sustainable electronics sector. This study offers insights tailored to TSS AB's temperature loggers and valuable lessons applicable to other firms in electronics manufacturing.

6.1 Limitations

The study investigating the sustainability of only two specific devices underlines critical insights while acknowledging several limitations. The study faced challenges accessing upstream manufacturing data, necessitating reliance on literature and the Ecoinvent database for inventory inputs. A more granular end-of-life treatment analysis could refine the results of this study. The study's reliance on the basic openLCA license and Ecoinvent 3.7.1 database necessitates the use of proxies for missing materials, suggesting an upgrade to a more detailed database for accuracy. Moreover, the study's temporal scope does not consider future technological, regulatory, or industry shifts that might impact the devices' sustainability.

Environmental focus marks the study's primary lens, sidelining other sustainability dimensions such as social or economic factors. Broadening this view to include a wider stakeholder perspective, including suppliers, consumers, and regulators, could enrich the sustainability assessment. Acknowledging these limitations underscores the necessity for ongoing research and adaptation of sustainability practices in the electronics sector, emphasizing the dynamic nature of environmental impact assessments.

6.2 Future Research

For future enhancements in the lifecycle analysis of temperature loggers, it is crucial to extend the examination across all lifecycle stages. Key improvements include optimizing material selection and enhancing the precision of primary data collection, which will sharpen insights into environmental impacts. Diversifying functional units and exploring different allocation strategies, particularly with an eye on the inclusion of recycled content, could refine evaluations of the product's life stages.

Acknowledgements. The authors would like to acknowledge the support and collaboration of the Research Institutes of Sweden (RISE), TSS AB, and Kungliga Tekniska Högskolan (KTH) within the RESPIRE project, funded by the Production2030 program and Sweden's Government Agency for Innovation VINNOVA [grant number 2021–03685]. The authors extend their gratitude to Martin Kurdve from RISE and Johanna Eriksson from TSS AB, for their assistance and resources. Thanks also to the team members of TSS AB, including Ingrid Szabo and Andras Kalmar, for their insights and collaboration.

References

1. IPCC: Climate change widespread, rapid, and intensifying. In: The Intergovernmental Panel on Climate Change (2021). https://www.ipcc.ch/2021/08/09/ar6-wg1-20210809-pr/
2. UNEP: UN report: time to seize opportunity, tackle challenge of e-waste. In: UN Environment (2019). https://www.unep.org/news-and-stories/press-release/un-report-time-seize-opportunity-tackle-challenge-e-waste
3. UNEP: Chemicals of Concern in Electronics Review of Legislative and Regulatory Approaches Geneva, 23/12/2020 2 Acknowledgements (2020)
4. Biswas, C., Lee, Y.H.: Graphene versus carbon nanotubes in electronic devices. Adv. Funct. Mater. **21**, 3806–3826 (2011). https://doi.org/10.1002/adfm.201101241
5. Reyes, P.M., Man, J., Jaska, P., et al.: Recycle system design for end-of-life electronics in developing countries. Int. J. Integr. Supply Manage. **14**, 101 (2021). https://doi.org/10.1504/ijism.2021.113565
6. Kuchta, R., Stefan, P., Barton, Z., et al.: Wireless temperature data logger. In: IEEE Xplore (2005). https://ieeexplore.ieee.org/document/1564541
7. Hsu, A., Hu, C.: Green supply chain management in the electronic industry. Int. J. Environ. Sci. Tech. **5**, 205–216 (2008)
8. Murphy, M.: Electronics can trigger a more circular, sustainable world – here's how. In: World Economic Forum (2021). https://www.weforum.org/agenda/2021/05/electronics-can-trigger-a-more-circular-sustainable-world-here-s-how/
9. Liu, J., Yang, C., Wu, H., et al.: Future paper based printed circuit boards for green electronics: fabrication and life cycle assessment. In: Research Gate (2014). https://www.researchgate.net/publication/265138290_Future_Paper_based_Printed_Circuit_Boards_for_Green_Electronics_Fabrication_and_Life_Cycle_Assessment
10. ISO (2006) ISO 14040:2006. In: ISO. https://www.iso.org/standard/37456.html

11. Andersen, O., Hille, J., Gilpin, G., Andrae, A.: Life cycle assessment of electronics. In: Research Gate (2014). https://www.researchgate.net/publication/267509635_Life_Cycle_Assessment_of_Electronics
12. Rigamonti, L., Falbo, A., Zampori, L., Sala, S.: Supporting a transition towards sustainable circular economy: sensitivity analysis for the interpretation of LCA for the recovery of electric and electronic waste. Int. J. Life Cycle Assess. **22**, 1278–1287 (2016). https://doi.org/10.1007/s11367-016-1231-513
13. Golsteijn, L.: LCA basics: life cycle assessment explained. In: PRé Sustainability (2020). https://pre-sustainability.com/articles/life-cycle-assessment-lca-basics
14. JRC EU: ILCD Handbook: Recommendations for Life Cycle Impact Assessment in the European context First edition (2012)
15. TSS AB Logger - webLogger II. In: www.tssab.com. https://www.tssab.com/logger/weblogger-ii
16. TSS AB Logger - TempTracer BLE. In: www.tssab.com. https://www.tssab.com/logger/temptracer-ble

Developing a Robust Multi-objective Optimization Model for Reverse Logistics of Electric Vehicle Batteries

Roozbeh Yaghoubi, Aysan Mahboubi(✉), Mina Kazemi Miyangaskary, and Samira Keivanpour

Department of Mathematical and Industrial Engineering, Polytechnique Montreal, Montreal, QC, Canada

aysan.mahboubi@polymtl.ca

Abstract. In response to the damaging impacts of greenhouse gas emissions, Electric Vehicles (EVs) have emerged as a sustainable alternative. However, the rise of EVs creates the challenge of disposing numerous retired batteries. Hence, designing and optimizing Reverse Logistics (RL) for Electric Vehicles Batteries (EVBs) can be an effective approach. Existing uncertainties, however, cause a gap between the outputs of exact optimization models and real-world conditions. Addressing this, the present study develops a robust multi-objective optimization model that incorporates uncertainty such as the return rates of retired batteries. Making it more applicable to real-world scenarios. This research compares the relative performances of deterministic and the proposed robust optimization models and validates the model by calculating violation probabilities within a robust optimization framework. This validation is conducted by allocating a specific budget for robustness to ensure that the model remains effective under various uncertain conditions. Additionally, this study explores adjusting the level of conservatism in decision-making by applying the price of robustness approach to manage conservatism in our decision-making process. Highlighting the importance of innovative sustainable practices, this work offers a practical route for stakeholders to collaboratively mitigate the challenges associated with the end-of-life management of EV batteries, thus contributing to environmental sustainability and resource efficiency in the BEV industry.

Keywords: Electric Vehicle Batteries · Reverse Logistic · Robust Optimization

1 Introduction

The transportation sector has become a major source of GHG emissions, with road transport accounting for 76% of all the EU's transport GHG emissions (European Environment Agency, 2023,Yarahmadi et al., 2023)). Therefore, Electric vehicles (EVs) present a viable solution for this problem. According to the International Energy Agency (IEA, 2023), EVs represented 18% of the total car sales, with Europe alone contributing 3.4

Published by Springer Nature Switzerland AG 2024
M. Thürer et al. (Eds.): APMS 2024, IFIP AICT 733, pp. 492–507, 2024.
https://doi.org/10.1007/978-3-031-71645-4_33

million EVs in 2023. However, the environmental benefits of EVs are limited by the lifespan of their batteries. When they lose 20% of their capacity, the batteries are retired and can start a second life. Spent EV batteries typically retain around 80% of their usable capacity, so recycling them generates additional value.

Several articles examine the treatment activities of EV batteries, highlighting different recycling methods including hydrometallurgy, pyrometallurgy, and direct recycling, as well as the detailed components of EV batteries (Chen et al., 2019; Harper et al., 2019). Additionally, the potential for repurposing EV batteries in less demanding applications, such as stationary energy storage, is discussed by Zhu et al. (2021). However, the question arises as to how to address this uncertainty in return rates within the EV battery supply chain models. This uncertainty can stem from various factors, including customer behaviors regarding battery returns, changes in policies, rapid technological advancements in battery design, and the diversity in battery lifespans.

This research aims to compare the efficacy of robust approaches in scenarios where the underlying distributions are unknown. To make the model align with reality, the "return percentage" of products can be estimated based on market populations. This approach makes the "return percentage" as a critical uncertain parameter in our analysis. Additionally, we will assume the absence of shortages and backlogs within the system.

Our primary focus is to address the robustness of the model with respect to changes in the model's parameters. Therefore, in this robust optimization approach, the focus on return percentage of retired batteries from EV markets is considered as a key technological coefficient. We apply a budget of robustness concept, where not all variables must reach their worst-case scenario simultaneously; this would be overly conservative. To achieve this goal, we introduce a parameter, Γ (e.g., $\Gamma = 5$), to represent scenarios where any five technological coefficients may deviate to their worst-case values. This ensures the model remains robust.

The goal of this study is to find an optimal robust solution viable under all uncertainty scenarios, hence this approach systematically addresses problems where uncertainty predominates within constraints. Unlike traditional robust optimization, which considers all worst-case scenarios as in the Ben-Tal's method, our model adopts a more balanced approach. This contribution enables the evaluation of different scenarios on a broader level, so that stakeholders in reverse logistics of EV batteries can get a better vision and understanding of the future supply of and demand for retired batteries and take actions if necessitates. With the proposed model, the impacts of other uncertainties, including the number of reusable cells or the demand for EVs, can also be evaluated. Therefore, this model facilitates a more expansive examination of the closed-loop supply chains for EVs and batteries.

The organization of the remaining sections is as follows: After a literature review in Sect. 2 that identifies the research gaps our study addresses, Sect. 3 details our methodology, including the proposed mathematical and robust models. Section 4 discusses the correlation between gamma (Γi) and the optimal outcomes of the robust model, offering insights into our findings. Finally, Sect. 5 summarizes the innovative contributions of our research, discusses its limitations, and suggests directions for future work.

2 Literature Review

In the context of optimizing closed-loop supply chains for electric vehicle (EV) battery recycling, a review of the literature reveals a variety of methodologies aimed at enhancing supply chain resilience and efficiency. Notably, a gap identified in these studies is the lack of focus on the uncertainties inherent within the supply chain, especially those related to environmental and technological.

This section reviews conducted studies on optimizing EV battery supply chain. However, a common limitation across these works is the insufficient exploration of stochastic models that can robustly account for the various sources of uncertainty affecting the EVB supply chain. While Hoyer et al., (2015) and Hendrickson et al., (2015) provide foundational models for technology selection and infrastructure optimization, their approaches do not explicitly model the stochastic nature of market demand or regulatory changes. While Gu et al., (2018), Scheller et al., (2020), and Wang et al., (2020) develop strategies for pricing, cooperative planning, and network design, the potential variability in key parameters like battery return rates, technological advancements, and carbon tax policies is not directly addressed through stochastic modeling. This limits the ability of these models to provide reliable and realistic solutions in the face of the EV battery supply chain's inherent uncertainties.

The works of Nguyen-Tien et al., (2022), Huster et al., (2022) and Li et al., (2018) proposed an approach by considering geospatial configurations and the remanufacturing potential, highlighting the importance of spatial and temporal dynamics. Nonetheless, these analyses remain deterministic, overlooking the stochastic fluctuations in EV adoption rates and technological shifts in battery composition. Gonzales-Calienes et al., (2022), Pamucar et al., (2022) and Zhang et al., (2022) contribute perspectives on reverse logistics, recovery center selection, and recycling mode decisions. However, the uncertainty in logistical costs, material recovery rates, and regulatory environments could impact the viability and optimization of recycling networks, a consideration that requires stochastic methods to fully understand.

Kazemi Miyangaskary et al., (2023) and Zhang et al., (2023) offer closer approaches to addressing uncertainties, with Zhang et al., (2023) exploring carbon emission reduction strategies under different collection scenarios and Kazemi Miyangaskary et al., (2023) employing fuzzy logic to navigate supply chain parameter vagueness. Yet, the adoption of stochastic models that capture the probabilistic nature of these uncertainties is absent. Robust optimization can consider a set of possible values each parameter can take. This creates solutions that are feasible under multiple potential futures.

The existing literature primarily focuses on deterministic models, which restrict applicability in real-world scenarios. A significant gap exists in understanding the nature of randomness in the first mile problem in EVB supply chains. This highlights the need for a deeper understanding of how uncertainty can impact the effectiveness and reliability of decision-making strategies.

2.1 Identified Research Gaps

Our literature review on EV battery recycling supply chains highlights critical areas needing further exploration:

- **Dominance of Deterministic Models:** Current studies heavily rely on deterministic models, missing the nuanced complexities of environmental and technological uncertainties.
- **Limited Use of Stochastic Models:** There's a notable deficiency in applying stochastic models to address uncertainties like market demand fluctuations, technological advances, and regulatory changes, limiting the realism and applicability of existing models.
- **Neglecting Uncertainty's Impact on Decision-Making:** Previous research seldom models the stochastic nature of critical variables, such as EV adoption rates and technological shifts, undermining the reliability of decision-making strategies in the face of uncertainty.

These gaps underscore the necessity for a shift towards stochastic modeling to capture the probabilistic nature of uncertainties in the EVB supply chain, promising more realistic and dependable analytical frameworks for decision-making.

3 Methodology

3.1 The Mathematical Programming Model

The configuration of the EVB closed-loop supply chain network is displayed in the Fig. 1.

Model's assumptions are as follows:

- The capacity of all facilities is determined.
- Quantity discounts are not considered in the purchase.
- Shortage of products to supply customers' demands is allowed and incurs a cost.

The following notation is used to formulate the problem:

Sets

I: Set of Suppliers (Mines), index by i.
B: Set of BEV Manufactories, index by b.
M: Set of EV Manufactories, index by m.
K: Set of EV Retailer, index by k.
C: Set of EV Market, index by c.
F: Set of Collection & Sorting Centers, index by f.
S: Set of Echelon Markets, index by s.
R: Set of Recycling Centers, index by r.
D: Set of Disposal Centers, index by d.
E: Set of Echelon Utilization Centers, index by e.

Parameters

PS_{bm}: Unit sale price of EV batteries from BEV manufactories b to EV manufactories m

PS_{mk}: Unit sale price of EVs from EV manufactories m to EV retailer k

PS_{kc}: Unit sale price of EVs from EV retailer k to EV market c

P_{ib}: Unit purchase cost of raw material from mine i to BEV manufactories b

P_{rb}: Unit purchase cost of raw material from recycling center r to BEV manufactories b

Db: The demand for BEV

DEV: The demand for EV

M: Required raw materials to produce an EV battery unit

C_g: The capacity of facilities

Pr_g: Unit Processing cost in node

T_{gh}: Unit Transport cost from node g to node h

shb: Unit Shortage cost of EV batteries in BEV manufactories

SHm: Unit Shortage cost of EV in EV manufactories

SHk: Unit Shortage cost of EV in EV retailer

R: Return percentage of retired batteries from EV markets

Rs: Return percentage of second retired batteries from echelon markets

Ps_{es}: Unit sale price of retired batteries from echelon utilization e to echelon markets s

Ps_{rb}: Unit sale price of raw material from recycling centers r to BEV manufactories b

Decision Variables

QR_{ib}: Quantity of raw material purchased by BEV manufactories b from mines I

QR_{rb}: Quantity of raw material purchased by BEV manufactories b from recycling centers r

QB_{bm}: Quantity of EV batteries transported from BEV manufactory b to EV manufactory m

QB_{mk}: Quantity of EV transported from EV manufactories m to EV retailer k

QB_{kc}: Quantity of EV transported from EV retailers k to EV market c

QB_{fr}: Quantity of retired batteries transported from collection centers f to recycling centers r

QB_{fe}: Quantity of retired batteries transported from collection centers f to echelon utilization

QB_{fd}: Quantity of retired batteries transported from collection centers f to disposal centers d

QB_{es}: Quantity of retired batteries transported from echelon utilization e to echelon market s

QB_{er}: Quantity of retired batteries transported from echelon utilization e to recycling center r

QB_{ed}: Quantity of retired batteries transported from echelon utilization e to disposal center d

QC_{rd}: Quantity of scraps transported from recycling centers r to disposal centers d

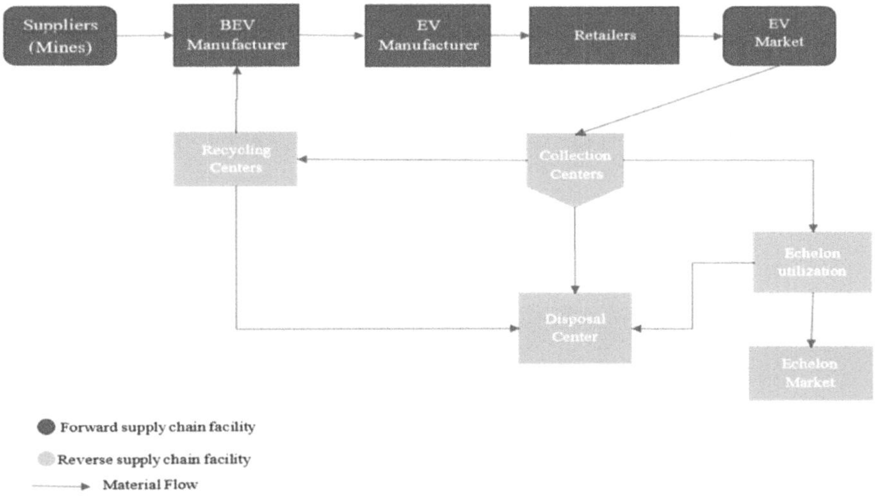

Fig. 1. Configuration of the EVBs supply chain

Mathematical Model

The following presents the multi-objective problem in this study:

$$
\begin{aligned}
\mathbf{MaxF}_1(\mathbf{X}) = & \Big[\Big(\sum_b \sum_m PS_{bm} \times QB_{bm}\Big) + \Big(\sum_m \sum_k PS_{mk} \times QB_{mk}\Big) \\
& + \Big(\sum_k \sum_c PS_{kc} \times QB_{kc}\Big) + \Big(\sum_e \sum_s Ps_{es} \times QB_{es}\Big)\Big] - \Big[\Big(\sum_i \sum_b P_{ib} \times QR_{ib}\Big) \\
& + \Big(\sum_r \sum_b P_{rb} \times QR_{rb}\Big) + \Big(\sum_i \underset{i}{\Pr} \times \sum_b QR_{ib}\Big) + \Big(\sum_b \underset{b}{\Pr} \times \sum_m QB_{bm}\Big) \\
& + \Big(\sum_m \underset{m}{\Pr} \times \sum_k QB_{mk}\Big) + \Big(\sum_k \underset{k}{\Pr} \times \sum_c QB_{kc}\Big) + \Big(\sum_f \underset{f}{\Pr} \times \Big(R \times \sum_c QB_{kc}\Big)\Big) \\
& + \Big(\sum_e \underset{e}{\Pr} \times \Big(\sum_f QB_{fe} + \sum_s Rs \times QB_{es}\Big)\Big) + \Big(\sum_r \underset{r}{\Pr} \times \Big(\sum_b QR_{rb} + \sum_d QC_{rd}\Big)\Big) \\
& + \Big(\sum_d \underset{d}{\Pr} \times \Big(\sum_f QB_{fd} + \sum_e QB_{ed} + \sum_r QC_{rd}\Big)\Big) + \Big(\sum_i \sum_b T_{ib} \times QR_{ib}\Big) \\
& + \Big(\sum_r \sum_b T_{rb} \times QR_{rb}\Big) + \Big(\sum_r \sum_b T_{rb} \times Qc_{rd}\Big) + \Big(\sum_b \sum_m T_{bm} \times QB_{bm}\Big) \\
& + \Big(\sum_m \sum_k T_{mk} \times QB_{mk}\Big) + \Big(\sum_f \sum_e T_{fe} \times QB_{fe}\Big) + \Big(\sum_f \sum_r T_{fr} \times QB_{fr}\Big) \\
& + \Big(\sum_f \sum_d T_{fd} \times QB_{fd}\Big) + \Big(\sum_e \sum_s T_{es} \times QB_{es}\Big) + \Big(\sum_e \sum_r T_{er} \times QB_{er}\Big) \\
& + \Big(\sum_e \sum_d T_{ed} \times QB_{ed}\Big) + \Big(Shb \times \Big(Db - \sum_b \sum_m QB_{bm}\Big)\Big) \\
& + \Big(Shm \times \Big(DEV - \sum_m \sum_k QB_{mk}\Big)\Big) + \Big(Shk \times \Big(DEV - \sum_k \sum_c QB_{kc}\Big)\Big)\Big]
\end{aligned}
\tag{1}
$$

$$
\mathbf{Max\ F}_2(\mathbf{X}) = \Big(\frac{\sum_{\mathrm{c}} \sum_{\mathrm{k}} \mathrm{QB}_{\mathrm{kc}}}{\mathrm{DEV}}\Big) + \Big(\frac{\sum_{\mathrm{k}} \sum_{\mathrm{m}} \mathrm{QB}_{\mathrm{mk}}}{\mathrm{DEV}}\Big) + \Big(\frac{\sum_{\mathrm{b}} \sum_{\mathrm{m}} \mathrm{QB}_{\mathrm{bm}}}{\mathrm{Db}}\Big) \tag{2}
$$

st:

$$\sum_{\mathrm{b}} QR_{\mathrm{ib}} \le C_i \qquad \forall \mathrm{i} \in \mathrm{I} \tag{3}$$

$$\sum_{\mathrm{i}} \mathrm{QR}_{\mathrm{ib}} + \sum_{\mathrm{r}} \mathrm{QR}_{\mathrm{rb}} \le C_b \qquad \forall b \in B \tag{4}$$

$$\sum_{\mathrm{b}} \mathrm{QB}_{\mathrm{bm}} \le C_m \qquad \forall m \in M \tag{5}$$

$$\mathrm{R} \times \sum_{\mathrm{c}} \mathrm{QB}_{\mathrm{kc}} \le \sum_f C_f \tag{6}$$

$$\sum_{\mathrm{f}} \mathrm{QB}_{\mathrm{fe}} + \left(\mathrm{Rs} \times \sum_{\mathrm{s}} \mathrm{QB}_{\mathrm{es}}\right) \le C_e \qquad \forall e \in E \tag{7}$$

$$\sum_{\mathrm{e}} \mathrm{QB}_{\mathrm{es}} \le C_s \qquad \forall \mathrm{s} \in \mathrm{S} \tag{8}$$

$$\sum_{\mathrm{f}} \mathrm{QB}_{\mathrm{fr}} + \sum_{\mathrm{e}} \mathrm{QB}_{\mathrm{er}} \le C_r \qquad \forall r \in R \tag{9}$$

$$\mathrm{R} \times \sum_k \sum_c QB_{KC} = \sum_r QB_{fr} + \sum_d QB_{fd} + \sum_e QB_{fe} \qquad \forall f \in F \tag{10}$$

$$\sum_{f} QB_{fe} + \left(\text{RE} \times \sum_{s} QB_{es}\right) = \sum_{r} QB_{er} + \sum_{d} QB_{ed} + \sum_{s} QB_{es} \quad \forall e \in E \quad (11)$$

$$\sum_{i} QR_{ib} + \sum_{r} QR_{rb} = M \times \sum_{m} QB_{bm} \quad \forall b \in B \quad (12)$$

$$\sum_{k} \sum_{c} QB_{KC} \leq \text{DEV} \quad (13)$$

$$\sum_{k} \sum_{m} QB_{Km} \leq DEV \quad (14)$$

$$\sum_{b} \sum_{m} QB_{bm} \leq Db \quad (15)$$

$$\text{All of variables } \geq 0 \quad (16)$$

The first objective function maximizes the profits of all supply chains. To do that, total income is subtracted from total cost. The second objective function maximizes the service level. Constraints 3 to 9 display facilities' capacities. Constraints 10 and 11 pertain to collection centers and echelon utilization centers, respectively, ensuring that the total input produced by these entities equals their respective outputs. Constraint 12 establishes a relationship between battery production quantity and raw material procurement. Constraint 13 mandates that the quantity of electric vehicles supplied by retailers to the market must not exceed market demand. Similarly, Constraint 14 dictates that the quantity of EVs supplied by manufacturers to retailers should not exceed the demand of each individual retailer. Constraint 15 asserts that battery demand must always surpass battery production. Finally, Constraint 16 specifies that all variables must maintain non-negative values.

3.2 Robust Model

Formulating the Robust-Counterpart (using budget of uncertainty)

In this research, our objective is not to adopt Ben-Tal's approach (Ben-Tal & Nemirovski, 1999), which is characterized by a high level of conservatism, as this could potentially limit its applicability. Instead, this research adopts a methodology that allows for a better balance between robustness and flexibility. This is achieved by employing the concept of the price of robustness. This approach ensures that the proposed robust optimization model maintains the characteristics of a linear optimization problem, so it can retain computational efficiency while addressing real-world complexities. The methodology aligns with the perspectives presented by Bertsimas & Sim (2004), where the emphasis is on controlling the level of conservatism.

This methodology states that the chance of all variables reaching their worst-case scenario simultaneously is low. In this framework, Bertsimas & Sim (2004) propose an approach where one can control the number of variables that may be vulnerable to their worst case. Consequently, as long as fewer than Γ coefficients shift to their worst-case scenario, the solution remains robust and viable.

The introduction of the uncertainty budget presents a risk: if the robust constraint is applied, there is a possibility of violation. To address this issue, Bertsimas & Sim have proposed upper bounds for the probability of such violations, which are probabilistic in nature. In this model, we do not specify which parameters should assume their worst-case values; instead, the model autonomously determines this. Maintaining robustness is costly, necessitating the introduction of flexibility to mitigate both the expense and the level of conservatism. The revised approach protects against violations by ensuring that the solution remains feasible as long as no more than a pre-specified number of coefficients, denoted by Γ, undergo changes. This method ensures feasibility by allowing changes in fewer than Γ coefficients.

Regarding the price of robustness (cordiality robust optimization), this aspect allows us to modulate the level of conservatism in our decision-making problems, particularly focusing on scenarios where the coefficients on the left-hand side are stochastic, while those on the right-hand side remain deterministic.

The robust multi-objective optimization model contains:

J_i: Set of random coefficients.

$|Ji|$: The cardinality of Ji.

$\Gamma\ i$: It changes between 0 and $|Ji|$ & the role of this parameter is to adjust the robustness of the proposed method, against the level of conservatism of the solution.

$$\eta ij = \frac{\widetilde{aij} - \overline{aij}}{\widehat{aij}}$$

where:

$\widetilde{aij}$ is the randome value

$\overline{aij}$ is the nominal value

$\widehat{aij}$ is the drift

However, we intend to impose a condition that this summation remains less than or equal to the value denoted as Γ for constraint No.6. This is what is referred to as the price of robustness or the budget of uncertainty. By introducing this Γ threshold, we are setting a limit on the number of parameters that can go to their worst-case scenarios. This allows us to develop a set of solutions and then estimate the probability that any one constraint will be violated.

$$\sum_{j=1}^{n} \eta ij \leq \Gamma i$$

In this scenario, certain parameters may exceed their nominal values, while others may fall short of their estimates.

If Γi equals zero, our model remains deterministic, representing the most optimistic scenario possible; but if Γi matches the cardinality of Ji (as Ben-Tal's approach), every coefficient reaches its worst-case scenario. Hence, we are navigating a spectrum between zero and the size of Ji. The purpose of this Γi parameter is to adjust the robustness of our proposed method relative to the solution's level of conservatism.

Uncertainty in the EV Battery Reverse Logistics Model
To implement the above-mentioned approach in the reverse logistics model of EV batteries, constraint No.6 has been chosen to contain the random coefficient "R_c". It has been modified to include its robust counterpart to necessitate the addition of a new constraint to the previously existing model. It is understood that "R_c" represents random values, each assumed to have a nominal value despite potential deviations above or below this level. However, it is important to note that not all coefficients are subject to randomness; in this research, we specifically focus on the "Return percentage of retired batteries from EV markets" constraint. Moreover, new dual variables q_k & r_{kc} are introduced, in relation to constraint No.17 to facilitate the development of the Robust-Counterpart (RC) model.

To develop the Robust-Counterpart (RC) model, we have modified the constraint No.6 with its following primary descriptions:

$$\mathrm{R} \times \sum_{\mathrm{c}} \mathrm{QB}_{\mathrm{kc}} \leq \sum_{f} C_f \tag{17}$$

- R: Return percentage of retired batteries from EV markets
- C_g: The capacity of facilities g ∈ I, J, B, M, K, F, S, R, D, E (Ton)
- QB_{kc}: Quantity of EV transported from EV retailors k to EV market c (Ton)

Now our new model is formulated as below:

$$\textit{Initial}\ \text{constraint}: \sum_{c} Rc * \sum_{c} QBkc \leq \sum_{f} Cf \qquad \forall k$$

So first we add a protection function to our constraint:

$$\begin{aligned}
&\sum_{c} \overline{R}_c * \sum_{c} QB_{kc} + \mathrm{Max}\{\sum_{c} R_c * \sum_{c} QB_{kc} * Z_{kc}\} \leq \sum_{f} C_f && \forall k \\
&\sum_{c} Z_{kc} \leq \Gamma_{kc} \quad (q_k\text{: Dual}) && \forall k \\
&0 \leq Z_{kc} \leq 1 \quad (r_{kc}\text{: Dual}) && \forall k,c
\end{aligned} \tag{18}$$

To enhance our new protection function, we have introduced a novel variable, *Zkc*, to maintain the model's linearity. The challenge lies in the "Maximum" term within a constraint, which introduces nonlinearity. To solve this, we have transformed the constraint into a linear format by integrating *Zkc*, a continuous decision variable constrained between 0 and 1.

Moreover, we have constructed the dual of this model to align the objective function of the dual with that of the original model. This duality is essential for linearizing the model, especially since it includes a "Maximum" term in one of its constraints, marking it as nonlinear. Through this approach, we convert the problem into one of "minimization" within the corresponding constraint. This constraint is then subject to a less than or equal to relationship on its right-hand side which allows us to simplify the constraint by removing the "minimum" term, thus achieving a simple linear constraint.

$$\xrightarrow{\text{Dual}} \text{Min } \Gamma k * qk + \sum_{c} rkc$$
$$\text{s.t.:} \tag{19}$$
$$\begin{array}{ll} qk + \text{r}kc \geq Rc + QBkc & \forall k, c \\ qk \geq 0 & \forall k \\ \text{r}kc \geq 0 & \forall k, c \end{array}$$

$$\xrightarrow{\text{Eliminaing "Min"}}$$

The final model would be as the following

$$\sum_{c} \overline{R}c * \sum_{c} QBkc + \Gamma k * qk + \sum_{c} rkc \leq \sum_{f} Cf \quad \forall k$$

$qk + \text{r}kc \geq Rc + QBkc \quad \forall k, c$

(new constraintsto be added below the constraint No.6)

$qk \geq 0 \quad \forall k$

$\text{r}kc \geq 0 \quad \forall k, c$

(new decision variables)

According to the strong duality theorem, given that problem (18) is both feasible and bounded, the corresponding dual problem (19) will also be feasible and bounded, sharing the same objective values. We establish primary constraints beneath this original constraint to form the robust counterpart; notably, we have removed the "Min" operation. This is because if the constraint holds for any given values of q_k & r_{kc}, then it satisfies for their minimum. Thus, we effectively convert the condition into a linear programming constraint.

4 Discussion

The issue with introducing a budget of uncertainty is the risk when it is implemented in the robust solution, as in reality the constraint may be violated. To address this, Bertsimas and Sim proposed probabilistic bounds for this probability of violation scenarios.

It can never be guaranteed that the solution will always remain feasible unless Γ takes its maximum value. Hence, we attempt to find some bounds for the probability that the constraint might be violated as the following situation occurs:

$$\sum_{j=1}^{n} \eta_{ij} \geq \Gamma_i$$

By changing the value of Γi, Berstimas and Sim develop a set of solutions. This allows to estimate the probability that was solved with a given Γi, recognizing that if Γi is not at its maximum, there is always a risk of violation. In method of Ben-Tal, assigning all values to their maximum ensures absolute robustness; for example, if there are 5 options and $\Gamma i = 5$, then the solution will always be feasible regardless of the situation (the absolute robust). The formula for calculating these probabilities is represented:

$$pr(\sum \eta_{ij} \geq \Gamma_i) = 1 - \phi(\frac{(\Gamma_i - 1)}{\sqrt{n}})$$

To elaborate more details about the functionality of the robust-counterpart model, a numerical example with its figures provided.

To start, the set $\Gamma i = 1, 2, 3, 4, 5$ is defined for Constraint No. 6, which contains uncertain parameters. This set represents the limit on the number of parameters that might take their worst-case values in any combination of the five parameters within the constraint (Table 1).

Figure 2 illustrates the probability of violation of constraints. Initially, with a small value for Γ_i, the probability of violation is high. However, as Γ_i increases, this probability decreases.

Figure 3 reports the optimal objective value on the vertical axis. When $\Gamma_i = 0$, the maximum objective function value is achieved, as there is no consideration for the worst-case scenario (Table 2). As Γ_i increases towards 5, the objective function value decreases, but only by 2% compared to when $\Gamma_i = 0$.

Table 1. Stochastic Model

Γ_i	n	(Γi - 1) / SQRT(n)	ϕ	1 - ϕ ((Γi - 1) / SQRT(n))	Optimal Robust Value
1	5	0.00	0.50	0.50	927,710
2	5	0.45	0.67	0.33	922,683
5	5	0.89	0.81	0.19	917,771
4	5	1.34	0.91	0.09	912,973
5	5	1.79	0.96	0.04	908,288

n: Number of random coefficients/ϕ: Cumulative distribution function of a standard normal.

Table 2. Deterministic Model

Γ_i	Optimal Robust Value
0	932,955

Figure 4 shows the marginal impact on the optimal value as the protection level increases. For up to a 46% chance of constraint violation, the objective function is reduced by only 2%. This indicates that by selecting a moderate value for Γ_i, we can achieve a balance between a small probability of violation and minimal loss in the objective function value.

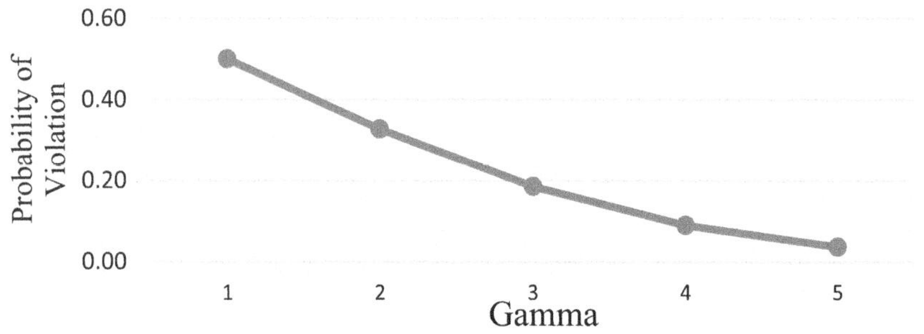

Fig. 2. Gamma Vs. The Probability of Violation

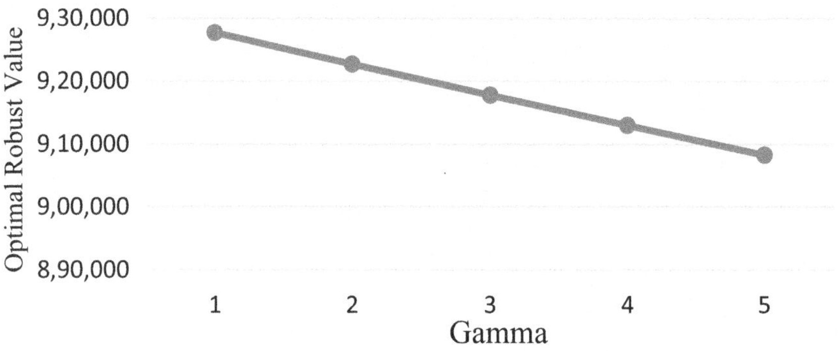

Fig. 3. Gamma Vs. The Optimal Robust Value

In this project, by introducing a set of constraints for the robust-counterpart model and establishing a computationally tractable representation, it is demonstrated that adopting the worst-case scenario for all options in terms of return percentage results in a low value for our maximization objective function. This outcome indicates that the price of robustness, when set to be highly robust (100% robust), is substantial. Therefore, we propose introducing some flexibility to reduce the cost associated with this robustness and to adjust the level of conservatism accordingly.

Increasing Γ_i can reduce the probability of constraint violations. This is because a larger Γ_i provides greater protection against infeasibility. However, as Γ_i increases, the maximization objective function tends to decrease. Hence, setting a large Γ_i can lead to overly conservative solutions with less favorable objective function values.

Therefore, determining the optimal value of Γ_i involves balancing the trade-off between the probability of infeasibility and the desired objective function outcome. In

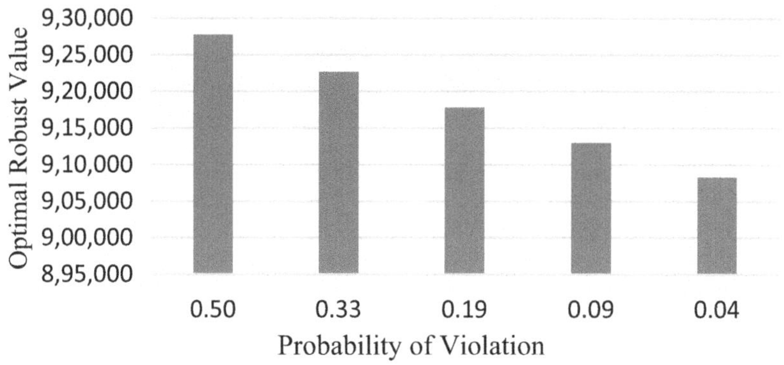

Fig. 4. The Optimal Robust Value Vs. The Probability of Violation

this project, we present graphs to illustrate this trade-off, aiming to achieve a high objective function while minimizing the chance of infeasibility. Therefore, we have estimated these probabilities to inform our final recommendations.

5 Conclusion

In this study, we propose a robust multi-objective optimization model for the reverse logistics of EV batteries. This model enables the exploration of the theoretical supply potential for various treatment activities, such as recycling, by incorporating the uncertain return percentage rates of the batteries. To do so, the price of robustness by adjusting the flexibility level of conservatism is proposed. This approach will guarantee that the model will remain feasible by allowing violations of the specific constraint under consideration.

To target the environmental goal of our model, there is always demand for retired batteries, so our aim is to address how remanufacturing or recycling EV battery centers can control the uncertainty nature of return batteries. In addressing robust scenarios, our objective is to determine the strategies for minimizing potential worst-case outcomes. This involves assuming a set of uncertain variables that vary within specified intervals, lack of known probability distributions. Therefore, our primary objectives are twofold: firstly, to assess the computational feasibility of robust counterparts for this uncertain optimization problem; and secondly, to identify the optimal structure for the uncertainty set, ensuring that the robust counterpart remains a manageable model. To our knowledge, our model is the first to implement this uncertain constraint in the closed-loop supply chain of BEV industry. However, the attractiveness of battery treatment to companies is not only determined by the amount of return. To assess other aspects such as the technical, economic feasibility and the social acceptance of retired batteries, further research is needed. Moreover, other uncertainties, such as determining whether retired batteries require remanufacturing, recycling, or other forms of treatment activities based on their condition, can also be addressed.

Disclosure of Interests. No potential conflict of interest was reported by the author(s).

References

1. European Environment Agency. Greenhouse gas emissions from transport in Europe. European Environment Agency (2023). https://www.eea.europa.eu/en/analysis/indicators/greenhouse-gas-emissions-from-transport
2. International Energy Agency. Electric Vehicles (2023). https://www.iea.org/energy-system/transport/electric-vehicles
3. Harper, G., et al.: Recycling lithium-ion batteries from electric vehicles. Nature **575**(7781), 75–86 (2019)
4. Chen, M., et al.: Recycling end-of-life electric vehicle lithium-ion batteries. Joule **3**(11), 2622–2646 (2019)
5. Ben-Tal, A., Nemirovski, A.: Robust solutions of uncertain linear programs. Oper. Res. Lett. **25**(1), 1–13 (1999)
6. Bertsimas, D., Sim, M.: The price of robustness. Oper. Res. **52**(1), 35–53 (2004)
7. Gonzales-Calienes, G., Yu, B., Bensebaa, F.: Development of a reverse logistics modeling for end-of-life lithium-ion batteries and its impact on recycling viability—a case study to support end-of-life electric vehicle battery strategy in Canada. Sustainability (Switzerland) **14**(22) (2022). https://doi.org/10.3390/su142215321
8. Gu, X., Ieromonachou, P., Zhou, L., Tseng, M.L.: Developing pricing strategy to optimise total profits in an electric vehicle battery closed loop supply chain. J. Clean. Prod. **203**, 376–385 (2018). https://doi.org/10.1016/j.jclepro.2018.08.209
9. Hendrickson, T.P., Kavvada, O., Shah, N., Sathre, R., Scown, C.D.: Life-cycle implications and supply chain logistics of electric vehicle battery recycling in California. Environ. Res. Lett. **10**(1) (2015). https://doi.org/10.1088/1748-9326/10/1/014011
10. Hoyer, C., Kieckhäfer, K., Spengler, T.S.: Technology and capacity planning for the recycling of lithium-ion electric vehicle batteries in Germany. J. Bus. Econ. **85**(5), 505–544 (2015). https://doi.org/10.1007/s11573-014-0744-2
11. Huster, S., Glöser-Chahoud, S., Rosenberg, S., Schultmann, F.: A simulation model for assessing the potential of remanufacturing electric vehicle batteries as spare parts. J. Clean. Prod. **363** (2022). https://doi.org/10.1016/j.jclepro.2022.132225
12. Kazemi Miyangaskary, M., Keivanpour, S., Lamghari, A.: Optimizing closed-loop supply chain in the electric vehicle battery industry: a fully fuzzy approach. In: Danjou, C., Harik, R., Nyffenegger, F., Rivest, L., Bouras, A. (eds.) PLM 2023, IFIP, pp. 38–49. Springer, Cham (2023). https://doi.org/10.1007/978-3-031-62582-4_4
13. Li, L., Dababneh, F., Zhao, J.: Cost-effective supply chain for electric vehicle battery remanufacturing. Appl. Energy **226**, 277–286 (2018). https://doi.org/10.1016/j.apenergy.2018.05.115
14. Nguyen-Tien, V., Dai, Q., Harper, G.D.J., Anderson, P.A., Elliott, R.J.R.: Optimising the geospatial configuration of a future lithium ion battery recycling industry in the transition to electric vehicles and a circular economy. Appl. Energy **321**, 119230 (2022). https://doi.org/10.1016/j.apenergy.2022.119230
15. Pamucar, D., Ebadi Torkayesh, A., Deveci, M., Simic, V.: Recovery center selection for end-of-life automotive lithium-ion batteries using an integrated fuzzy WASPAS approach. Expert Syst. Appl. **206**, 117827 (2022). https://doi.org/10.1016/j.eswa.2022.117827
16. Scheller, C., Schmidt, K., Herrmann, C., Spengler, T.S.: Decentralized planning of lithium-ion battery production and recycling. Procedia CIRP **90**, 700–704 (2020). https://doi.org/10.1016/j.procir.2020.01.075
17. Wang, L., Wang, X., Yang, W.: Optimal design of electric vehicle battery recycling network – from the perspective of electric vehicle manufacturers. Appl. Energy **275**, 115328 (2020). https://doi.org/10.1016/j.apenergy.2020.115328

18. Yarahmadi, A., Morency, C., Trepanier, M.: New data-driven approach to generate typologies of road segments. Transp. Transp. Sci. (2023). https://doi.org/10.1080/23249935.2022.2163206
19. Yarahmadi, A.: A Methodological Framework to Develop Optimal Sets of Driving Cycles for a Region (Doctoral dissertation, Polytechnique Montréal) (2023)
20. Zhang, C., Chen, Y.X., Tian, Y.X.: Collection and recycling decisions for electric vehicle end-of-life power batteries in the context of carbon emissions reduction. Comput. Ind. Eng. **175**(195), 108869 (2023). https://doi.org/10.1016/j.cie.2022.108869
21. Zhang, C., Tian, Y.X., Han, M.H.: Recycling mode selection and carbon emission reduction decisions for a multi-channel closed-loop supply chain of electric vehicle power battery under cap-and-trade policy. J. Clean. Prod. **375**, 134060 (2022). https://doi.org/10.1016/j.jclepro.2022.134060
22. Zhu, J., et al.: End-of-life or second-life options for retired electric vehicle batteries. Cell Rep. Phys. Sci. **2**(8), 100537 (2021). https://doi.org/10.1016/j.xcrp.2021.100537

Construction Logistics Conceptualization: A Comprehensive Framework and Existing Challenges

Uyen Ngo(✉), Erlend Alfnes, Fabio Sgarbossa, and Marco Semini

Norwegian University of Science and Technology, Trondheim, Norway
{thi.t.u.ngo,erlend.alfnes,fabio.sgarbossa,marco.semini}@ntnu.no

Abstract. Playing a key role in the nation's infrastructure development, construction industry has been studied extensively for several decades. However, there is a need of systemizing knowledge on construction logistics – a crucial element in improving performance of the industry. Combining multiple case study and literature review methods, this research offers a comprehensive framework for construction logistics with construction logistics activities positioned throughout construction project phases. Additionally, it explores practical challenges in on-site and prefabricated construction logistics operation, through which offering valuable insights into areas to be further developed and suggesting improvement ideas for practitioners and policymakers in construction and similar Engineer-to-Order industries.

Keywords: Construction logistics · Construction operation management · Engineer · to · order logistics · Engineer · to · order operation management

1 Introduction

Construction industry - the industry that performs necessary services to create long-lasting structures and building works [1] - is a crucial part of the country development process, as it fulfils the infrastructure demand of the social, cultural, and economic growth. The industry is significant to the growth of the economy along with the population explosion [2]. Together with customized machineries and ships manufacturing, construction is a typical engineer-to-order (ETO) industry [3]. It involves various tasks such as establishing new buildings, making modifications, conducting repairs, and carrying out demolitions [1]. A construction project is the product of the construction industry, which is normally done through a three-phase process, which is Design – Procurement – Installation [4] or Engineering – Procurement – Construction (EPC) [5]. The Design or Engineering phase transforms customer needs and technical requirements into a feasible model or design of the project. The Procurement phase helps acquire resources needed to complete the designed project from the previous phase. Lastly, the Installation phase contains assembly and installation of components integral to buildings or constructions.

Published by Springer Nature Switzerland AG 2024
M. Thürer et al. (Eds.): APMS 2024, IFIP AICT 733, pp. 508–521, 2024.
https://doi.org/10.1007/978-3-031-71645-4_34

In conventional construction, most of the construction activities happen on the construction site (on-site construction), while prefabricated construction is an approach in which majority of the on-site tasks are conducted in a factory setting away from the construction site [6]. Typically, construction projects are highly customized and implemented on a case-by-case foundation, characterized by limited instances of repeated orders, all while facing challenges related to cost and delivery time constraints [7]. In a few cases, multiple similar construction projects are undertaken, such as mountain cabins, beach cottages, student residences, military accommodations [8].

The construction industry is facing various performance problems [9] for which improvement opportunities could be initiated from logistics [10]. Generally, logistics is the process of coordinating, arranging, and managing the movement of materials and information from their source to the destination [11]. Appeared commonly in the literature, logistics in construction or construction logistics is a term covering the material and information flows to sustainably plan, execute, and control the right resources – humans, materials, equipment, and information - from the source of resources to the construction sites and during on-site operation at the right time, in the right quantity, with the right quality [12–14]. Figure 1 illustrates an overview of construction logistics, in which material and information flows link different key actors in the supply chain network such as suppliers, transporters, prefabrication factories, and operators on the construction site.

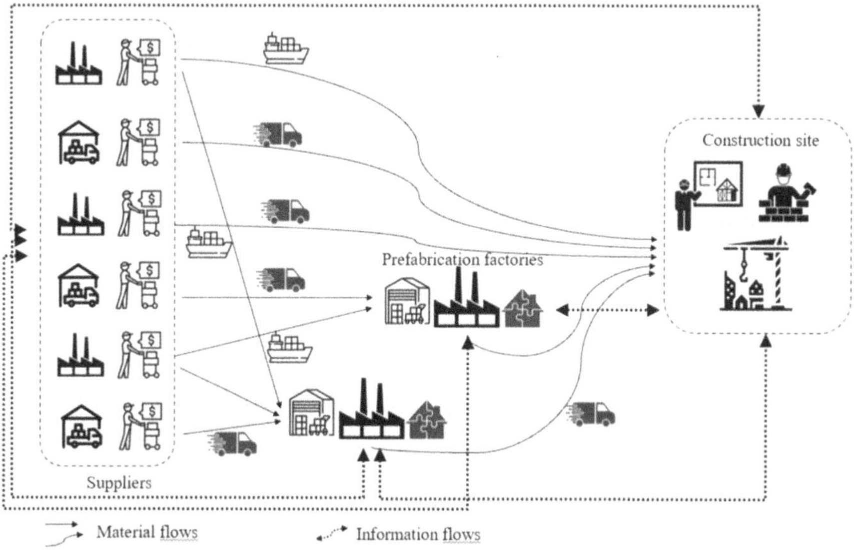

Fig. 1. Construction logistics

To ensure the efficiency and effectiveness of the flows in construction logistics, various tasks or activities should be undertaken, namely construction logistics activities. These activities are critical for maintaining the project schedule and budget, placing significant impact on the performance of the industry [15]. Because construction projects become more demanding and must adapt quickly to the dynamic economy, it is critical to ensure efficient and effective logistics. The unique characteristics of the construction combined with the extensive scale and lifecycle complexities contribute to the complex nature of logistics and supply chain management within the industry [16]. To improve construction logistics, it is important to build a comprehensive understanding of the topic. Similar to other ETO industries, construction is typically more experience-based, conservative and slow-moving than other industries [15, 17]. Therefore, there is a need for more formal, research-based knowledge on construction logistics. In order to be able to measure and improve their performance, the activities belonging to construction logistics need to be identified and clearly described. Furthermore, needs and challenges need to be mapped in order to define relevant paths for future research. This study attempts to fulfil these needs by answering the following research questions:

RQ1: What are activities of construction logistics along construction project phases?
RQ2: What are challenges and needs of construction logistics?

RQ1 aims to describe construction logistics activities and their positions corresponding to construction project phases. The expected answer to this question is a conceptual framework on construction logistics with logistics activities, aligned with the construction project phases and indicating their extent of overlap. Based on that, RQ2 helps explore existing challenges and needs of construction logistic, revealing opportunities to expectedly enhance the performance of logistics management.

The paper is structured as follows: chapter 2 outlines the methodology including research design and description of case studies. In the chapter 3 and 4, the research findings are synthesized and discussed to answer RQ1 and RQ2 accordingly. Lastly, chapter 5 finishes the paper by summarizing its contributions, acknowledging limitations, discussing implications, and suggestion of further research.

2 Research Method

2.1 Research Design

Both research questions are answered by a combination of literature review and multiple case studies. The literature review is used to gather theoretical knowledge from conference papers, journal articles, books, and other academic materials from SCOPUS database from the scope of the literature search also covered other ETO sectors than construction, such as shipbuilding and customized machinery production. Specifically, activities and challenges of construction logistics are gathered from the literature as a part of the answer for the RQ1 and RQ2 respectively.

The findings from literature search were then validated and expanded with case studies conducted at construction companies in Norway to enrich what has been discussed in the literature, fulfilling answers of the two research questions. The multiple case study method explores practical knowledge of the field with data collected from semi-structured interviews, industry documents, field notes, etc., Qualitative data analysis approach is then employed to analyze collected data from the literature search and the cases.

2.2 Data Collection

The data used for this study is collected through a combination of literature search and a multiple case study. The literature review is based on the guideline of [18], gathering relevant publications found from SCOPUS databases including journal articles, conferences papers, books, professional reports, etc.,. The search related to RQ1 was done for keywords combination of "Construction" OR "Yard" OR "Engineer-to-order" AND "Logistics" OR "Supply Chain Management", then adding "Challenges" OR "Barriers" OR "Problems" for RQ2. There are five construction companies involved in the study, with ten interviews completed and two site visits. Besides, other documents such as company workflows and internal documentations were also provided as inputs for the study analysis. The case companies are located in various places of Norway with varying working scopes. Two out of the five cases (case A, B) focus on prefabricated construction, while the scope of the other three is mainly on-site construction (case C, D, E). Table 1 presents the cases and the interviewees involved in each case. Interview recordings were transcribed and analyzed under tabular forms. The findings are presented in chapter 3 and 4.

Table 1. Case companies and corresponding interviewees

Case	Case description	Interviewees
A	Producing prefabricated elements for residential buildings, supplying prefabricated elements, constructing residential projects (cabins, houses, garages, and so forth)	Technical drawing engineer, Purchasing manager, Production leader, Sales manager
B	Engaging in the production and distribution of prefabricated concrete construction elements	Transportation engineer, Planning and transport team, Project manager
C	Constructing public infrastructures, industrial infrastructures, residential structures, and rehabilitation projects	Purchasing manager
D	Constructing public infrastructures, industrial infrastructures, residential structures, and rehabilitation projects, alongside the production of prefabricated elements	Project engineer
E	Constructing various types of structures including public infra-structures, industrial facilities, residential buildings, rehabilitation projects, and offshore installations	Production manager

3 Construction Logistics Activities and Construction Logistics Framework

3.1 Result from Literature Review

Adapting the views about construction project phases from [4, 5], this study refers construction project to be done through three-phase process of Design – Procurement - Fabrication and Construction, as it is compatible with the logistical perspective of the study. Construction industry possesses certain unique features that impact the logistical aspects, such as its distinctive project design, temporary location, irregular material supplies, specific sequence of the tasks [19], and many more. Therefore, it is essential for its actors to ensure the timely and precisely logistics activities. The specific features of the construction work can pose barriers to construction logistics activities, many of which are mentioned in the literature, and many await to be explored through practical experiences.

In the existing research works, different activities can be considered under construction logistics: planning of supply and planning of the tasks and personals [20–22]; specification and acquisition of supply resource [21, 22]; material receipt, handling and transportation [15, 23]; arrangement of machinery operations on the site [24]; storage and installation of materials on the site [15, 20, 24]; movements of materials, machineries, workers, and waste [15, 20]; creating the arrangement of the site and addressing conflicts among different construction teams on the site [22]; progress monitoring and reporting [23]. Popularly, logistics functions in construction can be grouped into supply logistics and on-site logistics [22, 25]. As can be seen, most of the scholars have been viewing construction logistics under the conventional on-site construction view. For prefabricated construction logistics, it is recommended to look at the manufacturing sector but not replicating it due to difference in work features [15]. Because construction logistics activities are fragmented in the literature, a framework is needed to organize these activities across different phases of a construction project. This will help connect and systemize our understanding towards construction logistics.

3.2 Result from Multiple Case Studies

Based on the knowledge about construction logistics from the literature, eight construction logistics activities are synthesized as following. The cases confirm the findings from the literature.

1. *Planning and coordination:* setting the objectives, schedules and allocation of necessary tasks and resources (personnel, materials, equipment, tools, funding) needed to achieve overall goal of the construction project.
2. *Resource acquisition:* obtaining the necessary personnel, materials, equipment, tools, funding required in alignment with the planned objectives and schedules. Long-term strategic activities such as personnel training, supplier relationship management, or long-term fundraising are excluded from the scope.
3. *Resource transportation:* moving, loading, and handling of necessary personnel, materials, equipment, and tools.

4. *Resource storage and handling:* receiving, handling, maintaining, and monitoring of construction materials and elements.
5. *Element fabrication:* production activities of construction elements ready for the on-site installation.
6. *Element installation:* integrating and assembling different construction components and elements to produce construction structures.
7. *Environmental management:* controlling environmental standards of the materials and production, handling, and disposal of various types of waste generated during element production and on-site construction activities.
8. *Progress monitoring and reporting:* recording, monitoring, and updating reports of the above activities.

Based on the findings from the cases and abovementioned construction logistics activities, a framework for construction logistics is proposed as illustrated in Fig. 2. In this framework, logistics activities are positioned through a three-phase construction project, with dashed arrows indicating feedback loops and communication between the activities. In particular, the "Planning and coordination" activity is highly connected to the "Progress monitoring and Reporting" activity, while each of them is in communication with the other activities. Additionally, the activities relating to the resource (Resource acquisition, Resource transportation, Resource storage and handling) are combined. Besides, the cases' findings separate two types of resource, namely project-specific resource and standard resource, which will be discussed in the later text.

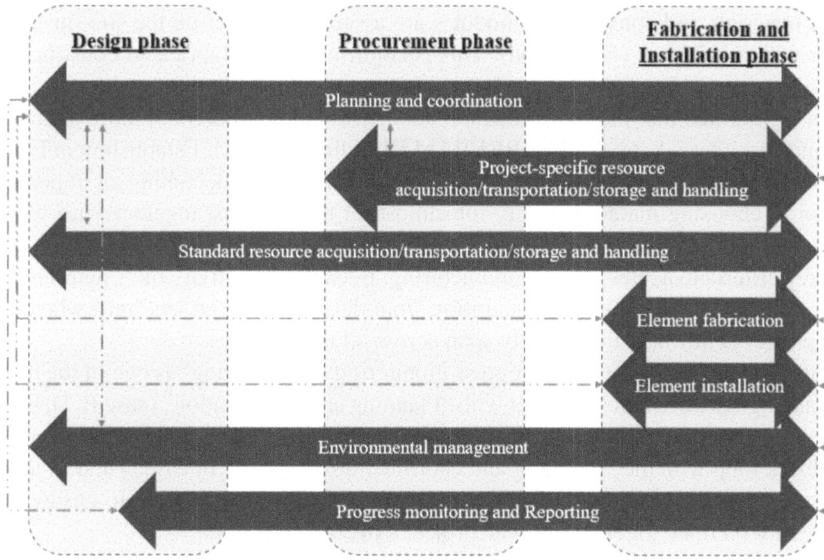

Fig. 2. Construction logistics framework

The cases studies suggest that "Project handling" and "Cleaning and finishing" could be added to the project phases (case C, D). However, they are considered as an ending part of the Fabrication and Installation phase and there is no specific logistics activity fitted particularly to them. Thus, they are not considered to be added to the framework.

The "Planning and coordination" activity should be initiated once the Design phase starts, even before contract signing (case B) so that resources and production capabilities are planned in alignment with the customer requirements of the project. Moreover, this activity must be repeatedly conducted along the project phases to timely adapt any changes in the design, resource control and production (case B, D, E), which is consistent with the discussion of [22].

The activity of "Resource acquisition" starting together with Procurement phase should only be applicable for project-specific materials, which are materials and components that suit specifically for a project (designed doors, structural steels, interior bricks, specialized equipment, etc.,.). For the standard materials and components that are consumed generally in every project (screw, hammer, drill, cement, tape, etc.,.), frequently procurement is executed based on standard reorder point method regardless of the project phase (case B). This separation of the resource types also happens to the "Resource storage and handling" and "Resource transportation" activities. These three resource-related activities should be conducted continuously until the last phase of the project is finished (case B, E). It is necessary to monitor and optimize resource movements as this is significant to the efficient logistics within and around the construction site, especially for the projects in urban areas with limited space and urban traffic congestion (case B, E).

The activity of "Element fabrication" and "Element installation" are completed in the third phase of the construction project – Fabrication and Installation phase. In this phase, the raw materials and construction modules are assembled, either on the site (installation) or within a factory (fabrication). This is confirmed by prefabricated construction companies (case A, B).

It is highlighted that the environmental concern is getting more and more attention in the industry (case A, C, D). The BREEAM (Building Research Establishment Environmental Assessment Method) certificate standard [26] is now becoming an important criterion in choosing material supply for almost all the projects, together with waste management (case C, D). For prefabricated construction (case A), reducing waste is one of the key criteria to achieve lean manufacturing. Because the activity of "Environmental management" is put under consideration from designing the project until when the whole project is finished, this activity span across all three phases.

Most of the cases agree that "Progress monitoring and reporting" is one of the most important logistics activities together with "Planning and coordination" (case B, D, E). It should start right after the contract is signed and be monitored in a tight communication or feedback loop with the "Planning and coordination" (case B). In case D and E, there are frequent weekly meetings, monthly meetings, and reporting scheme to ensure the activities are on track together with the project's overall progress.

"Quality inspection" is suggested to be added to the logistics activities (case B). Indeed, it is part of "Progress monitoring and reporting". An interesting finding is found that the proposed activities are positioned and planned by time sequence along the project phases. However, in the actual operation, they should be planned backward, starting from the expected delivery date of the whole project due the pull-based strategy of the industry (case A, C).

During the execution of the construction logistics activities, numerous challenges and needs must be identified and addressed to enhance construction logistics. Chapter 4 synthesizes this knowledge from previous research works and case studies.

4 Construction Logistics Challenges and Needs

4.1 Result from Literature Review

The construction industry is encountering critical challenges on performance [9, 27, 28]. Based on the logistics activities, the literature search identifies many characteristics of the construction industry that place substantial impacts on logistics, as mentioned by different scholars. It was mentioned that the construction industry is manual task based, conservative, lacking incentive to change and low-tech advancement [15, 29, 30]. The material supplies vary according to the construction phase, and any delay in one activity will affect the sequence of the other tasks and overall progress of the project [19]. Next, the work setting is one-off and temporary, thus hindering long-term optimization [15, 19, 30]. Employment is usually fragmented with the use of short-term and outsourced labor, thus reducing trust and hurting information sharing [15, 31]. Also, the communication between project owner, main contractors and subcontractors is poor [30, 32].

The characteristics of the construction industry are blamed to prevent it from successfully addressing the logistics [15]. In specific, various logistical problems have been mentioned in the previous research works. One of the problems is the poor management of construction equipment, tools, and materials [25]. Next, there is insufficient planning of material transportation, material deliveries and unloading [31, 33]. Planning is highly uncertain and affected by external factors such as weather changes and site topography [15]. Besides, there is lack of trust and coordination among actors due to inadequate information flow and communication [21, 30, 34]. High logistics costs due to traditional trading method is also another logistical problem [35], along with the subsequent delays and waste of materials due to delay in supply [36]. On the construction site, machineries are not utilized well [37] as well as vehicle arrivals are not optimized due to limited storage area in urban projects and uncontrolled vehicle arrangement at site [15]. Construction logistics is also facing lack of unified material acquisition procedure [33], lack of IT system used [38], and unsatisfactory in communication between material suppliers and subcontractors [31].

When it comes to the research methodology employed in previous studies, many of the challenges/barriers in construction logistics are discovered through reviewing literature, discussing concepts, or analyzing synthetic simulated data. However, it is crucial to gather insights from real-world experiences to fully understand these challenges. Moreover, specific challenges of the prefabricated construction have not been stated. The present study aims to contribute to filling these gaps with results from multiple cases presented in the next section.

4.2 Result from Multiple Case Studies

Based on the data collected from the semi-structured interviews and company visits, fifteen challenges in construction logistics have been grounded and presented in Table 2. Also, domains of challenges and possible consequences are presented in the table.

In general, there are four domains that might be used to categorize these construction logistics challenges: Logistics operations, IT infrastructure, Culture and working habits, and Externalities. These challenge domains are discussed further in Table 3. The challenges found from the case studies are accused of causing different consequences in logistics of the construction industry. Challenges #1 to #5 are blamed as a reason of information loss, thus creating inefficient planning and control of the logistics. Challenges #6 to #9 could lead to damages of materials and elements, while challenges #10 to #12 about poor planning activities are affecting construction progress in a negative way. The rest of the challenges are causing miscellaneous problems such as inappropriate performance measurement, repetitive human mistakes, and more cost incurred.

Most of the above challenges have been existing for a long time and consistent with what has been discussed by other scholars. For example, challenge #1 and #7 are consistent with findings of [21, 30, 34], which mentions the lack of trust and coordination due to poor information flow and communication among actors. [38] mentioned something similar to challenge #3 and #4 about a lack of unified system to share information and update changes among all actors, or challenge #9, #15 about limited storage in urban sites together with challenge #8 about weather condition have been mentioned by [15].

Noticeably, there are several challenges found from the case studies that have not been discussed widely in the literature, even though they have been existing for a while in the industry. Specifically, case company C mentioned the risk of extraordinary circumstances such as currency or exchange rate fluctuations, global disease, or wars that could affect material supply, causing supply uncertainties and possible supply chain disruptions (challenge #6). Next, the challenge #13 about inappropriate performance measurement is another obstacle, when some performance indicators are suitable for this logistics activity, or this project may be not applicable for the others. This could negatively impact performance assessment accuracy and demotivate construction logistics actors. For example, case company B is now using hour per ton to measure average time needed to produce a ton of finished good as a performance indicator for worker productivity. However, this indicator depreciates the difference in complexity level of different products or projects, making workers feeling that their performance is unfairly assessed.

Challenge #5 about human personalities and #10 about human resource planning are interesting. A change in staffing due to sick leave or unexpected circumstances could create subsequent delays in logistics activities, and misunderstandings caused by personality problems may severely affect the communication, long-term coordination, and hurt the trust among logistics actors. This might be further emphasized now due to an increased frequency of hiring, firing, and turnover among workers, greater reliance on foreign labor, and more frequent fluctuations in the industry.

Table 2. Challenges in construction logistics from case studies

Challenge	Challenge domain	Consequences
1. Inefficient information flow and coordination among the teams, or between internal teams and subcontractors	IT infrastructure, Culture and working habits	Loss of information, mismatch in client specifications and production capabilities, poor planning, overload/underload operation, progress delays
2. Traceability of materials due to inefficient inventory control	IT infrastructure, Logistics operations	Information loss leading to inefficient planning, material damages and wastes
3. Manual documentation	IT infrastructure	Loss of information
4. Fragmented tools, lack of unified system	IT infrastructure	Information loss, poor planning
5. Human personalities	Culture and working habits	Ineffective communication, loss of information
6. Supply uncertainties (supply lead time, market fluctuation, currency changes, war, etc.,)	Externalities	Supply chain disruption and resource (time and money) wastes
7. Conflict of interests between different departments	Culture and working habits	Inefficient decision-making, resource waste
8. Weather challenge (extreme weather conditions, changes in weather)	Externalities	Material damages, resource waste, delays in progress
9. Limited in storage capability and transportation in urban projects	Logistics operations, Externalities	Material damages, progress delays
10. Ineffective human resource planning	Logistics operations	Production inefficiency and delays
11. Sudden changes in operation due to poor planning	Logistics operations, IT infrastructure	Waste of resource, progress delays
12. Poor planning between delivery and on-site traffic	Logistics operations	Waste of resource, progress delays
13. Inappropriate performance measurement	Logistics operations	Ineffective assessment on the logistics performance

(*continued*)

Table 2. (*continued*)

Challenge	Challenge domain	Consequences
14. Experience-based working method	Logistics operations, Culture and working habits	Possible human repetitive mistakes
15. Special restriction on vehicle accessibility to the site	Logistics operations	More cost incurred

The construction logistics challenges found could be classified into four groups of challenge domains as shown in Table 3. These are the primary needs that should be considered by construction logistics practitioners and managers. The Logistics operation challenges are directly related to logistics operation (planning, resource acquisition, resource transportation, etc.,.). The IT infrastructure challenges are about the lack or inefficiency in technological systems and tools to support and enhance logistics operations. Next, the Cultural and working habit challenges are about human personalities and their working manners. Lastly, the Externalities challenges are related to extraordinary circumstances or uncontrollable external factors. These challenge domains are interdependent to each other. For instance, the lack of efficient IT infrastructure might create barriers in logistics operation, and Externalities challenges such as wars could result in worker psychological problems, thereby causing working habit challenges or logistics operation challenges. Therefore, it is important that the management pays attention to address all these four domains of challenges simultaneously, thereby sustainably improve the construction logistics.

The suggestive remedies to mitigate these challenges are also presented in Table 3. To overcome challenges from logistics operations, it is suggested to focus on improving information flows among stakeholders, standardize operating procedures, or consider utilizing third-party logistics services. To mitigate IT infrastructure challenges, construction companies should consider developing a unified IT system to support logistics activities. Next, improving communication, implementing standard operating procedures, and providing logistics training are recommended to address challenges in culture and working habits. Lastly, construction companies could address challenges from externalities by predicting them promptly, adopting a multi-sourcing strategy, and maintaining effective inventory control to handle sudden supply changes.

Table 3. Construction logistics challenge domains and suggested remedies

Challenge domain	Suggested remedies
Logistics operations	Improve information flows among all stakeholders Implement standard operation procedures to minimize reliance on experienced-based operations Consider using third-party logistics services
IT infrastructure	Develop comprehensive IT tools to support various logistics functions like supply management, warehouse operations, planning, and production management. Prefer integrated systems over fragmented ones
Culture and working habits	Improve communication between internal staffs and subcontractors Implement and adhere to standard operating procedures Provide logistics-specific training for the workforce
Externalities	Promptly predict and respond to external circumstances Adopt multi-sourcing strategy to mitigate risks Manage inventory effectively to handle supply chain disruptions

5 Conclusion

The study contributes to the knowledge of the field with two primary outcomes. Firstly, it presents a comprehensive framework on construction logistics describing the logistics activities and their alignment across construction project phases. Secondly, the study explores existing practical construction logistics challenges and needs, with four domains of the challenges and suggested remedies to mitigate them. These outcomes contribute not only to systemizing knowledge within the field of construction logistics but also to offering practical insights for practitioners and policymakers in the construction operation management in general. Furthermore, the findings are expected to be useful for similar ETO industries.

This scope of this study is limited by its reliance on a small number of case studies, which may hinder the generalization of findings. Additionally, some interviewees may not have comprehensive insights into all aspects of their company's logistics functions, potentially restricting their views. These limitations emphasize the need for cautions when applying the study conclusions to broader contexts and highlight opportunities for future research to address these gaps. Besides that, future research is suggested to put an emphasis on analyzing prefabricated construction logistics, proposing analytical models to quantitatively validate the found challenges, and applying new IT technologies in improving construction logistics.

References

1. Hillebrandt, P.M.: The nature of construction economics. In: Economic Theory and the Construction Industry, pp. 3–9. Springer, Cham (1974). https://doi.org/10.1007/978-1-349-01927-4_1

2. Zidane, Y.J.-T., Andersen, B.: The top 10 universal delay factors in construction projects. Int. J. Manag. Proj. Bus. **11**(3), 650–672 (2018)
3. Schulze, F., Dallasega, P.: Industry 4.0 application in ETO companies: an empirical comparison. In: Alfnes, E., et al. (eds.) IFIP International Conference on Advances in Production Management Systems. Springer, Cham (2023)
4. Kermanshachi, S., Pamidimukkala, A.: Uncertainty analysis of key schedule performance indicators in design, procurement, and construction phases of heavy industrial projects. J. Leg. Aff. Disput. Resolut. Eng. Constr. **15**(1), 04522042 (2023)
5. Force, U.o.T.a.A.C.I.I.C.S.C.T. and U.o.T.a.A.B.o.E.: Research. Model Planning and Controlling System for EPC of Industrial Projects, vol. 6. Institute, University of Texas at Austin (1987)
6. Han, Y., Yan, X., Piroozfar, P.: An overall review of research on prefabricated construction supply chain management. Eng. Constr. Archit. Manag. **30**(10), 5160–5195 (2023)
7. Cigolini, R., et al.: Supply Chain Management in Construction and Engineer-to-Order Industries, pp. 803–810. Taylor & Francis (2022)
8. Lawson, M., Ogden, R., Goodier, C.I.: Design in Modular Construction, vol. 476. CRC Press Boca Raton, FL (2014)
9. Fikri Mohamed, S., Anumba, C.J.: Potential for improving site management practices through knowledge management. Constr. Innov. **6**(4), 232–249 (2006)
10. Matouzko, V., Methanivesana, N.: Improving Construction Logistics: A Case Study of Residential Building Project (2012)
11. Sundquist, V., Gadde, L.-E., Hulthén, K.: Reorganizing construction logistics for improved performance. Constr. Manag. Econ. **36**(1), 49–65 (2018)
12. Janné, M.: Construction Logistics in a City Development Setting, vol. 2091. Linköping University Electronic Press (2020)
13. Lange, S., Schilling, D.: Reasons for an optimized construction logistics. In: Proceedings of the 23rd Annual Conference of the International Group for Lean Construction. 2015. IGLC 2015 Organizing Committee Perth, Australia (2015)
14. Guffond, J.-L., Leconte, G.: Developing construction logistics management: the French experience. Constr. Manag. Econ. **18**(6), 679–687 (2000)
15. Sullivan, G., Barthorpe, S., Robbins, S.: Managing Construction Logistics. John Wiley & Sons (2011)
16. AlMaian, R.Y., et al.: Supplier quality management inside and outside the construction industry. Eng. Manag. J. **27**(1), 11–22 (2015)
17. Khan, A.N., et al.: Paving the way for technological innovation through adoption of artificial intelligence in conservative industries. J. Bus. Res. **165**, 114019 (2023)
18. Moher, D., et al.: Preferred reporting items for systematic reviews and meta-analyses: the PRISMA statement. Ann. Intern. Med. **151**(4), 264–269 (2009)
19. Janné, M., Fredriksson, A.: Construction logistics governing guidelines in urban development projects. Constr. Innov. **19**(1), 89–109 (2019)
20. Fredriksson, A., et al.: WP1: Construction logistics scenarios and stakeholder involvement Deliverable 1.1: Scenarios of construction logistics. (2020)
21. Fadiya, O., et al.: Decision-making framework for selecting ICT-based construction logistics systems. J. Eng. Des. Technol. **13**(2), 260–281 (2015)
22. Jang, H., Russell, J.S., Yi, J.S.: A project manager's level of satisfaction in construction logistics. Can. J. Civ. Eng. **30**(6), 1133–1142 (2003)
23. Strandhagen, J.W.: Towards Next-Generation Yard Logistics (2022)
24. Brusselaers, N., Fufa, S.M., Mommens, K.: A Sustainability assessment framework for on-site and off-site construction logistics. Sustainability **14**(14), 8573 (2022)

25. Shigute, S., Nasirian, A.: The future of construction logistics consolidation centers in construction. Master of Science Thesis in the Master Program Design and Construction Project Management Chalmers University of Technology Sweden (2014)
26. Schweber, L.: The effect of BREEAM on clients and construction professionals. Build. Res. Inf. **41**(2), 129–145 (2013)
27. Todsen, S.: Produktivitetsfall i bygg og anlegg. 2018. https://www.ssb.no/bygg-bolig-og-eiendom/artikler-og-publikasjoner/produktivitsfall-i-bygg-og-anlegg.last. Accessed 2023/July
28. Ahmad, S.B.S., et al.: Labour productivity statistics: a reality check for the Norwegian construction industry. Int. J. Constr. Manag. **20**(1), 39–52 (2020)
29. Hossain, M.A. and A. Nadeem. Towards digitizing the construction industry: State of the art of construction 4.0. in Proceedings of the ISEC. 2019. (2019)
30. Kim, S.-Y., Nguyen, V.T.: Supply chain management in construction: critical study of barriers to implementation. Int. J. Constr. Manag. **22**(16), 3148–3157 (2022)
31. Behera, P., Mohanty, R., Prakash, A.: Understanding construction supply chain management. Prod. Plann. Control **26**(16), 1332–1350 (2015)
32. Maskuriy, R., et al.: Industry 4.0 for the construction industry: Rev. Manag. Perspective. Econ. **7**(3), 68 (2019)
33. Ying, F., Tookey, J., Roberti, J.: Addressing effective construction logistics through the lens of vehicle movements. Eng. Constr. Archit. Manag. **21**(3), 261–275 (2014)
34. Briscoe, G., Dainty, A.: Construction supply chain integration: an elusive goal? Supp. Chain Manag. Int. J. **10**(4), 319–326 (2005)
35. Vrijhoef, R., Koskela, L.: The four roles of supply chain management in construction. Eur. J. Purchas. Supp. Manage. **6**(3–4), 169–178 (2000)
36. Agapiou, A., et al.: The role of logistics in the materials flow control process. Constr. Manag. Econ. **16**(2), 131–137 (1998)
37. Dubois, A., Hulthén, K., Sundquist, V.: Organising logistics and transport activities in construction. Int. J. Logist. Manag. **30**(2), 620–640 (2019)
38. Salami, E., Aydinli, S., Oral, E.L.: Barriers to the implementation of supply chain management-case of small to medium sized contractors in turkey. Int. J. Science and Research **5**(9), 516–520 (2013)

Integrated Dynamic Flexible Job Shop and AIV Scheduling: Deep Reinforcement Learning Approach Considering AIV Charging and Capacity Constraints

Arman Hosseini[1], Mohammad Feizabadi[2], and Zakaria Yahouni[2](✉)

[1] Systems Engineering, University of Virginia, Charlottesville, VA 22903, USA
ufy5tc@virginia.edu

[2] Univ. Grenoble Alpes, CNRS, Grenoble INP G-SCOP, 38000 Grenoble, France
seyedmohammadghasem.feizabadi@etu.univ-grenoble-alpes.fr,
zakaria.yahouni@grenoble-inp.fr

Abstract. Scheduling Automated Intelligent Vehicles (AIV) in Dynamic Flexible Job Shops is a challenging problem due to its high level of stochasticity and dynamic nature. Various heuristic and exact methods have proven effective when the complexity of the problem is relatively low. Moreover, in the past decade, machine learning algorithms, particularly reinforcement learning, have been applied to sophisticated scheduling tasks and demonstrated their efficiency in solving such problems. In this study, a deep reinforcement learning approach is proposed to address the integrated Dynamic Flexible Job Shop and AIV scheduling. Aimed at optimizing two objectives: total lateness of jobs and total energy consumption of AIVs. The study takes into account the limitations of AIV transporters, such as charging consumption and loading capacity. To validate the proposed method, a case study of a flexible job shop is designed, and our approach is compared with a combination of existing heuristics.

Keywords: AIV scheduling · Flexible Job shop · Manufacturing optimization · Reinforcement learning · Deep Q-network

1 Introduction

Material Handling Systems (MHS) such as transportation robots, play a crucial role in enhancing manufacturing system performances [6]. These systems encompass various activities, including storage, packaging, and product movement. The emergence of industry 4.0 technologies has automated MHS functions, allowing human operators to focus on tasks with added value. Among these technologies, Automated Intelligent Vehicles (AIVs), stand out for their ability to autonomously transport products while considering workshop constraints and layout, making them ideal for flexible autonomous manufacturing [5].

M. Thürer et al. (Eds.): APMS 2024, IFIP AICT 733, pp. 522–536, 2024.
https://doi.org/10.1007/978-3-031-71645-4_35

In this context, developing control strategies for AIVs becomes imperative to align with manufacturing objectives, such as timely product delivery, workload balance among workstations, and energy consumption. These strategies entail scheduling rules that assign AIVs to transportation requests. Moreover, the dynamic and uncertain nature of manufacturing environments presents challenges such as machine breakdown, necessitating real-time scheduling and adaptation to dynamic events.

Within the aforementioned context, our study aims to investigate and propose a capacitated AIV scheduling approach that empowers decision-makers to effectively plan and manage their AIV fleets. The primary objective is to simulate a manufacturing and logistic process and to leverage Machine Learning (ML) algorithms, particularly Deep Reinforcement Learning (DRL) techniques to predict optimal/near-optimal scheduling rules for both products and AIV robots. We introduce a deep Q-network framework that simultaneously allocates machines and AIVs to jobs to optimize the total lateness of jobs and energy consumption of AIVs.

The contribution of our approach lies in its ability to address the challenges of dynamic scheduling in a manufacturing environment with machine's breakdown using DRL techniques. Specifically, our methodology integrates a deep Q-network with two types of agents to achieve efficient AIV scheduling. Furthermore, we demonstrate the effectiveness of our approach through its application to a flexible job shop case study and compare it with six pairs of heuristics.

The remaining of the paper is structured as follows: Sect. 2 delves into the literature background. Section 3 introduces a deep reinforcement learning approach using two separate agents to optimize product and AIV scheduling. Section 4 presents a case study of a flexible dynamic job shop problem with AIV transporters. Section 5 elucidates the detailed experimental protocol and results, and Sect. 6 presents the conclusions of this study.

2 Literature Review

Scheduling, particularly in the context of Job Shop Scheduling Problems (JSSP), has been a focal point of optimization research due to its NP-hard nature [23]. The challenge is further amplified in flexible job shop scheduling problems (FJSSP), where each job can be processed by multiple machines [4]. Moreover, in modern-day job shops, the majority are equipped with automated MHS for transferring raw materials and sub-assemblies between different machines. While the scheduling of transporter robots such as Automated Guided Vehicles (AGVs) has garnered attention in previous studies [7,24], it is noted that power supply considerations and capacity of AGVs have not been deeply explored in the literature [13,19]. To the best of our knowledge, only one paper has addressed capacitated AGV scheduling [16], unlike other studies that typically assume AGVs can transfer only one product at a time. The upgraded version of AGVs is represented by AIVs which utilize mapping and environment recognition to get to the destination [8]. Unlike AGVs, AIVs do not necessitate a fixed movement line and are therefore more adaptable to flexible manufacturing scenarios with variable layouts, such as open-plan manufacturing environments [5]. Our study is centered on integrated FJSSP and multi-AIV scheduling, taking into account factors such as charging consumption and the capacity of AIVs.

To address FJSSP, methods such as mixed integer linear programming (MILP) and metaheuristics have been commonly employed in previous studies [17]. However, it is worth noting that in a real-world FJSSP, unexpected events such as machine breakdowns or late arrival of jobs are common occurrences. These uncertainties contribute to the complexities and create a dynamic environment that is challenging to address solely with MILP and metaheuristics [21]. Dynamic flexible job shop scheduling problem (DFJSSP) necessitates real-time actions. One prominent approach involves the selection of priority dispatching rules at each decision state. These heuristic approaches select the first job to be processed based on different priorities such as Shortest Processing Time (SPT). While these methods are straightforward to apply and interpret, their drawback lies in their inability to guarantee an optimal solution [15]. With the advent of machine learning and the availability of data through the Internet of Things (IoT), researchers have turned their attention towards data-driven approaches. Machine learning methods have proven efficient in addressing production scheduling and rescheduling problems [14]. In this context, machine learning methods emerge as a viable alternative to classical operations research techniques, particularly for predicting scheduling rules based on manufacturing states. Within the literature, two primary machine learning paradigms are prevalent: supervised learning, which relies on historical data, and reinforcement learning, which is well-suited for dynamic environments and utilizes an agent-reward-policy paradigm for scheduling [18].

Concerning supervised learning, [14] applied classification methods such as Random Forest and Support Vector Machine to forecast the most appropriate times for rescheduling in FJSSP. The approach of [8] was different as they employed multiple linear regression to predict the optimal decision in each state with a focus on minimizing the makespan. This study served as a bridge between supervised learning and Markov Decision Processes, establishing decision states and actions. It is noteworthy that supervised learning methods necessitate extensive data collection and are not conducive to real-time decision-making processes. On the other hand, reinforcement learning (RL) methods excel at rapidly adapting to unexpected events by leveraging previously obtained information [10]. In job shop and AIV scheduling problems, Q-learning stands out as a widely employed method [22]. Authors [2] applied a Q-learning approach to optimize the makespan in a job shop scheduling problem. However, tabular reinforcement learning approaches like Q-learning and Sarsa are more suited for simpler environments with small state and action spaces. In more complex environments with larger state and action spaces, the memory and computational requirements of maintaining a tabular representation become impractical [20]. To address this dimensionality issue, researchers have shifted their focus to a combination of deep learning and reinforcement learning methods, known as deep reinforcement learning.

[11] employed a deep Q-network to schedule jobs on machines. Deep reinforcement learning becomes even more predominant when addressing multiple decision points within a system. In the study by Hu et al. [9], a deep Q-network is employed to tackle two primary decisions concurrently: Selecting a dispatching rule for the jobs between five heuristic candidates and Selecting an idle AGV for transferring the task. Results show that the proposed approach outperformed classical RL methods such as SARSA and Q-Learning. However, their work did not consider constraints related to the dynamic behavior of the environment.

[12] introduced a multi-action deep reinforcement learning approach for flexible job shops, incorporating two actions: job operation selection and machine selection, aimed at minimizing the makespan. Their proposed Multi-Proximal Policy Optimization (Multi-PPO) method showcased notable generalization capabilities, particularly for larger-sized problems. However, from a practical standpoint, solving any scheduling problem entails tackling a multi-criteria problem wherein various objectives must be simultaneously optimized. These objectives may include factors such as the energy efficiency of AIVs and the completion time of jobs [1].

In this study, a multi-objective Deep Q-Network approach is introduced to optimize the integrated dynamic flexible job shop and AIV scheduling, taking into account the charging consumption and transfer capacity of AIVs.

3 Methodology

3.1 Deep Reinforcement Learning Architecture

In this section, a deep reinforcement learning architecture is presented to address the dynamic FJSSP with AIV transporters. This architecture addresses two types of decisions: The first decision, managed by the machine agent, involves job-machine assignment. This decision is crucial as jobs can be processed on different machines within the FJSSP. The second decision, handled by the AIV agent, involves job-AIV assignment, considering multiple AIVs on the shop floor.

Figure 1 illustrates the proposed architecture. At the top level, a central cloud contains all recently arrived jobs and those awaiting the next operations. Jobs in the central cloud are sorted using the Earliest Due Date (EDD) policy. The utilization of the EDD policy aligns with one of the primary objectives of the study, which is to minimize the total lateness of the jobs. EDD has been chosen as it has proven to be a qualified heuristic for addressing this objective [3]. This policy is dynamically applied, updating the priority queue of the central cloud when new jobs arrive or operations are completed.

Moving to the second level, the machine agent assigns a machine to process each job in the queue. Once this assignment is made, the destination for each job becomes clear, indicating the need for job transfer to their designated machines. Finally, at the lowest level AIV agents determine which AIV should carry out the job transfer.

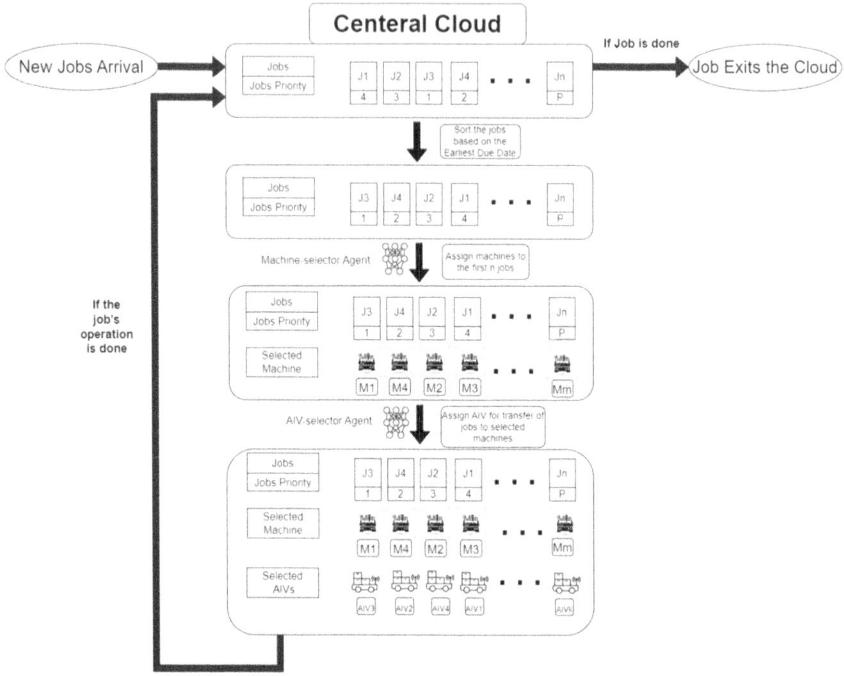

Fig. 1. Decision process

3.2 Deep Q-Network

Reinforcement learning approaches, rooted in the framework of Markov Decision Processes (MDP), excel in handling sequential decision-making problems. The fundamental principles of reinforcement learning methods include states, actions, and rewards. At each time step t, the agent observes the environment's state S_t, and upon taking action A_t, transitions to the next state S_{t+1}. Subsequently, the agent receives an immediate reward upon entering the new state S_{t+1}. In this section, we model dynamic FJSSP with AIVs as a Partially Observable Markov Decision Environment, including two types of reinforcement learning agents: the machine-selector agent and the AIV-selector agent each of which has a different observation from the environment based on which the agent performs an assignment. It is worth noting that these two agents are decentralized, with each equipped with a separate deep Q-network (DQN). The machine-selector network focuses on optimizing jobs' lateness, while the AIV-selector network works on minimizing the AIVs' energy consumption. It is worth noting that in each training process of agents, the configuration of the FJSSP problem (job due dates, distances between machines, machines' processing times) is fixed, and for different configurations, the agents require a new training process. More details of the FJSSP configurations are provided in Sect. 5.1.

Detailed descriptions of both agents are provided subsequently.

Machine-selection agent network, where states include the following:

- Queue length of machines
- Machines' busy time percentage
- Jobs' next processing times
- Jobs' remaining processing times
- Location of AIVs
- States (busy/idle) of AIVs
- Charging percentage of AIVs

The action of the machine-selector agent consists of selecting an appropriate machine for the available jobs in the central cloud. It should be noted that the jobs in the central cloud are sorted in real-time based on their due dates. In each state, the machine-selector assigns the first n jobs of the central cloud to the machines, where n is equal to the sum of capacities of all AIVs. For example, if there are 4 AIVs and each AIV has a capacity of 2, the machine-selector agents assign machines to the first 8 jobs of the central cloud. Whenever each of these jobs finishes its operation on one of the machines, the immediate reward will be updated based on the current lateness of the jobs.

It's essential to note that handling the variation in the dimension of actions at each step requires the use of a mask for the first n operations within the cloud. This mask indicates the potential actions (machines) available for each operation. For example, if job J1 can only be served on machines M1 or M3 for its first operation among all m machines, we mask out the other machines except M1 and M3. Consequently, the output of the network is adjusted to $-\infty$ where no viable action exists (masked options). This adjustment to $-\infty$ is beneficial as it allows the objective function, which uses an argmax, to ignore the masked options. The current lateness attribute is calculated as follows:

$$\text{Job current lateness} = \text{Job's due} - \text{Current simulation time} - \text{Job's remaining processing time} \quad (1)$$

At the end of each episode, when all the jobs are processed, a cumulative reward is given based on the total (sum of) lateness of all jobs. Due to the lack of optimal solution, to evaluate the performance of the agents, this reward is related to the network's performance in previous episodes and is calculated as the difference between the total lateness of the current episode and the minimum total lateness of previous episodes which is considered as the best solution explored. In this scenario, if the total lateness improves compared to the minimum of previous episodes, it will receive a positive reward; otherwise, it will receive a negative reward.

AIV-selector agent network includes the following states:

- Queue length of transfer requests for each AIV
- Location of each AIV
- State of each AIV (busy/idle)
- Battery charging percentage of AIVs
- Busy time percentage of AIVs
- Location of the first n jobs in the central cloud

The actions of this agent consist of assigning the proper AIVs to the first n jobs of the central cloud whose machines are already assigned by the machine-selector agent. Using the same example of having 4 AIVs, each with a capacity of 2, the AIV-selector agents decide which two jobs should be assigned to each AIV. After transferring each job to the desired machine, a -1 reward is allocated for each percentage of energy consumption during the transfer of one job on an AIV.

At the end of each episode, and after finishing the processing of all jobs, a global reward is given to the AIV-selector Q-network. Similar to the machine-selector network, this reward is based on the difference between the total charging consumption of AIVs in the current episode and the minimum total charging consumption of all previous episodes. If the consumption is less than the minimum, the reward will be positive; otherwise, it will be negative.

4 Case Study

The assumed shop floor is designed to evaluate and validate the proposed approach within a flexible job shop composed of four machines (M1, M2, M3, and M4), each capable of performing two different operations (though not simultaneously). In total, there are four types of operations in this workshop, represented as (Op1, Op2, Op3, Op4). For example, machine M1 is capable of performing operations Op1 and Op4. The list of machines and the operations they are capable of is provided in Table 1.

Table 1. Operations of each machine

Machine	M1	M2	M3	M4
Operations	Op1, Op4	Op2, Op3	Op1, Op3	Op2, Op4

Four final products (P1, P2, P3, P4) are processed through the job shop, with each product consisting of multiple jobs all arriving at the beginning of the simulation. The number of jobs is assumed to be equal between all products. For example, in a problem with a total number of jobs of 20, there are 5 jobs for each product. Upon arrival, each job will be stored in the storage. Once a job finishes processing, it will remain in the buffer of the last station, which is assumed to have an infinite capacity.

Each product follows a specific series of operations to become a final product. The list of required operations for each product is provided in Table 2, presented in the correct order of operations.

As an example, for jobs of product P1, they start with operation Op1, then proceed to operation Op3, and finish with operation Op4. Referring to Table 1, we know that operation Op1 can be performed on either machine M1 or M3, operation Op3 can be performed on either machine M2 or M4, and operation

Table 2. List of operations of each product

Products	P1	P2	P3	P4
Operations	Op1 -> Op3 ->Op4	Op2 -> Op3 -> Op1	Op1 -> Op2 -> Op3	Op3 -> Op4 -> Op2

Op4 can be performed on both machine M1 and M4. Therefore, a decision-making process is required for the selection of machines for each operation of jobs. This decision is handled by the machine selector agent. Figure 2 represents the decision tree of machine selection for product P1.

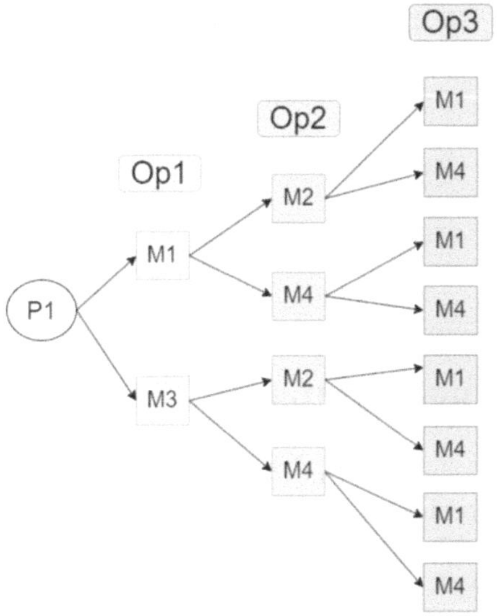

Fig. 2. Mahine selection decision tree for product P1. There are 8 possible routings for product P1

Two AIVs are exclusively responsible for transporting jobs between storage, machinery units, and a charging location. Each AIV can accommodate a maximum of two jobs per transfer. AIVs remain at their last destination if no requests are assigned to them. When an AIV's battery charge falls below 20%, it transfers the current jobs it carries and proceeds directly to the charging station. After charging its battery to 100%, it resumes its transportation activities. The energy consumption of AIVs correlates with the load of the jobs and traveling distance. The size of the jobs is assumed to be identical, and each job occupies one out of two capacity units of the AIVs. A distance matrix D defines the distances between machinery units, charging stations, and storage locations. This matrix is assumed to be fixed in all scenarios. Figure 3 illustrates the model of the flexible job shop with AIV transporters.

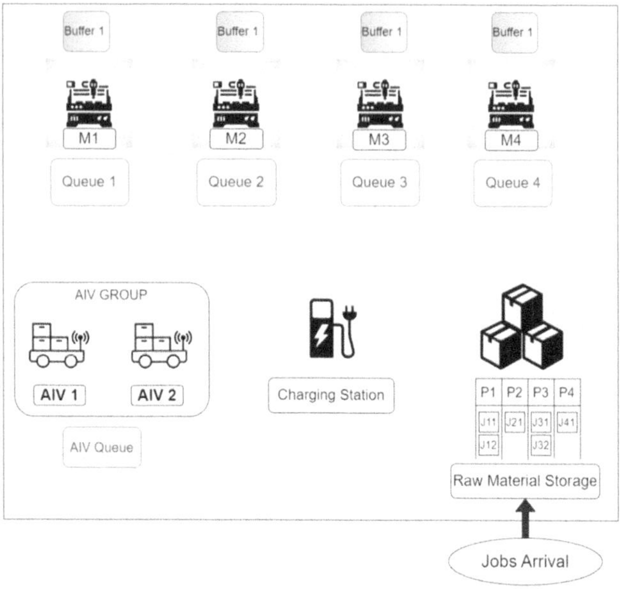

Fig. 3. Model of case study

5 Experimentation and Results

The flexible job shop is simulated in Python using OpenAI gym and agents are implemented using Pytorch library. Python provides flexibility and ease of implementation, while OpenAI Gym offers a standardized framework for developing and testing reinforcement learning algorithms. Experimentation protocols and scenarios are described in the following sections.

5.1 Protocol

Different configurations of the case study are simulated to assess the performance of the proposed approach. A configuration consists of a set of properties including the number of jobs for each product to be produced, the due date of each job, and the processing times for operations on each machine. All these properties are randomly generated, except for the total number of jobs, where seven values are tested: 20, 40, 60, 80, 100, 120, and 160. For example, 20 jobs mean 5 jobs for each of P1, P2, P3, and P4.

A machine breakdown is also incorporated as a dynamic behavior of the system. Each machine has a 20% probability of breaking down in every 200 time unit. The duration of each breakdown follows a normal distribution with $\mu = 125$ time units and $\sigma = 10$ time units.

As previously mentioned, two AIVs are employed for transferring products. Both AIVs are considered identical in terms of speed, charging consumption, and loading capacity. The speed of AIVs is assumed to follow a normal distribution with $\mu = 4$ m/s and $\sigma = 0.4$ m/s.

5.2 DQN Training

For each agent, machine-selector, and AIV-selector, a separate DQN is implemented. The machine-selector aims to minimize the total lateness of jobs, while the AIV-selector aims to minimize the total energy consumption of AIVs. For each selector agent, a fully connected deep neural network with 5 hidden layers and 20 nodes in each layer, along with ReLU activation functions, is proposed. Stochastic Gradient Descent with a learning rate of 0.1 optimizes the Mean Squared Error (MSE) loss function. Additionally, a learning rate scheduler is applied throughout the training process to reduce the learning rate by a factor of 0.9 based on the performance of the network. The discount factor for both networks is set to 0.9. Figure 4 illustrates the convergence of agents for 40 jobs (10 jobs for each product of P1, P2, P3, and P4) across 60 episodes. Each episode is conducted with a new configuration while maintaining the same total number of jobs.

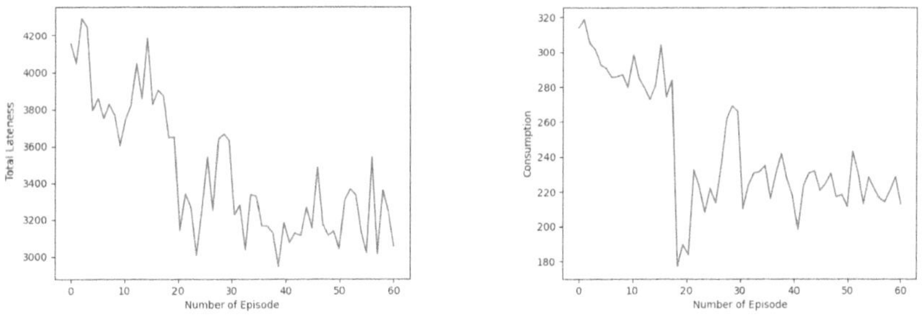

(a) Lateness (b) AIVs energy Consumption

Fig. 4. Convergence of DQN for an example of 40 jobs

5.3 Evaluation

For the evaluation, the proposed DQN is compared with heuristics derived from blending dispatching rules for machine selection and AIV selection. Three dispatching heuristics for machine selection and two heuristics for AIV selection are provided, resulting in a total of six rules generated from their combination.

- SPT: Machine with the Shortest Processing Time is selected for each job. It is worth noting that the processing time of each operation varies depending on the machine performing this process.

- Qlenght: Machine with the shortest queue Lenght has the priority.
- SWL: Machine with the Shortest Workload or shortest percentage of busy time has priority.

Two dispatching rule candidates for AIV selection:

- STT: Select AIV with the Shortest Transfer Time (from its current location to the location of the first job to pick up)
- SWL: Choose AIV with the Shortest Workload or shortest percentage of busy time.

The results of the proposed DQN implementation are compared with the performance of these six heuristic combinations. To achieve this comparison, we conducted simulations with increasing numbers of jobs (20, 40, 60, 80, 100, 120, and 160). For each configuration, the simulation was executed 10 times, and the average result was collected. Figures 5 and 6 demonstrate the performance of the proposed method against the heuristic combination methods.

Table 3. Average results of total lateness

Jobs	(STT, SPT)	(STT, Qlength)	(STT, SWL)	(SWL, SPT)	(SWL, Qlength)	(SWL, SWL)	DQN Agent
20	1442.2	1565.6	1765.9	2320.8	2607	2593.3	1053.5
40	3775.6	3602.8	3538.8	4766.1	5069.5	5253.5	3183.9
60	5412.6	4512.2	5629.9	7773.7	8176.6	7792.1	3760.1
80	7250.5	8192.8	8280.1	11393.9	12608.4	12120.3	5764.2
100	9093.6	9138.4	9391.7	12637	13239.4	14425.9	6178.9
120	12556.5	13755.9	14455.2	19832.6	20136.8	20005.5	9098.8
160	15979.6	16963.2	16982	21716.9	23784.8	24201.7	9717.2

The proposed method outperformed all the heuristics in the total lateness objective, and its superiority was amplified by increasing the number of jobs. In the case of 20 jobs, the difference between the mean total lateness of our method and the best heuristic (STT, SPT) is 388.7 units (as shown in Table 3), while with the worst heuristic (SWL, Qlength), it is 1,553.5 units. The difference grows significantly with 160 jobs, where the difference between the proposed method and the best heuristic (STT, SPT) is 6,262.4 units, and with the worst heuristic in this case (SWL, SWL), it is 14,484.5 units.

Increasing the number of jobs within a job shop environment significantly amplifies the complexity, posing challenges to optimization models. Our proposed approach adeptly navigates this complexity. The results demonstrate that our method exhibits superior convergence in optimizing total lateness as the number of jobs escalates.

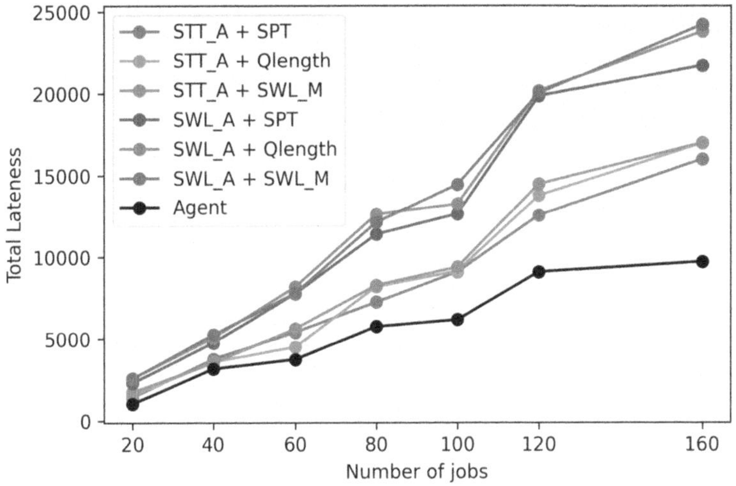

Fig. 5. Total lateness of jobs

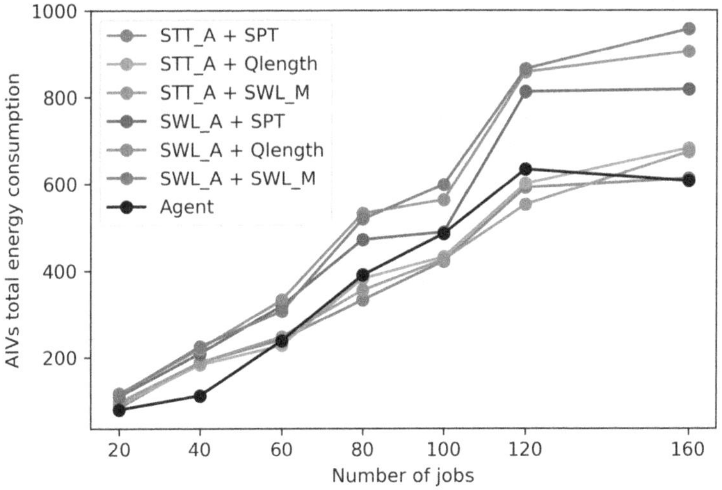

Fig. 6. AIVs energy consumption

While DQN agents outperform three heuristic candidates in terms of AIVs energy consumption, unsatisfactory results are observed for scenarios involving 60, 80, 100, and 120 jobs. This discrepancy may be attributed to the multi-objective nature of the problem. Conflicting objectives may lead to suboptimal decisions, where prioritizing total lateness may not necessarily coincide with optimal decisions for energy consumption.

6 Conclusion

In this study, we have proposed an integrated dynamic flexible job shop scheduling approach coupled with automated intelligent vehicle (AIV) routing using two agents of deep reinforcement learning (DRL). Our methodology addresses the challenges of dynamic scheduling in manufacturing environments by leveraging advanced machine learning techniques, specifically DRL. Through two agents architecture, we modeled the dynamic flexible job shop scheduling problem (FJSSP) with AIV transporters. Two distinct types of decisions were addressed: job-to-machine assignment and AIV-to-job assignment. Our approach utilized a Deep Q-Network (DQN) framework to optimize both total lateness of jobs and energy consumption of AIVs simultaneously.

During the training phase, the DQN agents demonstrated convergence after 60 episodes, with the reward function being profoundly influenced by total lateness throughout the simulation. Evaluation of the proposed method against heuristic approaches revealed promising results. Our DQN approach outperformed several heuristic candidates, particularly in optimizing total lateness as the number of jobs increased. However, challenges arose, especially concerning AIVs' energy consumption in environments with higher job complexity. The multi-objective nature of the problem led to conflicting decisions between optimizing total lateness and minimizing energy consumption.

Further research could focus on refining the DRL model to better balance conflicting objectives and address specific challenges in complex manufacturing environments. Several perspectives emerge from our study, aiming to generalize and extend the applicability of our approach beyond the scope of the current research.

Firstly, it is imperative to generalize our approach by testing it on various case studies with different configurations. This includes exploring scalability by considering larger numbers of machines, as well as varying capacities and layouts. By diversifying the testing scenarios, we can assess the robustness and adaptability of our methodology across different manufacturing environments.

Furthermore, our focus is on addressing challenges encountered within the multi-agent framework and multi-objective nature, particularly those stemming from the decentralized architecture. We plan to enhance communication between agents and explore methods for efficiently sharing parameter states and actions among all agents. Additionally, we aim to investigate the implementation of a shared reward system.

Eventually, the proposed methodology can be applied in a real world job shop using a fleet management software with the proposed approach implementation where at each decision making point the software is used to perform assignments.

References

1. Barak, S., Moghdani, R., Maghsoudlou, H.: Energy-efficient multi-objective flexible manufacturing scheduling. J. Clean. Prod. **283**, 124610 (2021)
2. Belmamoune, M.A., Ghomri, L., Yahouni, Z.: Solving a job shop scheduling problem using q-learning algorithm. In: International Workshop on Service Orientation in Holonic and Multi-Agent Manufacturing, pp. 196–209. Springer (2022)
3. Blackstone, J.H., Phillips, D.T., Hogg, G.L.: A state-of-the-art survey of dispatching rules for manufacturing job shop operations. Int. J. Prod. Res. **20**(1), 27–45 (1982)
4. Chan, F., Wong, T., Chan, L.: Flexible job-shop scheduling problem under resource constraints. Int. J. Prod. Res. **44**(11), 2071–2089 (2006)
5. Cronin, C., Conway, A., Walsh, J.: State-of-the-art review of autonomous intelligent vehicles (aiv) technologies for the automotive and manufacturing industry. In: 2019 30th Irish signals and systems conference (ISSC), pp. 1–6. IEEE (2019)
6. Esmaeilian, B., Behdad, S., Wang, B.: The evolution and future of manufacturing: a review. J. Manuf. Syst. **39**, 79–100 (2016)
7. Fu, J., Zhang, H., Jian, Z., Jiang, L.: Review on agv scheduling optimization. J. Syst. Simul. **32**(9), 1664–1675 (2020)
8. Hosseini, A., Yahouni, Z., Feizabadi, M.: Scheduling aiv transporter using simulation-based supervised learning: a case study on a dynamic job-shop with three workstations. IFAC-PapersOnLine **56**(2), 8591–8597 (2023)
9. Hu, H., Jia, X., He, Q., Fu, S., Liu, K.: Deep reinforcement learning based agvs real-time scheduling with mixed rule for flexible shop floor in industry 4.0. Comput. Ind. Eng. **149**, 106749 (2020)
10. Kayhan, B.M., Yildiz, G.: Reinforcement learning applications to machine scheduling problems: a comprehensive literature review. J. Intell. Manufacturing, pp. 1–25 (2021)
11. Lang, S., Behrendt, F., Lanzerath, N., Reggelin, T., Müller, M.: Integration of deep reinforcement learning and discrete-event simulation for real-time scheduling of a flexible job shop production. In: 2020 Winter Simulation Conference (WSC), pp. 3057–3068. IEEE (2020)
12. Lei, K., Guo, P., Zhao, W., Wang, Y., Qian, L., Meng, X., Tang, L.: A multi-action deep reinforcement learning framework for flexible job-shop scheduling problem. Expert Syst. Appl. **205**, 117796 (2022)
13. Li, M., Wang, G.G.: A review of green shop scheduling problem. Inf. Sci. **589**, 478–496 (2022)
14. Li, Y., Carabelli, S., Fadda, E., Manerba, D., Tadei, R., Terzo, O.: Machine learning and optimization for production rescheduling in industry 4.0. The Int. J. Adv. Manuf. Technol. **110**(9), 2445–2463 (2020)
15. Luo, S.: Dynamic scheduling for flexible job shop with new job insertions by deep reinforcement learning. Appl. Soft Comput. **91**, 106208 (2020)
16. Maoudj, A., Kouider, A., Christensen, A.L.: The capacitated multi-agv scheduling problem with conflicting products: model and a decentralized multi-agent approach. Robot. Comput.-Integrated Manuf. **81**, 102514 (2023)
17. Nouiri, M., Bekrar, A., Jemai, A., Niar, S., Ammari, A.C.: An effective and distributed particle swarm optimization algorithm for flexible job-shop scheduling problem. J. Intell. Manuf. **29**, 603–615 (2018)
18. Seeger, P.M., Yahouni, Z., Alpan, G.: Literature review on using data mining in production planning and scheduling within the context of cyber physical systems. J. Ind. Inf. Integr. **28**, 100371 (2022)

19. Tubis, A.A., Poturaj, H.: Risk related to agv systems-open-access literature review. Energies **15**(23), 8910 (2022)
20. Wang, L., Pan, Z., Wang, J.: A review of reinforcement learning based intelligent optimization for manufacturing scheduling. Complex Syst. Model. Simul. **1**(4), 257–270 (2021)
21. Wang, Y.: Flexible job shop scheduling rules mining based on random forest. In: 2021 2nd International Conference on Computing and Data Science (CDS), pp. 220–226. IEEE (2021)
22. Xue, T., Zeng, P., Yu, H.: A reinforcement learning method for multi-agv scheduling in manufacturing. In: 2018 IEEE International Conference on Industrial Technology (ICIT), pp. 1557–1561. IEEE (2018)
23. Yahouni, Z., Mebarki, N., Sari, Z.: Evaluation of a new decision-aid parameter for job shop scheduling under uncertainties. RAIRO-Oper. Res. **53**(2), 593–608 (2019)
24. Zhong, M., Yang, Y., Dessouky, Y., Postolache, O.: Multi-agv scheduling for conflict-free path planning in automated container terminals. Comput. Ind. Eng. **142**, 106371 (2020)

Multi-project Scheduling Problem with Limited Preemption and Ordinary Feeding Precedence Relations

Ali Fradi(✉) and Maha Benali(✉)

Department of Mathematics and Industrial Engineering, Polytechnique Montréal, Montréal, Canada
{ali.fradi,maha.ben-ali}@polymtl.ca

Abstract. Many companies operating in an Engineering-to-Order environment are under constant pressure to manage the complexity of production planning. This paper deals with the case study of a company offering highly customized products that are modular buildings. To support tactical production planning and capacity allocation, this paper formulates the production planning problem as a Preemptive Resource Constrained Multi-project Scheduling Problem and proposes two mathematical formulations to solve it. The first presented formulation performs preemption using continuous activity splitting whilst the second permits discontinuous energy distribution. With the presented formulations we aimed to curtail resultant makespans and delays of all set projects. Resource capacities are defined as accumulated weekly working hours of available and similar skills. We evaluated the performance of the proposed formulations, in terms of solving time and makespan, using three real-world datasets. Our findings indicate that the formulation using discontinuous energy allocation outperforms the formulation offering more control on the sub-activities.

Keywords: Linear programming · Preemption · Resource Constrained Multi-project Scheduling Problem · Ordinary Feeding Precedence Relations · Production Planning

1 Introduction

Industries often adopt an Engineering-To-Order (ETO) approach to serve a market characterized by customized products. ETO involves tailoring the product creation process to meet unique customer requirements, which adds complexity to production planning and capacity allocation. Tactical planning comes as a solution that divides the global problem into sub-problems that are easier to solve [1,2]. While on the medium-term horizon planners may have limited access to information such as performance details, resource availability, and required materials, activity aggregation and flexibility in resource allocation can simplify planning [3]. In the literature, this problem is usually presented as a Resource

Published by Springer Nature Switzerland AG 2024
M. Thürer et al. (Eds.): APMS 2024, IFIP AICT 733, pp. 537–551, 2024.
https://doi.org/10.1007/978-3-031-71645-4_36

Constrained Multi-project Scheduling Problem (RCMPSP), since each customer order can be considered as an individual project.

This article is motivated by the need of our industrial partner operating in an ETO context to develop a decision support system to support tactical production planning. Our partner manufactures modular buildings according to customer requirements. Production planning and capacity allocation for the medium-term horizon are challenging due to the large number of activities competing for available resources and the constraints imposed by each project. Since planners need to preempt some activities to deal with urgent projects, the problem can be formulated as a Preemptive RCMPSP (P-RCMPSP). A tool is needed to support production planning, especially when a new order is received or a customer requests additional changes.

This paper aims to develop and compare different mathematical formulations to solve this P-RCMPSP, taking into account the specificities of the production process of modular buildings. To this end, we propose two linear programming formulations that allow preemption and are adapted to our partner's real case study. While the first formulation focuses on continuous activity splitting, the second allows preemptions with discontinuous energy distribution. The performance of these two formulations in terms of solving time and global makespan is evaluated on three real project portfolios.

The rest of the paper is structured as follows: Sect. 2 presents a brief literature review of the RCPSP and the different extensions of this problem. While mathematical formulations are presented in Sect. 3, experiments are described in Sect. 4. Section 5 analyses the results. Section 6 concludes the major points of the current paper.

2 Related Literature Review

Resource Constrained Project Scheduling Problem (RCPSP) is well-studied and literature offers a variety of extensions to it. Most of the time, literature treats the planning of projects as an all-in-one mega-project. However, when each project has its own utility function (cost, priority, deadline...etc.), it is more convenient to define it as Resource Constrained Multi-Project Scheduling Problem (RCMPSP) [4]. In addition, some works consider preemptive activities as they promote curtailing costs linked to delays which define a new extension known as Preemptive RCMPSP (P-RCMPSP). Preemptions can be an integer, discrete, or continuous [5]. Integer preemptions are most common in literature and occur in integral points of time without an explicit bounded count. In contrast, discrete preemptions are limited and integral. Continuous preemptions can occur infinitely. Bounding the occurrence of preemptions during the activity execution is needed to keep a sense of productivity.

Literature about RCPSP highlighted different types of precedence relations. Some relations account for competing for resources such as concurrency and exclusiveness precedence relations. Others assume time intervals to consider between activities [4]. Generalized work relationships and generalized feeding precedence relations suggest a more complex definition of activity sequencing. While the latter includes all precedence relations features, Generalized work relationships define ordinary relations between initial and final parts of preceding and succeeding activities respectively [5]. For P- RCMPSP, traditional precedence relations need to be redefined to assist the sequential order of activities [3–5]. In this case study, an activity can not be launched until its predecessor is accomplished at some rate. This type of precedence relations is called ordinary feeding [3,5].

The literature highlights two different possibilities to perform preemptions [6]: by continuous splitting [7] or by discontinuous energy distribution [3]. The literature presents RCPSP as a sub-class of P-RCPSP when activities are not allowed to be suspended. Suspending an activity is possible through the allocation of zero energy during one or more time periods. It is also possible when an activity is satisfied through the accomplishment of disjoint sub-activities when resources are continuously engaged.

Studies of RCMPSP problems show an interest in different objective functions [4]. Objective functions can focus on resources (net present value, resource leveling, etc.), projects (time balance, project quality, project splitting), or delays (makespan, delivery delay, etc.). Combinations of different orientations of the objective function are also possible.

In ETO manufacturing systems, a decision-maker may accept a near-optimal or equivalently optimal solution with adjustments to fit his preferences. In this direction, [8] suggests investigating the near-optimal solution space to select alternatives solution that better responds to decision-maker preferences. [9] discusses a triangular algorithm with a variable locking feature to assist the decision-maker in his analysis of preferred solutions, while [8] uses bipolar heuristic and [10] uses Large Neighborhood Search (LNS) for the so-called Mixed-Initiative System (MIS). For instance, a solution space is required as input to such algorithms for MIS interaction. The solution space is generally pulled from a linear formulation afforded in advance.

As shown in Table 1, to the best of our knowledge, no study has addressed the P-RCMPSP considering discrete preemption and ordinary feeding precedence relations. This paper proposes two formulations inspired by two studies dealing with a single-project scheduling problem (P-RCPSP). The first formulation extends the model proposed in [7], which deals with a P-RCPSP and does not consider any constraint on overlapping activities, unlike our case study. The second formulation is inspired by the model proposed in [3], which also deals with a P-RCPSP and does not consider ordinary precedence relations as in this paper. In addition, this article has the advantage of being a real-world case study.

Table 1. Positioning in comparison to the related literature

Reference	Multi-project	Problem acronym	Preemption	Precedence relations	Data
[11]	No	P-RCPSP	Integral	Ordinary	Customized data
[6]	No	P-RCPSP	Integral	Generalized work	Modified PSPLIB
[12]	No	P-RCPSP	Integral	Ordinary	Modified PSPLIB
[3]	No	P-RCPSP	Integral	Generalized feeding	Real
[7]	No	P-RCPSP	Discrete	Ordinary	Generated
[13]	Yes	P-RCMPSP	Integral	Concurrency	Generated
[14]	Yes	P-RCMPSP	Integral	Exclusive	Generated
[15]	Yes	P-RCMPSP	Integral	Ordinary	Real
[16]	Yes	P-RCMPSP	Integral	Generalized feeding	Generated
[17]	Yes	P-RCMPSP	Integral	Ordinary	Generated
This paper	Yes	P-RCMPSP	Discrete	Ordinary feeding	Real

3 Mathematical Formulation

In this section, we first describe the case study. Then, we present the objective functions and two linear programming formulations (A and B) for the preemption. Formulation A uses continuous activity splitting, while formulation B permits discontinuous energy distribution.

3.1 Problem Definition

As presented in Fig. 1, the modular building production process includes six activities: sketching, design, machine work, welding, paintwork, and finishing. For each confirmed project (i.e., a confirmed customer order), production activities must be planned to meet the customer's requirements and the desired delivery date. Thus, the problem at a specific moment can be formulated as follows: the production planner has to plan production for a portfolio of $\mathcal{I}$ projects. This involves scheduling the six activities of each project i on a horizon $\mathcal{H}$. The goal is to minimize the makespan denoted as $E_{i|\mathcal{J}(i)|}$.

Tables 2 and 3 present shared sets and parameters between the two formulations A and B. It is worth noting that, in formulation A, a number of continuous sub-activities μ_{ij} is equivalent to the definition of $\mu_{ij} - 1$ as the limit of preemptions in formulation B.

Fig. 1. Modular building production process

Table 2. List of Sets

Sets	Definitions
$\mathcal{I}$	Set of projects
$\mathcal{J}(i)$	Set of activities in project i
$\mathcal{H} = \{1, \ldots, l\}$	Set of time buckets
$\mathcal{K}$	Set of resources
$\Psi = \bigcup_{i,j} \psi(i,j) = \bigcup_{i,j} \{1, \ldots, \mu_{ij}\}$	Set of possible sub-activities
$\bar{\psi}(i,j) = \{\mu_1, \ldots, \mu_{ij} - 1\}$	Set of possible preemption counts for activity j of project i
$succ(j)$	Set of successors of activity j

Table 3. List of Parameters

Parameters	Definitions
ε_{ij}	Minimum duration of a sub-activity j of project i
ρ_{ij}	Required progress ratio on activity j of project i to launch its successor
p_{ij}	Energy required to satisfy activity j of project i
ds_i	i^{th} customer desired delivery date
r_{kt}	Total amount of resource k available in time bucket t
η_m	Minimal non null energy to allocate
M_R	Relaxation constant

3.2 Formulation of the Objective Function

An objective function minimizing delays and makespan is usually used for linear programming models dealing with multi-project planning problems. In this paper, we propose two objective functions OF1 and OF2. The easiest option is to linearize the sum of possible delays and makespan of all projects as depicted by the objective function OF2. Constraints (2) and (3) activate the variable V_i indicating a delayed delivery of the project i.

Closely similar, the objective function OF1 does the same with proportional minimized terms for the makespan of each project delivered before its due date. Constraints (5) and (6) have to be added to take a linear form of the expression $max(E_{i|\mathcal{J}(i)|} - ds_i, 0)$, which corresponds to the resultant delay of project i.

OF1

$$min \Sigma_{i \in \mathcal{I}} 2E_{i|\mathcal{J}(i)|} - ds_i.V_i \quad (1)$$

$$l * V_i \geq E_{i|\mathcal{J}(i)|} - ds_i, \forall i \in \mathcal{I} \quad (2)$$

$$-l * (1 - V_i) \leq E_{i|\mathcal{J}(i)|} - ds_i, \forall i \in \mathcal{I} \quad (3)$$

OF2

$$min \Sigma_{i \in \mathcal{I}} E_{i|\mathcal{J}(i)|} + D_i \quad (4)$$

$$D_i \geq E_{i|\mathcal{J}(i)|} - ds_i, \forall i \in \mathcal{I} \quad (5)$$

$$D_i \geq 0, \forall i \in \mathcal{I} \quad (6)$$

In the following, two formulations are proposed to consider a limited count of activity preemption and ordinary feeding precedence relations.

3.3 Non-preemptive Splitting: Formulation A

The formulation A offers the possibility to control the minimal time of continued work on an activity j of project i before suspending it through ε_{ij}. In this formulation, having μ possible continuous sub-activity is equivalent to having $\mu - 1$ possible preemptions. $E_{i|\mathcal{J}(i)|\mu_{i|\mathcal{J}(i)|}}$ is used to denote the end date of the latest activity of project i (i.e., the makespan). Table 4 defines the variables of formulation B, then, the constraints are depicted.

Table 4. Variables definition of formulation A

Variable	Support	Definition
$E_{ij\mu}$	$\mathbb{N}^*$	End date of sub-activity μ of j in project i
N_{ijt}	$\mathbb{N}^*$	Hours of resource k allocated to activity j of i in t
$P_{ij\mu}$	$\mathbb{N}$	Duration of sub-activity μ of j in i
$S_{ij,\mu}$	$\mathbb{N}^*$	Start date of sub-activity μ of j in project i
$Y_{ij\mu}$	$\{0, 1\}$	Equals to 1 if the sub-activity μ is of non-zero duration, 0 otherwise
α_{ijt}	$\{0, 1\}$	Equal to 1 if the energy spent at t on activity j of project i has not yet satisfied the minimum progress ρ_{ij}, 0 otherwise
$X_{ij\mu t}$	$\{0, 1\}$	Equal to 1 if the sub-activity μ associated with activity j of project i is running at period t, 0 otherwise
Z_{ijt}	$\{0, 1\}$	Equal to 1 to execute activity j of project i is running at period t, 0 otherwise
V_i	$\{0, 1\}$	Equal to 1 if project i is late, 0 otherwise

$$S_{ij(\mu+1)} \geq E_{ij\mu}, \forall i \in \mathcal{I}, j \in \mathcal{J}(i), \mu \in \bar{\psi}(i,j) \tag{7}$$

$$P_{ij\mu} \geq \varepsilon_{ij} Y_{ij\mu}, \forall i \in \mathcal{I}, j \in \mathcal{J}(i), \mu \in \psi(i,j) \tag{8}$$

$$Y_{ij\mu} \geq Y_{ij(\mu+1)}, \forall i \in \mathcal{I}, j \in \mathcal{J}(i), \mu \in \bar{\psi}(i,j) \tag{9}$$

$$l.Y_{ij\mu} \geq P_{ij\mu}, \forall i \in \mathcal{I}, j \in \mathcal{J}(i), \mu \in \psi(i,j) \tag{10}$$

$$E_{ij\mu} = P_{ij\mu} + S_{ij\mu}, \forall i \in \mathcal{I}, j \in \mathcal{J}(i), \mu \in \psi(i,j) \tag{11}$$

$$E_{ij\mu} \geq S_{ij(\mu+1)} - l.Y_{ij(\mu+1)}, \forall i \in \mathcal{I}, j \in \mathcal{J}(i), \mu \in \bar{\psi}(i,j) \tag{12}$$

$$S_{ij\mu} - t \leq l.(1 - X_{ij\mu t}), \forall i \in \mathcal{I}, j \in \mathcal{J}(i), \mu \in \psi(i,j), t \in \mathcal{H} \tag{13}$$

$$t - E_{ij\mu} + \eta_m \leq l.(1 - X_{ij\mu t}), \forall i \in \mathcal{I}, j \in \mathcal{J}(i), \mu \in \psi(i,j), t \in \mathcal{H} \tag{14}$$

$$\Sigma_{t \in \mathcal{H}} X_{ij\mu t} = P_{ij\mu}, \forall i \in \mathcal{I}, j \in \mathcal{J}(i), \mu \in \psi(i,j) \tag{15}$$

$$\Sigma_{\mu \in \psi(i,j)} X_{ij\mu t} = Z_{ijt}, \forall i \in \mathcal{I}, j \in \mathcal{J}(i), t \in \mathcal{H} \tag{16}$$

$$\Sigma_i N_{ijt} \leq r_{jt}, \forall j \in \mathcal{K}, t \in \mathcal{H} \tag{17}$$

$$\Sigma_t N_{ijt} = p_{ij}, \forall i \in \mathcal{I}, j \in \mathcal{J}(i), t \in \mathcal{H} \tag{18}$$

$$N_{ijt} - M_R.Z_{ijt} \leq 0, \forall i \in \mathcal{I}, j \in \mathcal{J}(i), t \in \mathcal{H} \tag{19}$$

$$N_{ijt} \geq \eta_m - (1 - Z_{ijt}) * M_R, \forall i \in \mathcal{I}, j \in \mathcal{J}(i), t \in \mathcal{H} \tag{20}$$

$$\sum_{h=1}^{t} N_{ijh} \geq \rho_{ij}.p_{ij} - \alpha_{ijt} p_{ij}, \forall i \in \mathcal{I}, j \in \mathcal{J}(i), j \in \mathcal{K}, t \in \mathcal{H} \tag{21}$$

$$\sum_{h=1}^{t+1} N_{ij_{succ}h} \leq p_{ij_{succ}}(1 - \alpha_{ijt}), \forall i \in \mathcal{I}, j \in \mathcal{J}(i), j_{succ} \in succ(j), j \in \mathcal{K}, t \in \mathcal{H} \tag{22}$$

$$S_{ij1} \geq t * Z_{ij't}, \forall i \in \mathcal{I}, j \in \mathcal{J}(i), t \in \mathcal{H}, j > j' \tag{23}$$

$$E_{ij\mu_{ij}} \leq E_{i(j+1)\mu_{i(j+1)}}, \forall i \in \mathcal{I}, j \in \mathcal{J}(i) \tag{24}$$

Constraints 7 ensure precedence and non-overlapping between sub-activities. The set of constraints 8 prevent the model from preempting an activity j of a project i before expending ε_{ij} units of time. Constraints 9 and 10 imply that the zero-duration sub-activities decided by the model come at the end and not at the beginning or in the middle. Constraints 11 calculate the end dates of the sub-activities. Constraints 12 give start and end dates equal to the end date of the last non-zero duration sub-activity. Constraints 13, 14 and 15 define the time ranges associated with the sub-activities while those 16 establish the time ranges where activity j in a project i is running. Respecting resource limits for each period is governed by Constraints 17. Then those of 18 guarantee that distributed energy satisfies the completion of an activity. Constraints 19 disengage resources outside the working ranges. Constraints 20 define the minimum energy to be allocated per period during an activity. Constraints 21 and 22 allow overlapping between two consequential activities j and $succ(j)$ after having achieved a minimum progression of ρ_{ij} in j. Constraints 23 prevent the overlap of two activities that are not immediately consequential. Finally, Constraints 24 ensure to finish an activity j of a project i before the end date of the next activity.

3.4 Preemptive Work Allocation: Formulation B

Formulation B creates preemptions through the discontinuous distribution of workload. This offers the option to control the number of possible preemptions during each activity while no additional control options are allowed as in formulation A. $E_{i|\mathcal{J}(i)|}$ is used in formulation B to denote the end date of the latest activity of project i (i.e., the makespan), unlike formulation A that used $E_{i|\mathcal{J}(i)|\mu_{i|\mathcal{J}(i)|}}$. Table 5 defines the variables of formulation B, then, the constraints are depicted.

Table 5. Variables definition of formulation B

Variable	Support	Definition
S_{ij}	$\mathbb{N}$	Start date of project i activity j
E_{ij}	$\mathbb{N}$	End date of project i activity j
B_{ijt}	$\{0, 1\}$	1 indicates the start of activity j of project i at t
F_{ijt}	$\{0, 1\}$	1 indicates the end of activity j of project i at t
X_{ijt}	$[0, 1]$	Progress ratio for activity j of project i at t
T_{ijt}	$\{0, 1\}$	1 indicates that project i's activity j is strictly progressing at t

$$\sum_{i=1}^{\mathcal{I}} p_{ij} * X_{ijt} \leq r_{jt}, \forall j \in \mathcal{K}, t \in \mathcal{H} \tag{25}$$

$$X_{ijt} \leq 1, \forall i \in \mathcal{I}, j \in \mathcal{J}(i), t \in \mathcal{H} \tag{26}$$

$$X_{ijt} \geq 0, \forall i \in \mathcal{I}, j \in \mathcal{J}(i), t \in \mathcal{H} \tag{27}$$

$$\sum_{t \geq 1} X_{ijt} = 1, \forall i \in \mathcal{I}, j \in \mathcal{J}(i) \tag{28}$$

$$S_{ij} = \sum_{t \geq 1} t B_{ijt}, \forall i \in \mathcal{I}, j \in \mathcal{J}(i) \tag{29}$$

$$B_{ijt} \in \{0,1\}, \forall i \in \mathcal{I}, j \in \mathcal{J}(i), t \in \mathcal{H} \tag{30}$$

$$\sum_{t \geq 1} B_{ijt} = 1, \forall i \in \mathcal{I}, j \in \mathcal{J}(i) \tag{31}$$

$$F_{ijt} \in \{0,1\}, \forall i \in \mathcal{I}, j \in \mathcal{J}(i), t \in \mathcal{H} \tag{32}$$

$$\sum_{t \geq 1} F_{ijt} = 1, \forall i \in \mathcal{I}, j \in \mathcal{J}(i) \tag{33}$$

$$E_{ij} = 1 + \sum_{t \geq 1} tF_{ijt}, \forall i \in \mathcal{I}, j \in \mathcal{J}(i) \tag{34}$$

$$X_{ijt} \leq \sum_{h=1}^{t} B_{ijh}, \forall i \in \mathcal{I}, j \in \mathcal{J}(i), t \in \mathcal{H} \tag{35}$$

$$F_{ijt} \leq \sum_{h=1}^{t} X_{ijh}, \forall i \in \mathcal{I}, j \in \mathcal{J}(i), t \in \mathcal{H} \tag{36}$$

$$B_{ij_{succ}t} \leq 1 - \rho_{ij} + \sum_{h=1}^{t} X_{ijh}, \forall i \in \mathcal{I}, j \in \mathcal{J}(i), j_{succ} \in succ(j), j \in \mathcal{K}, t \in \mathcal{H} \tag{37}$$

$$\sum_{h=1}^{t} F_{ijh} \geq 0.1 - \sum_{h \geq t+1} X_{ijh}, \forall i \in \mathcal{I}, j \in \mathcal{J}(i), t \in \mathcal{H} \tag{38}$$

$$\sum_{h=1}^{t} X_{ijh} \geq 0.1 * \sum_{h=1}^{t} B_{ijh}, \forall i \in \mathcal{I}, j \in \mathcal{J}(i), t \in \mathcal{H} \tag{39}$$

$$\sum_{h=1}^{t} X_{ij_{succ}h} \leq \sum_{h=1}^{t} X_{ijh}, \forall i \in \mathcal{I}, j \in \mathcal{J}(i), j_{succ} \in succ(j), j \in \mathcal{K}, t \in \mathcal{H} \tag{40}$$

$$T_{ijt} \geq X_{ijt}, \forall i \in \mathcal{I}, j \in \mathcal{J}(i), t \in \mathcal{H} \tag{41}$$

$$T_{ijt} \leq 100 * X_{ijt}, \forall i \in \mathcal{I}, j \in \mathcal{J}(i), t \in \mathcal{H} \tag{42}$$

$$T_{ijt} \in \{0,1\}, \forall i \in \mathcal{I}, j \in \mathcal{J}(i), t \in \mathcal{H} \tag{43}$$

$$\sum_{h=1}^{t} T_{ijh} \leq l. \sum_{h=1}^{t} B_{ijh}, \forall i \in \mathcal{I}, j \in \mathcal{J}(i), t \in \mathcal{H} \tag{44}$$

$$\sum_{h \geq t} T_{ijh} \leq l. \sum_{h \geq t} F_{ijh}, \forall i \in \mathcal{I}, j \in \mathcal{J}(i) \tag{45}$$

$$E_{ij} - S_{ij} - \sum_{t \geq 1} T_{ijt} \leq \mu_{ij} - 1, \forall i \in \mathcal{I}, j \in \mathcal{J}(i) \tag{46}$$

Constraints 25 allow us to respect resource limits for each period. Constraints 26 and 27 define the progress ratios for all activities and projects while 28 ensure the completion of all activities. Through constraints 29, we define the start date of activity j of project i. Constraints 30 and 31 start the activity j of project i only one time. Similarly, constraints 32 and 33 finish activity j of project i only once. 34 calculate the end of activity j of project i. Constraints 35 announce the start of activity j of project i at the first strictly positive progress ratio when 36 prevent the indication of the end of activity j of project i before reaching 100%

progress. Launching the successor activity after having achieved a minimum progress defined by ρ_{ij} is ruled by constraints 37. Reaching 100% of accomplishment for the activity j of project i is given by constraints 38. Starting activity j of project i at its trigger (the first strictly positive ratio is greater than 10%) is ensured through constraints 39. The activities must finish in the appropriate chronological order 40. The set of constraints 41, 42 and 43 indicate the periods with a strictly positive progress ratio. This indication is exclusively guaranteed during the realization range of activity j of project i, which is inducted by 44 and 45. Finally, we limit the number of preemptions to $\mu_{ij} - 1$ using constraints 46.

4 Computational Experiments

For our case study, to avoid work-in-progress inventories, preemption is only allowed during sketching, design, and finishing. We propose to investigate the influence of some parameters of the production system on the solving time obtained with formulations A and B. As parameters, we consider the workload, the portfolio size (i.e., the number of projects, and the Maximum number of preemptions that are allowed respectively for the activities Sketching (MS), Design (MD), and Finishing (MF).

Formulations A and B are implemented with IBM ILOG CPLEX Optimization Studio version 22.1 and solved with the mathematical programming solver. All experiments are performed on an Intel® Core^{TM} i9-13900HX, 2200 Mhz 32 GB RAM machine. While keeping standard CPLEX settings, we tested the four linear programming models, obtained by combining formulations A and B with the function objectives OF1 and OF2, with three project portfolios (see Table 6) and different values of the Maximum number of preemptions that are allowed respectively for the activities Sketching, Design, and Finishing. In this case study, we suppose that MS, MD, and MF can take 0 (no preemption is allowed), 1, or 2.

The three project portfolios tested include real-world projects sampled from our partner database. As presented in Table 6, these portfolios correspond to different numbers of projects, different numbers of activities, and different levels of accumulated workloads.

Table 6. Project portfolios

Number of projects	Number of activities	Accumulated workload (man-day)
2 projets	12	1343.46
4 projets	24	2616.21
15 projets	90	3463.5

5 Results and Discussion

First, we analyze the solving time obtained with the four linear programming models (obtained by combining formulations A and B with the function objectives OF1 and OF2). To this end, we perform a linear regression, where the dependent variable is the solving time. Table 7 presents the significance of each feature including the objective function, the formulation, the maximum number of preemptions that are allowed respectively for the activities Sketching, Design, and Finishing (MS, MD and MF), the accumulated workload and the number of projects considered in the portfolio. The results indicate that in contrast to workload, the number of projects considered in the portfolio has a significant impact on the solving time. Formulation B outperforms formulation A since it offers a better relaxation to the problem. Among MS, MD and MF, limiting preemption during the finishing has a greater impact on the solving time.

Table 7. Significance of each feature on the solving time

	Estimate	Std. Error	t value	Pr(> |t|)
Intercept	–320,88	302,36	–1,06	0,29
OF2	88,387	126,82	0,69	0,48
Formulation B	–535,91	126,82	–4,22	3,12E-05*
MS	133,34	77,66	1,71	0,08
MD	151,40	77,66	1,94	0,05
MF	279,09	77,66	3,59	0,37E-3*
Accumulated workload	-0,31	0,15	–2,06	0,03
Number of projects	202,84	23,33	8,69	1,93E-16 *

Considering the 15-project portfolio, Fig. 2 illustrates the average solving time obtained with the four linear programming models (obtained by combining formulations A and B with the function objectives OF1 and OF2) considering different values for MS, MD and MF.

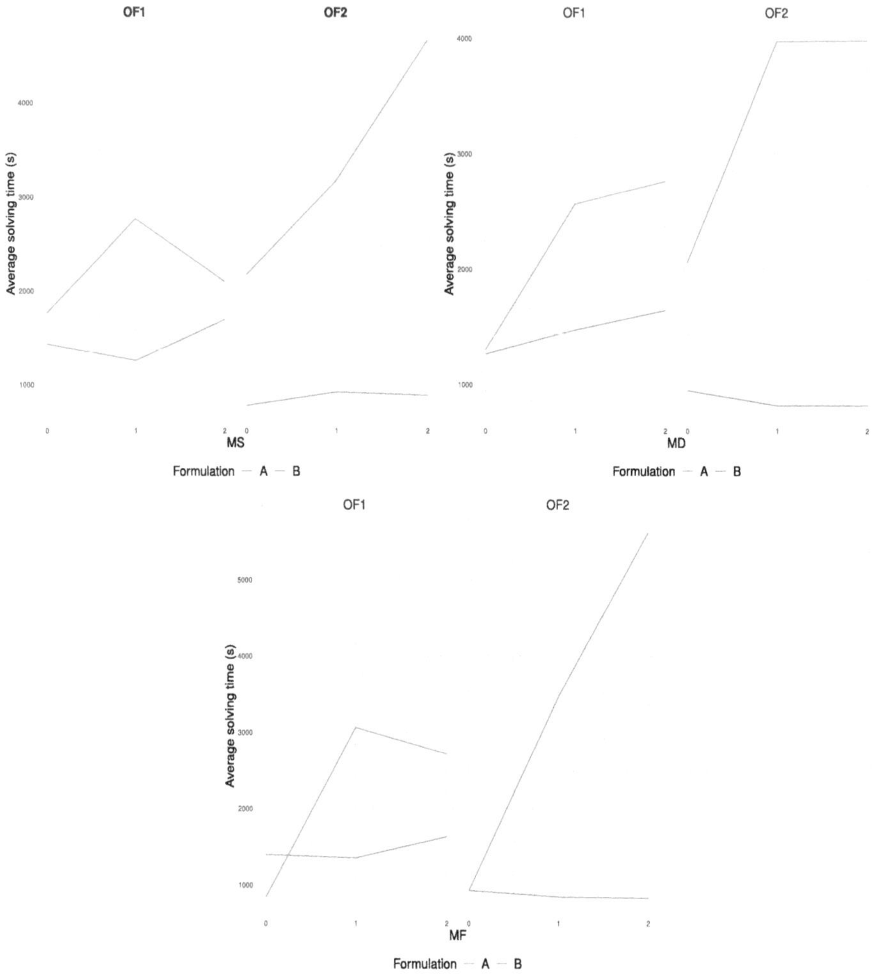

Fig. 2. Sensitivity of solving time to preemption limits

The values of MS, MD, and MF have remarkable effects on solving time for formulation A, regardless of the objective function used. In contrast, formulation B seems to be insensitive to the values of MS, MD and MF. This is due to the constant count of binary variables as the maximum number of preemptions allowed increases. Formulation A, however, is of increasing number of binary variables when limits of preemption is also increased. Formulation B achieves a shorter solving time, which is consistent with what has been shown in Table 7. Second, formulation A is not only more complex to solve, but also takes more time as MS, MD and MF increase (especially with OF2).

It is important to note that while limiting the number of possible preemptions has no direct control over the minimal time units of continuous work, it can help to prevent frequent preemption after a short time of the progress of the activity. Therefore, using the formulation B with OF2 is a good compromise in terms of utility, solving time and sensitivity to preemption parameters compared to formulation A. Another reason to prefer formulation B to A, as shown in Fig. 3, is that when activities require a small amount of energy to accomplish, formulation A forces activities of the same project not to be scheduled during the same time unit, whereas it might be feasible to do so, in contrast to formulation B.

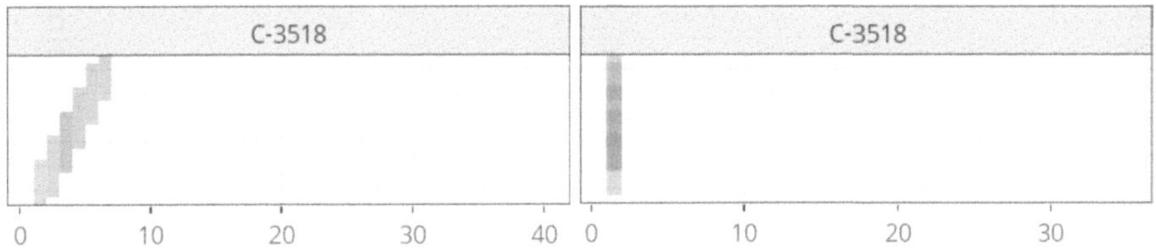

Fig. 3. Gantt chart of a project with Formulation A (left) and Formulation B (right)

Figure 4 illustrates the number of preemptions allowed and those decided by the model for the three portfolios. We can see that, with more projects to plan, the average number of preemptions proposed by the model increases. Also, this number increases each time we allow more preemptions in the different production activities. This confirms the usefulness of setting a limit on allowed preemptions, as in practice it is preferable to control preemptions.

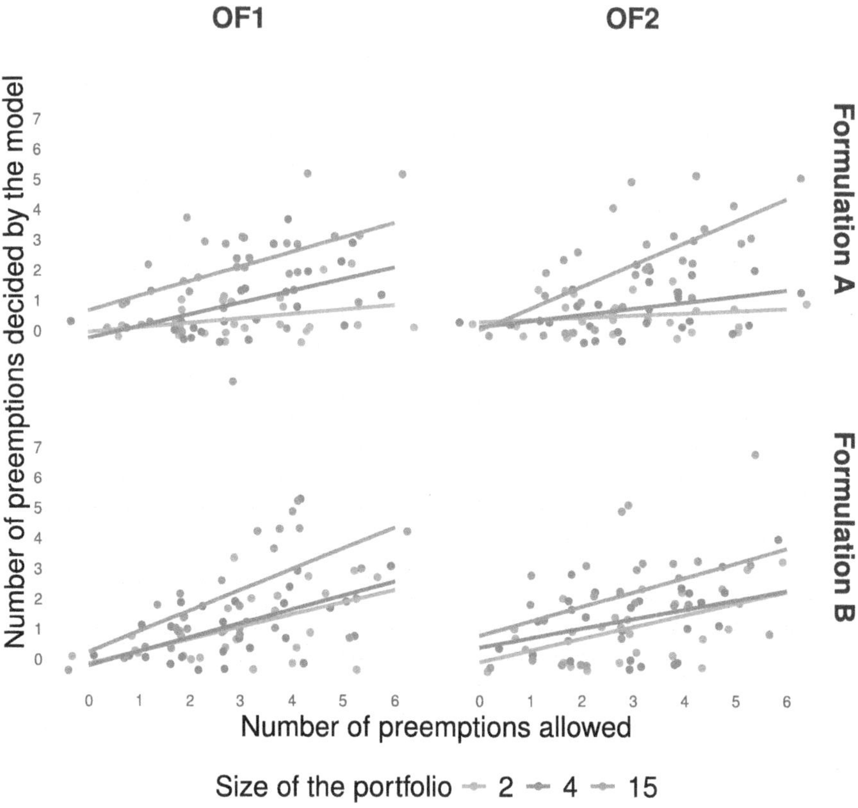

Fig. 4. Number of preemptions allowed vs. Number of preemptions decided by the model

6 Conclusion

This paper deals with a real-world case study and addresses a RCMPSP with preemption. To support the tactical planning of a modular building manufacturer, we propose different models allowing activity overlapping and a maximum number of allowable preemptions. Two linear programming formulations have been tested with two objective functions. The formulation allowing preemptions with discontinuous energy distribution outperforms the formulation using continuous activity splitting and offering more control over planning. This work contributes to our partner's efforts to develop a Mixed-Initiative System (MIS), which will integrate the decision-maker preferences.

Acknowledgment. This work was financially supported by MITACS (Canada) and our industrial partner.

References

1. Klein, R.: Scheduling of resource-constrained projects. Springer Science & Business Media (1999)
2. Neumann, K., Schwindt, C., Zimmermann, J.: Project scheduling with time windows and scarce resources: temporal and resource-constrained project scheduling with regular and nonregular objective functions. Springer (2002)
3. Alfieri, A., Tolio, T., Urgo, M.: A project scheduling approach to production planning with feeding precedence relations. Int. J. Prod. Res. **49**(4), 995–1020 (2011)
4. Sánchez, M.G., Lalla-Ruiz, E., Gil, A.F., Castro, C., Voß, S.: Resource-constrained multi-project scheduling problem: a survey. Europ. J. Operat. Res. **309**(3), 958–976 (2023)
5. Schwindt, C., Paetz, T.: Continuous preemption problems. In: Handbook on Project Management and Scheduling, vol. 1. pp. 251–295. Springer (2015)
6. Cheng, J., Fowler, J., Kempf, K., Mason, S.: Multi-mode resource-constrained project scheduling problems with non-preemptive activity splitting. Comput. Oper. Res. **53**, 275–287 (2015)
7. Quintanilla, S., Lino, P., Pérez, Á., Ballestín, F., Valls, V.: Integer preemption problems. In: Schwindt, C., Zimmermann, J. (eds.) Handbook on Project Management and Scheduling Vol.1. IHIS, pp. 231–250. Springer, Cham (2015). https://doi.org/10.1007/978-3-319-05443-8_12
8. Hamel, S., Gaudreault, J., Quimper, C.-G., Bouchard, M., Marier, P.: Human-machine interaction for real-time linear optimization. In: 2012 IEEE International Conference on Systems, Man, and Cybernetics (SMS), 673–680 (2012)
9. Bouchard, J., Gaudreault, J., Quimper, C.-G., Marier, P., Brotherton, E., Simard, N.: Mixed-initiative system for tactical planning allowing real-time constraint insertions. IFAC-PapersOnLine, Elsevier **50**, 15233–15240 (2017)
10. Quimper, C.-G., Rousseau, L.-M.: A large neighbourhood search approach to the multi-activity shift scheduling problem. J. Heuristics **16**, 373–392 (2010)
11. Issa, S., Tu, Y.: Integrated multi-resource planning and scheduling in engineering project. J. Project Manage. **2**(1), 11–26 (2017)
12. Naber, A., Kolisch, R.: MIP models for resource-constrained project scheduling with flexible resource profiles. Europ. J. Oper. Res. **239**(2), 335–348 (2014)
13. Pritsker, A.A.B., Waiters, L.J., Wolfe, P.M.: Multiproject scheduling with limited resources: a zero-one programming approach. Manage. Sci. INFORMS **16**(1), 93–108 (1969)
14. Hao, Q., Shen, W., Xue, Y., Wang, S.: Task network-based project dynamic scheduling and schedule coordination. Adv. Eng. Inform. **24**(4), 417–427 (2010)
15. Joo, B.J., Chua, P.C.: Multimode resource-constrained multi-project scheduling with ad hoc activity splitting. In: 2017 IEEE International Conference on Industrial Engineering and Engineering Management (IEEM), pp. 2261–2265. IEEE (2017)
16. Issa, S.B., Patterson, R.A., Tu, Y.: Solving resource-constrained multi-project environment under different activity assumptions. Int. J. Prod. Econ. **232**, 107936 (2021)
17. Afshar-Nadjafi, B.: A solution procedure for preemptive multi-mode project scheduling problem with mode changeability to resumption. Appl. Comput. Inform. **14**(2), 192–201 (2018)

Author Index

A

Adenipekun, Ebenezer Olatunde 229
Alão, Tiago 3
Al-Dulaimy, Auday 461
Alfnes, Erlend 508
Alieksieiev, Volodymyr 92
Alves, Lucas 49
Antons, Oliver 180
Apaydın, Tuğçe 19
Arlinghaus, Julia C. 180
Aslanidou, Ioanna 425
Ates, Aylin 19

B

Bai, Yutong 449
Bakhshi-Khaniki, Hessam 244
Baptiste, Pierre 49
Battini, Daria 138
Behnam, Moris 461
Benali, Maha 49, 537
Benderbal, Hichem Haddou 155
Bengtsson, Emanuel 476
Borade, Trupti 476
Bouzegag, S. Hocine 213
Buer, Sven-Vegard 65

C

Calzavara, Martina 138
Chavez, Zuhara Zemke 476
Chen, Qiqi 168
Chirumalla, Koteshwar 461
Christiansen, Atle M. 34

D

Ding, Jiyuchen 168
Dolgui, Alexandre 307

E

Erdeş, Hakan 274

F

Fattouh, Anas 461
Feizabadi, Mohammad 522
Fradi, Ali 537
Freitag, Mike 437
Fujii, Yu 410

G

Ghasemi, Mohammad 307
Goyal, Akshay 425
Große, Nick 394
Gustafsson, Christopher 338

H

Hämmerle, Oliver 437
Hanzalek, Zdenek 244
Hatvani, Leo 461
Holmström, Jan 353
Hosseini, Arman 522
Huang, George Q. 168

I

Iza-Mendez, Martin 325

J

Jahn, Carlos 92
Janecki, Luca 180

K

Karaoğlan, İsmail 291
Keivanpour, Samira 492
Kesen, Saadettin Erhan 213, 274
Khajavi, Siavash Haghighat 353
Konyalioglu, Aziz Kemal 19

L

Lagorio, Alexandra 78
Li, Congdong 381
Li, Weiyuan 111
Lim, Ming K. 111, 449

Published by Springer Nature Switzerland AG 2024
M. Thürer et al. (Eds.): APMS 2024, IFIP AICT 733, pp. 553–554, 2024.
https://doi.org/10.1007/978-3-031-71645-4

Limère, Veronique 229
Liu, Lei 381
Lödding, Hermann 92
Lopes, Isabel 3

M
Ma, Lin 381
Mahboubi, Aysan 492
Maier, Julian B. 123
Mangano, Giulio 78
Martignago, Michele 138
Meyer, Anne 394
Mirnezami, Seyed-Ali 307
Miyangaskary, Mina Kazemi 492
Mourtos, Ioannis 258
Mpourdakos, Ilias 258

N
Ngo, Uyen 508
Nonaka, Tomomi 410
Nygren, Christer 338

O
Oura, Fumihito 410

P
Papathanasiou, Dimitrios 258
Paton, Steve 19
Pinto, Roberto 78
Pires, Cláudia 3
Pletzer-Zelgert, Leonie Pauline 325
Pyeon, Hyeongoo 197

Q
Qu, Ting 381

R
Rafele, Carlo 78
Reh, Daniel 180

S
Salmi, Mika 353
Schiller, Christian 437
Schleifenbaum, Johannes Henrich 325
Schmid, Nico André 229
Semini, Marco 508
Sgarbossa, Fabio 508
Shahabi-Shahmiri, Reza 307
Shahrivar, Rafat Beigpoor 92
Shirasaka, Seiko 410
Sid-Lakhdar, Mokhtar Nizar 155
Skjelstad, Lars 65
Son, SookYoung 197
Souier, Mehdi 155
Steger-Jensen, Kenn 34
Sun, Mingyue 168

T
Tavakkoli-Moghaddam, Reza 244, 307
Thürer, Matthias 381
Tjaden, Greta 394

U
Unruh, Lukas 325

V
Vahedi-Nouri, Behdin 244
Vatikiotis, Stavros 258
Vidskjold, Kenneth 34

W
Wiendahl, Hans-Hermann 123
Wilson, James 111

Y
Yaghoubi, Roozbeh 492
Yahouni, Zakaria 522
Yamamoto, Yuji 425
Yilmaz, Nurşah 291
Yuan, Mingze 381

Z
Zenezini, Giovanni 78
Zhang, Kai 381
Zhao, Zhiheng 168

SPRINGER NATURE

GPSR Compliance

The European Union's (EU) General Product Safety Regulation (GPSR) is a set of rules that requires consumer products to be safe and our obligations to ensure this.

If you have any concerns about our products, you can contact us on ProductSafety@springernature.com

In case Publisher is established outside the EU, the EU authorized representative is:

Springer Nature Customer Service Center GmbH
Europaplatz 3
69115 Heidelberg, Germany

The manufacturer's authorised representative in the EU is Springer Nature Customer Service Centre GmbH, Europaplatz 3, 69115 Heidelberg, Germany. If you have any concerns regarding our products, please contact ProductSafety@springernature.com

Printed and bound by CPI Group (UK) Ltd, Croydon, CR0 4YY
08/10/2026
02229819-0003